Fruit Production

— Major Fruits —

The Authors

Dr. R.R. Sharma, born on March 14, 1962 in Hamirpur (Himachal Pradesh), Dr. Sharma completed his B.Sc. (Agri.) from HPKV, Palampur during 1983 and Post Graduation from IARI, New Delhi in 1994 and joined as Scientist in the Division of Fruits and Horticultural Technology, IARI, New Delhi in 1997. During his scientific career, he has served several research projects and has been associated with release of mango hybrids like Pusa Arunima, Pusa Surya, Pusa Lalima, Pusa Shresth, Pusa Pitamber etc., and standardized several fruit production and post harvest technologies. He is a prolific writer and has published 40 research articles in International journals, 55 in National journals, 160 popular articles in different magazines, and 25 book chapters. He has also authored 8 books on different aspects of fruit production. He has delivered several radio and T.V. talks on different aspects of fruit production and post harvest management of horticultural crops. He has received several honours and awards, of which, Dr. R.N. Singh award (twice) of IARI, New Delhi, Dr Rajinder Prasad award of ICAR (twice), New Delhi are worth mentioning. He has served as International Mango Registrar and Education award of MHRD, Govt. of India for about 5 years (1999-2004). He has been honoured with *Himachal Shri Award* by Govt. of Himachal Pradesh in 2008. He has also been honoured by International journals like Scientia Horticulturae (The Netherlands), International Journal of Fruit Science (U.K.), American Journal of Plant Sciences (US), Journal of Food Processing and Technology (US), and Stewart Postharvest Reviews (UK) by including him as a Member, Editorial Board. He is a reviewer of several International and national journals, and Chief Editor of International Journal of Processing and Postharvest Technology, India.

Dr. Hare Krishna, born on June 28, 1981, Dr Hare Krishna graduated from the Institute of Agricultural Sciences, B.H.U., Varanasi and completed his M. Sc. and Ph. D. from Indian Agricultural Research Institute, New Delhi. He has got an outstanding academic career. With his experience and vision, he contributed significantly to address the major obstacles of commercial micropropagation. Based on his work, a protocol for bio-hardening of tissue culture-raised plantlets was developed, for which he got IARI Gold Medal in 2004. He was invited for the creation of a Scirus Topic page on Mango Tissue Culture for the scientific community run by Elsevier, The Netherlands. He joined as Scientist in 2006 at Vivekanand Institute of Hill Agriculture, Almora, later transferred to Central Institute of Temperate Horticulture-Regional Station, Mukteshwar. Since 2010, he is working as a Senior Scientist at Central Institute for Arid Horticulture, Bikaner. His research works are primarily focused on physiological understanding of the problems of horticultural crops. During this short tenure as scientist, he has published 10 research articles in international and 15 in national journals. Presently, he has been vested with responsibility of the development of a DUS descriptor for *ber* fruit crop; besides, several other important research issues. He is also a reviewer of several international journals such as Scientia Horticulturae, Crop Protection, Journal of Environmental Management, Journal of American Society for Horticultural Science, Journal of Agricultural Sciences etc.

Fruit Production

— Major Fruits —

Authors

R.R. Sharma
Senior Scientist (Fruit Science)
Division of Post Harvest Technology
IARI, New Delhi

Hare Krishna
Senior Scientist (Fruit Science)
Central Institute of Arid Horticulture, Bikaner

2014

Daya Publishing House®
A Division of
Astral International Pvt. Ltd.
New Delhi – 110 002

Published by   : **Daya Publishing House**®
                          A Division of
                          Astral International Pvt. Ltd.
                          – ISO 9001:2008 Certified Company –
                          4760-61/23, Ansari Road, Darya Ganj
                          New Delhi-110 002
                          Ph. 011-43549197, 23278134
                          E-mail: info@astralint.com
                          Website: www.astralint.com

Laser Typesetting   : **Classic Computer Services**, Delhi - 110 035

Printed at   : **Replika Press Pvt. Ltd.**

PRINTED IN INDIA

Dedicated to
Our Respective Mothers

डा. एस. अय्यप्पन
सचिव एवं महानिदेशक

Dr. S. AYYAPPAN
SECRETARY & DIRECTOR GENERAL

भारत सरकार
कृषि अनुसंधान और शिक्षा विभाग एवं
भारतीय कृषि अनुसंधान परिषद
कृषि मंत्रालय, कृषि भवन, नई दिल्ली 110 001

GOVERNMENT OF INDIA
DEPARTMENT OF AGRICULTURAL RESEARCH & EDUCATION
AND
INDIAN COUNCIL OF AGRICULTURAL RESEARCH
MINISTRY OF AGRICULTURE, KRISHI BHAVAN, NEW DELHI 110 001
Tel.: 23382629; 23386711 Fax: 91-11-23384773
E-mail: dg.icar@nic.in

Foreword

Fruits are considered as 'protective foods' as they contain health promoting substances, such as essential vitamins, minerals and antioxidants, which protect us from several ailments. Our country is endowed with varied climatic conditions, which had made it possible to grow diverse group of fruits i.e., from temperate to tropical, to subtropical and even to arid niches. The cultivation of fruit crops is, presently, one of the most important and remunerative areas of agriculture. Growing fruit crops translate into realization of more yield/ income from a unit area of land than any of the agronomic crops. The art and science of cultivation has now transformed into one of the most skillful and intensive forms of land utilization practice.

Besides social and nutritive values, fruits also contribute in strengthening National economy by fetching foreign exchange. Annually, the country earns a lot of foreign money by export of major fruits like mango, grapes, banana, citrus, guava, apple, pineapple, sapota, etc. In view of the economic importance of these major fruit crops, the R&D efforts were focused for the last four decades, primarily, by the Indian Council of Agricultural Research, other Central Research Organizations and State Agricultural Universities, which resulted into a paradigm shift in fruit crop husbandry. However, our productivity of most of the fruit crops is dismally low in comparison to several developed countries of the world; hence there is an urgent need to considerably increase the yield and productivity of these crops, particularly under changing environmental conditions, to feed the millions of this country. This can be achieved through the adoption of proper improvement, production and protection practices, being generated by Pomologists of our country.

I congratulate *Dr. R.R. Sharma* and *Dr. Hare Krishna* for bringing out this University-level textbook. It is gratifying to note that the authors have painstakingly toiled to incorporate latest information on every aspect of production of major fruit crops in this book.

I am confident that this book will be very useful for teachers, scientists, researchers, UG and PG students, extension personnel, and even growers of our country.

New Delhi

S. Ayyapan

Preface

The history of fruit cultivation is as old as human civilization itself. In India, fruits are grown extensively under varying climatic conditions from temperate to arid zones. Fruits are regarded as protective food as they are rich source of essential minerals, vitamins and antioxidants, which protect us from several fetal diseases. In India, fruit crops are grown on an area of 6,329 (000' ha) area with an annual production of 71,516 (000' MT). Of total fruit production, 90% of share is being contributed by major fruit crops like banana, mango, citrus, papaya, guava, apple, pineapple, sapota, grapes, litchi and pomegranate. These fruits are unique in their own taste, palatability and nutritive value and are incomparable with any other food commodity. However, their production is encountered with several complex problems, which if not addressed in time, may result in complete crop failure.

Ample literature is available on various aspects of fruit cultivation, which have been published from time-to-time. However, keeping in view the enormously increasing and updated scientific knowledge, an immediate need to bring out a book on this topic in new apparel so as to elicit assenting and committed response in students, teachers, researchers and growers, was felt.

For writing this book, we have consulted several books, research periodicals, research reports, web content etc. in order to bring in the most recent and relevant information together. In addition, the input given by the fruit scientists was also taken into consideration. Utmost care has been taken to make the book comprehensive to the readers supported with the ample numbers of tables and plates. However, there are chances of errors and omissions, and therefore, we invite readers for their constructive suggestions to make this more comprehensive in future.

This book could not have come in its present form without the help received from various scientists, colleagues and students during different stages of its preparation. Hence, we profusely thank one and all for their enormous help.

We also acknowledge our family members who were ever there to extend unconditional support during the preparation of this book and we thank all of them from the core of our hearts.

In the last but not the least, we thank, Astral International Pvt. Ltd., New Delhi, who had faith in us and gave the responsibility of writing this proposal.

R.R. Sharma
Hare Krishna

Contents

1

Mango

Mango (*Mangifera indica*) is the one of the leading fruit crops of our country and considered to be the 'king of fruits'. Besides delicious taste, excellent flavour and attractive fragrance, it is rich in vitamin A and C.

India is a leading mango producing country in the world. Mango contributes 20.3 per cent to national fruit production (15.18 MT) and occupies 35.9 per cent of the total fruit area (22.97 million ha) of the country. Maharashtra occupies the largest area under mango cultivation (4.77 million ha) with 3.31 million tones production. Uttar Pradesh (U.P) is the main mango growing state of India sharing 23.9 per cent of total mango production in the country from an area of 2.67 million ha. The highest productivity has been recorded in U.P. (13.6 t/ha) as compared to national average of 6.6 t/ha (Table 1.1).

ORIGIN, HISTORY AND DISTRIBUTION

Mangoes have been cultivated for an estimated 4,000 years in India. However, it was virtually unknown to any Botanist until 1605 when Carol Clusius first mentioned about it in his writings. The name *Mangifera* was given for the first time by Bontius in 1658 when he referred it as *Arbor mangifera* (the tree producing mango). Later, it was mentioned in the literature as *Mangifera indica, Mangas domestica* or *Mangas sylvatica*. Linnaeus also referred to it as *Mangifera arbor* in 1747, prior to changing the name to its present form (*Mangifera indica*) in 1753, in his much quoted book, *Species plantarum*.

It is considered to be native to southern Asia, especially eastern India, Burma, and the Andaman Islands, where the mango has been cultivated, praised and even revered since time immemorial. Buddhist monks are believed to have taken the mango on voyages to Malaya and eastern Asia in the 4th and 5th Centuries B.C. The Persians are said to have carried it to East Africa in the 10th Century A.D. It was commonly grown in the East Indies before the earliest visits of the Portuguese who apparently introduced it to West Africa early in the 16th Century and also into Brazil. After becoming established in Brazil, the mango was carried to the West Indies, being first planted in Barbados about 1742 and later in the Dominican Republic. It reached Jamaica in 1782 and, early in the 19th Century, reached Mexico from the Philippines and the West Indies.

Besides the Indian subcontinent, mango is grown in several countries of the tropical and subtropical world, where it was introduced by the Muslim missionaries, Spanish voyagers, and Portuguese explores during the 15th to the 18th centuries. It has been reported in the literature that mango was being cultivated at the head of the Persian Gulf by the 16th century. It was introduced to the humid, tropical countries of the Malay Archipelago, and South-East Asia during the 4th and 5th centuries B.C. where it continued to develop and has become naturalized. It was introduced in the Philippines after 1600, in the Moluccas in 1665 and in Yemen in the later part of the 18th century. Further, it was introduced in Hawaii between 1800 and 1820. Mango introduction to Florida began in 1861 with the importation of No. 11, a polyembryonic, seed propagated cultivar from Cuba. The US Department of Agriculture introduced Mulgoa, an improved cultivar from India, in 1889. Later, several superior types were selected from Mulgoa and large numbers of mango cultivars were introduced in Florida, as a result, Florida became a contemporary secondary centre of diversity for mango germplasm.

Table 1.1: Statewise Area, Production and Productivity Scenario of Mango

State	Area (000'Ha)	Production (000'MT)	Productivity (MT/Ha)
Uttar Pradesh	267.2	3623.2	13.6
Andhra Pradesh	391.1	3363.4	8.6
Karnataka	161.6	1778.8	11.0
Bihar	147.0	1334.9	9.1
Gujarat	130.1	911.3	7.0
Tamil Nadu	148.0	823.7	5.6
Odisha	190.1	642.0	3.4
West Bengal	89.5	620.2	6.9
Jharkhand	38.9	427.9	11.0
Kerala	62.2	380.9	6.1
Maharashtra	477.0	331.0	0.7
Others	194.0	951.1	4.9
Total	2,296.8	15,188.4	6.6

Source: National Horticulture Board Database-2011.

Presently, it is cultivated commercially in several countries of world such as India, Pakistan, Bangladesh, Myanmar, Sri Lanka, Thailand, Cambodia, Vietnam,

Malaysia, the Philippines, Indonesia, the Fiji Islands, Tropical Australia, Egypt, Israel, Kenya, Tanzania, South Africa, Nigeria, Madagascar, Mauritius, the USA, Venezuela, Mexico, Brazil and The West Indies.

Centers of Mango Diversity

In 1910, a seedling of 'Mulgoba' came into production in Florida. Its fruits had a highly attractive red blush, and appeared to bear more heavily than its parent(s). This selection was named 'Haden'. Although 'Haden' was not superior with respect to fruit quality in comparison to the imported germplasm from India and its genetic base was much wider. During the 20th century, more introductions of mango germplasm into Florida occurred from South-East Asia (the Philippines, Cambodia), India and elsewhere. It was at one time believed that these introductions of mango germplasm created a secondary centre of diversity of the species. 'Eldon', 'Glenn', 'Lippens', 'Osteen', 'Parvin', 'Smith', 'Springfels', 'Tommy Atkins' and 'Zill' are progeny of 'Haden'. 'Saigon' seedling selection was made from 'Cambodiana', a polyembryonic introduction from Indochina. From 'Saigon' seedlings, 'Alice', 'Herman' and 'Florigon' were selected. Based upon more recent genetic analysis involving microsatellite markers, it is now estimated that the majority of Florida cultivars are descended from only four monoembryonic Indian mango cultivar accessions, *i.e.*, 'Mulgoba', 'Sandersha', 'Amini' and 'Bombay', together with the polyembryonic 'Turpentine' from the West Indies. The Florida mango cultivars have been found to be highly adaptable to many agro-ecological areas and bear regularly, whereas, many of the outstanding Indian cultivars have been unproductive outside their centre of domestication, and are alternate bearing. These selections also have a highly attractive red blush at maturity, firm flesh, a high flesh to seed ratio and a regular bearing habit. Some of the Florida cultivars, for example 'Tommy Atkins', 'Keitt', etc. are also moderately resistant to anthracnose, the most important production and postharvest problem of mango in many areas. In the latter half of the 20th century, plantings of Florida cultivars have been established in many countries and now form the basis of international trade of mangoes.

COMPOSITION AND USES

Ripe mango fruit is a rich source of several vitamins and minerals. Among vitamins, it is considered as a very good source of β-carotene, vitamin C and vitamin B_9. In addition, it contains appreciable amount of calcium and magnesium (Table 1.2).

Uses

The mango is generally sweet, although the taste and texture of the flesh varies across cultivars, some having a soft, pulpy texture similar to an overripe plum, while the flesh of others is firmer, like a cantaloupe or avocado, or may have a fibrous texture. Ripe mangoes are used as table fruits but are also used to develop several value added products. Ripe mangos may be frozen whole or peeled, sliced and packed in sugar and quick-frozen in moisture-proof containers. Green mangos are peeled, sliced, parboiled, and then combined with sugar, salt, various spices and cooked, sometimes with raisins or other fruits, to make chutney; or they may be salted, sundried and kept for use in *chutney* and pickles. Thin slices, seasoned with turmeric, are dried, and sometimes powdered, and used to impart an acid flavour to *chutneys*, vegetables and soup. Green or ripe mangos may be used to make relish. The fresh kernel of the mango seed (stone) constitutes 13 per cent of the weight of the fruit, 55 per cent to 65 per cent of the weight of the stone. The kernel is a major by-product of the mango-processing industry. In times of food scarcity in India, the kernels are roasted or boiled and eaten. After soaking to dispel the astringency (tannins), the kernels are dried and ground to flour, which is mixed with wheat or rice flour to make bread and sometimes also used in puddings.

Table 1.2: Nutrient Composition of Ripe Mango Fruit

Component	Per 100 g Edible Portion	Component	Per 100 g Edible Portion
Carbohydrates	15 g	Niacin (vit. B_3)	0.67 mg
Sugars	13.7 g	Pantothenic acid (B_5)	0.2 mg
Dietary fiber	1.6 g	Vitamin B_6	0.12 mg
Fat	0.38 g	Folate (vit. B_9)	43 µg
Protein	0.82 g	Vitamin C	36 mg
Vitamin A equivalent	54 µg	Calcium	11 mg
β-carotene	640 µg	Iron	0.16 mg
Thiamine (vit. B_1)	0.03 mg	Magnesium	10 mg
Riboflavin (vit. B_2)	0.04 mg		

The fat extracted from the kernel is white, solid like cocoa butter and tallow, edible, and has been proposed as a substitute for cocoa butter in chocolate. The peel constitutes 20 per cent to 25 per cent of the total weight of the fruit. It has been shown that the peel can be utilized as a source of pectin. Average yield on a dry-weight basis is 13 per cent. Immature mango leaves are cooked and eaten in Indonesia and the Philippines. Dried immature fruits can be converted into powder, commonly called as '*Amchur*' in India.

TAXONOMY AND BOTANICAL DESCRIPTION

Mango is one of the 73 genera of the family Anacardiaceae in order Sapindales. The family Anacardiaceae includes a large number of plants found within the tropics and a few growing in the Mediterranean region, Japan, and temperate North America. The best known relatives of the mango are, probably, the cashew (*Anacardium occidentale*), widely cultivated in the tropics for its edible fruit; the pistachio nut (*Pistacia vera*) of the Mediterranean region; several species of *Spondias*, which are grown for their edible fruits; cooking spice, *chironji* (*Buchanania lanjan*), which is used widely in preparation of sweets in India; the obnoxious poison ivy (*Rhus toxicodendron*) of the United States; and the pepper-tree, *Schinus molle*, familiar in the southern California. Of approximately 40 other species of *Mangifera*, a few are cultivated for their fruits (*e.g. Mangifera indica*) and several have been employed as rootstocks for the mango in Malaya such as *M. sylvatica* Roxb., *M. foetida* Lour., *M. caesia* Jack, *M. odorata* Griff.;

besides, some lesser-known species like *M. longipetiolata* King, *M. maingayi* Hook f., *M. kemanga* Blume and *M. pentandra* Hook f. (Table 1.3)

Table 1.3: A List of Species of *Mangifera* and their Locations.

1.	*Mangifera duperreana*	Cochinchina (Vietnam), Siam (Thailand)
2.	*M. pentandra*	Burma (Myanmar), Malaya (Malaysia), Indochina (Cambodia)
3.	*M. cochinchinensis*	Cochinchina
4.	*M. lanceolata*	Malaya (Malaysia)
5.	*M. indica*	Tropics of Old World
6.	*M. longipes*	Burma (Myanmar), Malaya (Malaysia), Sunda Archipelago, the Philippines
7.	*M. caloneura*	Burma (Myanmar), Siam (Thailand)
8.	*M. siamensis*	Siam (Thailand)
9.	*M. sylvatica*	India, Burma (Myanmar), Indochina (Cambodia)
10.	*M. oblongifolia*	Malacca, Siam, Indochina (Cambodia)
11.	*M. minor*	New Guinea, Celebes, the Solomon Island
12.	*M. zeylanica*	Ceylon (Sri Lanka)
13.	*M. khasiana*	Assam
14.	*M. gracilipes*	Malacca
15.	*M. camptosperma*	Burma (Myanmar), Siam (Thailand), COchinchina (Vietnam), Sumatra
16.	*M. gedebe*	Java
17.	*M. microphylla*	Malaya (Malaysia)
18.	*M. griffithii*	Malaya (Malaysia)
19.	*M. sclerophylla*	Malaya (Malaysia)
20.	*M. merillii*	The Philippines
21.	*M. baccarii*	Sarawak
22.	*M. similis*	Sumatra, Java
23.	*M. altissima*	The Philippines
24.	*M. rumphii*	Banda Island
25.	*M. philippinensis*	The Philippines
26.	*M. havilandi*	Sarawak
27.	*M. ridida*	Sumatra
28.	*M. maingayi*	Malaya (Malaysia)
29.	*M. longipetiolata*	Malaya (Malaysia)
30.	*M. quardrifida*	Malaya (Malaysia), Sumatra, Borneo
31.	*M. spathulaefolia*	Borneo
32.	*M. timorensis*	Timor, Banda, Sumatra
33.	*M. monandra*	The Philippines
34.	*M. andamanica*	The Andaman Islands

Contd...

Table 1.3–*Contd...*

35.	*M. lagenifera*	Siam (Thailand), Malaya (Malaysia), Sumatra
36.	*M. macrocarpa*	Malaya (Malaysia), Sunda Archipelago, The Anambas Island, Indochina (Cambodia)
37.	*M. foetida*	Malaya (Malaysia)
38.	*M. odorata*	Malaya (Malaysia), The Philoppines
39.	*M. kemanga*	The Malaya Peninsula and Archipelago
40.	*M. caesia*	Malaya (Malaysia)
41.	*M. superba*	Malaya (Malaysia)

On the basis of studies on certain *Mangifera* species (*M. indica, M. caloneura, M. sylvetica, M. foetida, M. caesia, M. odorata* and *M. zeylanica*), the basic chromosome number of 20 and of the cultivated varities are diploid with 2n=40. However, the variety Vellaicollumban is said to be tetraploid (2n =80). Further, the cultivated mango has been reported to be an allopolyploid.

BOTANY

The mango tree is erect, 30 to 100 ft (roughly 10-30 m) high, with a broad, rounded canopy which may, with age, attain 100 to 125 ft (30-38 m) in width, or a more upright, oval, relatively slender crown. In deep soil, the taproot descends to a depth of 20 ft (6 in), the profuse, wide-spreading, feeder root system also sends down many anchor roots, which penetrate for several feet. The tree is long-lived, some specimens being known to be 300 years old and still fruiting.

The mango inflorescence is primarily a racemose. The inflorescence may be terminal or axillary as per the position of panicles on the branches. Mango inflorescence can be grouped in two main groups *viz.* (a) Pure inflorescence. It contains only floral structure; (b) Mixed inflorescence. It has leaves or leafy bracts in addition to the floral buds. Thus, the panicles may be categorized into four groups:

1. *Pure terminal panicles:* Terminal panicles with floral structure only.
2. *Mixed terminal panicles:* Terminal panicles with floral structure and leaves or leafy bracts.
3. *Pure axillary panicle:* Terminal flower bud is undeveloped and a number of axillary panicles arise from the axil of leaves situated just below the apex of shoot, thus giving rise to a cluster of panicle. All such panicles contain only floral structure. Sometimes, the terminal bud may also develop into a purely floral panicle in addition to the axillary panicle, which are also included in this category.
4. *Mixed axillary panicle:* It has a simple origin as in the previous category but the panicles have both floral and leafy structures.

There are mainly two types of flowers in mango: (i) Male flowers, which contain male parts with viable pollens. It is also known as staminate flower, and (ii) hermaphrodite flowers, which are also called as perfect flowers and have androecium and gynoecium both.

Nearly evergreen, alternate leaves are borne mainly in rosettes at the tips of the branches and numerous twigs from which they droop like ribbons on slender petioles 1 to 4 inch (2.5-10 cm) long. The new leaves, appearing periodically and irregularly on a few branches at a time, are yellowish, pink, deep-rose or wine-red, becoming dark-green and glossy above, lighter beneath. The midrib is pale and conspicuous and the many horizontal veins distinct. Full-grown leaves may be 4 to 12.5 inch (10-32 cm) long and 3/4 to $2^1/_8$ inch (2-5.4 cm) wide. Hundreds and even as many as 3,000 to 4,000 small, yellowish or reddish flowers, 25 per cent to 98 per cent male, the rest hermaphroditic, are borne in profuse, showy, erect, pyramidal, branched clusters 2½ to 15½ inch (6-40 cm) high. There is a great variation in the form, size, colour and quality of the fruits. They may be nearly round, oval, ovoid-oblong, or somewhat kidney-shaped, often with a beak at the apex, and are usually more or less lop-sided. They range from 2½ to 10 inch (6.25-25 cm) in length and from a few ounces to 4 to 5 lbs (1.8-2.26 kg). The skin is leathery, waxy, smooth, fairly thick, aromatic and ranges from light-or dark-green to clear yellow, yellow-orange, yellow and reddish-pink, or more or less blushed with bright-or dark-red or purple-red, with fine yellow, greenish or reddish dots, and thin or thick whitish, gray or purplish bloom, when fully ripe. Some varieties have a 'turpentine' odour and flavour, while others are richly and pleasantly fragrant. The flesh ranges from pale-yellow to deep-orange. It is essentially peach-like but much more fibrous (in some seedlings excessively so-actually 'stringy'); is extremely juicy, with a flavour range from very sweet to sub-acid to tart.

There is a single, longitudinally ribbed, pale yellowish-white, somewhat woody stone, flattened, oval or kidney-shaped, sometimes rather elongated. It may have along one side a beard of short or long fibers clinging to the flesh cavity, or it may be nearly fiberless and free. Within the stone, is the starchy seed, monoembryonic (usually single-sprouting) or polyembryonic (usually producing more than one seedling).

SOIL AND CLIMATIC REQUIREMENTS

Soil

Mango being a deep rooted tree, needs soil profile of at least 2 m depth, though in the hilly regions, the seedling mango trees have been observed to be growing on only one meter deep soils. However, more the depth of soil, better is the suitability.

Hard soils, soils poor in depth or soils having hard pan in sub-soil should be avoided. Mango grows well on all types of soil provided they are deep and well drained. Red loamy soils are quite ideal. This crop can be grown on alluvial as well as lateritic soils. Clay and black cotton soils are unsuitable for mango plants due to poor drainage and aeration. The vigour and cropping behavior of a mango tree are affected by the soil type. Certain cultivars of mango can withstand slight soil salinity and alkalinity. The mango growing soils should preferably have a very low total water soluble salt content of 0.04 to 0.05 per cent. Mango is rated as moderately tolerant to salts with 4-6 dsm^{-1}. However, in general, it prefers slightly acidic soil. Soils with an appreciable amount of gravel ($CaCO_3$) too can grow good crop provided they are not very alkaline. A pH of 5.5-7.5 has been found suitable. The presence of small amounts of *kankar* in neutral or slightly alkaline soils upto pH of 7.5 may not harm the tree.

A constant water table is more preferred for good growth and development of mango crop. For mango, the water table should always be at a depth of 1.80 to 2.40 m. If water table is too high, then feeder roots will be submerged with water for a long time, leading to development of chlorotic patches on leaves.

Climate

Mango can be grown under both tropical and sub-tropical climate from sea level to 1,400 m altitude, provided there is no high humidity, rain or frost during the flowering period. However, for commercial cultivation of mango crop, 600 m altitude is ideal. As the altitude of the place increases over and above 1,000 m from MSL, the growth and productivity of the crop are poor. The altitude has a definite role on the time of mango flowering. It has been observed that an increase in every 12 m altitude, flowering is retarded by four days. Similarly, for each degree latitude, south or north of the tropics, flowering is delayed by four days.

For growing mango on a commercial and profitable scale, the temperature and rainfall have to be with in a clearly defined range. It does well within a temperature range from 24°C to 27°C, although it can endure even temperatures as high as 48°C during the period of fruit development and maturity. The annual mean temperature at which mango thrives best is around 26.7 °C. Higher temperatures during the period of fruit development hastens maturity, improves fruit size and quality. When atmospheric temperature is high, fruits exposed to direct sunlight are normally affected by spongy tissue disorder in Alphonso mango. Similarly, air temperature of over and above 40.5 °C develops spongy tissue in Alphonso mangoes. The limiting factors are low temperatures, and frost during the period of flowering. Temperature below 2°C even for few hours in consecutive days may damage the vegetative and reproductive organs.

The amount of rainfall is not so important factor as its intensity and distribution. It can do well in areas having an average rainfall of as low as 25 cm to as high as 250 cm. Localities, which experience bright sunny days and a relatively low humidity during flowering period are ideal for mango cultivation. One of the pre-requisites for successful growing of mango is the absence of rains during the flowering time. Rain at flowering not only washes away the pollen, which adversely affects fruit set, but also encourages greater incidence of mango hoppers, mealy bugs and diseases like powdery mildew and anthracnose, which sometimes damage the crop completely. Further, cloudy weather with resultant increased humidity in the atmosphere also encourages greater incidence of such pests and diseases. This also interferes with the activity of pollinating insects, thus adversely affecting fruit set.

Mango is adversely affected by frosts and freezes if not properly protected. The damage depends on several factors, such as the age of the tree, moisture content of the soil, condition of growth, timing, severity and duration of the frost. If the temperature is below 1.1 °C, the mango plants are adversely affected by frost. A short spell of –3.3 °C and consequent longdrawn out cold spell may lead to the drying out of the young shoots and leaves of mango plants, killing the tree from the top down. The young mango trees with vigorous growth may be injured seriously at 0°C. For protection against frosts, the young mango trees should be covered (thatched) fairly early and the thatching should be removed only when the danger of forst is over.

Exposure to strong winds, whether hot or cool, are harmful to mango crop. Strong winds during the fruiting season cause many fruits to fall prematurely. Sometimes, high velocity of wind may uproot the whole tree. This problem is commonly observed in areas frequently affected with heavy cyclone, especially, in Nellore, Krishna, Ranga Reddy and Viskhapatnam districts of Andhra Pradesh and Kanyakumari district of Tamil Nadu. Dwarf mango varieties like Amrapali, Rumani and Kerala Dwarf are less prone to wind damage compared to tall and huge trees like Pairi and Langra. Seedlings of *Casuarina*, silver oak or *Accasia* can be grown as windbreaks around the mango orchards. However, these windbreaks may compete with the mango crop for nutrients and moisture. Therefore, a trench of half a meter width and half-to-one meter depth should be dug open between the mango tree rows and windbreak to minimize the competition for natural resources.

Sometimes, occurrence of hailstorm may also cause reduction in yield. The hailstorms occur sporadically, particularly, during the pre-monsoon showers. The damage caused to fruit is by physical hitting of hailstorms, which leads to rupture of tissue and such areas get discoloured and start rotting. Anti-hail net can be employed to reduce the impact of hail storms if trees are dwarf.

IMPORTANT VARIETIES

Two hundred years ago when vegetative methods of propagation were not in practice, all the mango trees were seedling ones. This has resulted in the accumulation of a large number of varieties with different taste, flavour, sweetness and other qualities. There are more than one thousand varieties available in India. However, the popular varieties grown in different states are Alphonso, Bangalora, Banganpalli, Bombai, Bombay Green, Dashehari, Fazli, Fernandin, Himsagar, Kesar, Kishen Bhog,Langra, Mankhurd, Mulgoa, Neelum, Alphonso, Chausa, Suvarnarekha, Vanaraj and Zardalu (Table 1.4).

Classification of Mango Cultivars

On the basis of horticultural traits, mango varieties have been classified in different ways as under :

Vigour of the Tree

On the basis of growth habit, mango varieties may be dwarf, semi-dwarf or vigorous. Singh (1954) grouped mango varieties as dwarf if plants having 13 to 15 ft (4 to 5 m) height and 14-16 ft (4.5 to 5.5 m) spread, while those having 32-38 ft (10 to 12 m) height and 30-39 ft (10-12 m) spread were grouped as vigorous varieties.

Maturity of Fruit

The varieties have been classified as early, mid-season and late as per the maturity or availability of fruits. Bangalora and Banganpalli were reported very early in producing the fruits under south Indian condition. Zardalu, Bombay, Ranipasand, Gulabkhas are earlier in eastern and northern part of the country; Langra, Dashehari,

Khasulkhas, Himsagar etc. are mid-season and Fazli, Chausa, Mallika, Amrapali, Kaitki, Neelum etc., are late-season varieties.

Table 1.4: Some Important Mango Varieties Grown in Different States of India

State	Varieties being Grown
Andhra Pradesh	Allumpur Baneshan, Banganapalli, Bangalora, Cherukurasam, Himayuddin, Suvernarekha, Neelum, Totapuri
Bihar	Bathua, Bombai, Himsagar, Kishen Bhog, Sukul, Gulab Khas, Zardalu, Langra, Chausa, Dashehari, Fazli
Goa	Fernandin, Mankurad
Gujarat	Alphonso, Kesar, Rajapuri, Vanraj, Jamadar, Totapuri, Neelum, Dashehari, Langra
Haryana	Dashehari, Langra, Sarauli, Chausa, Fazli
Himachal Pradesh	Chausa, Dashehari, Langra
Jharkhand	Zardalu, Amrapali, Mallika, Bombai, Langra, Himsagar, Chausa, Gulabkhas
Karnataka	Alphonso, Bangalora, Mulgoa, Neelum, Pairi, Baganapalli, Totapuri
Kerala	Mundappa, Olour, Pairi
Madhya Pradesh	Alphonso, Bombay Green, Langra, Sunderja, Dashehari, Fazli, Neelum, Amrapali, Mallika
Maharashtra	Alphonso, Mankurad, Mulgoa, Pairi, Rajapuri, Kesar, Gulabi, Vanraj
Odisha	Baneshan, Langra, Neelum, Suvarnarekha, Amrapali, Mallika
Punjab	Dashehari, Langra, Chausa, Malda
Rajasthan	Bombay Green, Chausa, Dashehari, Langra
Tamil Nadu	Banganapalli, Bangalora, Neelum, Rumani, Mulgoa, Alphonso, Totapuri
Uttar Pradesh	Bombay Green, Dashehari, Langra, Safeda Lucknow, Chausa, Fazli
West Bengal	Bombai, Himsagar, Kishen Bhog, Langra, Fazli, Gulabkhas, Amrapali, Mallika

Fibre Content of Fruit

The varieties may also be fibrous (*e.g.*, Sukul, Baramasia) about fibreless like Langra, Dashehari, Zardalu, Ranipasand etc. and moderately fibrous such as Fazli, Sipia, Taimuria etc.

Shape of Fruit

Burns and Prayag (1920) classified the mango fruits as round, long and indefinite on the basis of shape of fruit.

Firmness of the Pulp

Mango varieties for table types purpose should have soft pulp (*e.g.*, Langra, Dashehari etc.) while few have firm pulp like Bangalora, Mallika etc., and some have loose pulp like Sukul, Baramasia etc.

Bearing Behaviour

On the basis of bearing habit, mango varieties are classified as regular (*e.g.* Amrapali, Neelum, Bangalora) or irregular bearer (*e.g.* Langra, Chausa etc.)

Number of Embryos

On the basis of emergence of seedling from the stone, mango varieties are polyembryonic in nature producing more than one seedling from a stone (*e.g.*, Mylepalium, Goa, Kurukkan, Olour, and other are monoembryonic in nature, producing single seedlings from a stone.

Colour Development

On the basis of colour of fruit, mango varieties are coloured varieties like Sinduria, Kesar, Gulabhas, Zafrani etc. while others are waxy green in colour.

Monoembryonic and Polyembryonic Mangoes

There are two distinct races in mango *viz.*, monoembryonic and polyembryonic. Monoembryonic mangoes consist of a single embryo which gives rise to a single hybrid plant not true to the type (*i.e.* do not have the characteristics of parents); therefore, they are raised by asexual propagation. Almost all the varieties of mango grown in India are monoembryonic. But there are mango varieties, wherein seed contains several embryos, of which, only one will be the result of crossing of two different varieties (sexual progeny), while the remaining ones will be breeding true to type like parents (asexual progenies). The asexual progenies in mango are derived through the nucellar tissues and; hence, they are true to mother plants. Zygotic plantlet in polyembryonic varieties is screened out as the one, which is the closest to the basal side of the seed and it degenerates or, do not develop well. On the other hand, nucellar plantlets are those which develop very well and become the most vigorous in diameter and height. Polyembryonic types are now-a-days used for raising genetically uniform clonal rootstock, which are presently lacking in mango. These are mostly concentrated in the South especially Malabar. However, they are also found in the moist tropics of S.E. Asia such as Malaysia, the Philippines and Indonesia. Some of the polyembryonic varieties are Bappakai, Chandrakaran, Goa, Kurukkan, Olour, Bellary, Kasargod, Mazagoan, Nileshwar Dwarf and Salem.

Salient characters of some important commercial cultivars of mango are as follow:

Langra

It is the most popular variety of India growing mostly in U.P., Bihar and Punjab as mid-season crop. The average fruit weight is about 200 g with oblong oval in shape and waxy light-green in colour. It has a wide range of adaptability and has excellant sugar/acid blend with pleasant flavour. It has very high pulp content (more than 78.5 per cent). The pulp is very sweet. It is a heavy yielder but irregular in bearing. The keeping quality of the ripe fruit is also poor.

Dashehari

It is also one of the most important and popular varieties of northern India. The fruit is medium sized (about 150 g per fruit), attractive in shape (oblong oval) and colour. The pulp is also soft and sweet with pleasant flavour. It is heavy a yielder but irregular in bearing. The keeping and canning quality of the fruit is good. It comes in bearing even after four years of planting.

Gulabkhas

This is also one of the most important, popular and early varieties of North and East India. The fruit is famous for its rose flavour, which is also very sweet in taste. Fruits mature in last week of May and are medium-sized (about 180 g per fruit). It is oblong oblique jn shape, has deep to shallow sinus with prominent raised beak. The ripe fruits are attractive yellow in colour with reddish blush on the base and the side.

Bombay Green

It is also called as Malda in U.P. and Sehroli in Delhi. The fruit is obliquely elongated and green in colour. It has very attractive pleasant flavour with very sweet pulp. It resembles Bombay Yellow except for the development of yellow colour on the shoulder. It is an early variety and susceptible to malformation. Bearing is good but highly irregular.

Zardalu

It is originated in Bengal but has established itself in Bihar especially in Bhagalpur region as an early variety. It has excellent fruit quality and bears moderately to heavily. The fruit is oblong oval, medium in size, attractive yellow in colour and has very soft pleasant flavoured sweet pulp of moderate quantity.

Alphonso

A famous variety of western India. It grows on a large scale in Ratnagiri and Bulsar regions on the West coast. It has limited range of adaptability and does best on the coastal area. It is believed to have originated in Goa. Keeping quality of the fruit is very good (3 weeks from the date of picking). It is also used for canning and export. The fruit is medium sized, ovate oblique and cadmium yellow in colour. The bearing is medium but irregular in habit. Its flesh develops spongy tissue due to reduced hydrolysis of starch into sugar.

Pairi

It is another leading variety of Mumbai region and is extensively cultivated all over the Decan planes of the Western Ghats. The fruit quality is much inferior to that of Alphonso but popular in South India and thrives better in humid region. The medium sized ovate fruit had a prominent red blush and is very attractive. It is an early variety with heavy but irregular bearing. The keeping quality of the fruit is fair.

Neelum

This is also a famous cultivar of South India. It has a wide range of adaptability and excelled in yield per unit area in northern India. The fruit is medium to small in size and ovate in shape. It has medium deep sinus with prominently raised beak. The quality of the fruit is not so good but due to its regular bearing and dwarf growth habit, mango breeders used it in many combinations and succeeded in obtaining these characters in the hybrids as an additional advantage.

Hybrids

Attempts are being made to obtain varieties with maximum desirable characters. For this purpose, some hybrids have been developed in different research stations. Salient characteristics of some important mango hybrids are as follows:

Prabha Shankar

This hybrid has been developed at A.R.I. Sabour from the parents of Bombay x Kalapady. The tree is semi-dwarf and regular in bearing. The pulp is orange in colour, sweet with strong pleasant flavour, fiberless and soft. Fruits become pale green when ripe. This variety has a strong tendency to bear fruits in adverse weather conditions. The fruit is of good keeping quality and ripens a month later then cv. Bombay.

Mahmud Bahar

This is the another hybrid with same parents *i.e.*, Bombay x Kalapady developed at A.R.I., Sabour. The tree is semi-dwarf and regular in bearing. The average fruit weight is 210.0 g, pulp is 70.8 per cent, and Brix is 22 °Brix. The fruit matures from the last week of June to first week of July. The fruits become pale green when ripe. Its pulp is yellow, sweet with mild pleasant flavour, and has a few thick fibres and the pulp is semi-soft.

Jawahar

This hybrid has also been developed at A.R.I., Sabour as a result of cross between Gulabkhas x Mahmud Bahar. The tree is semi-dwarf and highly regular. The average fruit weight is 215.0 g, pulp is 79.5 per cent and Brix is 22.5°.

Mallika

This hybrid has been developed at I.A.R.I., New Delhi. It is a cross between Neelum and Dashehari. Fruits are medium sized, cadmium coloured with good quality, reported to be a regular bearer.

Amrapali

This hybrid has been developed at I.A.R.I., New Delhi. It is a cross between Dashehari and Neelum. It is a dwarf, vigorous, regular and late bearing hybrid. It yields on an average 16 t/ha and about 1600 plants can be accommodated in one hectare.

Manjeera

It is a cross between Rumani and Neelum. It is a semi-vigorous type with a regular bearing habit. Fruits are medium-sized with light yellow coloured skin, firm and fibreless flesh and sweet to taste.

Ratna

It is a cross between Neelum and Alphonso. It is a regular bearer and free from spongy tissue. Fruits are medium-sized with excellent quality. Flesh is firm and fibreless, deep orange in colour with high TSS (19-21 Brix).

Arka Aruna

It is a hybrid between Banganapalli and Alphonso with regular bearing habit and dwarf in stature. About 400 plants can be accommodated per hectare. Fruits are large sized (500-700 gm) with attractive skin colour. Pulp is fibreless, sweet in taste (20-22 Brix). Pulp percentage is 73 and the fruits are free from spongy tissue.

Arka Puneet

It is a regular and prolific bearing hybrid of the cross between Alphonso and Banganapalli. Fruits are medium-sized (220-250 g) with attractive skin colour, having red blush. Pulp is free from fibre, pulp percentage being 70 percent. Fruits are sweet in taste (20-22 Brix) with good keeping quality and free from spongy tissue. It is a good variety for processing also.

Arka Anmol

It is a semi-vigorous plant type from the cross between Alphonso and Janardhan Pasand. It is also a regular bearing and free from spongy tissues. Fruits ripen to uniform yellow colour. Keeping quality of the fruit is very good and it is suitable for export. It has got excellent sugar and acid blend and fruits weigh on an average about 300 g. Pulp is orange in colour.

Other mango hybrids *i.e.*, Nelphonso (Neelum x Alphonso), Neelshan (Neelum x Baneshan), Neeleshwari (Neelum x Dashehari), Swarna Jahangir (Swamrekha x Jahangir), Sabri (Bombay x Gulabkhas), Alfazli (Alphonso x Fazli) and Sunder Langra (Langra x Sunder Pasand), Neeluddin (Neelum x Himauddin) are also released from Fruit Research Stations, Paria, Sangareddy and Sabour and these are performing well.

Recently, some mango hybrids and varieties have been released for cultivation by different Institutes like IARI, New Delhi, CISH, Lucknow and PAU, Ludhiana. A brief introduction to such varieties is presented below:

Pusa Arunima

It is cross between Amrapali and Sensation. The tree is semi-vigorous, regular, attractive red fruit colour with medium TSS, high beta-carotene and good shelf-life.

Pusa Surya

An exotic cultivar 'Eldon' has been renamed as Pusa Surya after assessment for 22 years at IARI, New Delhi. The trees are semi-vigorous, regular with attractive apricot yellow peel fruit colour, medium TSS, and good shelf-life

Pusa Pratibha (Amrapali x Sensation)

The fruits of Pusa Pratibha are attractive in shape, bright red peel develops at ripening and has orange pulp. The fruit size is medium (181 g) with higher pulp content (71.1 per cent). The fruit contains 19.6 per cent total soluble solids and is rich in vitamin C (34.9 mg/100 g pulp) and β-carotene content (11,474 mg/100 g pulp). It has pleasant flavour with improved shelf life (7 to 8 days) at room temperature after ripening.

Pusa Shreshth (Amrapali x Sensation)

It is an exclusive mango hybrid with attractive elongated fruit shape, red peel and orange pulp. The fruit size is about 228 g with attractive red peel colour and higher pulp content (71.9 per cent). The total soluble solids are 20.3 per cent, rich in vitamin C (40.3 mg/100 g pulp) and β-carotene content (10.964 mg/100 g pulp), excellent sugar:acid blend, and has pleasant gustatory aroma with enhanced shelf life (7 to 8 days) at room temperature after ripening.

Pusa Lalima (Dashehari x Alphonso)

The fruits of Pusa Lalima are attractive in shape and having jasper red peel and orange pulp. The average fruit weight is about 209 g, with higher pulp content (70.1 per cent). The total soluble solids (19.7 per cent), vitamin C (34.7 mg/100 g pulp) and β-carotene content (13.028 mg/100 g pulp), and has approving flavour with enhanced shelf life (5 to 6 days) at room temperature after ripening.

Pusa Pitamber (Amrapali x Lal Sundari)

The fruits of Pusa Pitamber are attractive oblong shaped with bright yellow peel at full ripening. It is moderately resistant to mango malformation and major insect-pests of mango. The average fruit weight is about 213 g with higher juicy pulp (73.6 per cent). It has medium total soluble solids (18.8 per cent), vitamin C (39.8 mg/100 g pulp) and β-carotene content (11.737 mg/100 g pulp), and appealing flavour with good shelf life (5 to 6 days) at room temperature after ripening.

Ambika

This variety has been developed at CISH, Lucknow. It is a cross between Amrapali x Janardan Pasand. Fruits have attractive skin colour with firm flesh and high TSS.

CISH M-2

It has also been released from CISH, Lucknow. It is a hybrid between Dashehari x Chausa. The fruits are dark yellow in colour with firm flesh and scanty fibres. It is a late season variety.

Rajendra Mango-1

It is a cross between Langra and Sunderprasand. Tree is medium and spreading. Bearing is moderate and regular. Fruits are large and similar to those of Langra in physical appearance. Pulp is yellow, firm, sweet and mildly flavoured. Keeping quality is fair. It matures in the last week of June.

Rajendra Mango-2

It is a cross between Alphonso and Fazli. Tree of this hybrid grows somewhat tall and erect and bears a medium crop regularly. Fruits are greenish and bigger in size like Fazli. Pulp is fibreless but of average quality. It ripens in the end of June, about a month earlier than Fazli.

Gangiyan Sindhoori 19

It is a sucking type variety released from PAU, Ludhiana.

PLANT PROPAGATION

Seed Propagation

In old days, mango was primarily propagated through sexual means of propagation *i.e.* by seeds. But, as it is heterozygous, amphidiploid and highly cross-pollinated crop, the seedling trees never come true to type and had any resemble to the mother plant; besides, extended juvenility. During the Mughal period, many of the choice seedling trees were designated as distinct varieties and attempts were made to propagate them vegetatively. However, with the advancement made in the propagation techniques, it has now become possible to propagate mango by different vegetative methods, easily.

Seed propagation is still the major method of raising rootstocks in mango. Germination rate and vigour of seedlings are highest when seeds are taken from fruits that are fully ripe, not still firm. Also, the seed should be fresh, not dried. If the seed cannot be planted within a few days after its removal from the fruit, it can be covered with moist earth, sand, or sawdust in a container until it is planted, or kept in charcoal dust in a desiccators with 50 per cent relative humidity. Seeds stored in the latter manner have shown 80 per cent viability even after 70 days. Higher rates of germination are obtained if seeds are stored in polyethylene bags but the seedling behaviour may be poor. Sprouting occurs in 8 to 14 days in a warm, tropical climate and 3 weeks in cooler climates. Seedlings generally take 6 years to fruit and 15 years to attain optimum yield for evaluation. Regarding the production of seedling from polyembryonic varieties, largest number of seedling per stone was noticed in Moovandan and Villai Columban followed by Bappakai, Kurukkan and Olour.

Vegetative Propagation

Inarching, veneer grafting, side grafting and epicotyl grafting are the popular methods of propagation in mango. These methods are briefly destributed hereunder.

Inarching

It has been a common method of propagating mango cultivars and is still followed by the nurserymen all over the country. However, it has the following major disadvantages:

(i) It is a cumbersome method since the rootstock seedlings have to be carried to the mother plant for grafting.

(ii) It is a laborious and time consuming method since raised structures are to be fabricated around the mother tree for placing the rootstock seedlings for grafting.

(iii) The inarched plants have to be irrigated and looked after for at least 2 to 3 months at various odd places around the tree.

(iv) It is an uneconomical method since only one plant can be obtained by grafting a long scion-shoot.

(v) Inarched plants usually get pot-bound and hence mortality on transplanting is rather high, the ideal age of potted seedlings for grafting being about a year.

The method consists of uniting the selected shoot (scion) of a desired parent tree (mother plant) with the potted or transplanted seedling (rootstock) by approach grafting. For this purpose, about one-year-old seedlings are most suitable when they attain a height of about 30-45 cm and thickness ranging from 0.75 to 1.5 cm. These seedlings are either grown in pots or under the mother plant from which the grafts are to be prepared, depending upon the availability of suitable branches. Generally, a one-year-old twig of the scion tree about 60 cm in length and nearly of the same thickness as that of the stock is chosen for grafting. Young and non-bearing trees should not be selected as mother plants.

Inarching should be done during the growing period when the tree is in active sap flow condition termed as active growth period. A hot and very dry period, as well as heavy rainfall during the inarching period is not suitable. The end of the monsoon in heavy rainfall areas and early monsoon in the light rainfall areas is the best period for inarching. In North India, July is the best month for inarching. In the more equitable climate of South India, the operation can be done in any time between July and February.

A thin slice of bark and wood, about 5 cm in length, 7.5 mm width and 2 mm deep, is removed by means of a sharp grafting knife from the stem of the stock as well as from the scion branch. The dimensions can be proportionately increased or decreased according to the thickness of the stock and scion. The cuts thus made should be absolutely flat, clean, boat shaped, even and smooth. The ends of these cuts should be round and not angular. The cut surfaces of both, *i.e.,* stock and scion are made to coincide facing each other so that there remains no hollow space between the two. These are then tightly tied by polythene/alkathene strips of about 1.5 cm in width and preferably of 200 gauge thickness, which has proved to be a good tying material. After about one month of operation, the scion below the graft union and stock above the graft union should be given light 'V' shape cuts at weekly intervals so that grafts can finally be detached while giving the fourth cut. In the last stage, the top of the stock above graft union should also be removed completely.

Veneer Grafting

In veneer grafting, the scions or budsticks are detached completely from the mother plant at the time of grafting. Thus, scion may be procured from distant places and brought to the nursery for grafting. The scion should be a healthy 22-25 cm long terminal or axillary shoots of the previous season growth (3-4 months old) with some activated buds. Its leaf blades are cut off, leaving petioles about 10 days earlier and the buds should be ready to sprout. This is the most important factor for the success of this method. A slanting cut of 5 cm long on one side of the stem is made and the bark along with wood is removed with an oblique cut. A slanting cut of the same size on one side of the scion base is made, which will just fit with the notch of the stock. The scion is then placed in position on the stock in such a way that the two cambium rings of both stock and scion come in close contact. It is then wrapped tightly with alkathene sheet keeping the terminal end free. When the scion begins to grow at the top (after about 3 weeks), the upper part of the stock is removed, thus forcing the buds to grow more rapidly. The plastic wrap is removed after 2-3 months.

Softwood Grafting

In softwood grafting method, which was first developed in India on mango at Gujarat Agricultural University, Anand in 1971, one month old seedlings with bronze leaves, are propagated by wedge grafting with 3-4 months old scion sticks. Lamina of scion leaves is cut off about a week before detachment without disturbing apical bud. The grafting is then tied with 200-300 gauge polythene tape.

Stone Grafting

Stone grafting or epicotyl grafting is done on germinating seedlings (stone being attached with seedling). Splice and wedge grafting techniques are used in this method in rainy season when humidity is high in atmosphere. This is a predominantly followed method of propagation in Konkan region.

'T' Budding

Shield or 'T' budding and patch budding are also practiced in mango; though, at smaller scale. Activated buds give better success of budding than dormant buds.

Rootstocks

Stionic effects have been studied in mango and in this regard, Taimuria, Olour, Rumani, Moreh and Nakkare Kurukkan appeared to be dwarf rootstocks, whereas, Dashehari, Chausa, Moovandan and Sakarchina showed vigorous nature. On the other hand, Goa has been noted as vigorous rootstock followed by Bappakai and Chausa. In India, double-grafting has been found to dwarf mango trees and induce early fruiting. Naturally, dwarf hybrids such as 'Julie' have been developed. The polyembryonic Indian cultivars, 'Olour' and 'Vellai Colamban', when used as rootstocks, have a dwarfing effect; so has the polyembryonic 'Sabre' in experiments in Israel and South Africa. In Peru, the polyembryonic 'Manzo de Ica', is used as rootstock; in Colombia, 'Hilaza' and 'Puerco'. 'Kaew' is utilized in Thailand.

FLOWERING, POLLINATION AND FRUIT SET

Mango trees must undergo a period of rest or dormancy, a period without any externally observable growth activity in mature shoots after flushing, for about 3 – 5 months, depending on the variety. This is a pre-condition required for the successful flowering. This dormancy may be naturally induced by low temperatures in most sub-tropical environments and soil moisture deficits under tropical conditions and may be by both ways under some situations. Low N supply also tends to promote this rest in mature shoots. Though, the term dormancy or rest is used to explain this stage of growth, plants are not actually in a dormant stage physiologically. At this time, plants very actively photosynthesize and produce carbohydrates for storage to be used in the following season.

High temperature, high soil moisture and high N levels in the soil or in tree tissues might break this rest and force the tree to flush again and again. This will affect the physiology of trees by changing the food reserve base and will lead to a further vegetative bias. Trees under such conditions produce poor crops during the following fruiting season. This happens more often when varieties used to rest under low temperature climates are planted under high temperature environments.

During the rest period, tree carbohydrate reserve levels increase as against tree N level. Thus, C:N ratio of the trees increase. Also at the same time, shoots require to make flower primordia on the terminals of mature shoots. When all these conditions are accomplished, the mature shoots become physiologically ready for flowering. After this stage is achieved, trees are ready to flower when the environment is appropriate for flowering.

Flower bud differentiation has been reported to start in October. It has been found that corolla and androecium develop almost simultaneously. And neither the high humidity and rain at the time of bloom nor the late rain appeared to influence flower bud differentiation indirectly during the following year.

Mango trees less than 10 years old may flower and fruit regularly every year. Thereafter, most mangoes tend toward alternate, or biennial bearing. A great deal of research has been done on this problem, which may involve the entire tree or only a portion of the branches. Branches that fruit one year may rest the next, while branches on the other side of the tree will bear. Leaves in mango play an important and immediate role in flower bud differentiation and their effect might be localized, more particularly branch wise. There was a strong indication that the leaves supply some kind of substances, which might be vital for flower bud differentiation. In the biennial varieties, deblossoming of shoots or of a small branch in limited number failed to induce flower bud formation in those shoots during the following year. Deblossoming in the regular bearing varieties induced immediate axillary flower bud formation as well as leafy shoots, which were able to differentiate flower bud in the following year. In the Philippines, various methods are employed to promote flowering: smudging (smoking), exposing the roots, pruning, girdling, withholding nitrogen and irrigation, and even applying salt. In the West Indies, there is a common folk practice of slashing the trunk with a machete to make the tree bloom and bear in 'off' years. Deblossoming (removing half the flower clusters) in an 'on' year will induce at least a small crop in the next 'off' year. Almost any treatment or condition that retards vegetative growth will have this effect. Spraying with growth-retardant chemicals has been tried, with inconsistent results. Potassium nitrate has been effective in the Philippines. It has been found that a single mechanical deblossoming in the first bud-burst stage, induces subsequent development of particles with less malformation, more hermaphrodite flowers, and, as a result, a much higher yield of fruits. Cultivar, 'Neelum', in South India that bears heavily every year, apparently because of its high rate (16 per cent) of hermaphrodite flowers. The average for 'Alphonso' is 10 per cent.

Presence of leaves in shoots is essential for overall development of plant including flower bud differentiation. To support the growth of a single fruit to a normal size of a mango fruit, it requires 60-90 fully expanded leaves if the assimilate were to be met from current photosynthesis. The mango inflorescence is primarily a racemose. The inflorescence may be terminal or axillary as per the position of the panicle on the branches. The size of a panicle may vary from a few centimeters to more than half meter. The panicle takes 3 weeks to attain its full size, usually.

The period of blooming varies from 10 days to several weeks depending upon varities and the weather conditions. Generally, the anthers dehisce within 5 to 8 hours after the opening of flowers. Anthesis starts before the completion of the panicle growth and the total flowers in a panicle open within 20-35 days. It has been noted that pollen for cross-breeding can be stored at 32° F (0° C) for 10 hours. If not separated from the flowers, it remains viable for 50 hours in a humid atmosphere at 65° to 75° F (18.33° –23.09° C). The stigma remains receptive for 18 hours before full flower opening and, some say, for 72 hours after.

The mango trees do not flower simultaneously in all directions; at least two distinct flushes are noticed. The number of flowers per panicle depends on the variety, periodicity, age of the plant and size of panicle, and the prevailing climatic temperature. In general, it ranges from 197 to more than 9,000 flowers in different varieties and locations. Mango flowers are visited by fruit bats, flies, wasps, wild bees, butterflies, syrphids, moths, beetles, ants and various bugs seeking the nectar and some transfer the pollen but a certain amount of self-pollination also occurs. However, houseflies (*Musca domestica*) seem to be major pollinator for mango. Although, honeybees, *Melipona* sp. (a small Dipterous fly) and *Syrphidae* sp. (a king of hover fly) also reported to be responsible for pollination. Soon after pollination and fertilization of the ovary with pollen, fruits start to develop. Then the plant become active in translocating carbohydrates and other food reserves from the storage tissues such as roots and trunks to the developing fruits for their growth. Then the size of the crop a tree can bear, is primarily determined by the tree reserve level. Excess fruits drop may happen at various stages of fruit development. Fruit drop is governed physiologically by the size of reserve food base in the tree and also by climatic conditions prevailing during this time. Even if trees have sufficient reserves, lack of K or soil moisture deficits during fruit development phase or pest and disease attacks on flowers and developing fruits may act as limiting factors controlling the size of the final crop. Therefore, at this stage, trees have to be managed very well providing all necessary inputs correctly.

Self Incompatibility

Although some workers had suspected the existence of self-sterility in mango, it was not until some years later its existence in 'Dashehari' was reported. Later on, it was proven that 'Dashehari', 'Langra', 'Chausa' and 'Bombay Green' are self-unfruitful. Initial fruit set following self- pollination is negligible (0.0-1.68 per cent) in these varieties compared with that after cross-pollination (6.4-23.4 per cent). The majority of the selfed fruitlets drop within 4 weeks of pollination and none attains even half- grown size. An analysis of its causes reveals that the processes culminating in ovule fertilization are the same as after compatible cross-pollination. The difference between self-and cross-pollination becomes apparent from the fifteenth day after pollination, when degeneration of the endosperm and the surrounding nucellus is widespread after selfing. 'Himayuddin' as the polliniser resulted in 50 per cent more fruit set and increase in fruit size compared with that of the open-pollinated 'Rumani' fruits with 'Jahangir' as the pollen parent, however, fruit size was reduced. This emphasizes the need for study on suitable pollinizers for mango cultivars. Likewise, 'Dashehari' is cross-incompatible with 'Chausa' and 'Safeda Malihabad' but it is cross-compatible with 'Langra', 'Rataul' and 'Bombay Green'. 'Langra' is cross-incompatible with 'Alphonso', 'Bombay Green', 'Chausa'. 'Fazli', 'Rataul', 'Safeda Malihabad' and 'Surkha Burma'. 'Chausa' is cross- incompatible with 'Bombay Green' and 'Rataul' but cross-compatible with 'Langra' and 'Safeda Malihabad'. Cultivars like 'Bombay Green' and 'Dashehari' appear to be the best pollinizers for 'Dashehari' and 'Chausa', respectively. 'Neelum' is partially self-compatible, whereas 'Mallika' is self-incompatible.

FRUIT GROWTH AND DEVELOPMENT

Fruit growth and development in mango fruit is characterized by sigmoid curve. Development of fruit in 'Langra' and 'Dashehari' starts in the last week of March and is completed by the end of second week of June. The percentage increase in growth in 'Dashehari' and 'Samar Behisht Chausa' as expressed in terms of length, breadth and thickness, is maximum in April, followed by May and March, respectively. It is least in June. However, maximum increase in weight and volume of fruits is recorded in May, followed by April and June, respectively. It is almost negligible in March. Many characters of the pericarp like cell size, laticiferous canals, intercellular spaces, etc., in different tissues of the fruit contribute to the increase in length, breadth, thickness and volume of the fruit, whereas increase in weight in the later stages is associated with the accumulation of starch grains in the cells. In mango, the period of rapid growth is directly associated with the period of maximum activity of auxin and gibberellin like substances in the seed. The second period of rapid development of fruit may be due to rapid initiation of development of seed and decrease in the inhibitor content of the pericarp. Further, the slowing of growth after 64 days in 'Dashehari' and 29 days in 'Chausa' may be due to the lignification and development of endocarp, as it results in competition for food substances in the formation of the endocarp and the fleshy part of the fruit. In the later stages of the growth and development of mango fruit, the exocarp region develops into a leathery protective skin, the mesocarp into a fleshy and pulpy region and the endocarp into a hard, stony and non-edible region. The laticiferous canals meant for storage of fluid, called mango *chenp*, secreted by a well-defined layer of secretary cells surrounding the canal, appear in the endocarp region at anthesis and in the outer regions 7 days after anthesis. They originate schizogenously and grow throughout the period of fruit growth. Percentage increase in the growth of these canals is found highest in March, followed by April, in all the regions of fruit. The growth in the exocarp is due to cell division, increase in cell size and increase in size and number of laticiferous canals. Cell division in this region continues up to 35 days in 'Dashehari' and 42 days in 'Chausa'. Thereafter, the growth is controlled by other factors such as cell size and size and number of laticiferous canals.

The increase in the thickness of the mesocarp region is initially by cell division, cell size and size of laticiferous canals. But cell division stops after 42 days in 'Dashehari' and 49 days in 'Chausa'; thereafter the growth-contributing factors are cell size, size of laticiferous canals and increase in intercellular spaces. In the endocarp, region, the increase in thickness is due to cell division, cell size and size of the laticiferous canals in the early stages. But cell division stops after 4-5 weeks and further increase in growth is due to increase in cell size, size of laticiferous canals and increase in the reticulate fibrous structures. The hardening of endocarp region starts 64 days after anthesis in 'Dashehari' and slightly later in 'Chausa'. The cell division in the inner epidermis, however, continues for a longer period than in the main endocarp region. The accumulation of starch grains adds to the increase in fruit weight in later stages.

The developing mango fruits is a rich source of many growth substances such as cytokinin, auxin and gibberellins. Lack of pollination and fertilization results in the cessation of the active growth of the ovule, thereby eliminating the sources of growth hormones, which are required for the enlargement of fruits. Also, the occurrence of natural parthenocarpy has not generally been reported in mango. Observations show that growth substances produced in the seeds are necessary for the development of mango fruits. In a study, fully developed parthenocarpic fruits were obtained by spraying N^6 benzyladenine (250 ppm) at the time of anthesis and later on a combination of 'Symbol'-naphthoxyacetic acid (10 ppm) and GA_3 (250 ppm) at fortnightly intervals. The fruits although small in size were superior in quality to the normal seeded fruits.

PLANTING AND ORCHARD ESTABLISHMENT

Planting should be done when the growth flushes in the plants have matured. Leaves should be cut to half its size to minimize transpirational water loss and to reduce transplanting shock. It is better to establish plants in the orchard after heavy rains are over. In case appropriate drainage and irrigation facilities are available, planting may be done at any time of the year. It is advisable to paint the stems of plants with a white water based paint to protect plant stems from direct sunlight and heat.

Plant spacing may be adjusted depending on soil type, variety and management systems to be adopted. Tall varieties require 10-12 m space between plants and rows as mango is conventionally planted in a square system. While the dwarf varieties require only 2.5-3 m or more distance between plants and rows. A number of modifications by adopting rectangular and quincunx systems of planting are not uncommon in mango orchards with an anticipation of accommodating more number of plants per unit area and higher fruit production. But the introduction of high density planting with dwarf variety has surpassed the previous records of production. A hectare of land accommodating 1,600 Amrapali trees can produce 11.5-22.0 tonnes of fruit during 5th and 9th year of age as compared to an average of 3.67 tonnes of ftuits/ha from the trees of traditional varieties with conventional spacing.

The planting operation for propagules starts from digging of pits (0.6-1.0 m cube) at suitable distance one month earlier, to dry up the soil and to destroy the soil borne insects and diseases. The dug out soil mixed with manures, fertilizers, and plant protection chemicals is used to fill half of the pit. The root ball of the propagule is placed on the soil mixture at the centre of the pit, gradually covered with soil mixture and pressed tightly. A prop should be used and the plant should be tied on 2-3 position with the prop and copiously irrigated. Planting should be carried out in such a way that the stock-scion union is about 4 – 6 inches above the ground level. Scion should never touch the ground at all.

High Density Planting

High density orcharding appears to be the most appropriate answer to overcome low productivity and long gestation period for early returns and export of mangoes. High density planting helps increase the yield/unit area. In mango, three different methods of high density planting *viz.*, low density, moderate density and high density planting are followed. The low density planting at a spacing of 10 × 10 m accommodates 100 plants/ha (40 plants/acre), the moderate density at a spacing of 7 × 7 m accommodates 204 plants/ha (82 plants/acre) and high density planting at a spacing of 5 × 5 meter accommodates 400 plants/ha (160 plants/acre). Planting system could be paired row planting to facilitate mechanization in mango orchards. For mid-season and late-season varieties, 10 × 5 × 5 m spacing in paired row planting with 222 plants/ha is found to be an ideal population. In North India, mango Amrapali is found amenable for high density planting with a spacing of 2.5 m × 2.5. Soil drenching with paclobutrazol (2ml/tree) induces flowering during off year. It has become a commercial practice in Konkan region of Maharashtra. If coupled with pruning, it helps increase production/unit area in Dashehari. The polyembryonic mango Vellaikolumban when used as rootstock, imparts dwarfing in Alphonso. At present, high density plantation of mango is followed over hundreds of hectares in states like Maharashtra, Goa, Andhra Pradesh, Karnataka and Tamil Nadu.

Comparing traditional and high density planting, Mahammed and Wilson (1984) made the following observation (Table 1.5).

Table 1.5: Comparison between Traditional and High-Density Planting Systems in Mango.

Attribute	Traditional System	High-Density
Basic need	Dwarf cultivars	Vigours cultivars
Tree number	A far large trees : 150-200 trees h⁻¹	Many small trees : 500-1600 trees h⁻¹
Bearing	Long juvenile period (6 to 8 years)	Precocious bearing (2 to 3 years)
Production	Low : 15 to 25 t/ha	High : 30 to 50 t/ha
Management	Difficult to manage due to large size of the tree	Easy-to-manage due to small size of the tree
Labour	Requires more labour	Requires less labour
Production cost	High	Low
Harvesting date	Normal	Easy hastens or delayed
Fruit quality	Large canopy, poor sunlight penetration and poor quality	Small canopy, better sunlight penetration and better quality
Establishment cost	Low	High
Machinery	Does not require expensive machine	Requires expensive machine
Chemical or growth substances	Not essential to use growth substances to control growth, flowering and fruiting	Requires the use of growth substances to control flowering and fruiting
Utilization of natural resources profits	Low	Efficient high

It has to be worked out in different locations with the different planting distances to get the appropriate number of plants for a particular region with a particular cultivar of mango grafts.

Hints for Growing 'Amrapali' under HDP

The development of Amrapali hybrid of mango opened the doors for high density planting in a vigorous fruit trees like mango. The package of practices for growing Amrapali mango is more or less similar to those followed for other commercial varieties but there are certain important points, which are necessary to get fruitful yield from high density orchard of Amrapali as under:

Establish *'in situ'* Orchard of Amrapali

It is always advisable to establish Amrapali orchards *in situ* for high success. For the establishment of *in situ* high density orchard, it is advisable to sow 5 stones in each pit during July at 2.5 x 2.5 m distance in triangular system for raising seedlings *in situ*. Side veneer grafting in the germinated seedlings should be done in the next July. If it is not possible to sow stones *in situ*, alternatively three germinated seedlings can be planted per pit. Out of these, two healthy seedlings in a pit are grafted in July by veneer grafting technique. At last, only one plant per pit should be retained.

Train Plants to a Bush by 'Pinching Off'

To train the plant to a bush, it is advisable to pinch off the terminal vegetative growth in the first and 2nd year of planting to encourage formation of branches. Afterwards, the tree is maintained dwarf by regular bearing habit of the variety.

Apply Manure and Fertilisers Timely and Adquately

To the young plants up to the age of 3 years, apply fertilisers and organic manure in March and April and to the bearing plants, apply all FYM in October and fertilisers in two split doses in July (immediately after harvest) and in October along with FYM.

Irrigate at a Right Time

Irrigation should be given at right time and as and when needed . However, stop irrigation in the bearing plants, which have developed more than 15 shoots, from October onward till flowering. Irrigation after October will disturb C:N ratio, which may change the rythum of regularity.

Maintain Optimum Number of Leaves per Fruit

It is important to maintain certain number of leaves/fruit. It is absolutely necessary in Amrapali as it requires lesser number of leaves per normal sized fruit as compared to other varieties. But because of dwarfed nature and continuous fruiting, the leaf area sometimes falls below the optimum level, thereby reducing the fruit size.

Avoid Intercropping

Farmers usually go for inter-cropping in commercial orchards of mango during early years of planting. This practice should be avoided in Amrapali high density orchards.

Pruning for Sustained Production

Normally, high density Amrapali orchards start giving low yields after the fruitful production of 12-14 years, primarily due to overcrowding or intermingling of tree of canopies. Pruning is thus absolutely necessary in such orchards. Pruning should be done immediately after harvesting of fruits (3rd week of July). It will help in maintaining the productivity and vitality of high density Amrapali orchards for many more years.

Reduce Malformation Incidence

Amrapali is susceptible to 'mango malformation'. To control floral malformation, single spray of 200 ppm NAA or 40 ml planofix in 9 litres of water during first week of October and deblossoming at bud burst stage in January should be practised regularly for all bearing plants.

Keep Control on Mealy Bugs

In areas where 'mealy bug' is a problem, grease collar banding or alkathene banding should be started from December besides soil raking with some insecticides in December end. Similarly, the crawling bugs can be killed by 0.2 per cent spray of metasystox.

Control Hopper Timely

Mango hopper is a serious pest of Amrapali and can cause heavy losses. In the event of hopper attack, spray the plants with diazinon (0.05 per cent) or rogor (0.05 per cent) or phosphamidon (0.05 per cent) in January and repeat it again after 20 days.

Manage Powdery Effectively

Like other mango varieties, Amrapali is seriously affected by 'powdery mildew'. There is expected attack of powdery mildew in March either due to rains or cloudy weather. Thus, a prophylactic spray of karathane (0.02 per cent) in March is highly beneficial. It may be repeated after 15-20 days. In extreme cases, a third spray is also necessary, which, may be given 15-20 days after the second spray.

Harvest Fruits at a Right Time

Many farmers do not harvest Amrapali fruits in time. Normally, they harvest it by 3rd or 4th week of June, which is not the right time of harvesting Amrapali. It is a late variety and is ready for harvesting during the 3rd week of July. For judging the maturity of fruits, farmers can use a simple technique. Take fully mature fruits from the plant and put them in one jug full of water. If the fruit sinks in water that means fruits are ready for harvesting otherwise not.

Thus, for higher production and productivity, it important that the above mentioned practices should be followed as and when required. Farmers are therefore advised to follow these recommendations strictly to make their Amrapali orchard much more productive.

Advantages of High Density Planting

☆ Pre-bearing period is reduced if dwarf hybrid varieties are used.

☆ Higher yield from small area due to more number of trees (400 trees/ha at 5 × 5 m) than conventional method (100 trees/ha at 10 × 10 m).

☆ Optimum utilisation of available natural resources *viz.*, land, water and sunshine.

☆ Management of canopy by manipulating agronomic practices like irrigation, fertigation, system of training and pruning, application of growth regulators. This requires more labour leading to employment generation. Canopy is maintained nearer to the trunk, resulting in more efficient use of water and nutrients. Otherwise barren shoots also take nutrients but do not flower.

☆ Mango bears flowers on the previous year's shoot growth. However, adoption of scientific pruning methodology allows manipulating the vegetative growth of the plant and regulating flowering and fruiting. Pruning of one year old shoots at the base followed by manuring, fertilizer application and irrigation along with plant growth regulator treatments like NAA (Planofix) induced increase fruit set, reduce fruit drop and results in better fruit growth.

☆ Due to pruning, the height of the trees remains under control, facilitating mechanized harvesting of the matured fruits.

☆ Handling of fruits in the field is easy and smooth.

☆ Regular pruning could overcome the problem of alternate bearing.

INTERCULTURAL OPERATIONS

Training and Pruning

Young plants should be pruned early to allow development of a strong open frame. Remove top to force 3-4 side branches about 60-90 cm above the ground. The main stem can be allowed thereafter, spaced at 20-25 cm apart in such a way that they grow in different directions. Thereafter, when the side branches are ready to prune, remove top of those branches to force further multiple shoots to get a spreading growth habit. To train branches as spreading limbs, staking may also be useful. By pruning this way, plants must be trained to have an open, well spaced and spreading canopy.

This pruning and training process must be a continuous activity. All this can not be done in a single pruning or two. Therefore, growers must regularly visit the plants and prune them as and when required. Also, pruning must be commenced when the plants are well established after transplanting, but before any major growth has been started.

In bearing orchards, trees must be pruned after harvesting. The objectives of pruning are to remove dead or diseased wood, to remove additional growth flushes, to allow more light penetration into the leaf canopy and also to control tree height to facilitate cultural management practices. Erect branches are less fruitful compared with spreading branches and these must be removed. At the same time, lower branches are pruned to about 1 meter from ground level to facilitate cultivation practices such as weed control, fertilizer application and irrigation. A light pruning is also recommended after main flush matures, in order to remove suckers and excess shoots.

Top-Working of Inferior Seedling Trees

One of the main handicaps that had led to slow pace of evolution of the mango industry in India is the preponderance of seedling trees, which are mostly of inferior type. Young seedling trees below 20 years can conveniently be top-worked with scion woods of commercial varieties. The technique of veneer grafting can be used to top-work the seedling tree with elite scion. For top-working, new shoots are encouraged by heading back. Heading back can be done early in the spring season. After heading back, new shoots arise in the spring, which are ready to be veneer grafted during the rainy season. If the tree is older, it is better not to head back all the main branches simultaneously, to avoid a sudden shock to the tree, which could result in the splitting of the bark of the main stem. The tree can be converted into a commercial variety in stages. The main trunk of this tree should be wrapped with hessian cloth or gunny bag during winter to avoid bark splitting. Such top-worked trees start giving commercial crop in the third year of the top-working.

Rejuvenation of Senile Orchards

In general, after 30 years of age, mango trees exhibit declining trend in fruit yield because of dense and overcrowded canopy, especially in areas where the tree growth is very vigorous. New emerging shoots are weak and are unsuitable for flowering and fruiting. The population of insect-pests builds up and the incidence of diseases increases in such mango orchards. These unproductive mango trees can be converted into productive ones by pruning. Intermingling, diseased and dead branches should be removed.

Undesirable branches of unproductive trees should be headed back from 1.5 to 2.0 m from the distal end after harvest in South Indian conditions. The cut portion of branches is applied with copper-oxychloride paste to avoid infection of diseases. During March-April or October-November, a number of new shoots emerge around the cut portion of the pruned branches. Only 8 to 10 healthy and outward growing shoots may be retained at proper distance to develop a good frame work in the following years. These pruned trees need to be fertilized with adequate quantity of N, P and K during June-July and September–October after soil test. The plants need to be irrigated at an interval of 5 days especially during summer.

Farm yard manure at the rate of 50 – 100 kg per tree may be applied. Unwanted new shoots should be regularly removed to maintain the tree canopy and to avoid re-crowding of branches. This helps in getting proper nourishment to retained shoots. After two years of pruning, new shoots come into bearing and the fruit yield of tree starts increasing gradually. By this technique, old and unproductive trees may be transformed into productive ones.

Irrigation

Irrigation is an effective management tool to improve tree productivity in bearing trees. Irrigation is also important in enhancing the fertilizer use efficiency of trees. Young trees must be frequently irrigated until the plants are well established in the field. Thereafter, up to about 3-4 years, plants must be regularly irrigated as determined by soil moisture status. In case of grown up trees, irrigation at 10 to 15 days interval from fruit set to maturity is beneficial for improving yield. However, irrigation is not recommended for 2-3 months prior to flowering as it is likely to promote vegetative growth at the expense of flowering.

Manuring and Fertilization

Application of plant nutrients economically at correct time with right amounts in a way that nutrients could be taken up by plants efficiently with minimum losses, covers the whole issue of nutrient management. Proper nutrient management has significant effects on tree productivity. The main purpose in nutrient management is to keep mineral nutrient levels in the tree with in the desired range to have the growth and development effects and fruiting of trees as desired by the grower. A general recommendation cannot be made for manuring mango trees, because it is affected by many factors such as age of the plants, variety, soil etc. But there are certain facts, which apply in case of fruit plant for manuring of an orchard. For example, the concentration of the feeder roots of the tree plays an important role in absorption of nutrients. In general, 170 g urea, 110 g single super phosphate and 115 g muriate of potash per plant per year of the age from first to tenth year and thereafter 1.7 kg, 1.1 kg, and 1.15 kg, respectively of these fertilizers per plant per year can be applied in two equal split doses (June-July and October). Foliar spray of 3 per cent urea is recommended before flowering in sandy areas. Lime application is an integral component in mango nutrient management. Mango prefers a soil pH of around 6.5. At this pH, most soil nutrient elements become available to plants. If the soil is acidic, finely ground lime or dolomite may be applied to adjust the soil pH. These liming materials may be applied as bands along the tree row or just around the trees on the root zone. The exact lime requirements may be estimated by using soil analysis results.

For young plant, basin method of application is suitable, while for older trees, furrow method of application is convenient and economic. The manure should be well dug into the soil in trench 60 cm broad, 15 cm deep and 30 cm away from the trunk in case of one-year-old trees. The trench should be widened about 15 cm and its inner edge taken 15 cm further from the tree every year. In older trees, furrow bottom placement method is applied. In this method, the fertilizers are placed in a continuous band on the bottom of the furrow in the process of ploughing. Each band is covered as the next furrow is turned. This method of application is helpful for not only maximizing the utilization of fertilizers by the tree but also minimizing the weed population during ploughing.

Intercropping

Intercrops such as vegetables (tomato, onion, cauliflower, bean, radish, palak etc.), legumes (pea, lentil, moth bean, black and green gram), short duration and dwarf fruit crops like papaya, guava, peach, plum, etc. depending on the agro-climatic factors of the region can be grown. Heavy feeders like cereals and cucurbits should preferably be avoided for cultivation in the mango orchards. In fully grown bearing orchards, where shading is a problem, shade-loving crops like turmeric may be raised.

The water and nutrient requirements of the intercrops must be met separately. But during the cultivation of intercrops, some principles should be kept in consideration. These crops should not interface with the growth of the main crops, *i.e.* fruit trees. At bearing stage, irrigation for intercrops should be arranged in such a way to facilitate the growth and production of fruit trees. In other words, priority should be given to the fruit crops and not to the intercrops, and it will not create any problem if irrigated.

MAJOR PRODUCTION PROBLEMS

A. Fruit Drop

A large number of fruits drop down from the tree before they reach maturity. The shedding of blossoms and developing fruits in mango starts by the end of March and continues throughout the fruiting season. The fruit drop in mango can be divided into 3 distinct phases: (a) pin head drop, (b) post setting drop or April drop, and (c) May drop. However, economically the most important drop is the May drop than other two fruit drops. It is also possible that if during the first two drops, more fruits are retained on tree, then subsequently, the third, pre-harvest fruit drop (May drop) becomes more important. Lack of pollination, low stigmatic activity, defective perfect flowers, poor pollen transference, occurrence and extent of incompatibility, competition between developing fruitlets, lack of moisture in the soil, unfavourable climatic conditions during fruit development period (*viz.*, wind and hail storm), high incidence of serious diseases and insect-pests are the main causes of fruit drop. Adopting suitable cultural practices such as thinning, pruning and maintaining adequate mineral and moisture regime helps in reducing fruit drop. Application of growth regulators such as triple spray of NAA at 5 ppm first at full bloom, second at pea stage and third at marble stage, very effective in controlling fruit drop in mango. Similarly, spray of urea (4 per cent), KNO_3 (3 per cent) and NAA (40 ppm) also increased the fruit number per plant in mango.

B. Biennial Bearing

The production of crop in alternate year or irregularly is known as alternate or biennial or irregular bearing. In fact, in mango tree, bearing is not observed exactly in alternate year, so irregular bearing is more appropriate term for this kind of behaviour. Most of the commercial varieties of mango show the same pattern of bearing (biennial); but 'Baramasi' may exhibit erratic and off-season bearing; and 'Totapari Red Small', 'Neelum' and 'Bangalora', show distinct regularity; particularly the latter. Even the regular- bearing types, if they carry a heavy load of crop in 1 year, show a tendency towards reduced yield in the following year. Hence, the basic tendency of bienniality exists even in the so-called regular-bearing varieties of mango. There are various factors, which are largely held responsible for this phenomenon in mango.

Factors Governing Bienniality in Mango

1. Growth Pattern

Early initiation and cessation of shoot growth in mango, followed by a definite dormant period is considered to help the shoots to attain proper physiological maturity, which is essential for fruit bud initiation in them. However, there is now enough evidence to show that the growth of mango shoots is purely a varietal feature and that fruit-bud differentiation in regular bearing varieties is an annual feature. On the other hand, in biennial bearing varieties, it is governed by 'on' and 'off' year phase of the trees rather than by the time of origin and cessation of growth of shoots. The shoot depending upon the variety, may stop putting forth extension growth after May or continue till September or onwards. The potentiality of these shoots to form flower buds will depend on the floriferous condition of the tree, which in turn will be determined by the amount of fruit load carried by the tree in the previous year.

2. Nitrogen and Carbohydrate Reserves

In almost all the varieties excepting in 'Baramasi', it has been found that higher starch reserve, total carbohydrates and C:N ratio in the shoots favoured flower initiation in mango. Studies have shown that the total nitrogen content was higher in the stem and leaves of trees, which were expected to initiate flower buds irrespective of the varieties they belonged to. The available evidence indicates that nitrogen and carbohydrate reserves play an important role, if not the primary role in flower-bud initiation. Possibly, the accumulation of these compounds may create a favourable condition for the synthesis and action of the substances actually responsible for flower induction in these plants.

For the normal development of a mango fruit, it requires 60-90 fully expanded leaves for development, if the assimilates are to be met from current photosynthesis. As a consequence, the developing fruit withdraws large amount of reserve metabolites from vegetative organs during the 'on' year. This might contribute to biennial and erratic bearing in mango.

3. Hormonal Control of Flower Formation

Studies have indicated that the shoots of 'Dashehari' during 'on' year and from 'Totapari Red Small' trees, which initiated flower buds, were found to contain higher level of the growth-promoting substances during the period of flower-bud initiation than those from 'Dashehari' of 'off' year tree, which remained vegetative. It is possible that the auxin is only active in combination with a receptor complex. Auxin appeared to play major role in the induction of flowering in mango. However, if endogenous auxin is the only factor favouring flowering in mango, external application of synthetic auxins would have initiated flowering in the 'off' year. The results so far obtained do not support it. This indicates that factors other than auxins are also involved in the initiation of flowering in mango.

Further, results of different studies have shown that gibberellins are antagonistic to flowering in mango. The endogenous levels of gibberellins were higher in shoot tips of 'off' year than in those of 'on' year. This suggests that the failure of flowering in an 'off' year may be accompanied by a higher level of gibberellins in the shoot tips.

Similarly, upon investigation, the endogenous levels of cytokinins were noted to be higher in the shoot tips of 'on' year trees than of 'off' year ones at the time of flower-bud differentiation. Cytokinin concentration in panicle of mango was highest, 5-10 days after full bloom and decreased rapidly thereafter. The amount of cytokinins increased from 10 days after full bloom, reaching a peak at 40 days after full bloom.

Certain inhibitors have also been found to play role in biennial bearing habit of mango. The presence of inhibitors similar to abscisic acid, in the shoots of mango trees has been observed. The shoots of 'Dashehari' from 'on' year trees and of 'Totapari Red Small' trees contain relatively higher level of this inhibitor during flower-bud initiation than those of 'Dashehari' from 'off' year trees. This indicate that it may also be involved in the initiation of flowering in mango. Since the inhibitor is antagonistic to both GA_3 and auxin effects on cell elongation, it may perhaps check vegetative growths of mango, thereby providing conditions suitable for flower-bud initiation.

4. Climatological Factors

Climatic factors are associated with the biennial bearing of mango trees in 2 ways: (*i*) by damaging the crop directly by destroying the fruit bud, blossoms and fruits, or (*ii*) by creating conditions that indirectly affect the production of flower or fruit on the tree adversely. Frost, high temperature accompanied by low humidity and hailstorm fall under the first category, as they directly damage the fruit bud and developing fruits'respectively and thereby reduce the crop considerably. Cloudy weather and rains during blossoming period reduce the crop indirectly by creating favourable conditions for the spread of mango hoppers and of diseases such as powdery mildew and anthracnose.

Besides, mango trees in extreme humid places and under milder climatic conditions may not fruit at all due to their increased tendency toward vegetative growth. Even if they flower, owing to excessive humidity, the pollen is never in a suitable condition to be transferred by insects for pollination purpose. These factors are contributory in nature and bring about biennial bearing only indirectly. At times, severe attack of mango hopper, blossom blight or early spring frost, destroys completely the fruit blossoms and this converts an 'on' year into an 'off' year condition. The following year, which would have been normally be an 'off' year becomes an 'on' year, in which, an excessive amount of fruit buds are produced, resulting in heavy bearing. Consequently, the following year is turned into a completely 'off' year and thus the biennial rhythm is initiated.

5. Cultural Practices

It appears that phenomenon of biennial bearing in mango is inherent in the tree from the very beginning and becomes more and more operative as the tree grows in age and in that it may be aided indirectly by environmental factors. Cultural practices like optimum uses of fertilizers, irrigation, deblossoming etc. can only reduce the impact of biennial bearing on production to only some extent.

6. Crop Load

Generally, moderate blossoming is one of the chief conditions of annual fruit bearing in fruit trees. The biennial habit can be minimized by undertaking some measures that reduce the number of fruit buds setting into fruits. Fruit load on the tree appears to be the main factor governing production of shoots, their fruit-bud differentiation and ultimately the biennial bearing. Increased crop load seems to have a systematic depressing effect on many manifestations of tree growth. Apparently, in mango, the fruiting is an exhausting process. The number of fruits retained till harvest is a varietal feature. However, the total number of fruits that are harvested is important because of their deleterious influence on the production of new shoots and their subsequent fruit-bud differentiation. Therefore, the load of fruits appears to be the main conditioning factor for 'on' or 'off' year in mango. The fruit development not only depends on current assimilates but also to a great extent on the food reserve.

The utilization of reserve metabolites from vegetative organs during the 'on' year could contribute to biennial or erratic bearing. The load of crop in its very initial stages of fruit growth (within a fortnight of full bloom) inhibits the flower-bud formation for the crop of the following year. Non-bearing and bearing units on the same tree are growing under the same environmental conditions with their alternating potentials for flower-bud formation. In the bearing units of a tree, even when new shoots are initiated, which attain physiological maturity quite in time, these fail to differentiate flower buds if such a unit carries an optimum load of crop.

Management Practices to Overcoming Bienniality

1. Use of Regular Bearing Varieties

Regarding genetical improvement, some hybrids have been developed, which were claimed to be regular bearer. Regularity in bearing is supposed to be controlled by recessive gene or genes, which is linked with poor quality of the fruit. The variety Neelum, besides being regular in rearing, was relatively superior in fruit quality and it was therefore possible to obtain some exceptionally good hybrids by crossing it with other commercial but biennial varieties like Dashehari, Langra and Chausa. The hybrids like Amrapali, Ambika, Arka Puneet, Pusa Arunima, Pusa Pitamber, Pusa Pratibha, Pusa Shresth have been reported to be regular bearer.

2. Cultural Practices

Cultural operations like irrigation, manuring, timely disease and pest control etc. are also found to be effective in controlling irregularity in bearing to some extent. By good annual cultural treatments, trees can be kept reasonably healthy, and total failure of crops, which usually occurs between 2 heavy cropping can be averted.

3. Use of Growth Regulators

Growth regulators are utilized with two objectives *i.e.*, reducing the flowering in 'on' year and increasing the same in 'off' year. It was possible to ensure regularity through deblossoming by spraying some chemicals like ethephon (50 and 100ppm), M and B-25105 (50, 500, 1,000 and 2,000ppm) and carbaryl (5,000 and 10,000 ppm) on Dashehari and Chausa. In Alphonso, 7 sprays of 100 ppm of TIBA at an interval of 15 days from second week of September to second week of December resulted in maximum induction of flowering on vegetative shoots by passing the vegetative growth during expected 'off' year followed by CCC 500 and 1,000 ppm.

Some good results have also been obtained when nutrient elements were sprayed. Early, uniform and profuse flowering of Carabao mango at Philippines was obtained when sprayed with KNO_3 at 12.5 g/liter. Some workers also claimed to overcome irregular bearing by spraying urea (2-4 per cent).

In recent years, application of paclobutrazol (PP_{333}) and Mcpiquat Chloride (MP-Cl) achieved some success in including or stimulating flowering in mango. Soil and foliar application of these chemicals promoted flowering and regulated the time of flowering in mango.

C. Malformation: A Threat to Mango Industry

Mango, the 'king fruit' of India, is affected by many biotic and abiotic stresses, which hinder its production greatly. Among various problems, mango malformation is considered as one of the most burning problems, which renders mango orchards unproductive throughout the subtropical areas of the world. Malformation is an intricate disorder of mango, which was first reported by Watt in 1891 in Darbhanga, Bihar, India. Now it is widely reported in mango-growing countries such as India, Egypt, South Africa, Brazil, Sudan, USA, Israel, Mexico, Bangladesh and Pakistan. The frequency of infestation is, however, variable being very heavy to slight. Although it first appeared in a tropical region, yet it is comparatively less reported in this region. In Pakistan, there is no region and no variety, which is free from this disorder.

Symptoms of the Malady

Mango malformation usually appears in two forms: vegetative malformation and floral malformation. The symptoms of the malady are discussed herehunder:

(a) Vegetative Malformation

Vegetable malformation appears mostly in the seedling plants or new grafts in the nursery than in the old plants of an orchard. However, it may appear in the older plants occasionally, particularly in those, which have suffered from floral malformation. The affected seedlings or grafts produce small shootlets, bearing small scaly leaves with a bunch like appearance on the shoot apices. The vegetative buds in the axils or shoot apices swell, affecting the apical growth of the plant, as a result, numerous vegetative buds sprout. More often, the shootlets and their branches are not distinguishable due to overcrowding and the whole mass of rudimentary leaves gives a bunch like appearance, resembling 'witch's bloom' structure, which is also called as *'bunchy top'*.

The root system of the affected seedlings does not develop properly, which checks the seedlings growth. These seedlings are also not suitable for their use as rootstock, because the grafting success is very poor. Seedlings infected at an early stage finally die. Those infected later, may continue growth, but at a very slower rate. Grown up trees in the orchard may also show vegetative malformation in the leaf axils and at the apex of the branches. Bunches of small scaly leaves are commonly seen on shoot tips bearing malformed panicle along the malformed buds, or along the internodes. These structures, however, soon dry up and remain hanging to the branches or shoots as dry masses. The development of vegetative malformation usually continues throughout the year.

(b) Floral Malformation

Malformation of inflorescence manifests as deformed, suppressed and clustered inflorescence, thickened rachis, shortened primary and secondary axis. The panicles bear males with seldom a hermaphrodite flower. The manifestation is intricate in respect that from the same terminal, few buds are normal others malformed. On the same panicle, a part may be normal and the rest malformed. And on the same healthy panicle only one lateral is malformed while on a completely malformed inflorescence only one lateral is healthy. It indicates the intricacy of the problem or that there is some disorder in the internal system which either stops proper functioning or promotes abnormal functions at a certain site and from there the normal growth of the panicle is disrupted. This abnormality hence could be only localised. Floral malformation appears in the bearing trees when they start flowering. The inflorescence/panicles get deformed due to enlargement of the flowers. The primary, secondary and tertiary rachises become short, thickened and are much enlarged or hypertrophied, giving the flower a characteristically clustered appearance. Such panicles are usually greener and heavier than the normal panicles. A malformed panicle usually produces much larger number of flowers than healthy panicles, but most of the flowers remain unopened. Malformed flowers are usually male and rarely hermaphrodite. The ovary of the malformed hermaphrodite flowers is usually large, non functional with very poor stigmatic receptivity. The malformed hermaphrodite flowers had 1-4 ovaries/flowers as compared to normal hermaphrodite flowers, which had one or rarely two ovaries. The malformed flowers also show abnormal development of stigma, style and ovary. Further, embryo abortion takes place at a faster rate in the malformed flowers. Malformed panicles rarely set fruit and continue to grow on the tree for a long period. However, such panicles have been reported to produce normal flowers and sometime may also set fruits in off-season (September-October) under north Indian conditions. In general, floral malformation affects the bearing potential of the tree adversely. In acute cases, the whole tree may be rendered unfruitful.

The severity of malformation may vary from shoot to shoot and from panicle to panicle. Hence, both healthy and malformed panicles may appear on the same shoot. In general, recovery of the malformed panicles is almost impossible. As reported earlier, it is the most dreaded malady of mango in subtropical climate than the tropics. The incidence may be cent percent on a particular variety in a particular climate. In general, almost all the varieties in tropical climate are free from this malady, but when grown under sub-tropical climate, they are severely affected by malformation.

The incidence of vegetative malformation may be upto 80 per cent in the seedlings or grafted plants than the bearing plants. However, the incidence of floral malformation may be cent percent in some varieties under certain locations.

Causative Factors

The causes of this malady have been studied exhaustively by a number of workers, but the actual cause of the malady has been very controversial and still inconclusive.

However, the various factors associated with this malady have been briefly discussed below:

Varietal factors

Almost all the varieties of mango, except 'Bhadauran', 'Abib', 'Amin', 'Dhudhiya Langra' in India, are susceptible to malformation. In Egypt, 'Zebba', 'Hiendi' and 'Anshas' have been reported to be resistant to malformation. Susceptibility of the varieties varies from one percent to cent percent, depending upon the age of tree, climatic condition and the time at which flowering has taken place. For example, when flowering takes place at a temperature lower than 25⁰C (December-January) under north Indian conditions, the trees tend to produce more malformed panicles. Although, all the varieties in the tropical climate are free from malformation, but when grown under sub-tropical climate, they are severely affected by the malformation.

Mites

Various investigators have reported the association of mites like, *Aceria mangiferae* and *Tyrophagus castellanii* (*T. asiaticus*) with the cause of mango malformation. However, later the possibility of mite as a cause of this malady was ruled out, because the correlation between bud mite population and incidence of malformation could not be established and acaricidal sprays under field conditions also did not check the disease. Several workers also reported the role of mites as a carrier of fungal pathogens as a cause of this malady, which was also ruled out by other workers. Hence, from the several reports, which appeared in the literature to support or refute the role of mites in causation of mango malformation by direct feeding or as vector, it is not conclusive enough to state that mites are directly or indirectly involved in causing mango malformation.

Fungi

Several reports which have appeared have indicated the association of a number of *Fusarium* species like *F. moniliforme*, *F. moliniforme* var. *subglutinans*, *F. oxysporum* and *F. solani* with the cause of mango malformation. Of these, association of *F. subglutinans* with mango malformation is much more accepted than other species, because both vegetative and floral malformed tissues contain hyphae of this fungus in higher proportion than healthy tissues. It has also been isolated from cortex, phloem and parenchymatous pith cells of the malformed tissues. Several workers have also reported its association with malformation from the countries, like the USA, Israel and South Africa. Though, indifferent reports have appeared in literature and no fungicide has yet been recommended for its control, yet its involvement with the malady can't be entirely ruled out.

Viruses

As early as 1946, Sattar, a famous Virologist suggested that mango malformation might be of viral origin. This view was based on the inconsistent transmission and failure to isolate any pathogens from the malformed tissues. This view was further strengthened because of occurrence of high incidence of malformation in trees severely infested by hoppers (*Amaritodus alkinsoni*). However, later many workers could not transmit the disease in healthy trees by hoppers or other sap-sucking pests like mealy bugs, thrips or aphids. Some workers also reported the transmission of this disease though infected bud wood, but the other failed to transmit it thorough different grafting techniques. Similarly, some workers have used modern techniques like electron microscopy and serological tests and have ruled out the possibility of involvement of viruses being the cause of this malady.

Nutritional Factors

Association of nutrition imbalance with mango malformation has also been a matter of controversy among the scientists. For example, some earlier investigators have reported that the malformed tissues had lower level of N than the healthy tissues, but the others reported higher N, K and low Zn in malformed tissues. However, enhanced application of N curtailed malformation, whereas addition of P and K significantly increased the incidence. Tripathi (1992) reported that leaves of vegetative malformed seedlings had higher proportion of ash, silica, and calcium than the healthy seedlings, but spraying either with B, Ca, Cu, Fe, Mg, Mn, K, Na and Zn had failed to control malformation. Further, some workers have achieved reduction in malformation incidence by spraying sulphate of cobalt, cadmium and nickel, but none proved beneficial elsewhere. Hence, from the foregoing discussion, it appears that either excess or deficit of nutrients does not appear to have any relation with the cause of this malady.

Physiological Factors

Some Horticulturists in association with Plant Physiologists have reported the association of the following physiological factors with the cause of mango malformation.

(*i*) Chlorophyll Content

Contradictory information is available on chlorophyll content of the leaves producing healthy and malformed panicles. For example, some workers have reported higher amount of chlorophyll 'a, b' and total chlorophyll in the leaves from malformed shoots, but *vice versa* by the others.

(*ii*) Phytohormones

A low level of auxins and high level of gibberellins, cytokinins, ethylene and ABA has been reported by various workers in malformed tissues than in the healthy ones. However, a few have reported low gibberellins in the malformed tissues. The higher level of gibberellin like substances may account for the production of solely male flowers and continuous growth of the malformed panicles. Low level of auxins and high levels of gibberellin, zeatin, ABA and ethylene in the malformed seedlings caused inhibition in their apical growth, which affects the orientation and development of lateral organs, resulting in the production of numerous small shoots. Short internodes in the affected shoots might have been caused by sub-optimal levels of endogenous gibberellins, while the production of small leaves, showing epinastic curvature might be the result of an increased ethylene production. Hence, the cumulative evidence suggests that mango malformation seems to be linked with the imbalance between growth promoters and inhibitors.

(*iii*) Carbohydrates

Some scientists have reported the level of the carbohydrates to be higher in shoots bearing malformed panicles than the healthy ones. They emphasized that the starch in the malformed panicles was not hydrolyzed into sugars to meet their energy requirements. Yet another group of scientists have reported that the malformed panicles or the shoots bearing them had lower level of starch than the healthy ones, ruling out the possibility that starch in the malformed panicles was not hydrolyzed into simple sugars to meet their energy requirements. In contrast, he postulated that low level of starch and high level of sugars in malformed panicles suggests the hydrolysis of starch into simple sugars, which provides sufficient energy for excessive growth of the malformed panicles. Some authors, however, have proposed that malformation of the tissues might result from the degradation of cellulose and lignin as they observed lower level of saccharides in the malformed tissues. Hence, from the present day knowledge, it can't be concluded whether low or high levels of carbohydrates are related to the cause of malformation.

Biochemical Factors

On the basis of the information available in the literature, the following factors seem to have some association with mango malformation:

(*i*) C/N Ratio

Many researchers had been of the view that in malformed panicles failure to fruit set is caused by improper C/N ratio, which was considered to be an important factor responsible for differential growth and disturbed sex ratio in the malformed panicles. However, it is almost impossible to conclude whether low or high C/N ratio is responsible for the cause of mango malformation. For example, some researchers have observed higher C/N ratio in malformed panicles and leaves bearing malformed panicles, respectively. In contrast, other group reported low C/N ratio in malformed panicles or the shoots bearing malformed panicles than shoots bearing healthy panicles. From this discussion, it appears difficult to conclude whether high or low C/N ratio is responsible for the cause of malformation in mango.

(*ii*) Malformin

Ram and Bist (1984) for the first time isolated malformin like substances from the malformed panicles and seedlings, which was later confirmed by various other workers. Thus, they concluded that the accumulation of mangiferin results in an excessive vegetative growth, which helps in the continuous emergence of rudimentary leaves, inflorescence bearing florets intermixed with leaflets and florets transformed into leaf structures probably due to hormonal imbalance. However, the mechanism of synthesis of such substances in the plant system is still not yet understood.

(*iii*) Magniferin

Some workers have isolated mangiferin, a non-toxic polyphenol from the malformed tissues, establishing the fact that high content of phenols and steroids have strong correlation with the intensity of malformation. Different opinion of other workers led to conclude that these polyphenols are the effect of malformation rather than the cause, because application of any of these compounds could not induce malformation in mango.

(*iv*) Fusicoccin: A Possible Cause

Fusicoccin, a phytotoxin produced either by the inject injury or fungi, has been reported to be one of causes of malformation by the scientists of IARI, New Delhi.

(*v*) Enzyme Activities

Some workers have correlated the malformation with the activities of enzymes like IAA oxidase and polyphenol oxidase. The activities of these enzymes were reported to be higher than the healthy panicles. In contrast, the activities of amylase and catalase have been reported to higher in healthy panicles than the malformed ones.

(*vi*) Nucleic Acids, Amino Acids and Proteins

It has been observed by some workers that healthy panicles contain higher content of RNA, DNA, soluble proteins and total amino acids than the malformed panicles. In contrast, others have reported, higher content of RNA and DNA in malformed panicles than in healthy ones in the initial stages, but the trend is different in later stages. They also reported that malformed seedlings had higher protein and total amino acids than healthy seedlings, suggesting that their higher contents must be responsible for excessive growth, converting the healthy ones into malformed ones.

Environmental Temperature

Among various factors responsible for the development of malformation, environmental temperature seems to play an important role. It is a known fact that malformation occurs frequently in subtropical climate, but tropical areas are almost free from it. Moreover, the panicles, which appear early in the season even under subtropical climate tend to be malformed and unproductive than those, which appear late in the season. Because, the temperature in the tropics remain almost the same and higher than in the subtropics, where the temperature goes quite low in winter and higher in summer. Hence, it can be concluded that environmental temperature appears to be most closely related with the causation of malformation.

Control Measures

Several workers have made various attempts to control malformation by using plant growth regulators, deblossoming, nutrients, phenolics, malformation antagonists, pesticides and pruning etc., but none of the recommended measures could get commercial viability because of one or the other lacunae. The different measures recommended/used by the workers are briefly described hereunder:

Cultural Practices

Removal of Panicles

It has been reported earlier that panicles, which appear early in the season (December-January) when the temperature is low, tend to malformed and unproductive. Hence, it is advisable to remove such panicles as and when they appear. Experiences have shown that if such panicles are removed in time, healthy panicles appear in the shoots during the normal flowering time (March), which are healthy and fruitful.

Mulching

Some experiments have been conducted with the use of black polyethylene as mulch or covering the whole plant with it. Authors of this book is of the view that above all causes of the mango malformation; environmental temperature is the most vital. Hence, any manipulation done to raise the temperature to some extent would also reduce the incidence of malformation. The small but useful experiment conducted by Dr. P.K. Majumder and his associates at IARI, New Delhi in the early sixties reveal that if temperature is raised to some extent, the malformation does not appear. Hence, use of polythene either as mulch or to cover the plant can be an effective practice for reducing malformation.

Deblossoming

Deblossoming at bud burst stage alone or in combination with the application of NAA (200 ppm) during flower bud differentiation is perhaps the most effective treatment to reduce the malformation significantly. Various workers, whosoever had worked on the control of this malady, have recommended this practice for reducing the incidence of malformation. However, hand deblossoming is a tedious, cumbersome and expensive task. Hence, chemical deblossoming has been recommended. To achieve this, ethrel (200 or 500 ppm) or cyclohexamide (200 ppm) should be sprayed at panicle emergence stage for successful Deblossoming in mango.

Application of Nutrients

On the basis of the work done that nutritional imbalance may be the cause of mango malformation, various workers have conducted different experiments in the past. On the basis of information available, it can be concluded that neither soil application nor foliar sprays of macro- or micronutrients could reduce the incidence of malformation. However, some workers claim that application of nitrogen (200 g/ plant) reduced it to some extent, but addition of P and K enhanced it.

Selective Pruning

Selective pruning of the malformed parts or panicles helps in reducing the incidence of malformation in the subsequent years. However, some workers are of the view that after removing malformed parts/panicles, application/spray of insecticide and/or fungicide (captan or diazinon) is essential. The pruned malformed panicles should be destroyed or buried deep in the soil to avoid the spread of inoculum.

Raising Seedling in Polyhouse

Vegetative malformation is a serious problem in seedling mango plants in the nursery. Experiences of the author of this book and that of other workers have shown that seedlings raised for grafting in the polyhouse are relatively free from malformation. Hence, it is all time advisable to raise mango nursery in a low cost polyhouse. The seedlings raised so are not only healthy, but have very high grafting success as well.

Use of Malformin Antagonists

Use of malformin antagonists and antimalformins has been proved to be beneficial in some cases. For example, Ram and Bist (1984) have recommended to give three spays of glutathione (2,250 ppm), ascorbic acid (2,110 ppm) and silver nitrate (600 ppm) when the panicles are 4-6 cm long for taking fruitful yield from the malformed panicles. However, later, Singh and Dhillon (1990) proposed to spray these antagonists during first week of October rather than in February when panicles are 4-6 cm long, for getting fruitful results.

Use of Chemicals

Pesticides

Several investigators have attempted for the control of malformation by using various insecticides, fungicides or acaricides, but the results have been highly contradictory and inconclusive. Some workers have claimed that application of diazinon (0.04 per cent), bavistin (0.01 per cent), phosphamidon (0.03 per cent), methyl demeton (0.1 per cent), fytolan (0.2 per cent) and sulphur (0.25 per cent) minimize the incidence of malformation, but others have contradicted by saying that neither fungicides nor insecticides or acaricides are effective for controlling mango malformation. Spray of fungicides like Topson-M and captan were applied during panicle pruning in April and after harvesting during the month of July. The trees sprayed with Topson-M particularly during July produced less malformed panicles as compared to the control trees

Use of Plant Growth Regulators

The first recommendation for the control of mango malformation was suggested by Manumder *et al.* (1970) which was confirmed by several other workers with small modifications. On the basis of various recommendations, it can be concluded that application of NAA or planofix (200 ppm) during October (prior to flower bud differentiation) is most effective in reducing the incidence of floral malformation in almost all the cultivars of mango. In addition, some workers had been of the view to spray ethrel (400 ppm) at bud inception stage during February. Application of NAA (200 ppm) during October, followed by a spray of ethrel (200 ppm) during February has also been quite effective to reduce the floral malformation effectively (Table 1.6). GA @ 30ppm reduced incidence of malformation when sprayed prebloom, NAA @ 150ppm and paclabutrazol @ 1000ppm at post harvest stage proved significantly effective to control the incidence.

Use of Phenolics Compounds

It has been suggested by some workers that polyphenols also minimize the destruction of auxin and recommended that the application of catechol (1,000 ppm) during the first week of October to be very effective measure for reducing the floral malformation to a greater extent (Table 1.6).

Table 1.6: Some Recommended Plant Growth Regulators, Phenolics and Malformin Antagonists for the Control of Malformation

Chemical	Cultivar	Concentration (ppm)	Time of Application	Remarks
1. Plant growth regulators				
NAA	Dashehari, Chausa, Bombay Green	100 or 200	October	Reduced malformation remarkably
NAA	Langra	200	October	Reduced malformation remarkably
NAA	Bombay Green	200	October	Reduced malformation remarkably
Planofix	Bombay Green	250 (4 sprays at weekly interval)	After 20th October	Reduced malformation remarkably
NAA and Ethrel	Dashehari	200 + 500	1st week of October and at bud inception in February	Reduced floral malformation significantly
Ethrel	Dashehari	400	At bud inception in February	Reduced malformation significantly
2. Phenolics				
Catechol	Dashehari	1000	1st week of October	Quite effective to reduce malformation
3. Malformin antagonists				
Glutathione + Ascorbic Acid	Dashehari	2,240-2,110	3 times when malformed panicles are 4-6 cm long	Malformed panicle set fruits
Ascorbic Acid	Dashehari	2000	1st week of October	Subsequently reduced malformation

Growing Resistant Varieties

At present none of the commercial varieties of mango are free from malformation. Resistant varieties, like Bhadauran, Abib etc., don't bear fruits of commercial value. Hence, development of resistant variety (ies) is the only solution for this long lasting chronic problem. However, when 'Bhadauran', an immune variety to malformation, had been used as one of the parents in hybridization programme at IARI, New Delhi, all the resultant hybrid seedlings were not observed to be resistant to malformation. In addition, 'Bhadauran' also transmitted all its undesirable characters to the hybrid seedlings. Later, 'Lal Sundari' variety has been used as one of the parents in hybridization programme. Fortunately, the resultant hybrids of cross combination, 'Amrapali' and 'Lal Sundari', have been observed to be free from malformation. At present, hybrids like, H-8-11, H-2-6, H-2-10 etc., having 'Amrapali' and 'Lal Sundari' parents are at final stage of their testing, which are regular in bearing, free from malformation and powdery mildew, which are considered as the most serious problems of mango industry in north India.

CROP PROTECTION

A. Major Diseases and their Management

Powdery Mildew

The disease is caused by *Oidium magniferae*. The fungus attacks the leaves, the flower scales, buds of tender flower heads, axils, stalks and fruits. It manifests itself by the appearance of white superficial, powdery fungal growth on the affected parts. Later, whole surface is covered with the powdery substance, which is blown away by even a slight disturbance caused by the winds. The affected fruits do not grow in size and may drop before attaining pea size, which causes great losses to the growers.

Management Practices

1. Before flowering, a prophylactic spray of wettable sulphur (0.2 per cent) or karathane (0.1 per cent) or benlate (0.1 per cent) should be given.
2. Three sprays of bavistin (0.1 per cent) or microsul (0.2 per cent) has been found effective in controlling the disease. Spray with fungicides karathane, wettasul, baycor, calixin, bayleton, tridemorph and dinocap, anvil 5 sc, spotless 12.5 WP and bayleton 25 WP have been found effective against the disease.

Anthracnose

Anthracnose, also known as 'blossom blight' or 'leaf spot' or 'fruit rot', is a very common and widespread disease of mango in all mango-growing areas. The causal organism is *Colletotrichum gloeosporioides*. It affects tender twigs, shoots, flower and flower stems, and fruits and hence it produces symptoms of leaf spots, blossom blight, wither tip, twig blight, fruit rot. Young leaves are most susceptible and numerous oval or irregular grayish brown spots appear on the leaf lamina which coalesces to cover larger areas at later stage. The affected dead leaf tissue is blown away giving shot hole type of appearance. Symptoms of wither tip or die-back starts from the tip of young branches which defoliates first. Black necrotic areas are formed on tips and the twigs dry downwards. The inflorescence is also affected and minute black spots appear on the flowers which dry and shed. The severity of the disease varies according to prevailing weather conditions. Half mature to mature fruits show slightly sunken black spots which may be converted into cracks under severe infections. The spots are often concentric at the stem end but sometimes in streaks on one side of the fruit, suggesting the spread of disease through spores washed down by rain water from the stem end.

Management practices

Growers should adopt the following measures to avoid the incidence of anthracnose in mango:

1. All the infected twigs should be pruned and burnt along with fallen leaves. Proper irrigation and fertilization is essential to maintain tree vigour, which helps in avoiding twig infection.

2. Spraying of captan (0.3 per cent) along with zineb (0.25 per cent), Bordeaux mixture (3:3:50), blitox or bavistin (0.1 per cent) or topsin-M (0.1 per cent) thrice a year *i.e.*, during February, April and September.

Die-Back

Die-back is considered as one of the most destructive diseases of mango in India and many other mango growing countries of the world. The disease is destructive both on young and old plants and the incidence varies from variety to variety ranging from 30 to 96 per cent. It is caused by *Botryodiplodia theobromae*. Besides other fungi such as *Pestalotia magniferae, Phoma* sp. *Colletotrichum gloeosporioides, Sclerotium rolfsi* and *Rhizoctonia solani* are also associated with dieback symptoms.

The disease is characterized by dying back of the twigs from top downward, particularly in old trees, followed by complete defoliation. It gives an appearance to the trees as if they have been scorched by fire. In addition, discolouration and darkening of the bark at a certain distance from the tip is also seen. The upper leaves loose their colour, turn brown and curl upward. Such leaves nay shrivel and fall down. The bark of the twigs may split length wise, from which gum exudes before they die.

Management Practices

Following control measures are absolutely necessary for the control of dieback in mango:

1. Select scion wood from healthy trees for raising new plants.
2. Sterilize the grafting knife properly before use.
3. Cut and burn the affected branches.
4. These practices should be supplemented with spray of carbendazim (0.1 per cent) or Bordeaux mixture (5:5:50) or copper oxychloride (0.3 per cent) at fortnightly intervals during monsoon season.

Bacterial Blight and Canker

It is also known as bacterial spot, leaf spot, black spot, mango blight and bacterial black spot. It occurs in all parts of India and some parts of the world, but causes serious losses in humid weather and may cause 40-60 per cent losses in certain years. The disease is prevalent in Maharashtra, Tamil Nadu, Andhra Pradesh, Karnataka, Delhi, Haryana and Uttar Pradesh. Besides being pathogenic on several varieties of mango, the organism is capable of infecting wild mango, cashew nut and weeds as well. It is caused by *Xathomonas campestris* pv. *mangiferae indicae*.

The symptoms of the diseases appear on leaves, petioles, branches, trunk and fruits as spots and cankers. Initially, group of minute, water-soaked lesions measuring 1-5 mm in diameter appear towards the tip of the leaf, which are surrounded by a chlorotic halo. These lesions increase in size, coalesce and dry up, which are often rough and raised. These spots are small on younger leaves, but such leaves are drop off. In severe infestation, older leaves also fall down. Later on, such lesions also appear on petioles, fruits and tender shoots. On young fruits, water-soaked lesions are developed, which turn dark brown to black. Cracks may also appear on the affected fruits. Badly infected fruits drop off prematurely. Sometimes cracks appear on the skin of infected fruits releasing gummy ooze containing bacterial cells. On branches, twigs and stem, the fresh lesions are water-soaked, which later become raised and dark brown having longitudinal fissures. The vascular tissues beneath these lesions are filled with gum, which oozes out giving a sticky appearance. Cankers also appear on the flower stalks, which result in dropping of the flowers and just set fruits.

Management Practices

1. Keep the orchard clean to avoid further spread of the disease.
2. Seedling certification, inspection and orchard sanitation are important. Planting material should be taken from certified nurseries.
3. Grow resistant varieties like, Bombay Green, Fazli and Suvarnrekha etc.
4. Spray streptocycline (300 ppm) with copper sulphate (0.3 per cent) or cuprimicin 500 twice during May at 10 day's interval.
5. *Bacillus coagulans*, a bacterium antagonist, provides effective biocontrol measure against *Xanthomonas campestris* pv. *mangiferae-indicae*. Preharvest application of *Bacillus licheniformis* alone and alternated with Cu sprays (Kocide and Nodox) applied at 3-weekly interval from flowering until harvest controlled moderate levels of bacterial blight.
6. Integrated disease management practices that will help to keep the pathogen away prevent its dispersal or minimize the initialion of infection are: use of disease free planting stock, copper sprays + pruning to remove infected branches and improve aeration with in tree+ practices hygine such as sterilization of pruning and harvesting implements+ providing wind breaks to minimize wind damage + resistant cultivars.

Soft Rot

Soft rot of mango is caused by *Erwinia carotovora*, which also infests many vegetables and causes rotting. Bacterium enters the fruits through wounds or lentlcels and dissolves the middle lamella by producing the enzyme pectinase. As a result, some brownish lesions appear on the fruits. Later, these lesions increase in size, coalesce to form big patches and causes complete rotting of the fruits.

Management Practices

Following measures are considered necessary to control soft rot of mango:

1. Avoid mechanical injury to the fruits.
2. Dip the fruits in streptocycline sulphate solution (0.03 per cent) before storage.
3. Avoid cultivation of vegetables like tomato, chillies, peas, beet, cauliflower, brinjal etc., near orchards.
4. Spray streptocycline (100 ppm) during fruit development stage in May.

Phoma Blight

Phoma blight is caused by *Phoma glomerata* and first reported at Central Mango Research Station, Lucknow. The disease was later detected in mango growing belt around Lucknow region. It is now gaining economic importance.

The symptoms of the disease are noticeable only on old leaves. Initially, the lesions are angular, minute, irregular, yellow to light brown, scattered over leaf lamina. As the lesions enlarge, their colour changes from brown to cinnamon and they become almost irregular. Fully developed spots are characterized by dark margins and dull grey necrotic centres. In case of severe infection such spots coalesce forming patches measuring 3.5-13 cm in size, resulting in complete withering and defoliation of infected leaves.

Management Practices

The disease could be kept under control by spray of copper oxychloride (0.3 per cent) just after the appearance of the disease and subsequent sprays at 20 day intervals.

Red Rust

Red rust disease, caused by an alga *Cephaleuros virescens*, has been observed in some mango growing areas. The algal attack causes reduction in photosynthetic activity and defoliation of leaves, thereby lowering vitality of the host plant. The disease can easily be recognized by the rusty red spots mainly on leaves and sometimes on petioles and bark of young twigs and is epiphytic in nature. The spots are greenish grey in colour and velvety in texture. Later, they turn reddish-brown. The circular and slightly elevated spots sometimes coalesce to form larger and irregular spots.

The disease is more common in closely planted orchards. Fruiting bodies of the alga are formed in humid atmosphere. The zoospores formed by the sporangia initiate fresh infections. Stem entry is achieved by way of cracks. The affected areas crack and scale off. In severe infection by the zoospores the bark becomes thickened, twigs get enlarged but remain stunted and the foliage becomes sparse and finally dries up.

Management Practices

Two to three sprays of copper oxychloride (0.3 per cent) are effective in controlling the disease.

Sooty Mould

Sooty mould is caused by *Meliola mangiferae*. The disease is common in the orchards where mealy bug, scale insect and hopper are not controlled efficiently. The disease in the field is recognised by the presence of a black velvety coating, *i.e.*, sooty mould on the leaf surface. In severe cases, the trees turn completely black due to the presence of mould growth over the entire surface of twigs and leaves. The severity of infection depends on the honeydew secretion by the hoppers and scale insects. Honeydew secretions from insects sticks to the leaf surface and provide necessary medium for fungal growth. The fungus is essentially saprophytic and is non-pathogenic because it does not derive nutrients from the host tissues. Although, no direct damage is caused by the fungus, the photosynthetic activity of the leaf is adversely affected due to blockage of stomata.

Management Practices

☆ Pruning of affected branches and their prompt destruction prevents the spread of the disease.

☆ Spraying of 2 per cent starch is very effective.

☆ It could also be controlled by spray of nottasul + metacin + gumacasea (0.2 per cent + 0.1 per cent + 0.3 per cent).

Postharvest Diseases

The mango fruit is susceptible to many postharvest diseases such as anthracnose (*C. gloeosporioides*) and stem end rot (*L. theobromae*) during storage under ambient condtions or even at low temperature. Aspergillus rot is another postharvest disease of mango.

Management Practices

Preharvest sprays of fungicides could control the diseases caused by latent infection of these fungi. Postharvest dip treatment of fruits with fungicides could also control the diseases during storage. The following treatments are suggested:

(*i*) Three sprays of carbendazim (0.1 per cent), orthiophante-methyl (0.1 per cent) at 15 days interval should be done in such a way that the last spray falls 15 days prior to harvest.

(*ii*) Postharvest dip treatment of fruits in carbendazim (0.1 per cent) in hot water at 52±1°C for 15 minutes.

B. Major Insect-Pests and their Management

More than 492 species of insects, 17 species of mites and 26 species of nematodes have been reported to be infesting mango trees, about 45 per cent of which have been reported from India. Almost a dozen of them have been found damaging the crop to a considerable extent causing severe losses and, therefore, may be termed as major pests of mango. These are hopper, mealy bug, inflorescence midge, fruit fly, scale insect, shoot borer, leaf webber and stone weevil. Of these, insects infesting the crop during flowering and fruiting periods cause more severe damage. The insects other than those indicated above are considered as less injurious to mango crop and are placed in the category of minor pests. A brief description of the biology and control of major pests of mango is given below.

Plant Hopper

Of all the mango pests, hopper is considered as the most serious and widespread pest throughout the country. *Idioscopus clypealis* Lethierry, *Idioscopus nitidulus* (Walker) and *Amritodus atkinsoni* Lethierry are the most common and destructive species of hoppers, which cause heavy damage to mango crop. Large number of nymphs and adult insects puncture and suck the sap of tender parts, thereby reducing the vigour of the plants. Heavy puncturing and continuous draining of the sap cause curling and drying of the infested tissue. They also damage the crop by secreting a sweet sticky substance, (honeydew) which encourages the development of the fungus *Meliola mangiferae*, commonly known as sooty mould, which affects adversely the photosynthetic activities of the leaves. A low population of hoppers has been recorded in mango orchards throughout the year but it shoots up during February-April and June-August. Shade and high humidity conditions are favourable for their multiplication. Such conditions usually prevail in old, neglected and closely planted orchards. The female hoppers lay 100-200 eggs on mid rib of tender leaves, buds and inflorescence. In summers, the total life cycle occupies 2-3 weeks.

Management Practices

a) Chemical

Three sprays of 0.15 per cent carbaryl or 0.04 per cent monocrotophos or 0.05 per cent phosphomidon or 0.05 per cent methyl parathion have been found very useful in controlling the pest population. First spray should be given at the early stage of panicle formation. The second spray at full length stage of panicles but before full bloom and the third spray after the fruits are set and have attained pea stage are recommended.

b) Biological

Biological control agents such as the predators *Mallada boninensis* and *Chrysopa lacciperda,* the egg parasite *Polynema* sp. and a preparation of the fungus *Beauveria bassiana* are the important useful bioagents to control this pest

c) Integrated Pest Management (IPM)

The continuous use of pesticides though control the pests but pose some other serious problems like killing of pollinators and natural enemies, development of resistance to insecticides and residues, which are on fruits hazardous to human population. Besides, the high cost of pesticides, labour and maintenance of equipments are other limiting factors in pest control. Integrated pest management is gaining momentum to take care of these problems. To manage mango hopper pest, avoid dense planting and keep the orchard clean by regular ploughing and removal of weeds. Pruning of overcrowding and over lapping branches should be done in the month of December. Chemical spray is to be minimized. Neem products may be included in the management schedule of the pest. The use of insect growth regulator Buprofezin (0.0125 per cent) is also suggested as one of the sprays.

Mealy Bug

It is another major pest of mango in India and is widely distributed all along the Indo-Gangetic plain. *Drosicha mangiferae* Green is the most common mealy bug and causes severe damage to mango crop throughout the country. Nymphs and adults suck the plant sap and reduce the vigour of the plant. Excessive and continuous draining of plant sap causes wilting and finally drying of infested tissue. They also secrete honeydew, a sticky substance, which encourages the development of a fungus *Maliola mangiferae,* termed as sooty mould.

The adult male is winged and small, female is bigger and wingless. The female, after copulation, crawl down the tree in the month of April-May and enters in the cracks in the soil for laying eggs in large numbers encased in white egg sacs. The eggs lie in diapause state in the soil till the return of the favourable conditions in the month of November – December. Just after hatching, the minute newly hatched pink to brown coloured nymphs crawl up the tree. After climbing up the tree, they start sucking the sap of tender plant parts. They are considered more important because they infest the crop during the flowering season and if the control measures are not taken timely, the crop may be destroyed completely.

Management Practices

(i) Mechanical

Polythene (400 gauge) bands of 25 cm width fastened around the tree trunk have been found effective barrier to stop the ascent of nymphs to the trees. The band should be fastened well in advance before the hatching of eggs, *i.e.,* around November – December.

(ii) Chemical

Application of 250 g per tree of methyl parathion dust (2 per cent) or aldrin dust (10 per cent) in the soil around the trunk kills the newly hatched nymphs, which come in contact with the chemical. Spraying of 0.05 per cent monocrotophos or 0.2 per cent carbaryl or 0.05 per cent methyl parathion has been found useful in controlling early instar nymphs of the mealy bug.

(iii) Biological

Menochilus sexmaculatus, Rodolia fumida and *Sumnius renardi* are important predators in controlling the nymphs. The entomogenous fungus *Beauveria bassiana* is found to be an effective bioagent in controlling the nymphs of the mealy bug.

(iv) Integrated Pest Management (IPM)

The IPM schedule of mealy bug is very important and useful if timely operations are done. Flooding of orchards with water in the month of October kills the eggs. Ploughing the orchards in the month of November exposes the eggs to sun's heat. In the middle of December, 400 gauge alkathene sheet of 25 cm width may be fastened to the tree trunk besides raking the soil around the tree trunk and mixing of 2 per cent methyl parathion dust. The dust may also be sprinkled below the atkathene band on the tree. The congregated nymphs below the band may be killed by any of the suggested insecticides. The above IPM schedule holds promise to control the mealy bug but spraying of neem product and the spores of the fungus *Beauveria bassiana* will further ensure the reduction of the pest population.

Inflorescence Midge

The mango inflorescence midge, *Erosomyia indica* Grover (Diptera : Cecidomyiidae) is another major pest of mango. Recently, this pest has become very serious in certain pockets of Uttar Pradesh causing serious damage to mango crop by attacking both the inflorescence and the small fruits. The adult midge are harmless minute flies, which are short-lived and die within 24 hours of emergence after copulation and oviposition. The flies lay eggs singly on floral parts like tender inflorescence axis, newly set fruit or tender leaves encircling the inflorescence. The eggs hatch within 2-3 days. Upon hatching, the minute maggots penetrate the tender parts on which the eggs have been laid and feed on them. The floral parts finally dry up and are shed. The larval period varies from 7-10 days. The mature larvae drop down into the soil for pupation. The pupal period varies from 5-7 days. There are 3-4 overlapping generations of the pest spread over the period from January-March. Thereafter, as the weather conditions turn unfavourable, the mature larvae undergo diapause in the soil instead of pupating. They break diapause on the arrival of favourable conditions in following January.

The midge infests and damages the crop in three different stages. The first attack is at the floral bud burst stage. The eggs are laid on newly emerging inflorescence, the larvae tunnel the axis and thus destroy the inflorescence completely. The mature larvae make small exit holes in the axis of the inflorescence and slip down into the soil for pupation. The second attack of the midge takes place at fruit set. The eggs are laid on the newly set fruits and the young maggots bore into these tender fruits, which slowly turn yellow and finally drop. The third attack is on tender new leaves encircling the inflorescence. The most damaging one is the first attack in which the entire inflorescence is destroyed even before flowering and fruiting. The inflorescence shows stunted growth and its axis bends at the entrance point of the larvae. It finally dries up before flowering and fruit setting.

Management Practices

☆ Plough the orchard in summer. Ploughing of the orchards expose pupating as well as diapausing larvae to sun's heat, which kills them.

☆ Soil application of methyl parathion also kills pupating as well as diapausing larvae in the soil. The insecticide in the soil should be applied after monitoring larval population on white sheet below the tree.

☆ Spraying of 0.05 per cent fenetrothion or 0.045 per cent dimethoate or 0.04 per cent diazinon at the bud burst stage of the inflorescence has been found effective in controlling the pest population.

Fruit Fly

The oriental fruit fly is one of the most serious pests of mango in the country, which has created problem in the export of fresh fruits. *Daccus dorsalis, D. zonatus* and *D. correctus* are the most common fruit flies which cause serious damage to mature mango fruits. The adult flies are dark brown in colour and measure 7 mm in length and 4 mm across the wings. The females have tapering abdomen, which ends in an ovipositor. The female punctures the outer wall of the mature fruits with the help of its pointed ovipositor and insert eggs in small clusters inside the mesocarp of mature fruits. After hatching, the larva feeds on the pulp of fruit, which appears normal from outside, but drops down finally. The mature maggots fall down into the soil for pupation. The emergence of fruit fly starts from April onwards and the maximum population is recorded during May-July, which coincides with fruit maturity. The population declines slowly from August to September after which it is non-existent up to March.

Management Practices

(i) Chemical

The adult fruit flies can be controlled by bait sprays of carbaryl (0.2 per cent) + protein hydrolysate (0.1 per cent) or molasses starting at pre-oviposition stage (first week of April), repeated once after 21 days. Another method to control these flies is to hang traps containing a 100 ml water emulsion of methyl euginol (0.1 per cent) + malathion (0.1 per cent) during fruiting (April to June). About 10 such traps are sufficient for one hectare of orchard.

(ii) Integrated Pest Management (IPM)

☆ Collect and destroy of the infested and dropped fruits.

☆ Plough the orchards and expose the diapausing pupae to sun's heat

☆ Releasing of parasite and predator during December to February is helpful in reducing the pest population.

☆ Use methyl euginol traps in the orchards.

☆ Early harvesting of mature fruits.

☆ Selective and need based bait sprays.

☆ Hot water treatment or vapour heat treatment (VHT) of fruits before storage and ripening for killing the larvae.

Scale Insects

Scale insects were not considered serious pests on mango in any part of the country till recently, but of late, they have assumed the status of serious pest in certain parts of the country. *Pulvinaria polygonata, Aspidiatus destructor, Ceroplastis* sp. and *Rastococus* sp. are some of the most common scale insects infesting mango crop. The nymphs and adult scales suck the sap of the leaves and other tender parts and reduce the vigour of the plants. They also secrete honeydew, which encourages the development of sooty mould on leaves and other tender parts of the mango plant. In case of severe scale infestation, growth and fruit bearing capacity of the tree is affected adversely. Among the above scale insects, *P. polygonata* is posing a serious threat to mango industry of western Uttar Pradesh.

Management Practices

Pruning of the heavily infested plant parts and their immediate destruction followed by two sprays of monocrotophos (0.04 per cent) or diazinon (0.04 per cent) or dimethoate (0.06 per cent) at an interval of 20 days have been found very effective in controlling the scale population.

Shoot Borer (*Chlumetia transversa*)

This pest is found all over the country. Larvae of this moth bore into the young shoot resulting in dropping of leaves and wilting of shoots. Larvae also bore into the inflorescence stalk. The adult moths are shining grey in colour and measure about 17.5 mm with expanded wings. Hind wings are light in colour. Female moths lay eggs on tender leaves. After hatching, young larvae enter the midrib of leaves and then enter into young shoots through the growing points by tunnelling downwards. The full grown larva is dark-pink in colour with dirty spots and measures about 22 mm in length. There are four overlapping generations of the pest in a year and it overwinters in pupal stage.

Management Practices

The attacked shoots may be clipped off and destroyed. Spraying of carbaryl (0.2 per cent) or quinalphos (0.05 per cent) or monocrotophos (0.04 per cent) at fortnightly intervals from the commencement of new flush gives effective control of the pest. A total of 2-3 sprays may be done depending on the intensity of infestation.

Bark-Eating Caterpillar (*Indarbella quadrinotata*)

This pest is found damaging a variety of plants including a number of fruit trees, forest trees and ornamentals all over India. The old, shady and neglected orchards are more prone to attack by this pest. Larvae of this moth feed on the bark and weaken the tree. The moth is light-grey in colour with dark-brown dots and measures about 35-40 mm with expanded wings. A single female lays about 300-400 eggs in batches on the bark. The full grown caterpillar is dirty-brown in colour and is about 35-45 mm in length. The caterpillar spins brown silken web on the tree, which consists of their excreta and wood particles. Larvae also make shelter tunnels inside the stem in which they rest. Larvae actually feed from April to December. There is only one generation in a year.

Management Practices

Remove the webs from tree trunks and put emulsion of monocrotophos (0.05 per cent) or DDVP (0.05 per cent) in each hole and plug them with mud.

Stem Borer (*Batocera rufomaculata*)

Stem borer is widely distributed pest in India and attacks a variety of fruit trees including mango. Damage is caused by the grub of this beetle, which feeds inside the stems boring upward resulting in drying of branches and in severe cases attacked stem also dies. Adult beetles, 35-50 mm in size, are stout and greyish-brown in colour with dark-brown and black spots. Eggs are laid either in the slits of tree trunk or in the cavities in main branches and stems covered with a viscous fluid. Fully grown grubs are cream-coloured with dark-brown head and 90 × 20 mm in size. Pupation takes place within the stem. Beetle emerges in July-August. There is only one generation of the pest in a year.

Management Practices

The pest can be effectively controlled by following the recommendations given for the control of bark eating caterpillar.

Shoot Gall Psylla (*Apsylla cistellata*)

It is a very serious pest of mango in many parts of India, particularly in Terai region of U.P., North Bihar and West Bengal. This pest creates green conical galls in leaf axis. The activity of the pest starts from August. The galls dry out after emergence of psyllid adults in March. The females lay eggs in the midribs as well as in lateral axis of new leaves. Nymphs emerge from eggs during August-September and crawl to the adjacent buds to suck cell sap. As a result of feeding, the buds develop into hard conical green galls. The galls are usually seen during September-October. Consequently, there is no fruit set.

Eggs are white while nymphs are flat and pale-yellow in colour. Adults are 3-4 mm long with black head and thorax and light-brown abdomen. Female lays approx. 150 eggs during March-April and nymphs pass the winter inside the galls. There is only one generation of the pest in a year.

Management Practices

The galls with nymphs inside should be collected and destroyed to prevent carryover of the pest. The pest can effectively be controlled by spraying monocrotophos (0.05 per cent) or dimethoate (0.06 per cent) or quinalphos (0.05 per cent) at 2 week intervals starting from the middle of August. The use of same chemical for every spray should be avoided.

Leaf Webber (*Orthaga euadrusalis*)

The pest is attaining serious proportions in our country. Its infestation starts from the month of April and goes up to December. Eggs are laid singly or in clusters within silken webbings on leaves. Upon hatching, the caterpillars feed on leaf surface by scrapping. Later, they make web of tender shoots and leaves together and feed within. Generally, 1-9 larvae are found in a single web. Pupation takes place inside the webs in silken cocoons. However, the last generation (December-January) pupates in the soil. The adult moths are medium sized and sombre coloured. Fully grown caterpillar measures 2.5 to 3 cm. They are brownish in colour with brown spots and whitish striation on the dorsal surface. The pupae diapause for about five to six months. The infestation is severe in shady conditions. Old orchards with lesser space between tree canopies have more infestation than open orchards.

Management Practices

☆ Pruning of infested shoots and their burning in the month of April to July is found effective.

☆ Raking of the soil around the base of the trees in January, after the last generation has pupated, helps in checking the pest population.

☆ Three sprays starting from the last week of July at 15 days interval with carbaryl (0.2 per cent) or monocrotophos (0.05 per cent)) or quinalphos (0.05 per cent) have been found effective in controlling the pest.

Stone Weevil (*Sternochetus mangiferae*)

This insect is widely distributed in tropics. It is a common pest of mango in southern India. Another species, *S. frigidus*, of the pest is found in Assam and Bengal. Sweeter varieties such as Alphonso, Bangalora, Neelum, *etc.* are more prone to attack by this pest.

Female lays eggs on the epicarp of partially developed fruits or under the rind of ripening fruits. Newly emerged grubs bore through the pulp, feed on seed coat and later cause damage to cotyledons. Pupation takes place inside the seed. Discolouration of the pulp adjacent to the affected portion has been observed. Eggs are minute and white in colour. Adult weevils are 5 to 8 mm long, stout and dark-brown in colour. Life-cycle is completed in 40 to 50 days during June-July. Adults hibernate until the next fruiting season. There is only one generation in a year.

Management Practices

☆ Collect and destroy the affected fruits.

☆ Exposing the hibernating weevils by digging the soil

☆ Spraying the trees with fenthion (0.01 per cent)

C. Physiological Disorders and their Management

Several physiological disorders cause serious losses in mango, which have been described briefly hereunder:

Spongy Tissue

A non-edible sour patch developing in the mesocarp of mango fruit is termed as spongy tissue. Spongy tissue can be detected only after cutting the ripe fruit. The fruits become unfit for human consumption. Alphonso cultivar has been recorded most susceptible to this malady and about 30 per cent loss in this variety occurs due to this disorder. Fruits lower in calcium content are affected with spongy tissue. Convective heat arising from the soil is the main cause of damage. It is a physiological disorder in which fruit pulp remains unripe because of unhydrolyzed starch due to physiological and biochemical disturbances caused by heat in mature fruit at pre- and postharvest stages. Calcium chloride has been found most effective in reducing the spongy tissue in ripe Alphonso fruits. Sod culture, green vegetation, leguminous crop cover or mulching at pre-harvest stage helps in saving the crop from the disorder.

Black Tip

Black tip is a serious disorder, particularly in the cultivar Dashehari. The affected fruits become unmarketable and reduce the yield to a considerable extent. The damage to the fruit gets initiated right at marble stage with a characteristic yellowing of tissues at distal end. Gradually, the colour intensifies into brown and finally black. At this stage, further growth and development of the fruit is retarded and black ring at the tip extends towards the upper part of the fruit. Black tip disorder has generally been detected in orchards located in the vicinity of brick kilns. It has been reported that the gases like carbon monoxide, sulphur dioxide and ethylene constituting the fumes of brick kiln are known to damage growing tip of fruits and give rise to the symptoms of black tip. Apart from these factors, irrigation, condition of the tree and management practices also play important role in deciding the severity of the disorder.

Planting of mango orchards in North-South direction and 5 to 6 km away from the brick kilns may reduce incidence of black tip to a greater extent.

The incidence of black tip can also be minimised by the spray of borax (1 per cent) or other alkaline solutions like caustic (0.8 per cent) or washing (0.5 per cent) soda. The first spray of borax should be done positively at pea stage followed by two more sprays at 15 days interval.

Clustering Disorder (*Jhumka*)

A fruiting disorder, locally known as '*Jhumka*', is characterized by the development of fruitlets in clusters at the tip of the panicles. Such fruits do not grow beyond pea or marble stage and drop down after a month or so of fruit set. These fruits do not contain seeds when they are cut open. The disorder seems to be due to lack of pollination/fertilization, which may be attributed to many reasons. Among them, absence of sufficient population of pollinators in the orchards is the major reason. Surveys conducted in the mango belt of Lucknow revealed that the more problem is in Malihabad area and Dashehari is the most affected cultivar. The other reasons causing the disorder are old and overcrowding of trees, indiscriminate spraying against pests and diseases, use of synthetic pyrethroids for spraying, monoculture of Dashehari, and bad weather during flowering. Some of the remedial measures are suggested below:

☆ Insecticides should not be sprayed at full bloom to avoid killing of pollinators.

☆ Pests and diseases should be controlled in time by spraying the recommended pesticides only.

☆ Introduction of beehives in the orchards during flowering season for increasing the number of pollinators.

☆ The practice of monoculture of a particular variety may be avoided. In case of Dashehari, 5-6 per cent of other varieties should be planted in new plantations. In old orchards, where monoculture of a particular variety like Dashehari is followed, a few branches may be top worked with pollinizing varieties.

☆ Pruning of old trees may be done to open the canopy.

☆ Spraying of 300 ppm NAA may be done during October-November.

Leaf Scorch

Leaf scorching is common in mango during winter months. Scorching of leaves is identified by the development of brick red colour towards the tip and along the margins of old leaves and subsequent collapsing of these tissues. However, the new leaves are not affected. These symptoms resemble K deficiency but it is caused by chloride toxicity, which render K unavailable to the plants. In severe cases, the leaves may fall down and the tree vigour and yields are reduced significantly. This disorder is common in saline soil or where brackish water is used for irrigation and muriate of Potash is used as a source of K fertilizer.

Leaf scorch can be checked effectively by collecting and burning the affected leaves, using sulphate of potash as a source of K and by avoiding brackish water for irrigation. In case of acute problem, 4-5 foliar applications of potassium sulphate (5 per cent) on newly emerged growth flushes should be given at fortnightly intervals.

Internal Brown Necrosis

In this disorder, there is development of dark green colour in the immature fruits of mango on their lower halves, followed by the browning of the stone and mesocarpic tissues. These tissues may turn in to brown black necrotic area, which may extend up to the epicarp of the fruit. There is the development water-soaked spots on the fruit surface, exuding gummy substance below the green tip. In severe cases, the whole lower portion may turn necrotic, resulting in the cracking of the fruits. Among different varieties of mango, Langra is almost free from this malady. It can be effectively reduced by spraying boric acid (0.4 per cent).

Stem End Rot

It is a disorder of mango, wherein the discolouration of fruit tissues starts from the pedicel end of the ripening fruit. The affected peel of the fruit turns dark brown to purple black, while the pulp tissues are soft and watery. Deficiency of calcium is considered as the major factor for its causes. A single spray of calcium sulphate (0.1 per cent) helps to check it considerably.

Soft Nose

Soft nose, jelly seed and stem end rot seem to be the different stages of development and expression of the same disorder in mango because in all these disorders, there is disintegration or damage or injury to the vascular tissues at the stem end, while fruit is still attached to the tree.

Tip Pulp

Tip pulp is a serious problem in some varieties of mango, which usually remain unnoticed in many cases. The first visual symptom of this malady is the yellowing of the fruit tip, which may turn grey in colour during later stages. The tip of the fruit is pulpy, while the remaining part is compact, unripe and hard. The affected fruits are slightly sweeter than the normal fruits but are different in taste. Such fruits also don't ripen properly during storage. The causes of this malady are unknown. However, timely irrigation of the orchards in peak summer months tends to reduce the problem significantly.

Girdle Necrosis

Girdle necrosis has been reported from different mango growing belts as a serious disorder of mango fruits. The initial symptom of girdle necrosis starts with the disfiguring of the lower half of the fruit with small-etiolated spots and appearance of brown dotted etiolated areas. These spots collapse and form a necrotic girdle of the tissues around the sinus region of the fruit, leaving the green tip healthy in the initial stages. There may be exudation of brownish gummy substances from the fruit. The mesocarpic tissues disintegrate and form a cavity below the necrotic spots. The name girdle necrosis has been given to this disorder because the necrotic area encircles more or less completely the green tip of the fruit. The causes of this disorder are unknown, however, sprays of micronutrients like Zn, Cu, and B reduce the problem to a greater extent.

Taper Tip

This is a common disorder of Dashehari mangoes. In this disorder, the distal end of the fruits shows the intensification of the normal green colour, tapering abruptly and often curved, causing the affected fruit to remain smaller than the normal ones. The causes of this disorder are still unknown.

Jelly Seed

Jelly seed is a serious disorder of mango fruits and like spongy tissue, it may cause heavy losses to the growers. In this disorder, there may be complete disintegration of the pulp surrounding the stone in to a jelly like mass. In severe cases, the entire pulp may disintegrate and there may be the development of an internal cavity. Similarly, the affected fruits emit bad odour on cutting. Usually, it is very difficult to detect this disorder without cutting the fruit because fruits appear normal externally. However, in severe cases, there may be softening of the beak. Faulty nutrition, especially Ca, is the major cause of this malady. Thus, in nutrient management programme, due consideration should be given for calcium.

HARVESTING AND YIELD

Grafted plants start bearing at the age of 3 – 4 years (10-20 fruits) to give optimum crop from 10-15th year which continues to increase upto the age of 40 years under good management. Mangos normally reach maturity in 4 to 5 months from flowering. When the mango is full-grown and ready for picking, the stem will snap easily with a slight pull. If a strong pull is necessary, the fruit is still somewhat immature and should not be harvested. The operation of harvesting may be accomplished manually or with the help of machine depending on the height of the plants and proportion of maturity of fruits for the operation, in the context of the number of operations required to complete the same.

Hand picking is possible irrespective of height of the plants, although it is labour intensive and liable to induce mechanical injury to the fruits. Generally, fruits with stalk intact, are harvested with the help of a bag affixed on long pole from tall trees. An improvised harvester developed by KKV, Dapoli, Maharashtra, India is useful for this purpose. Harvested fruits are assembled in a platform lined with sufficient cushioning materials available from the locality. In most of the cases, mango leaves and straw are used for the purpose prior to taking the fruits for grading.

The yield of mango varies with the variety, periodicity of flowering, growing conditions influencing the size of the plants etc. The local commercial varieties are seen to produce 8-10 tonnes while the hybrids produce 15-25 tonnes of fruits per hectare during the 'on' year. Depending on the type of variety, the number of fruit per tree may vary between 1,000-3,000 during 'on' year.

POSTHARVEST MANAGEMENT

Fruit Maturity

If picked immature, fruits develop white patches or air pockets and show lower amounts of Brix: acid ratio, taste and flavour, whereas over-mature fruits lose their storage life. Such fruits present numerous problems during handling. It is therefore desirable to pick the fruits at the correct stage of maturity to facilitate ripening, distant transportation and maximum storage life, and thus to increase their quality and market value.

Appearance of waxy coating, dots on the fruits, relative size of the fruit are some of the external appearances of the fruits, which indicate the maturity. However, some ripened fruits when start dropping (_tapka_ stage), it is said that maturity of that particular variety is reached. Specific gravity (1.01–1.02), acidity, starch and starch/acid ratio are some of the reliable indices for predicting maturity. Starch/acid ratio having 4 or more in Langra fruits may be said as full maturity. Maturity reaches after 114 to 116 days of fruit set, the critical period, in Baneshan mango fruits.

Grading and Packing

Grading is not practiced for local market. However, for long distant markets, grading is essential. Mango fruits can be grouped under three categories namely, Grade-I, _i.e._ fruit weighing over 320 g, Grade-II, _i.e._ 170 to 320 g and Grade-III, _i.e._ fruits under 170 g.

Likewise, Dashehari mango fruits can be graded in 4 groups _viz._, 130 g, 131 to 170 g, 171 to 200 g and above 200 g and termed as poor, average, good and excellent grades, respectively. While grading between varieties, other characters like size, shape, pulp weight, edible: non-edible ratio etc., should be considered in addition to fruit weight.

The grading of fruits may be helped by designing a suitable grader. For local market, the mango fruits, either mature or ripe, are sent in bamboo or mulberry baskets. For long distance transport, the mature fruits are individually wrapped in tissue paper and packed in single layers of 12 fruits in wooden trays lined with wooden wool. Five such trays are tied together. Wooden and metal crates are used in limited quantity. Now-a-days, mango fruits are packed in specially designed corrugated fibre board (CFB) boxes, having 5 kg capacity.

Ripening

Mango is a climacteric fruit and its period of ontogeny is characterized by a series of biochemical changes initiated by autocatalytic production of ethylene and increase in respiration. The methods of ripening do not affect the quality of fruit if harvested after full maturity. For local consumption, the fruits are spread over paddy or wheat straw in single, double or triple layers at a temperature of 9.4° to 21.1 °C in a ventilated store. In Alphonso fruits, ripening is very much hastened when the mature fruits are kept in a chamber saturated with ethylene gas released from 10,000 ppm ethrel solution containing sodium hydroxide pellets.

Postharvest Treatments and Storage

Washing the fruits immediately after harvest is essential, as the sap, which leaks from the stem burns the skin (Sapburn) of the fruit making black lesions, which lead to rotting. To avoid sapburn, dip the harvested fruits in 1 per cent lime solution for 2 minutes and remove their stalks in the solution. Remove the fruits and dry under fan. Mango fruits are able to respond metabolically to the environment under which it is stored. Various methods are employed to extend the storage life of mangoes. They are low temperature storage, sub-atmospheric pressure storage, controlled atmospheric storage, irradiation and use of chemicals. Some cultivars, especially 'Bangalora', 'Alphonso', and 'Neelum' have much better keeping quality than others. In Bombay, 'Alphonso' has kept well for 4 weeks at 11.11° C; 6 to 7 weeks at 7.22 °C. Storage at lower temperatures is detrimental as mangos are very susceptible to chilling injury. The fruits show chilling injury symptoms after 10 days of storage at 4 and 8 °C.

Green seedling mangos, harvested for commercial preparation of *chutneys* and pickles as well as for table use, are stored for as long as 40 days at 5.56-7.22 °C with relative humidity of 85 per cent to 99 per cent. Some of these may be diverted for table use after a 2-week ripening period at 16.67–18.13 °C.

Pre-and postharvest application of benomyl is recommended for control of latent infection of *Colletotrichum*. Further, the anthracnose of mango can be controlled by dipping the harvested fruits in a 1,000 ppm suspension of thiabendazole or benomyl at 55 °C prior to storage. Similarly, mangoes may be immersed in hot water (at 53°C for 5 minutes, and or 51-55 °C for 30 minutes) before storage or marketing to control diseases, particularly anthracnose. Another technology, vapour heat treatment (VHT) has been developed to control infestation of fruit flies in fruits after harvest. A recommended treatment of mangoes is 43°C in saturated air for 8 hours then holding the temperature for a further 6 hours. Similarly, the individual fruits of mangoes can be wrapped with cling film (modified atmospheric packaging) or stored under controlled atmosphere with 5 per cent CO_2 and 5 per cent O_2 at 13 °C.

Storage

The mango is a climacteric fruit and unless the fruits are stored properly, one cannot be sure of the condition in which the fruits will reach the desired market.

Proper storage is absolutely essential during the year of glut. Also, the processing units cannot utilize the entire produce at a time and proper storage conditions become vital. In general, green but mature fruits store better than those harvested ripe from the trees. Experimental results indicate that low temperature prolongs the storage life of different cultivars from 4 to 7 weeks (Table 1.7).

Table 1.7: Storage of Mango Fruits at Low Temperature

Cultivars	Temperature (^oC)	Relative humidity (per cent)	Duration (weeks)
Alphonso	7-9	90	7
Bangalora, Safeda	5.5-7	90	7
Neelum, Raspuri	5.5-9	90	5-6
Haden, Keitt	12-14	90	2
Julie	11-12	90	2
Zill	10	90	3

However, very low temperature may cause chilling injury. Chilling injury (skin pitting, staining and browning) in mango may occur after 15 days at 2°C and after about a month at 5°C.

Processing

Mango is perhaps one of the most important fruits of the world, which can be utilized by the processing industry during the different stages of its growth, development, maturity and ripening. The products prepared both from ripe and green mangoes are highly popular in India and abroad. Various processed products, which can be prepared from both green and ripe mangoes such as green mango powder (*amchur*), green mango drink (*panna*), mango pulp/puree, mango leather, mango pickle, *chutney*, frozen slices, raw slices in brine, concentrate, squash, nectar etc. In 2010-11, the country exported 2,072,014.8 metric tons of mango pulp and earned 3,85,562.8 Lakh Rupees. Export of processed mango products is continuously increasing.

CROP IMPROVEMENT

Breeding Problems

Mango breeding is beset with many problems as follows:

1. Mango is a out-breeding species and thus highly heterozygous
2. Pre-bearing period is very long
3. It is difficult to get large populations of hybrids because only one stone id produced by a fruit
4. Due to incompatibility, backcross progeny is a problem.
5. Selection of desirable parents is also a problem in mango because parents having desirable characters are only a few

6. Most of the desirable characters (*e.g.* regularity in bearing, dwarfness, beak shape etc.) are governed by recessive genes

7. Presence of polyembryony and self-incompatibility

Inheritance Pattern

Mango is not a convenient plant for general analysis due to:

1. Long life cycle
2. Cross-pollination
3. High degree of heterozygosity
4. Lack of detailed information on inheritance pattern
5. Intricate arrangement of sexes on the panicle, and
6. Excessive fruit drop.

However, the inheritance of some of the characteristics is as follows:

Growth Habit of Tree

Experiments have shown that upright growth habit (*e.g.*, Totapuri Red Small) of plant is dominant over spreading (*e.g.*, Dashehari, Langra, Chausa), which is dominant over dwarfness (*e.g.* Neelum).

Bearing Habit

Biennial bearing is dominant over regular bearing. Similarly, precocity is bearing (*e.g.*, Totapuri Red Small and Neelum) and regularity in bearing are governed by recessive genes.

Colour of New Leaves and Panicle

Inheritance pattern shows clear dominance of purple colour of leaves over light green it was inherited from Totapuri Red Small.

Leaf Flavour

Leaf flavour of Langra and Chausa was dominant in the hybrids having either of the variety as one of the parents. Thus, leaf flavour has distinct correlation with fruit flavour. This character is more useful in initial screening of hybrids from fruit quality point of view.

Fruit Size

Fruit size is an important character, which is governed by polygenes. Observations on fruit size in five parental combinations involving Neelum, Totapuri Red Small as parents indicated that fruit size of the hybrids, in general was inferior, though some hybrids in all combinations showed increased fruit size.

Fruit Colour

The genetics of fruit colour has not been studied in detail but the available reports indicate that it is governed by a number of genes as different combinations resulting in different colours. However. One report indicated that red colour is dominant over green, because when coloured variety, Janardhan Pasand was crossed with some green fruited variety, a wide variety of colour was observed in progeny.

Pulp Colour

Light yellow colour has shown some dominance to orange, but the gene action was primarily additive both within and among loci.

Inheritance of Beak

Presence of beak with marked sinus is considered as undesirable character in table varieties. For example, Dashehari and Langra have no beak; Chausa and Neelum have slight beak and Totapuri Red Small has more prominent beak. This undesirable character has shown dominance in the hybrid progeny.

Fruiting in Bunches

Dashehari has got pronounced bunching habit; Neelum bears fruits in moderate bunches and Langra, Chausa bear in single fruits. Whereever, Dashehari has been used as one of the parents, more especially with Neelum, the hybrids have shown tendency to bear fruits in bunches.

Juvenility

The inheritance of duration of juvenility period is yet to be examined critically. Totapuri Red Small has been found to have short juvenile phase and thus it can be used as male parent in hybridization programme to reduce juvenile period. This is because no effect of the female parameter has been found on the distribution of juvenile period or fertility.

Embrony

Polyembryouy has been observed to be governed by a single dominant gene .

Resistance to Diseases and Disorders

Resistance to floral malformation is perhaps controlled by recessive genes because when crosses were made with Bhaduran, a highly resistance variety to malformation. All hybrids, so produced had susceptibility to malformation. Susceptibility to bacterial canker is transmitted through cytoplasmic inheritance because whenever Neelum (highly susceptible to Bacterial canker) was used as female parent, almost all hybrids inherited this character indicating probably cytoplasmic inheritance. Bombay green is only resistant to canker. Spongy tissue, a serious physiological disorder of fruits has also been found to be governed by recessive genes.

Pre-Selection Criteria for the Prediction of Hybrid Characteristics

Certain characteristics can be used for the selection or rejection of hybrids at nursery stage. These are as follows:

- ☆ Leaf flavor: Leaf flavor has been reported to have direct correction with fruit flavor.
- ☆ Emergence of new growth flushes simultaneously with fruiting or immediately after harvesting is indication of regularity in bearing.

☆ High phloem: xylem ratio is associated with dwarfing nature. So if ratio exceeds 1.0, it is least vigorous; 0.6-1.0, it is moderately vigorous, and if it is less than 0.6, it would be most vigorous.

☆ High phenolics is apical bud has been shown to be associated with reduction in vigour and dwarfing habit.

☆ Lower stomatal density is associated with dwarfing.

☆ Higher polyphenol oxidase (PPO) activity at juvenile phase has direct and invert relationship with malformation.

Breeding Objectives

Ideal Mango Variety

A number of mango varieties have been developed but still, we lack in having an ideal mango variety, which should have the following characteristics:

☆ It should bear regularly

☆ It should be dwarf

☆ It should produce of medium size (200-250g)

☆ It fruits should have attractive red or golden apricot yellow colour

☆ It should be precocious in bearing

☆ Frutis should have high pulp : stone ration, fibreless

☆ It should be highly tolerant to various diseases including malformation

☆ Fruits should have good flavor with high shelf-life

☆ It should have a high ratio (3.31-4.00) of edible and non-edible matters

☆ It should have high quality of fruits

However, it is very difficult to incorporate all the desirable characteristics is one variety and hence, it is not advisable to start a breeding programme with all these objectives. Thus, these objectives vary from place to place.

Breeding Methods

Introduction and Assessment

The best method for mango improvement is introduction of superior types and their assessment in the introduced locality. Several varieties in different fruits in India have been introduced, which were assessed fro several years and then released for commercial cultivation. For example, Mulgoa was introduced by the USA at Florida, which excelled other varieties and became main variety of Florida. Later, several selections (*e.g.*, Eldon, Haden, Sensation, Tommy Atkins, Edward) were made from it, which revolutionized the mango industry in the world.

Selection

Most of the mango varieties in India (*e.g.*, Dahshehari, Langra, Chausa, Iphonso Baneshan) owe their existence of chance seedlings and their perpetuation through vegetative propagation. However, in the recent times, few selections have been made. For example, a superior clone of Dashehari, D-51 has been selected at CISH, Lucknow, which is a regular bearer and produces fruits of high quality, virtually free from malformation. Two clones of Banarasi Langra from Varanashi has been made. A clone, Paiyur-I has been selected from Neelum at Fruit Research Station, Paiyur. Sunderraja has been selected at Rewa (MP). In Punjab, selection of sucking mango type with abundant juice, less fibre, small stone and red blush on checks have been identified as GN-1 to GN-7. In north-eastern region, Manipur-I and Manipur-II have been identified, which are dwarf, precocious, polyembryomic and regular bearer. At BAC, Sabour, Menaka has been selected from Zardalu. In Maharashtra, off season selection 'Niranjin' has been made at Parbhani. An exceptionally superior clone 'Cardozo Mankurad' has been selected from 'Goa Mankurad', which is a regular bearer, bears attractive (red), large sized fruits of better quality. In china, 'Rumang', a chance seedling of 'Xiangmang' has been selected, which is high yielding, bears uniform fruits of better quality, weighing about 250g each. 'Panxi Hongmang' has been selected from open pollinated seedlings of Luzonmang. In south Africa, Neldawn, Neldica, Heidi and Cerise were selected from open-pollinated population. In Australia, 'Celebration' has been selected from unknown parents, which produces glossy brilliant red fruits with yellow background.

Hybridization

Intervarietal hybridization work was initiated first in India by Burns and Prayag 1910 but it did not result is any useful contribution. However, the crosses at Miami (USA) between Haden x Carabao resulted in the development of Simmonds and Edward and Saigon x Amini resulted into Samini. These hybrids produced superior quality fruits but all these fail in prolific fruiting, which is considered necessary for commercial cultivation. Later, crosses were made between Haden x Saigon and Haden x Malvina but the hybrids were irregular bearers and producers of poor quality fruits. However, Fascell (Brooks x Haden) was a superior. At present, around 50 hybrids have been produced/developed in India (Table 1.7).

Mutation Breeding

Mutation breeding plays a vital role as it gives better and quicker results than hybridization. Experiments have shown that irradiation doses beyond 5 Krads are lethal to mango. The LD-50 for Neelum, Dashehari, Amrapali and Mallika were 3.9, 2.9, 3.4 and 2.4 kR. The effective doses for ethyl methane sulphonate (EMS) and N-nitrosomethyl urea (NMU) were found to be 1.5 and 0.05 per cent, respectively. The resultant plants were dwarf with narrow and smaller leaves. Some natural mutants have also been selected in mango. For example,

Davis-Haden from Florida (larger fruits, one month late than Haden

↓

Haden

Rosica in Peru, a bud mutant from (Regular, precocious, large fruits, good size of high quality)

↓

Rosado de Lca

Table 1.7: Mango Hybrids Developed in India

Institute	Name of Hybrid	Parents	Chief Characteristics
BAC, Sabour	1. Mahmood Bahar	Bombai x Kalapadi	Study, medium-stature, regular, good quality.
	2. Prabha Sankar	Bombai x Kalapadi	Medium ht, regular, matures 15 days later than Bombai.
	3. Sunder Langra	Langra x Sundar Pasand	Semi-vigours, regular, fruit similar to Langra but larger.
	4. Alfazli	Alfhonso x Fazli	Tall, regular, fruit quality better than Fazli.
	5. Sabori	Gulabkhas x Bombai	Regular, colour like Gulabkhas, very sweet and fibreless.
	6. Jawahar	Gulabkhas x Mahmood Bahar	Precocious, regular, high pulp (79 per cent), fibreless and very sweet.
FRS, Sodur (AP)	1. Neeludin	Neelum x Himayuddin	Medium ht, regular, small fruit (200g), fibreless and aromatic fruits (300 g).
	2. Neelgoa	Neelum x Mulgoa	Bears medium crop every year, shape and colour of Neelum size of Mulgoa fibreless and juicy.
	3. Neeleshaw	Neelum x Baneshan	Fruits like Baneshan, regular, suitable for canning.
	4. Swarnajehangir	Chinna Suvamarekha x Jehangir	Fruits of good colour and quality (Jehangir) and profile bearing, sweet, juicy and fibreless.
	5. No 212	Neelum x Bangangalli	High yield with better quality.
	6. No. 212	Chinneswarnrekha x Neelum	High yield with better fruit quality.
HETC Saharanpur	1. Gaurav	Dashehari x Totapuri Hyderabad	High yield with better quality.
	2. Saurabh	Dashehari x Fazali Zafrani	High yield with better fruit quality.
	3. Varun	Dashehari x Rumani	High yield with better fruit but little turpentive flavour.
IARI, New Delhi	1. Mallika	Neelum x Dashehari	Semi-vigours, large fruit (400-500g), excellent taste, fibreless, less stone, doing well in south India.
	2. Amrapali	Dashehari x Neelum	Regular, precocious, distinctly dwarf, variability in fruit size, poor colour (green) on maturity, excellent sugar (24.5 per cent), fibreless, late maturity.
	3. Pusa Arunima	Amrapali x Sensation	Semi vigorous, regular, attractive fruit size (250g) and red colour improved shelf-life, suitable for export.

Contd...

Table 1.7–Contd...

Institute	Name of Hybrid	Parents	Chief Characteristics
	4. Pusa Pratibha	Amrapali x Sensation	Semi-vigorous, regular, attractive red fruits with moderate size (181 g), high pulp content (71 per cent) and medium TSS (19.6 per cent) and good shelf life (7-8 days).
	5. Pusa Lalima	Dashehari x Sensation	Early bearing, regular, highly attractive red fruits with moderate size (207 g), high pulp content and medium TSS (19.7 per cent) and good shelf life (5-6 days).
	6. Pusa Pitamber	Amrapali x Lal Sundri	Apricot yellow fruits, medium plants, which bear medium sized fruit, having medium TSS. Very less incidence of malformation.
	7. Pusa Shresth	Amrapali x Sensation	Unique hybrid, semi-vigorous plants, regular bearing highly attractive red fruits with moderate size (227 g), high pulp content (71.9 per cent) and medium TSS and good shelf life (7-8 days).
FRS, Sanga-reddy (AP)	1. A.U. Rumani	Rumani x Mulgoa	Regular and prolific bearer, round fruits, firm pulp, fibreless, juice and sweet.
	2. Manjeera	Rumani x Neelum	Precocious, regular, heavy yielders, firm pulp, fibreless and sweet.
HCRT, Periyakulam	1. PKM-1	Chinnasuvarnarekha x Neelum	Regular, medium sized fruits (200g), good quality.
	2. PKM-2	Neelum x Malgoa	Regular bearer, good quality fruits.
FRS, Paria	1. Neelphonso	Neelum x Alphonso	Dwarf, superior TSS and vit C.
	2. Neeleshan	Neelum x Baneshan	Regular, dwarf, good fruit quality.
	3. Neeleshwari	Neelum x Dahhehari	Regular, dwarf, high TSS and vit. C.
KKV, Dapoli (RRS, Vengurla)	1. Ratna	Neelum x Alphonso	Semi-vigorous, precocious, attractive shape and colour of fruits, free from spongy tissue, excellent quality, like Alphonso.
	2. Sindhu	Ratna x Aphonso	Regular, thin stone, fruits in clusters, free from spongy tissue pulp (83 per cent), fibreless, deep orange in colour with good aroma.
	3. Ruchi	Neelum x Alphonso	Regular bearer, good for pickling, av. fruit weight 350g.

Contd...

Table 1.7–*Contd...*

Institute	Name of Hybrid	Parents	Chief Characteristics
IIHR, Bangalore	1. Arka Aruna	Banganpolli x Alphonso	Dwarf, precocious, fibreless, free from spongy tissue, good flavor Av. Fruit wt (450g), fibreless.
	2. Arka Punit	Alphonso x Banganpalli	Vigours, regular, fruit (225g), fibreless, good aroma, free from spongy tissue, good keeping quality.
	3. Arka Anmol	Alphonso x Janardan Pasand	Semi-vigour, regular, high keeping quality.
	4. Arka Neelkiran	Alphonso x Neelum	Semi-vigours, regular, attractive red colour at maturity, pulp deep yellow, medium TSS.
CISH Lucknow	Ambika	Amrapali x Janardan Pasand	Bright yellow peel with dark red blush, scanty fibre, fruit wt 250g, TSS 21ºC.
MPKV Rahuri	Sai Sugandh	Kesar x Totapuri	Regular, fruit weight (320g), turmeric yellow pulp, peelable skin, fibreless free from spongy tissue.

Hongmang-6 has been selected from Zill in China. It bears red fruits, high yield of better quality, juicy, 15.8 per cent TSS). Similarly, IAC 100 Barerbonhas been selected from Bourbon.

Sources of Desirable Traits in Mango

Dwarfness

Taimuria, Olur, Rumani, Neelumn, Kerala Dwarf, Janardhaw Pasand, Creeping and Latara are useful varieties to impart dwarfness in the progeny.

Regularity in Bearing

In India, Bangalora (Totapuri Red Small) and Neelum have been extensively used in breeding programmes.as these are highly regular varieties, and they inpart dwarfness in the progeny. Bangalora has now been discontinued because it has strong tendency to impart undesirable fruit shape and poor fruit quality to the progeny. Tommy Atkins, Sensation, Fazli, Kalapady, Khas-ul-Khas, Banganpalli, Kurd, Allampur Baneshan are also regular in bearing.

Red Fruit Colour

Floridan cvs. like Sensation, Tommy Atkins, Kensington, Irwin, Haden, Julie, Kent and Zill have attractive red peel colour. At IARI, Sensation is being extensively used in hybridization. At CISH, Lucknow, Tommy Atkins has shown good promise, although some reports reveal that it imparts susceptibility to fruit fly in the progeny.

Resistance to Malformation

Bhadauran has been reported to be virtually free from malformation but when it is used in hybridization, none of the hybrids were free from malformation, which lead to conclude that this trait is controlled by recessive genes. Latest reports have shown that Elaichi, Bhadayam, Smar Bahist Rampur, Mian Saheb and H-8-11 (Amrapali x Lal Sundari) appear to be free from malformation.

Powdery Mildew Resistance

Lalif and Hurr cvs. are highly resistant to powdery mildew and can be used for future breeding programme.

Bacterial Canker Resistance

Neelum is highly susceptible and Bombay green highly resistant to Bacterial Canker and thus it holds promise for its use in future breeding programme.

Juvenility

Totapuri Red Small is highly precocious is bearing followed by Amrapali, Madhulica, a variety released by ICAR Res Complea, Barephai is also highly precocious, as reported.

Problem of Incompatibility

Mango is a highly cross pollinated crop. The out breeding is encouraged by dicling and entomophily. Many established varieties have been reported to be self incompatible or self sterile, and thus require pollination from other varieties. For example, most of the cultivars like Dashehari, Langra, Chausa, Bombay Green, Himsagar have been found to be self incompatible and these should not be planted in isolation. In addition, some varieties have also been found to be cross incompatible.

Variety	Compatible Pollen	Incompatible Pollen
Dashehari	Bombay Green, Rataul, Langra	Chausa, Safeda-Malihabad
Chausa	Dashehari, Langra Safeda Malihabad	Bombay Green and Rataul
Langra	Dashehari, Totapari	Bombay Green, Chausa, Fazli, Rataul, Safeda Malihabad

2

Banana

Banana is one of the oldest and most popular fruits. Bananas are likely to have been first domesticated in Papua, New Guinea. Banana is probably native to tropical area of South-East Asia. Today, bananas are cultivated throughout the tropics. It is widely used as a fresh fruit throughout the world. Besides, the central core of the pseudostem is used as a vegetable. The banana pseudostem is also used for manufacturing paper and boards.

India ranks first among the banana growing countries of the world, followed by Brazil and United Republic of Tanzania. The total area under banana in India is about 8.3 million ha with an annual production of about 29.78 million tones. Among the various states in India, Tamil Nadu, Karnataka, Maharashtra and Andhra Pradesh account for major share in area and production of banana (Table 2.1).

ORIGIN, HISTORY AND DISTRIBUTION

The Indo-Malayan region is considered to be the place of its origin, where so many varieties of wild bananas still grow there. Bananas have later travelled with human population. The first Europeans to know about bananas were the armies of Alexander, the Great, while they were campaigning in India in 327 BC. The Arabs brought them to Africa. Africans are credited to have given the present name, since the word banana has been derived from the 'Arab finger'. The Portuguese brought them to the Canary Islands. Bananas changed during all these trips, gradually losing its seeds, filling out with flesh and diversifying.

Table 2.1: State-wise Area, Production and Productivity Scenario of Banana in India

State	Area (000'Ha)	Production (000'MT)	Productivity (MT/Ha)
Tamil Nadu	125.4	8,253.0	65.8
Maharashtra	82.0	4,303.0	52.5
Gujarat	64.7	3,978.0	61.5
Andhra Pradesh	79.3	2,774.8	35.0
Karnataka	111.8	2,281.6	20.4
Madhya Pradesh	38.1	1,719.6	45.2
Bihar	31.9	1,517.1	47.6
Uttar Pradesh	32.4	1,346.1	41.5
West Bengal	42.0	1,010.1	24.0
Assam	47.6	723.6	15.2
Others	175.3	1,873.1	10.7
Total	830.5	29,779.9	35.9

Source: National Horticulture Board Database-2011.

When Spaniards and Portuguese explorers went to the New World, the banana travelled with them. In 1516, when Fiar Tomas de Berlanga sailed to Santo Domingo, he brought banana roots with him. From there, bananas spread to the Caribbean and Latin American countries. Bananas started to be traded internationally by the end of 19th century. Before that date, Europeans and North Americans could not enjoy them because of the lack of appropriate transport for bananas. The development of railroads and technological advances in refrigerated maritime transport, allowed for bananas to become the most important world traded fruit.

COMPOSITION AND USES

Composition

Banana is considered as a very good source of carbohydrates, which gives energy. Interestingly, banana is also a very good source of vitamin C. It contains substantial amount of minerals like potassium, phosphorus, calcium and sodium (Table 2.2).

Table 2.2: Nutrient Composition of Ripe Banana Fruit

Component	Per 100 g Edible Portion	Component	Per 100 g Edible Portion
Energy	103 K calories	Iron	0.6 mg
Water	73.3 g	Sodium	29.0 mg
Protein	1.3 g	Potassium	241.0 mg
Fat	0.4 g	Carotene	300 µg
Carbohydrate	23.6 g	Vit B$_1$	0.07 mg
Fibre	0.5 g	Vit B$_2$	0.08 mg
Calcium	11.0 mg	Niacin	0.7 mg
Phosphorus	28.0 mg	Ascorbic acid (vit. C)	173 mg

Uses

Banana fruit is normally consumed fresh. Some varieties can be processed into puree, fruit cocktail, jam, jelly, chips, crisp and powder. In addition to the fruit, the flower of the banana plant is used in soups and curry preparations. The tender core of the banana plant's trunk is also used in A.P., W.B. and Kerala for cooking purposes. Banana fritters can be served with ice cream as well.

The leaves of the banana plant are large, flexible, and waterproof. Banana leaves are also used to serve food and to wrap food for cooking or storage in India and other SE Asian countries.

Banana chips are a snack produced from dehydrated or fried banana or plantain slices, which have a dark brown color and an intense banana taste. Bananas have also been used in the making of jam. In some parts of India, juice is extracted from the corm and used as a home remedy for the treatment of jaundice, sometimes with the addition of honey, and for kidney stones.

TAXONOMY AND BOTANICAL DESCRIPTION

Botany

Banana does not have a true stem. The stem is constructed of clasping leaf stalks and called as pseudostem or trunk. Clump in banana is a cluster of several shoots or plants. The main growing point of a shoot is called as '*heart*' and buds are referred to as '*eyes*'. The shoot that grows from a plant, yields the 'plant crop.' A production unit is comprised of a mother, daughter (follower) and sometimes a grand-daughter. An offshoot from the parent corm is known as 'sucker'. There are four types of suckers namely 'sword sucker', 'water suckers' or shoot, 'maiden' and 'peeper'. Sword sucker is defined as young sucker on which the first narrow leaves have begun to unfold, whereas water suckers are small suckers of superficial origin with broad leaves and growing around the main shoot, which are unsuitable for followers and should be removed. In Queensland, sword suckers are called as 'spears', while water suckers are referred as 'umbrellas' and any small sucker of superficial origin is called as 'sitter'. A large but non-fruiting ratoon is called as 'maiden sucker'. If it has a corm big enough, which can be cut into piece/bits with a bud, bits can be employed for plantings. Any very young sucker bearing scale leaves, as it appears above the soil, is called as 'peeper'. A sucker, which has been set to produce fruit is called a 'follower' and later a 'ratoon'. The whole plant is called a 'mat' or 'hill'.

Corm is an underground portion of stem of the plant; portion of this forms the planting material. Any large corm is called as 'head' and form a plant, which has flowered a bullhead. The piece of pseudostem discarded form the head is called as 'cabbage'. Banana has a large expanded leaves, about 20 m long and 50 cm wide. It has entire leaf margin. The leaf stalk is elongated to form the leaf sheath.

The inflorescence of banana during its ascent up the pseudostem is called as *bull*. When first visible at the top, it is said to be *peep* and when fully emerged, it is said to be *shot*. The whole inflorescence bearing '*hands*' of several '*fingers*' of fruit. The 'count bunch' has nine hands, although there may be 8, 7 and 6 count bunches having 8, 7 and 6 hands, respectively. The hand is borne upon a protuberance, the crown or cushion. The axis of inflorescence is called the 'stalk'. The basal end by the first pistillate hand being the big end, while the other apical extremity the small end. The male flower at the end of the stalk is called the 'bell' flowers or 'navel'.

The fruit, which is botanically called as a *berry*, turns from deep-green to yellow or red, or, in some forms, green-and white-striped, and may range from 2½ to 12 in (6.4-30 cm) in length and 3/4 to 2 in (1.9-5 cm) in width, and from oblong, cylindrical and blunt to pronouncedly 3-angled, somewhat curved and hornlike. The flesh, ivory-white to yellow or salmon-yellow, may be firm, astringent, even gummy with latex, when unripe, turning tender and slippery, or soft and mellow or rather dry and mealy or starchy when ripe. The flavour may be mild and sweet or subacid with a distinct apple tone. Wild types may be nearly filled with black, hard, rounded or angled seeds $1/8^{th}$ to $5/8^{th}$ inch (3-16 mm) wide and have scant flesh. The common cultivated types are generally seedless with just minute vestiges of ovules visible as brown specks in the slightly hollow or faintly pithy centre, especially when the fruit is overripe. Occasionally, cross-pollination by wild types will result in a number of seeds in a normally seedless variety such as 'Gros Michel', but never in the Cavendish type.

Taxonomy

The genus name *Musa* is thought to be derived from the Arabic name for the plant (*mouz*) which, in turn, may have been applied in honour of Antonius Musa (63 – 14 BC), physician to Octavius Augustus, first emperor of Rome. The name 'banana' is derived from the Arabic banan = finger and was thought to be used in Guinea (West Africa) concomitant with the introduction of the fruit by the Portuguese. The name then spread to the New World. The genus *Musa* is a member of the family Musaceae, which includes at least one other genus (*Ensete*) and, depending upon the affiliations of the taxonomist, may also include the monotypic genus *Musella*. The plant family Musaceae, composed of bananas, plantains, and ornamental bananas, originally evolved in South-East Asia and surrounding tropical and subtropical regions (including New Guinea). Africa is considered as secondary centre of diversity.

Taxa in the Musaceae

Ensete

The genus *Ensete* is found throughout Africa and southern Asia. The *Ensete* species are monocarpic, unbranched herbs, which produce suckers rarely and are used for food, fibre, and as ornamentals. The plants of *Ensete* species resemble banana plants, but their wide spreading and immensely long, paddle-shaped leaves with usually crimson midribs, are unmistakable. Their fruits are similar in appearance to those of banana, but they are dry, seedy, and inedible. The entire plant dies after fruiting.

The important species are *Ensete gilletii* (De Wild.) Cheesman; *Ensete glaucum* (Roxb.) Cheesman. *Common names*: Wild banana, Seeded sweet banana, 'Virgin' banana, or Virgin (Philippines); *Ensete homblei* (Bequaert) Cheesman; *Ensete perrieri* (Claverie) Cheesman; *Ensete superbum* (Roxb.) Cheesman and *Ensete ventricosum* (Welw.)

Cheesman. *Common names*: Enset, Ensete, Abyssinian banana or Plantain, Ethiopian, Black, Bruce's or wild banana.

Musa

The centre of origin for genus *Musa* is Asia, primarily, southern and south-eastern Asia. A great number of important plants are found in the genus, which bear edible fruits. In addition to fruit, bananas and plantains are used for making medicines, beverages, fibres, edible floral parts, dyes, fuel, steam for cooking, cordage, wrapping materials, etc. Most cultivated varieties (cultivars) of edible banana originated from two species, *M. acuminata* and *M. balbisiana*.

Historically, five sections have been recognized in *Musa*, *viz.*, *Australimusa*, *Callimusa*, *Musa* (formerly known as *Eumusa*), *Rhodochlamys* and *incertae sedis* . Recent molecular analyses indicate a reduction to two sections, but much further study is required before the above system is abandoned. These sections of *Musa* have been describted briefly here under.

i. AUSTRALIMUSA (Chromosome Number, x =10)

Seeds are sub-globose or compressed, smooth, striate, tuberculate or irregularly angled. Contains the Fei bananas, which are important in the Pacific. Their origins are complex and may involve as many as three species, *M. lolodensis*, *M. maclayi* and *M. peekelii*. Also included in the section is an important source of fiber, abacá (*M. textilis*).

Plants in the section Australimusa are generally tall, with seeded fruit, and with distinctive green or greenish-yellow buds (if present). Their seed structure is important for classification purposes, *viz.*, either sub-globose or compressed, smooth, striate, tuberculate, or irregularly angled. Fei are robust plants bearing erect bunches of brilliant orange-gold fruit, which are delicious and nutritious when baked or boiled. *M. textilis* (abacá or Manila hemp) is particularly important in Philippine culture, and to a certain extent in some traditional islands of Micronesia, as a source of fibre. In the outer islands of yap, islanders still use hand-looms to weave abacá fiber into women's wrap-around skirts or lavalavas (pareus). Abacá produces a shiny, apple-green bud.

M. jackeyi W. Hill. (Common name: Johnstone River banana)

This has only a small range in North Queensland, Australia. It greatly resembles a Fei, with upright fruit stalk, Fei-like bananas, an enormous green bud pointing skyward, and 'bloody' sap. May be synonymous with *M. maclayi* subsp. *ailuluai*.

M. lolodensis Cheesman

It is a possible precursor of the Fei bananas. Native to the West Sepik region of Papua, New Guinea and parts of Indonesia.

M. maclayi von Muell. ex Mikl.-Maclay

It is also a possible precursor of the Fei bananas. The fruits are rounded, oppressed together in tight bunches, and in some varieties, partly joined together laterally.

M. peekelii Lauterb

This could also be a possible precursor of the Fei bananas. A very tall plant (>10 m [33 ft]), with a bunch of fruits tipped with a narrow green bud, which is pendent, rather than reaching skyward. Its geographical range includes Papua, New Guinea, and it was found in the Philippines (Palawan) in 1960.

M. textilis Née. (Common names: abacá, Manila hemp, amukid; Synonyms: *M. formosana* Hayata, *M. textilis* Née. var. tashiroi Hayata)

Before the advent of synthetic textiles, *M. textilis*, was the source of one of the world's premier fibres–soft, silky, glistening, and fine-textured. Its fibres are also suitable for manufacturing rayon, cellophane, and newsprint. It was even a constituent in some European paper money. Commercial production was greatest in the Philippines and Central America, but has now all but disappeared. Niche markets now cater to intricately woven floor mats, fancy place mats, and specialty paper.

ii. CALLIMUSA (Chromosome Number, x =10)

Bracts plain, firm, shiny on the outer surface, rarely glaucous and strongly imbricate when closed. These plants are most important as ornamentals. Most bear upright flower stalks, variously colored buds and flowers, and small seedy fruit.

M. beccarii Simmonds

This species bears a narrow, elliptical, bright scarlet bud, with green-tipped bracts. Fruits are green and skinny.

M. coccinea Andrews [Common Names: Red (flowering)]

These are called as Thai banana, scarlet banana, Thai red banana, coccinea, okinawa torch, okinawan banana flower, red ornamental banana.

The other species included in this group are *M. exotica* R. Valmayor, sp. nov.; *M. flavida*, *M. hotta*; *M. gracilis* Holttum, *M. salaccensis* Zoll. Common name: Javanese wild banana; *M. suratii* G. C. G. Argent; *M. violascens* Ridley.

iii. MUSA/EUMUSA (Chromosome Number, x =11)

Most cultivated varieties (cultivars) of edible banana originated from two species in this section, *M. acuminata* and *M. balbisiana*.

Musa acuminata Colla. (Synonyms: *M. chinensis* Sweet, *M. corniculata* Kurz, *M. rumphiana* Kurz, *M. simiarum* Kurz)

Recent genetic studies have identified whose subspecies were probable parents of some important edible cultivars. *M. acuminata*'s native habitat ranges throughout SE Asia (West to Myanmar) and Papua, New Guinea.

M. banksii F. Mueller

This is an edible, hybrid bananas, which arose within the Pacific. It is characterized in part by 15–20 cm (6–8 in), sausage-shaped fruit with rounded tips.

are the primary clone represented in Oceania, ranging from the western Pacific >6400 km (>4000 mi) eastwards to the Marquesas Islands.

Subspecies *burmannica* Simmonds

It is found in Burma, southern India and Sri Lanka.

Subspecies *burmannicoides* DeLanghe

Found in southern India.

Subspecies errans Argent. (Synonyms: *M. errans* Teodoro, *M. troglodytarum* L. var. *errans*, *M. errans* Teodoro var. Botoan)

It is a very pretty subspecies, with a blue-violet pendent bud and very pale green immature fruit.

Subspecies *malaccensis* (Ridley) Simmonds (Synonym: M. malaccensis Ridley)

It is distributed in Peninsular Malaysia and Sumatra. Paternal parent of 'Silk' AAB, the 'true apple' banana, common in the West Indies (not to be confused with 'apple' bananas of Hawaiii). The clone 'Pisang Lilin' is a derivative of this subspecies.

Subspecies *microcarpa* (Beccari) Simmonds

It is found in Borneo. This subspecies has given rise to the clone 'Veinte Cohol'.

Subspecies siamea *Simmonds*

It is found in Cambodia, Laos and Thailand.

Subspecies *truncate*

It is found in Peninsular Malaysia (highlands).

Subspecies *zebrina* (Van Houtte) R. E. Nasution. Java [Common Name: Blood banana]

Synonyms

M. acuminata Colla subsp. sumatrana {(Becc.) A.N., *M. sumatrana* 'Rubra'}

M. balbisiana Colla. (Common names: Balbis banana, Starchy banana, Mealy banana, Seedy banana, Wild (starchy) banana, Devil banana and Seeded apple banana (Maui, Hawaii)}

This species is extremely robust, fast-growing, and drought-resistant. The wild, seedy forms are much less variable than *M. acuminata*, although five morphotypes have been described. It is an useful windbreak. *M. balbisiana* is one of the parents of many edible seedless bananas. It is a native to Southeast Asia from Sri Lanka to the Philippines.

M. basjoo Sieb. {Common names: Japanese (fibre) banana; Synonym: *M. Japonica*}

It is used for fibre and as an ornamental. Native to Japan (including the Ryuku Islands), this is probably the world's most cold-hardy banana. It is a medium-sized plant (to 5 m [16 ft]) similar to abacá, with a beautiful, rounded, large green and yellow, shiny bud and inedible fruit. It is also used for fibre, elegant fabrics, and as an ornamental.

The other species include *M. cheesmani* Simmonds, *M. flaviflora* Simmonds, *M. halabanensis* Meije, *M. itinerans* Cheesman, *M. nagensium* Prain, *M. ochracea* Shepherd, *M. schizocarpa* Simmonds and *M. sikkimensis* Kurz.

iv. RHODOCHLAMYS (Chromosome Number, x =11)

Many highly ornamental species are found in this section such as *M. aurantiaca* Mann ex Baker, *M. laterita* Cheesman (Common name: Indian dwarf banana), *M. ornata* Roxb. and *M. mannii* H. Wendl. ex Baker.

The other species included in this group are *M. rosacea* Jacq., *M. rosea* Baker, *M. rubra* Wall. ex Kurz, *M. sanguinea* Hook. f., *M. velutina* H. Wendl. and Drude (Common names: Fuzzy (pink) banana, self-peeling banana, pink banana, hot pink banana, Velutina).

M. velutina H. Wendl. and Drude

Native to northern India, it is widespread in tropical botanical gardens, and is becomingly increasingly available for homegardens. Its upright hot pink bud, whose bracts are crowded with bright yellow flowers, produce small, fuzzy, flat, bright pink bananas. Their white inner flesh is packed with black seeds, which germinate readily, although rather slowly. This banana's species name, velutina, means velvety, as indeed it is. When ripe, the banana's flesh bursts through its skin at the apex, then proceeds to peel itself, true to its alternate common name, 'self-peeling' banana.

v. INCERTAE SEDIS

This is a taxa, which is considered to have uncertain taxonomic positions.

M. boman Argent

A tall plant from New Guinea with a glossy yellow bud, resembling abacá (*M. textilis*).

M. ingens Simmonds (Chromosome Number, x = 7)

This is the world's largest herb, and can reach 15 m (49 ft) in height and 2.5 m (8 ft) in circumference at the base. It is found on the island of New Guinea between 1,000 and 2,100 m (3,300–6,900 ft) in elevation.

The unresolved taxonomy at the family level continues down to the genus level and there are inconsistencies in the number of sections and number of species proposed for inclusion in the genus *Musa*. This has largely been brought about by the domestication of the fruit-bearing cultivars and the subsequent temporal and genetic separation from the original species, as well as the widespread vegetative reproduction in the genus and natural occurrence of many hybrids.

Before Simmonds and Shepherd's system, cultivated bananas were classified using the binomial nomenclature system developed by Carl Linneaus. Linneaus is the one who gave the name *Musa paradisiaca* to the banana. Later, Ernest Cheesman noticed that the model for *Musa paradisiaca* was in fact a type of plantain. When it was later realized that *Musa paradisiaca*, like *Musa sapientum* which Linnaeus had also added to the genus, and which turned out to be a Silk banana, and was hybrid between *Musa acuminata* and *Musa balbisiana*.

Over the years, several authors worked on the taxonomy of bananas on *Musa paradisiaca* and *Musa sapientum*. Sometimes, *Musa sapientum* was treated as a subspecies of *Musa paradisiaca*, but at other times, botanical priority was ignored and *Musa paradisiaca* was treated as a subspecies of *Musa sapientum*. Moreover, since *Musa paradisiaca* is seedless, the subspecies *seminifera* was created in order to accommodate the wild seeded forms.

It was eventually recognized that most cultivars (except for certain types such as the Fei bananas) are derived from either *Musa acuminata* alone or hybridized with *Musa balbisiana*. Some of these cultivars are, like their wild relatives, diploids, *i.e.* they have two sets of chromosomes (one inherited from each parent). The majority, however, are triploids, *i.e.*, they have three sets. This means that at one point, the reproductive cells of one of the parents did not undergo the normal halving of its genome and produced unreduced gametes. The other parent contributed a normal haploid genome.

This complexity made it difficult to devise a Latin name-based taxonomy, which could cope with all possible permutations. Cheesman realized that the use of Latin names for cultivars would have to be abandoned. Later, two of his young colleagues, Norman Simmonds and Kenneth Shepherd presented an alternative way to classify banana. The nomenclature system used to classify banana cultivars was developed by Norman Simmonds and Kenneth Shepherd in 1955. This system eliminates almost all the difficulties and inconsistencies of a taxonomy based on *Musa paradisiaca* and *Musa sapientum*.

Simmonds and Shepherd's Genome-Based System

In this system, bananas, at least the ones that are related to *Musa acuminata* and *Musa balbisiana*, are classified according to the relative contribution of these species, designated by the letter A, for *acuminata*, and B, for *balbisiana*. A cultivar is assigned to a genome group according to the number of chromosome sets in its genome (its ploidy) and the species that donated them. For example, *M. acuminata* and *M. balbisiana* are diploids, with genome AA and BB, respectively, and AA and AB clones are cultivated. Hybrid triploids are classified as AAA, AAB, or ABB. Tetraploid bananas (mostly products of breeding programs) may be AAAA, AAAB, AABB, or ABBB.

Genome groups are further divided into subgroups, defined as a set of closely related cultivars derived from a single original clone. On the basis of this system, cultivar names are put between inverted commas and preceded by the name of the genus and when known, the name of the group and subgroup. For example: *Musa* AAA (Cavendish subgroup) 'Robusta' (Table 2.3).

Table 2.3: Genome Nomenclature of some Banana Cultivars

Genome Group	Subgroup	Example of Common Cultivar Names
AA	Sucrier	'Sucrier'; 'Lady's Finger'
	Inarnibal	'Inarnibal'
	Lakatan	'Lakatan'; 'Lacatan'
	Pisang Lilin	'Pisang Lilin'
AB	Ney Poovan	'Lady's Finger'
	Kamarangasenge	'Sukari Ndizi'
AAA	Cavendish	'Giant Cavendish' (*e.g.* 'Williams', 'Mons Mari') 'Grande Naine' 'Dwarf Cavendish' 'Extra Dwarf Cavendish' (*e.g.* 'Dwarf Parfitt') 'Pisang Masak Hijau' 'Double'
	Gros Michel	'Gros Michel' 'Cocos' (Honduras), 'Highgate' (Jamaica) 'Lowgate' (Honduras)
	Mutika/Lujugira sub-group Synonym: *Musa brieyi* De Wild	'Beer', 'Musakala' 'Nakabululu' 'Nakitembe' 'Nfuuka'
	Ibota	'Yangambi Km 5'
	Red	'Red Dacca' 'Green red'
AAB	Maoli-Popoulu	'Pacific Plantain' 'Manini' 'Ele'ele' 'Hopa' 'Fa'i Samoa' 'Hai'/'Haikea' 'Maoli'
	Iholena	'Fa'i mamae' 'Iholena Iholena' 'Iholena Kâpua' 'Iholena Iele' 'Ore'a'
	Mysore	'Mysore'
	Pisang Raja	'Pisang Raja'
	Plantain	'French', 'Pisang Ceylan', 'False Horn', 'Horn'
	Pome	'Lady's Finger', 'Prata Aña', 'Pacha Naadan'
	Silk	'Sugar'
AAB	Bluggoe	'Bluggoe', 'Dwarf Bluggoe', 'Silver Bluggoe', 'Mondolpin'
	Monthan	'Nalla Bontha Bathees', 'Monthan'
	Klue Teparod	'Kluai Tiparod'

Contd...

Table 2.3–*Contd...*

Genome Group	Subgroup	Example of Common Cultivar Names
	Ney Mannan	'Blue Lubin'; 'Blue Java'
	Pelipita	'Pelipia'
	Saba	'Benedetta', 'Cardaba', 'Saba'
	Pisang Awak	'Ducasse'; 'Kluai Namwa Khom'
AAAA		'FHIA-02' (aka 'Mona Lisa'), 'FHIA-17', 'FHIA-23
AAAB		FHIA-01d ('Goldfinger'); FHIA-18 ('Bananza')
AAAB		'FHIA-20', 'FHIA-21'
AABB		'FHIA-03'
BB		'Tani'
BBB		'Kluai Lep Chang Kut'

Simmonds and Shepherd's Scoring System

The system is based on 15 characters that were chosen because they are different in *Musa acuminata* and *Musa balbisiana*.

Character	Musa acuminata	Musa balbisiana
Pseudostem colour	More or less heavily marked with brown or black blotches	Blotches very slight or absent
Petiole canal	Margin erect or spreading, with scarious wings below, not clasping pseudostem	Margin inclosed, not winged but clasping pseudostem
Peduncle	Usually downy or hairy	Glabrous
Pedicels	Short	Long
Ovules	Two regular rows in each loculus	Four irregular rows in each loculus
Bract shoulder	Usually high (ratio<0.28)	Usually low (ratio>0.30)
Bract curling	Bracts reflex and roll back after opening	Bracts do not reflex
Bract shape	Lanceolate or narrowly ovate, tapering sharply from the shoulder	Broadly ovate, not tapering sharply
Bract apex	Acute	Obtuse
Bract colour	Red, dull purple or yellow outside; pink, dull purple or yellow inside	Distinctive brownish-purple outside; bright crimson inside
Colour fading	Inside bract colour usually fades to yellow towards the base	Inside bract colour usually continuous to base
Bract scars	Prominent	Scarcely prominent
Free tepal of male flower	Variably corrugated below tip	Rarely corrugated
Male flower colour	Creamy white	Variably flushed with pink
Stigma colour	Orange or rich yellow	Cream, pale yellow or pale pink

Each character is scored on a scale from one (typical *Musa acuminata*) to five (typical *Musa balbisiana*). The possible total scores range from a minimum of 15 to a maximum of 75. The expected scores are 15 for AA and AAA, 35 for AAB, 45 for AB, 55 for ABB and 75 for BB.

IMPORTANT VARIETIES

The chief characteristics of the commonly grown banana varieties/clones in India are described briefly hereunder:

Poovan (AAB)

Foremost commercial variety in Tamil Nadu, A.P. and West Bengal. 11 months duration, average bunch weight 15-0 kg. Fruits medium in size, held firmly in the bunch and have a distinct mammillary tip, fruit rind thin, pulp cream coloured, sub-acid taste, ripe rind golden yellow and keeps well, resistant to panama wilt.

Monthan (ABB)

A leading commercial culinary banana of India. Plant is hardy and drought resistant. Duration is 12-14 months, each bunch weighs 10 kg, fruits are long with good girth in the middle, plump, angulate, slightly curved with blunt or knobbed apex, thick green rind. In some parts of Tamil Nadu, ripe fruits are eaten, but mostly for culinary purpose.

Dwarf Cavendish (AAA)

Dwarf type, Cavendish group, widely cultivated throughout the banana growing tract of the world, durations 10 months, each bunch weighs 15 kg, fruits large, curved, thick, greenish, flesh soft and sweet, retains green colour to some extent even after ripening, keeping quality poor, susceptible to bunchy top but resistant to Panana-wilt.

Harichal (AAA)

A semi-tall sport of Dwarf Cavendish, duration is 12 months, bunch weighs 25-30 kg, fruits large, skin thick, greenish to dull yellow, sweet and delicious, better keeping quality than Dwarf Cavendish. Commercially grown in Maharashtra, now gaining commercial importance in Tamil Nadu and Kerala.

Rasthali (AAB)

Choice table variety, duration 15-16 months, each bunch weighs 12 kg fruits are medium sized, similar to Poovan in appearance, skin thick, ivory yellow in colour, flower firm sweet with a pleasant flavour. Ayiranka Rasthali is a sport of this variety, which has only pistillate flowers upto the length of the axis.

Hill Banana (AAB)

This variety is a speciality of Tamil Nadu grown in lower Palney hills from 1,000-1,500 m as perennial shade crop for banana. The duration is 12 months, each bunch weighing 12 kg with 80-90 fruits per bunch. Two main types *viz.*, Sirumalai and Virupakshi are known to exist, the fruits of former one are having a thick rind, which comes off clean, while in the latter, threads of inner cushiony substance sticks to the pulp. Besides, fruits of Sirumalai are tastier, with more mellow pulp with fine flavour than those of Virupakshi, which are slightly sub-acid.

Nendran (AAB)

Fruits are always cooked to make it more palatable and hence known as plantain. Commercially grown in Kerala and parts of Tamil Nadu, 11-12 months in duration, bearing 12- 15 kg of bunch, each bunch has 6 hands, 8-15 fingers per hand. Fruits are relatively longer and thicker than most banana fruits, rind is thick and buff yellow when ripe, flesh is firm, yellowish with a characteristics good keeping quality (15 days even after ripening).

Sevazhai (TN) AAA

This cultivar is grown largely in Kerala. The characteristic feature of this variety is that pseudostem, midrib and fruit rind is purplish red, hence the name red banana. Fruit is of good size, slightly curved with a blunt apex, thick red and rind on ripening develops a characteristic strong flavour.

Kunnan (AB)

A popular dessert variety of Kerala, fruits have thin rind but with firm flesh.

CO-1 (AAB)

It was developed at TNAU, Coimbatore, involving multiple crossing between Ladan, *M. balbisiana* and Kadali. It was evolved by Ladan (AAB) × *M. balbisiana* (BB) ! AB × Kadali (AA) ! CO-1 (AAB). CO-1 was akin to pome group and resembles to Virupakshi (AAB) and has typical acid and apple flavour of Virupakshi. It is suitable for cultivation both in plains as well as hills (upto 1,220 m). The main crop duration is 14-15 months. The fruits contain a TSS of 22.6 °brix and an acidity of 0.58 per cent. The mean bunch weight is 10.57 kg.

Udhayam

It is a single plant selection from Kanthali at National Research Centre for Banana, Trichy, Tamilnadu. The average bunch weight is 37 kg having a potential upto 45 – 50 kg, which is 40 per cent higher than local Karpuravalli. Crop duration is 13 months and produces high yield in ratoons also. Exhibits field tolerance to Sigatoka leaf spot disease and nematodes.

BRS-1

This hybrid was developed through hybridization between Agniswar × Pisang Lilin and was released in 1999 from Banana Research Station, Kannara, Kerala. It is a triploid (AAB) hybrid of pome type. It has a short cropping cycle, resistance to leaf spot, Fusarium wilt and burrowing nematode (*Radopholus similis*). Medium statured variety and proudes about 14-16 kg bunch. Elongated fruits turn attractive golden yellow on ripening.

BRS-2

It is a hybrid between Vannan × Pisang Lilin, developed at Banana Research Station, Kannara, Kerala. It belongs to Mysore subgroup, Poovan type. It is a medium stature plant, growing upto 7-8 ft. Crop cycle is short with bunch coming to harvest in 11-12 months. Average weight of the bunch ranges from 15-20 kg with short, stout, dark green Poovan like fruit.

Besides, from AAU, Assam, a variety 'Bhimkal' was released, which was found to be resistant to nematodes. At CBRS, Kannara, Vannan, Agniswar and Harichal were crossed with Pisang Lilin, resulting in release of H-1 and H-2. H-1 has short crop cycle and have resistance to leaf spot, *Fusarium* wilt and burrowing nematode, while H-2 was found to resistance against leaf spot and nematode; besides, being suitable for subsistence cultivation. The characteristics of some other important banana varieties grown in India are as under:

Varietiy	Chief Characteristics
Berangan (AA)	Use for fresh. Each bunch has 8-12 hands and weighed 12-20 kg. Every hand has 12-20 fingers. It has medium to large size fingers ranges from 12-18 cm in length and 25-35 cm in thickness Fruit skin is thick, smooth and yellow in colour when ripen It has yellowish orange, pleasant aroma and sweet flesh.
Mas (AA)	Use for fresh. Each bunch has 5-9 hands and weighed 8-12 kg. Every hand has 14-18 fingers. It has small fingers ranges from 8-12 cm in length and 2-3 cm in thickness. Fruit skin is thin, smooth and golden yellow in colour when ripen. It has golden orange, pleasant aroma and very sweet flesh.
Embu (AAA) (Cavendish)	Use for fresh. Each bunch has 15-25 hands and weighed more than 20 kg. It has large size fingers ranges from 15-22 cm in length and 35-45 cm thickness. Fruit skin is thick, smooth and yellow in colour when ripen. It has white, fine textured, pleasant aroma and sweet flesh.
Rasthali (AAB)	Use for fresh. Each bunch has 5-9 hands and weighed 10-15 kg. Every hand has 12-16 fingers. It has medium size fingers ranges from 10-15 cm in length and 3-4 cm in thickness. Fruit skin is very thin and yellow in colour when ripen. It has white, pleasant aroma and sourish sweet flesh
Tanduk (AAB)	Use as fried banana (pisang goreng). Each bunch has 2 hands which weighs 7- 10 kg. Every hand has 5 fingers. It has a very large size fingers ranges from 25-35 cm in length and 6-7 cm in thickness. Fruit skin is very thick, smooth and yellow in colour when ripen. It has creamy orange and sweet flesh.
Nipah (BBB)	Use as fried banana (pisang goreng). Each bunch has 12-18 hands, and weighs 15-28 kg. Every hand has 12-20 fingers. It has large, angular fingers ranges from 10-15 cm in length and 3-5 cm in thickness. Fruit skin is very thick, smooth and yellow in colour when ripen. It has creamy white and sweet flesh.
Raja (AAB)	Use as fried banana (pisang goreng) or eaten fresh. Each bunch has 6-9 hands and weighed 10-15 kg. Every hand has 12-16 fingers. It has a large size fingers, 15 cm in length and 3-4 cm in thickness. Fruit skin is very thick, smooth and yellow in colour when ripen. It has creamy orange, coarse textured and sweet flesh.
Nangka (AAB)	Use as fried banana (pisang goreng). Each bunch has 6-8 hands and weighed 12-14 kg. Every hand has 14-24 fingers. It has large, angular shape with medium to large fingers.
Awak (ABB)	Use as fried banana (pisang goreng) or eaten fresh. Each bunch has 8-12 hands and weighed 18-22 kg. Every hand has 10-16 fingers. It has medium size fingers ranges from 10-15 cm in length and 3-5 cm in thickness. Fruit skin is thick, smooth and yellow in colour when ripen. It has creamy white and sticky flesh. In some cases, seed are present.

Distribution Pattern of Banana Cultivars in India

State	Cultivars
Andhra Pradesh	Dwarf Cavendish (AAA), Robusta (AAA), Amritpani (Rasthali, AAB), Thella Chakrakeli (AAA), Karpoora Chakrakeli (Poovan AAB), Monthan (ABB).
Assam	Jahaji (AAA), Dwarf Cavendish, Borjahaji (AAA, Robusta), Malbhog (AAB), China (AAB), Manohar (ABB), Kanchkol (AAB), Chini Champa (AB), Bhimkol (BB).
Bihar	Alpan (AAB), Chini Champa (AB), Malbhog (Rasthali, AAB), Muthia (ABB), Kothia (ABB), Basrai (AAA), Batheesa (ABB).
Gujarat	Dwarf Cavendish (AAA), Lacatan (AAA), Harichal (Lokhandi, AAA).
Karnataka	Dwarf Cavendish (AAA), Robusta (AAA), Poovan (AAB), Rasabale (AAB, Rasthali), Marabale (Pome, AAB), Monthan (ABB), Elakkibale (AB, Ney Poovan).
Kerala	Nendran (AAB, Plantain), Playankodan (AAB, Poovan), Kunnan (AB), Rasthali (AAB), Monthan (ABB), Red Banana (AAA).
Maharashtra	Basrai (Dwarf Cavendish AAA), Safed Velchi (AB), Rajeli (AAB, Plantain), Robusta (AAA), Monthan (ABB).
Tamil Nadu	Virupakshi (AAB), Poovan (AAB), Rasthali (AAB), Monthan (ABB), Dwarf Cavendish (AAA), Robusta (AAA), Peyan (ABB), Nendran (AAB Plantain).
West Bengal and Orissa	Champa (AAB), Morthaban (AAB, Rasthali), Amrit Sagar (AAB), Giant Governer (AAA), Lacatan (AAA), Monthan (ABB).

SOIL AND CLIMATIC REQUIREMENTS

Soil

Banana can be grown on a variety of soils, ranging from clay to sandy clay loam. However, the best soil is medium textured soil, uniform, reasonably deep and fertile, having good internal drainage. Soil pH between 5.5–8.0 is considered ideal for commercial banana cultivation.

Climate

The major banana-growing areas of the world are geographically situated between the equator and latitudes 20° North and 20° South. Banana is essentially a humid tropical plant, coming up well in regions with a temperature range of 10 °C to 40 °C with an average of 23 °C. In cooler climate, the duration is extended, sucker production is affected and bunches are smaller. Low temperatures (less than 10 °C) are unsuitable since they lead to a condition called 'choke'/'November dump' or impeded inflorescence and bunch development and the trapped bunches, which are exposed to the sun are said 'sun lookers'. Banana comes up well from sea level upto an altitude of 1,800 m above sea level. It enjoys an annual rainfall ranging from 100 to 325 cm. Banana is well adapted to areas with an annual rainfall between 1,000-2,000 mm. Banana should be planted in areas free from high wind/storms. The major field losses are caused by wind and flood. A wind storm causing plants to fall over is referred as 'blowdown'. Losses from storm damage are classified as 'doubling', when pseudostem collapses, uprooting when the plants fall over with the rhizome attached and snap off when rhizomes breaks near ground-level when the plant falls down.

Losses from 'break neck' occur when peduncle breaks prior to emergence from the pseudostem or soon after emergence.

PLANT PROPAGATION

Bananas are propagated from offshoots (suckers or keikis) or corms (bullheads). If enough buds are present, large bullheads can be halved or quartered. Planting material should be treated for nematodes as under:

(1) Cut off bottom half of corm and, if discolored, trim off up to 2/3rd of the bottom of the corm until only clean white tissue remains.

(2) Trim off about 1/2 inch of tissue around the sides of the corm.

(3) If bullheads are used, cut off the pseudostem 3-4 inches above the top of the corm.

(4) Either, (a) immerse the trimmed corms in a hot water bath at 50-52°C (122-126°F) for 15-20 minutes. Before planting, place the corms in a transparent plastic bag at room temperature until new roots begin to appear, or (b) coat the corms with parafilm wax prior to shipment or storage.

PLANTING AND ORCHARD ESTABLISHMENT

Preparation of Land

The selected field must be ploughed 4-6 times and allowed to weather for two weeks. Then the field is leveled by passing a blade harrow, plough furrows may be formed length wise and breadth wise of the field at the required spacing and its intersection of the plough furrows, pits of size 0.6m × 0.6m × 0.6m are dug sufficiently ahead at points fixed for planting. Usually, well decomposed compost or FYM are mixed with top soil for filling the pits at the time of planting. In areas where nematode problem is prevalent, nematicides and fumigants are also added to pits before planting. Suckers are normally used as planting material. There are many types of suckers namely sword suckers, water suckers, maiden sucker and peepers. Sword suckers, which are conical with very narrow leaves, are preferred because of their robustness. Suckers of 3 months old well developed, disease-free corms are separated from the mother plant and planted for starting a new plantation. The average planting distance is 3.0 m × 3.0 m. for tall cultivars and 2.0x2.0 m for dwarf varieties. Planting holes of 60 cm × 60 cm × 30 cm are dug and into each hole are placed 100 grams of phosphate (TSP) before planting. Robusta banana spaced at 2.4 m × 1.8m and Dwarf Cavendish banana at 2m × 2m or 2.5m × 2.5 m gave the highest yield. Although, higher yield is obtained in high-density planting but the growth of plant is slower and shooting is delayed. Finger tip disease is also severe in close planting. In high-density plantation nearly 30 percent of the plants can't be harvested in one time.

Season of Planting

The season of planting of banana varies from state to state. Keeping in view the divergence of climatic and soil conditions in our country, bananas are grown all through the year, while the peak seasons vary in different parts of the country. In most parts of the country, the colder seasons of the year are unsuitable for planting. On the contrary, in West Coast, planting is done from September-to-November with secured irrigation. Planting is done all the year round in order to fetch better prices in the market during the off season. In other areas, planting is done during South West monsoon in May-June, and continues thereafter till November. Planting in cold season is a troublesome and huge attention should be given for irrigating the crop in summer. Furthermore, it exposes the plants to high winds or cyclone damage during bunch formation time season. In Kerala, where Nendran is being grown as a pure crop, planting is executed in September-October. On the lower Palneys hillls, including Sirumalai, April planting is preferred. February-March is the best planting season in wet lands along the Cauvery bank as in Trichy. But in the perennial plantations in Tanjore, planting is done from January-to-June. The best time for planting in Karnataka, Andhra Pradesh and Odisha is by the end of June and in West Bengal, Bihar and Assam, planting could be done at any time during the South West monsoon, when the rains are not too heavy. The planting should not be taken up during very cold and very hot months. Similarly, the planting season should be so adjusted that during the period of high winds, banana should not be in flowering or near flowering stage. Similarly, the period of planting should be so adjusted that active growth phase of the plants can continue unhampered during flower bud initiation stage.

Planting System and Spacing

Square, rectangle and triangle systems of planting are recommended for planting. For mono-cultured cropping system, the recommended planting distance is 3.0 × 1.5 m. When intercropped with other permanent crops, the recommended planting distance is 2.4 × 2.4 m.

INTERCULTURAL OPERATIONS

Thinning of Suckers

Thinning of suckers aims at maintaining good vigour of the tree, obtaining desired number of plants per clump and enhancement in production of good quality fruits. Thinning involves removal of unwanted suckers; normally weak unhealthy, mainly water suckers using a sharp knife at the ground level, leaving one bearer, one follower and one sucker per clump at any time.

Deleafing

After the stem emerges, no new leaves are produced but some of older leaves and the bract leaves may be touching the recently emerged hands. This would lead to leaf scarring. Leaves that touch fingers, are cut off with pseudostem and bract leaves are pushed back away from the upper hands of the stem. Usually, no more than one or two leaves are removed, otherwise fruits will not be protected.

Thinning of Hands, Flowers and Bunch

De-handing refers to the removal of the false hand alone or the false hand plus one or two of the last apical hands at the time of bagging. Deflowering refers to removal of the tepals and pistil of the female flower in the field prior to bagging or the dried remnants of the flower at the packing plant prior to de-handing. To have a good quality fruit, bagging of fruit bunches is recommended. For bagging, strong polyethylene bag with size 75 × 120 cm is used. Bagging or sleeving consists of placing a perforated polythene bag over the bunch after all the female hands are exposed. There are three classes of bagging: normal, early and semi-early. Normal bagging is done when all the hands have emerged and early bagging when the buds have started to curve downward but before the bracts have lifted to expose the first hand. Semi-early bagging is done when 2-3 hands have been exposed. Early bagging is done in areas where certain peel scarring insects such as *Colaspis* sp., enter the hands when first exposed. It is also effective in reducing peel scratches from nectar-feeding bats. In addition to preventing peel blemishes, bagging of fruits creates greenhouse effect around the fruit by raising the temperature about 0.5 °C, which results in early maturity of fruits by 3-4 days.

Likewise, removal of the bunch soon after emergence to speed up the growth of follower is said to be de-bunching. Banana plants need a combined operation of bagging, removal of male buds (de-naveling/de-budding) and tying the plant with twine to adjacent plants or to an erect pole or to an overhead wire, which is called as fruit protection.

Weeding

During early years of growth, weeds should be controlled manually or using weedcide. Weeds between rows can be controlled using contact herbicides or by planting cover crops. As banana roots are superficial, care is taken during weeding to ensure that root damage does not occur. Weeds in between rows are controlled with herbicides using sprayer with a protective cone to the spray nozzle.

Manures and Fertilizers

Banana requires high fertilization due to its rapid and vigorous growth and high fruit yield. The nutrient uptake studies also reveal that the uptake of nutrients per unit area is more than any other crop. The nutrient uptake pattern analysis conducted in different countries showed that for a crop of 40-60 t/ha, removes nearly

Table 2.4: Recommended Doses of Manures and Fertilizers for Banana in India

Time	Type of Fertilizer	Fertilizer Rate/ha
At time of planting	TSP	110 kg
1 month	NPK 15:15:15	275 kg
3 months	NPK 15:15:15	385 kg
5 months	NPK mg 12:12:17:2	385 kg
1 month after first harvesting	NPK mg 12:12:17:2	385 kg
3 months after	NPK mg 12:12:17:2	385 kg
1 month after	NPK mg 12:12:17:2	385 kg
2 months after	NPK mg 12:12:17:2	385 kg
3 months after first harvesting	NPK mg 12:12:17:2	385 kg

250-300 kg N, 25-40 kg P and 800-1,200 kg K, 150-180 kg Ca, 40-60 kg Mg and 14-20 kg S per hectare. This reveals that the fertilizer applied should contain more nitrogen and potassium. The recommended fertilization programmes in India is shown in Table 2.4.

Water Management

Banana should be irrigated to encourage development and healthy growth especially in the early years of growth. Micro-sprinkler or drip irrigation system is recommended. Areas with frequent flash flood, construction of in-field drainage is recommended.

CROP PROTECTION

A. Major Diseases and their Management

Several diseases cause losses to banana world over. However, in India, the following diseases have been reported to be serious, which should be controlled by following the measures suggested for their effective management.

i. Fungal Diseases

Panama Wilt

Panama wilt, also known as Fusarium wilt was first reported by Higgins (1904) in Honolulu and now the disease is widespread and reported to occur in all the countries wherever bananas are grown. In India, the disease was first reported from Chinsurah in 1911 (Basu, 1911) and is presently prevalent in Tamil Nadu, Kerala, Karnataka, Bihar, Assam and Andhra Pradesh.

Causal Organism

The causal organism is *Fusarium oxysporum* f. sp. *cubense*. The fungus is mainly intracellular in the vessels, though intercellular hyphae may be observed in cortex of roots and parenchymatous tissues. Sporodochia appear through stomatal openings at late stage on the petioles and leaves of infected plants.

Symptoms

The fungus enters the roots through wounds, grows there and blocks the vascular system, thereby causing the plant to wilt. The infected plants show characteristic yellowing of leaf blades. The yellowing develops as a band along the margin and spreads towards mid rib. The leaves wilt, the petiole buckles and the leaf hangs between the pseudostem and the middle of lamina, while the leaf is still growing. All leaves except the youngest, eventually collapse and heart alone remains upright. New leaves, which appear are blotchy and yellow and often with wrinkled lamina. The pseudostem may show the longitudinal splitting of the outer leaf bases above the soil level. Young plants in heavily infected soil may exhibit dwarfing or stunting symptoms. In some cases, bunches also show various abnormal developments. For example, banana fruits become bottle necked, ripen irregularly and rapidly and flesh becomes pithy, acrid and yellow.

Internal symptoms are readily observed when diseased rhizomes or pseudostems are cut open. In a transverse section of diseased rhizomes, the disease symptoms are seen localized in the vascular strands. Individual diseased strands appear as yellow with red brownish dots and streaks. The diseased strands are distributed almost uniformly within and around the periphery of the stele. The pattern of discoloured streaks varies according to the region affected and severity. If the attack is severe, the yellow and red strands can be observed in the centre of the trunk. In advanced infections, the diseased vascular bundles are more numerous and deeply strained with dark-red and reddish brown pigments. When the infected pseudostem is cut through transversely, the yellow colour is evident in youngest leaf sheath, dark red in older and brown in outer leaf sheath. The outermost sheaths of the pseudostem are usually the first to show discoloured strands and the innermost sheaths the last. With advancing age, the affected strands darken in colour and can be traced upto the crown of the plant and petioles.

Management Practices

Growers should adopt the following measures to keep this disease under check:

1. Use disease-free planting material for raising new plantations.
2. Rougue out the infected plant material and destroy it.
3. Deep soil ploughing up to 60-90 cm thereby the pathogen could be destroyed by biological and other soil factors.
4. Follow crop rotation with sugarcane to reduce the level of the disease.
5. Flood fallowing of infested soil in 2 to 5 feet deep water for a period of 2 to 6 months help in reducing the disease.
6. Soil drenching with carbendazim (0.1 per cent) twice at monthly interval after 6 months of planting is effective in controlling the disease.
7. Integrated management programme utilizing disease free planting material, pre-planting dipping of rhizomes in carbendazim (0.2 per cent) for 45 minutes and pre-planting application of lime or neem cake (1 kg/pit) or application of urea (200 g/plant) plus sugarcane trash mulch at 5 and 7 months were recommended for effective disease control.
8. Varieties such as Dwarf and Giant Cavendish, Lacatan, Valery, Manzano, Robusta, Chiquita, Enaro, Congo, Walha, Bhimkol, Basrai Local, Red Banana, D. Local, Kasturi, Bombay Green, Kabuli and No. 15 varieties/accessions are resistant to Panama disease.

Sigatoka Leaf Spot

It is also a serious foliage disease of banana and causes great economic loss, wherever it appears. Sigatoka disease or banana leaf spot occurs throughout the world and is one of the most destructive diseases of banana. The name Sigatoka has been given, because it appeared in epidemic form in Sigatoka (Fiji Islands) in 1913. The pathogen causes losses by reducing the photosynthetic leaf surface, which results in production of small bunches having, uneven bananas that fail to ripen and may fall. Sigatoka disease is the reason of collapse of banana industry in Central and

South America. In India, it is one of the major diseases occurring in different banana growing regions of the country.

Causal Organism

It is caused by the fungus, *Cercospora musae*, with its perfect stage *Mycosphaerella musicola*.

Symptoms

The early symptoms of Sigatoka appear as small lesions on 3rd or 4th leaf from the top. These spots are indistinct, longitudinal, light-yellow, parallel to side veins and are clearly visible on both sides of the leaf surfaces. Spots increase in size, coalesce to form large dead area on leaf. Center of the spot dries out, becoming light gray with dark border, giving a look of 'eye spot'. In severe infections, the entire leaf dies; giving scorched appearance and leaf petiole collapses, which hangs down from the pseudostem. As a result of loss of leaf, the immature fruit bunches fail to fill out and ripen and fall prematurely. Individual fingers appear undersized and angular in shape. Pattern or distribution of spots over the surface of leaf lamina is governed by the kind of fungal spores causing infection. Generally, the spots produced by conidial infection are distributed over the whole of the leaf lamina, but they show characteristic linear distribution. Spots arising from ascospore infection may be distributed over the leaf, but heavy concentration of infection is towards the drooping leaf tip. In fact, black tip symptoms are the result of infection caused by ascospores.

The effect of disease on fruit yield and quality depends upon the severity of attack and duration of the epidemic. Under severe attacks, immature bunches fail to fill out and in case the fruits are approaching maturity at that time, these tend to remain undersized and pulp ripening begins unevenly. In case of very heavy infection in earlier stages of fruit development, relatively thin bunches can be seen to be ripening on the plant and fruit flesh develops a buff pinkish colour and store poorly.

Management Practices

Growers are advised to adopt the following measures to keep this disease under control:

1. Removal and destruction of infected leaves, improved drainage, efficient weed control, removal of suckers and proper spacing is recommended to check the spread of the disease.

2. Application of zineb or copper oxychloride or chlorothalonil, or mancozeb, or propiconazol, thiophanate methyl, or carbendazim, or benomyl, or derosal in combination with mineral oil gives better and less expensive control.

ii. Bacterial diseases

Moko Wilt

Bacterial wilt or Moko disease was first reported from Trinidad, West Indies during 1890. It is a serious disease of banana in many countries of world wherever banana is cultivated, but in India, it is not so serious. In India, the disease was first reported by Chattopadhya and Mukhopadhyay (1968) from West Bengal.

Causal Organism

Moko disease is caused by a bacterium, *Ralstonia (Pseudomonas) solanacearum* (Smith) Smith.

Symptoms

It is a wilt disease, more often confused with Panama wilt of banana. The disease, in general, is recognized by rapid wilting and collapse of leaves, pre-mature ripening of fruits, discoloration of vascular strands, and wilting and blackening of the suckers. The symptoms start on rapidly growing young plants. The youngest three leaves turn pale green or yellow and collapse near the junction of lamina and petiole. Most leaves collapse within 3-7 days. The characteristic symptoms appear on young suckers, which are blackened, stunted and twisted. In severe case, plants collapse before the emergence of bunches. The presence of yellow fingers in otherwise green stem often indicates the occurrence of Moko disease. Initially, a firm dry rot is found within the fruits of infected plants. Fruit stalk also shows discolored strands.

Management Practices

1. Grow resistant varieties like, Poovan and Montha etc., in the infected fields. Some of the diploids and hybrids such as F2P2, 1319-01, 1741-01, SH 3362 and Babi Yadefana are resistant to moko disease.
2. Eradicate the infected plants and destroy them.
3. Use disease-free planting material for raising new plantations.
4. Follow 3 year's crop rotation of banana, sugarcane and rice.
5. Disinfect the planting holes with chloropicrin.
6. The pruning machetes and other farm tools used in infested area should be disinfected with formaldehyde.
7. The male flower buds should be removed after emergence of female hands to avoid infection by insects.
8. Microbial antagonist, *Pseudomonas fluorescence*, is quite effective against the causal agent of Moko disease.

Bacterial Soft Rot of Rhizome

Soft rot of rhizome can cause severe losses in particular localities. A massive soft odorous rot of the centre or a portion of a rhizome characterizes the disease. It progresses upward the pseudostem and destroys the growing point, causing internal decay and discolouration of the vascular tissues. Leaves show yellowing, which wilt very soon. Sometimes, it is confused with Fusarium wilt.

Causal Organism

It is caused by *Erwinia chrysanthemi var. paradisiaca*

Management Practices

1. Uproot and destroy the affected plants.
2. Soil and plant drenching with bleaching powder at an interval of 10-15 days in quite effective for controlling the disease.

Bacterial Leaf Spot

Bacterial leaf spot is a serious bacterial disease in some varieties of banana in certain areas.

Symptoms

The disease affects only the leaf blades, in which it appears as chlorotic linear streaks along the veins. Later, these steaks coalesce to form large, irregular chlorotic patches. In severe infestations, the leaves may roll and dry up in the affected portions, presenting a scorched appearance. The affected plants become stunted, which affects the size of bunch and fingers, reducing the yield and quality invariably.

Causal Organism

It is caused by *Xanthomonas musicola*.

Management Practices

Try the following measures to control this disease effectively:

1. Cut and burn the affected leaves.
2. New plantings should be established from disease free planting material.
3. Grow resistant varieties and avoid Poovan, Rasthali, Peyan and Monthan cultivars in affected areas.

iii. Viral Diseases

Bunchy top (BBTV)

Bunchy top is considered as the most destructive disease of banana in Asia, Australia, Egypt and Pacific Islands. It was first noticed in Fiji during 1891 and later from other banana growing countries of the world. In India, its occurrence varies from state-to-state, and it may cause up to 40-50 per cent losses. It's primary spread is through the use of infected rhizomes/suckers as planting material and the secondary spread is through banana aphid (*Pentalonia nigronervosa*)

Symptoms

The symptoms of the disease may become apparent at any stage of plant growth. The first evidence of the bunchy top is seen in the leaves. Green streaks appear on the secondary veins on the underside of the lamina, on the mid-rib and petioles. In some leaves, these dark green steaks are found along many veins and several green dots or lines along the mid-rib and petiole. Successive leaves become more abnormal, and stunted in growth. In badly infected plants, the leaves are typically bunched together at the apex, forming dense rosette, hence the name '*bunchy top*'. The young infected leaves and plants are usually stunted and show more erect leaves than normal plants. Leaves are brittle and sometimes becomes darker green with upward rolled margin. Affected plants don't produce bunches, and if produce, the fingers are smaller in size with poor quality.

Management Practices

Phytosanitation is of paramount importance in the management of bunch top, hence:

1. Uproot and destroy the affected plants.
2. Don't use infected rhizomes or suckers for planting.
3. Keep banana aphid under control by spraying recommended insecticides like, metasystox or rogor (0.5 per cent).
4. Remove alternate host plants in the vicinity of plantation.
5. Treat the suckers/rhizomes with hot water or drench the soil or dip suckers in 500-1,000 ppm solution of tetracycline hydrochloride or 2 per cent solution of benomyl for 48 hours.

Infectious Chlorosis (ICV)

Infectious chlorosis or heart rot or mosaic of banana is caused by cucumber mosaic virus, which was first reported from New South Wales during 1930. Now, it is a widespread disease of banana in almost all banana-growing countries of the world. In India, it is more virulent on Poovan variety in which it may cause even up to 80 per cent loss. It is primarily transmitted through infected suckers and then by aphids like *Aphis gossypii* and *Pentalonia nigronervosa*.

Symptoms

Mosaic plants can easily be recognized by their stunted and mottled growth, distorted leaves with discontinuous linear streaking in bands extending from margins to mid-rib. Erratum of veins on the leaves, partial unfurling of leaf lamina, twisting and bunching of leaves at the crown and a rigid erectness in newly emerged leaves are also observed. The presence of dead or dying suckers is noticed in advanced cases referred as '*heart rot*', resulting from rotting of heartleaf and central portion of pseudostem due to association of some bacteria. In some cases, the diseased leaves taper rapidly from a broad base to an almost filiform tip. Similarly, sometimes, there is abnormal thickening of leaf veins.

Management Practices

Following measures have been considered necessary for the control of infectious chlorosis in banana effectively:

1. Uproot and destroy the infected plants.
2. Never use infected suckers or planting material for new plantings.
3. Keep the plantations free from weeds and alternate host plants such as cucumber.

4. Keep rigid control on aphids by spraying recommended insecticides, like metasystox or rogor (0.05 per cent).

Banana Steak Virus (BSV)

Banana streak virus is also of worldwide occurrence and can cause problems to the plantations. Infected suckers and mealy bugs like, *Planococcus citri* and *Saccharicoccus sachhari* mainly transmit it to the newer areas or plantations.

Symptoms

The disease is characterized by yellow streaking of the leaves, which becomes progressively necrotic, producing a black streaked appearance in the older leaves. The affected plants appear stunted and bear less bunches of poor quality.

Management Practices

1. Eradicate the infected plants.
2. Use disease-free planting material.
3. Follow quarantine measures strictly.
4. Control the vectors in time with suitable insecticides.

Banana Bract Mosaic Virus (BBMV)

This disease was first noticed in Philippines during 1990 and now it infests the banana plantations in many parts of the world.

Symptoms

Development of black streaks on the petiole, yellow or pinkish coloration of pseudostem and mosaic like purple and spindle-shaped streaks on the brackets are some of the symptoms of this disease. The affected leaves get clustered at the crown and represent a 'Travellers palm' appearance. The peduncle becomes elongated and the fruits remain undersized with poor quality.

Management Practices

Adopt the following measures for the control of this viral disease:

1. Rogue out the diseased plants and destroy them.
2. Use disease-free planting material for establishing new plantations.
3. Keep the plantations free from weeds.

B. Major Insect-Pests and their Management

The following insect-pests have been reported, which can cause severe losses to banana.

Rhizome Weevil (*Cosmopolites sordidus*)

Adult weevils feed on rotting banana tissue but they do not cause any significant damage. Damage to the banana plant is entirely the result of larval feeding. Their attack can prevent crop establishment, cause significant yield reductions in ratoon cycles and contribute to shorten the plantation life. The grubs attack pseudostems, which are riddled with holes and tunnels. Attacked pseudostem gets rotten and turn into a black mass. In severe cases, tunnels may extend up to stem, corm decays and injury prevents upward flow of nutrients, as a result, leaves turn yellow and die. The plants show premature withering, leaves become scarce and fruits become undersized. Severe infestation results in the toppling of plant. Adults and grubs also attack the suckers, which are killed.

Control Measures

1. Pin point the shelter and breeding places of weevils and destroy them.
2. Adults can be trapped on pieces of old rhizomes or pseudostems, which should be killed.
3. Cut the pseudostems very close to the ground.
4. The pseudostems from which bunches have been cut, should be chopped into small pieces so that they rot quickly.
5. The weevil spreads through infected material. Hence healthy suckers should be used for planting.
6. Trimming the rhizomes or suckers and dipping in monocrotophos (0.1 per cent) or methyl demeton (0.05 per cent) for 30 minutes protects rhizomes from weevil attack.
7. Soil application of phorate (10g/plant) or castor cake (250g) or carbaryl dust (50g) at planting time and after three months of planting is very useful.
8. In case of severe attack, methyl demeton (0.025 per cent) or fenitrothion (0.05 per cent) or rogor (0.03 per cent) may be sprayed around the collar region of plants.
9. Quarantine measures should be strictly followed to prevent further spread of weevil into non-affected areas.

Banana Pseudostem Weevil (*Odoiporus longicollis*)

In India, it causes considerable damage to most of the banana cultivars in eastern and north-eastern parts. The pest is active during summer and monsoon months. The grubs after hatching, start feeding on tissues near the leaf sheath and then bore into the pseudostems. Many grubs may be found boring a single plant. Exudation of plant sap is the initial symptom of its attack and blackened mass comes out from the holes made by the grubs. The pseudostem riddled with holes become weak, starts rotting and finally dies. The adult weevils feed on living and decomposing banana leaf tissues but eat little and are not considered as pests. Damage is done by the larval stage. The larvae attack the pseudostems and stem of banana plants, although they will occasionally feed within the rhizome.

Management Practices

1. Uproot and burn the infested plants.
2. Follow clean cultivation.

3. Place celphos tablets (3tablets/pseudostem) inside the pseudostem and then plaster the slip of pseudostem with mud.

4. Apply carbofuran granules (3g/stool) before the expected attack of the weevil.

5. After removing the dry outer sheaths of the pseudostem of all infested and un-infested plants in endemic areas, apply quinalphos (0.05 per cent), chlorpyriphos (0.04 per cent), carbaryl (0.2 per cent) or endosulfan (0.05 per cent). Repeat the treatment after 3 weeks, if the infestation persists.

Banana Aphid (*Pentalonia nigronervosa*)

Both adults and nymphs cause damage. Large number of adults and nymphs congregate under the outer leaf base on the pseudostems and around the crown of the plants. In the case of severe attack, there is progressive leaf dwarfing and leaf curling, fruit bunches become small and distorted. Whole plant looks stunted. Such plants may not bear at all, and if bear, they produce small fruits of poor quality. This aphid also acts as vector for causing the bunchy top disease of banana.

Management Practices

1. The insect can be controlled by spraying dimethoate (0.03 per cent), malathion (0.05 per cent) or oxydemeton methyl (0.025 per cent) at 10-15 days interval.

2. Granular insecticides like phorate @ 1 kg a.i./hectare should be applied to prevent the recurrence of this pest.

Nematode (*Radopholus similis*)

Nematodes cause serious loss to bananas. Nematode infected root turns reddish-brown and later become black. The root becomes short, blackened and reduce in number and thus, susceptible to wind damage. Larva feeds on the fruit. Evidence of attack is indicated by black spot on the skin. Good sanitation, hot water treatment to the planting material, drenching with fenamiphos and wrapping of fruits with polyethylene bugs are some effective measures to reduce damage caused by nematodes.

C. Physiological Disorders and their Managment

Chilling Injury

Symptoms include surface discoloration, dull or smokey anal color, subepidermal tissues reveal dark-brown streaks, failure to ripen, and, in severe cases, flesh browning. Chilling injury results from exposing bananas to temperatures below 13 °C (56 °F) for a few hours to a few days, depending on cultivar, maturity, and temperature. For example, moderate chilling injury will result from exposing mature-green bananas to one hour at 10 °C (50 °F), 5 hours at 11.7 °C (53 °F), 24 hours at 12.2 °C (54 °F), or 72 hours at 12.8 °C (55 °F). Chilled fruits are more sensitive to mechanical injury and decay casued by pathogens.

Skin Abrasions

Abrasions result from skin scuffing against other fruits or surfaces of handling equipment or shipping boxes. When exposed to low (<90 per cent) relative humidity conditions, water loss from scuffed areas is accelerated and their color turns brown to black.

Impact Bruising

Dropping of bananas may induce browning of the flesh without damage to the skin.

HARVESTING AND YILED

Maturity

Depending on the variety, banana plant starts to bear fruit 6-8 months after planting. Fruit maturity can be indicated by colour code or bunch covers tied on the bunch after shooting using a different colour weekly. In our country angularity of fingers is also used as maturity index. Maturity can also be indicated by measuring the grade with callipers. This system is called age-grade control. Calipers are caliberated in thirty seconds in inch in most of the American tropics and in millimeters elsewhwere. Grade specifications for harvest depend upon the market location, with lower grades harvested for distant markets. Fruits are said to be over grade and under grade when they do not fall between the maximum and minimum grades specified for the particular harvest. Hands are numbered consecutively from the first large basal hand downward to the last small female apical hand.

Harvesting

The banana tree starts to produce flowers in 7-12 months after planting and is ready for harvesting about 7-11 weeks later, depending on the variety. Most varieties flowers in 7-8 months after planting and are ready for harvesting 7-9 weeks later. The 'follower' plant will produce 3-4 months later, thus about four harvests from I clump per year is possible.

Yield

The average yield per year for a 3-year-cycle is Approx. 10 tonnes per hectare for several varieties, 12.0 tonnes per hectare for Rasthali and 20–40 tonnes per hectare for Embun. The general average yield is about 7 tonnes for the first year, 12 tonnes for the second year and 10 tonnes for the third year per hectare.

POSTHARVEST MANAGEMENT

Postharvest Handling

Individual plants are felled and the individual fruit are carefully cut and taken to packing shed. The hand is separated from the stalk and diseased and spoilt fingers are removed, cleaned, treated, graded and then packed into a suitable containers like corrugated fibre board or baskets.

Storage

To lengthen the shelf-life, banana should be stored in controlled environment, with a temperature of 13-14°C.

CROP IMPROVEMENT

Objectives of Improvement

The modern banana breeding objectives includes development of pure *acuminata* type to meet export demand, resistance to abiotic (salinity, heavy metals, wind) and biotic (*Fusarium* wilt, sigatoka leaf spot, weevil, borers, nematodes) stresses, dwarf stature with enhanced productivity and good agronomic qualities.

In India, banana breeding first started at Central Banana Research Station, Aduthorai (1950) and later it was continued at Agricultural College and Research Institute, Coimbatore. Presently, banana breeding is being pursued at Trisshur (Kerala), Kovvur (A.P.), Jorhat (Assam), Navsari (Gujarat), Hajipur (Bihar), Rahuri (Maharastra), UAS (Banglore), NRC Banana, Trichy and its research stations..

Natural diversity has been observed for wild banana species in India. In N-E states of country, *Musa flaviflora, M. velutina* and *M. cheesmani* are spread in Manipur and Meghalaya. India is major centre of biodiversity for AB, AAB and ABB group of banana.

The maximum accessions are available for AAB (84), ABB (84) and ABB (74). At CBRS, Aduthorai, *M. balbisiana, M. chiyocarpa, M. rosea* and *M. acuminate* were used to make some interspecific hybrids but most of the attempts failed.

Achievements

From AAU, Assam, a variety 'Bhimkal' was released, which was found to be resistant to nematodes. At CBRS, Kannara, Vannan, Agniswar and Harichal were crossed with Pisang Lilin resulting in release of H-1 and H-2. H-1 has short crop cycle and have resistance to leaf spot, *Fusarium* wilt and burrowing nematode, while H-2 was found to resistance against leaf spot and nematode; besides, being suitable for subsistence cultivation. At TNAU, Coimbatore, multiple crossing involving Ladan, *M. balbisiana* and Kadali. It was evolved by Ladan (AAB) x *M. balbisiana* (BB) → AB x Kadali (AA) → CO-1 (AAB). CO-1 was akin to pome group and resembles to Virupakshi (AAB) and has typical acid and apple flavour of Virupakshi.

Inheritance Pattern in Banana

The inheritance pattern of the important characters in banana is as follow;

1. **Parthenocarpy**: Controlled by series of complementary dominant genes (P_1, P_2 and P_3).
2. *Fuarium* **wilt resistance**: Resistance to *Fusarium* wilt race 1 is controlled by single dominant gene.
3. **Yellow sigatoka**: Controlled by multiple genes with dominant behaviour.

4. **Black leaf spot**: Controlled by multiple genes with dominant behaviour.

5. **Moko Disase**: Governed by dominant genes.

6. **Insect and nematode resistance**: Host-plant response to weevil is controlled by gene(s) exhibiting partial dominance towards the resistant parent and modifier genes with additive and dosage effects for susceptibility in plaintain parent. The resistance to burrowing nematode is controlled by one or more dominant alleles.

7. **Apical dominance**: Controlled by recessive gene (*ad*). The dominant allele *Ad*, which is probably fixed in AA bananas improves the suckering.

8. **Dwarfing**: Dwarfing in plant stature is governed by a single dominant gene in AAA bananas, though some modifier genes affect the degree of dwarfness.

9. **Suckering habit**: Governed by additive alleles.

10. **Bunch orientation**: Controlled by oligogenic trait regulated by epistatic interaction of at least three dominant loci with the threshold effect of dominant genes.

Constraints in Banana Breeding

Banana breeding is complicated with a variety of problems and over seven decades of breeding resulted in development of few cvs. superior to those prominent in world trade. Some of the associated banana breeding problems are enlisted hereunder;

1. High level of both male and female sterility and existence of vegetative parthenocarpy.

2. High level of heterzygosity.

3. Existence of polyploidy forms.

4. Existence of several interspecific species and natural variants.

5. Existence of meotic irregularities, morphological errors in both male and female gametes at post-meotic stages and physiological dysfunction during pollination and fertilization.

6. Poor seed set and exclusive vegetative propagation.

3

Citrus

Citrus comprise a group of fruits belonging to family Rutaceae. The members of this group are well spread over the areas ranging from tropical and subtropical regions of the world. A few species are frost tolerant and have well adapted the temperate conditions. After banana, it is the most extensively grown fruit crop globally. The total production of citrus fruits in the world is 96,325 thousand metric tonnes, of which, sweet orange (62,321) mandarin (6,351) lemons and limes (12,350) and grapefruit and pummelo (5,260) are the major types.

ORIGIN, HISTORY AND DISTRIBUTION

Citrus fruit have been in cultivation for over 4,000 years and they occupy the regions lying within 40°N and 40°S latitude. Wide scale dissemination of citrus is associated with several historical explorations and as a result, many species were introduced to the New World. Citrus species are believed to have originated in South-Eastern Asia including areas extending from eastern Arabia to the Philippines and from the southern Himalayas to Indonesia and Australia. Earlier, North Eastern India and Northern Myanmar (Burma) were believed to be the centre of origin but recent evidences suggest that Yunnan province in South Central China may be the primary centre of diversity, depending upon the number of species moved to West Asia before Christ. Citron (*Citrus medica* L.) is believed to have originated in the region extending from South China to northern eastern India. Limes (*C. aurantifolia* Swingle) originated in the east of Indian Archipelago. Lemon (*C. limon* Burmann) is believed to be a possible hybrid between citron and lime and spread to N. Africa and Spain. Sour orange (*C. aurantium* L.) is known for its south-east Asian origin, which later spread to N. Africa. Sweet orange (*C. sinensis* L.) Osbeck originated in southern China or

Indonesia. *Citrus grandis* (*Pummelo*) is believed to have originated in Malaysia, Indian Archipelago and Fiji Islands. Grapefruit (*C. paradisi* Macf.) originated in Barbados, West Indies, Mandarin (*C. reticulata* Blanco) probably originated in Indo-China and South China while, *C. deliciosa* is of typical Chinese origin. The others related minor genera *i.e. Poncirus trifoliata* (L.) Raf. (trifoliate orange) and *Fortunella margarita* (Lowi) Swingle, are known to have originated from southern China and both are freeze hardy species.

COMPOSITION AND USES

Juice present in the vesicles is the edible portion of the citrus fruit. The juice contains 12-14 per cent sugars, 0.5 to 1.5 per cent titrable acidity, 1 per citric acid, about 50 mg L-ascorbic acid (Vitamin C) per 100 ml juice and about 70-90 per cent water. Citric acid is the characteristics of citrus fruits and species like acid lime, lemons, grapefruits and sour orange contain about 1.5 to 3 per cent acid. The vitamin C present in juice prevents scurvy. The juice is also used for controlling rheumatism, dysentery, diaorrhea, cough, cold, fever and bronchitis.

The leaves, flowers and fruits contain essential oil, which has several flavanoids. 'Hesperidin' is the universal flavonoid present in most of the citrus fruits. Other flavonoids identified are neohesperidin, naringin, aurantamarin, limonin, narirutin, mobiletin, tangeretin etc. The bitter principles present are 'naringin' in grapefruit and 'limonin' of navel orange. The colour pigments in juice and rind are due to carotenoids. The peel oils are known for their anti-microbial and anti-fungal activities.

TAXONOMY AND CLASSIFICATION

Citrus belongs to the family Rutaceae, which contains more than a thousand species of South-East Asia, Africa and Australia. The botanical classification of citrus is as:

Order	:	Geraniales
Suborder	:	Geranea
Family	:	Rutaceae
Subfamily	:	Aurantoidae
Tribe	:	Citrae
Subtribe	:	Citrinae
Group C	:	True fruit trees (6 genera)
Genus	:	Poncirus (1 species), Fortunella (4 species), Citrus (2 subgenera
		– (a) Eucitrus, (b) Papeda

Currently, these are designated as Fortunella, Eremicitrus, Poncirus, Clymenia, Microcitrus and Citrus. The three genera *viz.*, *Clymenia, Eremicitrus* and *Poncirus* are monotypic with 1 species only. *Fortunella* and *Microcitrus* comprise of 4 and 6 species, respectively.

More than a dozen scholars from USA, Japan and India have tried to classify citrus. Earliest classification was attempted by W.T. Swingle, T. Tanaka and W. Hodgson. In India, S.C. Bhattacharya and S. Dutta classified the citrus species of Assam in 1950's, later Prof. Ranjit Singh attempted to classify citrus fruits and evolved a key for identification of different citrus fruits.

Swingle's system involves a total of 3 genera, 21 species (16 in citrus) and 9 botanical varieties. In Tanaka's system, there were 2 subgenera, 8 selections and 144 species. Hodgson made a more elaborate attempt and gave species status to closely related species and the strain of them, which was not justified. Of late, Prof. Ranjit Singh made the compromise between the Swingle and Tanaka systems and accepted forms of hybrid origin as independent species of hybrid origin. The classification proposed by him is listed below:

Citrus : 2 subgenera, Eucitrus and Papeda

Subgenus : Eucitrus (8 sections)

Section A : Decumana (7 species)

1. *C. pennivesiculata*, 2. *C. pennivesiculata* var. assamensis, 3. *C. semperflorens*, 4. *C. paradisi*, 5. *C. grandis*, 6. *C. megaloxycarpa* and 7. *C. megaloxycarpa* var. keem.

Section B : Medica Group (4 species)

1. *C. medica*, 2. *C. medica* var. Ertog, 3. *C. medica* var. *Sacrodactylus*, 4. *C. limonimedica*.

Section C : Limonoid Group (10 species)

1. *C. limon*, 2. *C. kama*, 3. *C. pseudolimon*, 4. *C. limonia*, 5. *C. limonia* var. *Kusai*; 6. Soh Thalia, 7. Acidless Rough lemon, 8. *C. jambhiri*, 9. *C. jambhiri* (intermediate type) and 10. *C. jambhiri* (Katajamir type).

Section D : Aurantium Group (5 species)

1. *C. sinensis*, 2. *C. aurantium* (bitter sweet orange), 3. *C. aurantium* (sour orange); 4. *C. aurantium* var. *myrifolia* and 5. *C. aurantium* (Natsudaidai).

Section E : Aurantoid Group (3 species)

1. *C. regulosa*, 2. Jenuru tenga, 3. *C. madaras* patna (Kichili).

Section F : Acrumen Group (6 species)

1. *C. nobilis*, 2. *C. unshiu*, 3. *C. deliciosa*, 4. *C. reticulata*, 5. *C. reticulata* var. *austere* and 6. *C. lycopersicae formis*

Section G : Limonellus Group (3 species)

1. *C. limettoides*, 2. *C. limettoides* var. *latifolia* and 3. *C. aurantifolia*

Section H : Pseudo Fortunelloa Group (3 species)

1. *C. indica*, 2. *C. tachibana* and 3. *C. madurensis*.

Section I : Subgenus Papeda

1. *C. inchangensis*, 2. *C. latipes*, 3. *C. micrantha*, 4. *C. micrantha* var. *microcarpa*, 5. *C. hystrix*, 6. *C. macroptera* and 7. *C. combara*.

Besides, Kumquat (*Forlunella* spp.) has three different species *viz., F. margarita; F. crassifolia* and *F. japonica* with fruit size ranging from 2 to 3 and trifoliate orange (*Poncirus trifoliate* (L.) Raf.), which is used as dwarfing rootstock.

Distribution

As stated earlier, Citrus is successfully grown between 40°N to 40°S latitudes. Brazil is the largest citrus producer country followed by USA, China, Spain, Mexico, etc. Spain is the largest producer of mandarin and oranges; USA is the largest producer of grapefruit while India is the largest producer for lemon and limes. Since last decade, the production in tropical regions is fast declining due to several threats such as extreme weather conditions, wide occurrence and spread of diseases *viz.,* canker, tristeza, greening etc. Some of the principal countries growing citrus are the USA, Brazil, Mexico, Spain, Italy, Algiers, Morocco, Israel, Egypt, Pakistan, China, India, Australia and Japan.

SOIL AND CLIMATIC REQUIREMENTS

Soil

Citrus in India is grown on a variety of soils having good drainage capacity. Sub-soil drainage is the most important consideration as poor aeration hinders root functions. Soil depth should be around 2-3 m to allow proper root growth as soils with hard-pan beneath, leads to short life span of the plants and thereby hamper the proper growth of roots and shoots. At present, commercial citriculture in India is done on soils ranging from coarse sands to heavy clays. High calcium carbonate content in soil leads to Zn, Fe and Mn deficiencies and in soils with more than 5 per cent $CaCO_3$, planting of limes should be avoided. Though citrus has been reported to be grown in soils with pH ranging from 4.0 to 8.5 but ideal pH ranges from 5.5 to 6.0. Lower soil pH leads to leaching of lime and Mg while, higher pH (alkaline range) leads to micronutrient deficiencies. Soils with low organic matter content or high sodium content require FYM or gypsum for reclamation. In regions with high water table, the planting should be done on raised beds.

Climate

The growth and development in citrus is optimum in temperature regimes ranging from 25 to 30°C to minimum of 13°C. Some species can tolerate extreme temperature of 50°C, while some can tolerate freezing point. High temperatures cause poor pigment development in fruits and also sun burn or sun scalding is occasionally met. Trifoliate orange is the most cold tolerant species followed by kumquats. Amongst citrus species, *C. unshiu* is the most cold hardy followed by mandarins, sour orange, sweet orange, pummelo, grapefruit, lemon, lime and citron (*C. medica*). Low relative humidity favours proper fruit colour development, while high RH leads to development of juicy fruits with thin rind thickness. Extreme high humidity present in the tropical regions favours the attacks of pests and diseases. Partial shade helps in improvement of fruit quality. Citrus plant is day neutral, however, long day conditions favour vegetative growth. Annual rainfall of 800-900 mm, well distributed throughout the year, is optimum for citrus growth and production.

In our country, mandarins are cultivated as irrigated crop in Maharashtra whereas, in southern and hilly region of north-eastern states, it is grown as a rainfed crop. Grapefruit can tolerate a temperature as high as 48°C, while lemon suffers severely beyond 37°C. Unconditional rainfall causes fruit splitting and hence regular irrigation is always desired.

IMPORTANT CULTIVARS

Distinction and nomenclature of citrus cultivars is also problematic as their taxonomical classification. Accidental mixtures, natural hybridization and occurrence of bud sports make difficulties in distinguishing varietal clones. Over long years in cultivation, no variety is stable and is bound to develop and certain changes and accordingly several local strains are also identified. Local names sometimes contribute to the chaos in nomenclature and distinction of citrus cultivars. Some of the important citrus cultivars are listed below:

Sweet Oranges (*Citrus sinensis* Osbeck)

These are tight skinned orange and have been classified into four broad groups *viz.*, (i) Spanish, (ii) Mediterranean, (iii) Blood Red and (iv) Navel.

(i)   Spanish Oranges

These have round fruits *e.g.*, Mosambi, Satthgudi, etc.

(ii)   Mediterranean Oranges

These have oval shaped fruits *e.g.*, Valencia, Pineapple, Jaffa, Hamlin, Shamouti, etc.

(iii)   Blood Red Oranges

These have red or red streaked pulp *e.g.*, Malta, Malta Blood Red, Ruby, Double Fine, Moro, Moscato, etc.

(iv)   Navel Oranges

These have large fruits with distinct navel opposite the stalk end, *e.g.* Washington Navel, Frost, Washington, Navelina, Gillett, etc.

Some of the popular sweet orange cultivars grown in India are described below:

Hamlin

It is an early maturing introduction from Florida. In India, it is grown in the states of Punjab, Haryana and Uttar Pradesh. Fruits are medium-to-small, globose to slightly oblate, cadmium-yellow or orange-red; rind, medium thick, tight, glossy, finely pitted; pulp colour golden yellow, sweet and juicy, excellent flavour; highly productive, seeds usually 2-6.

Jaffa

It is a mid-season introduction from Palestine and is cultivated in Punjab, Haryana and Uttar Pradesh. Fruits, medium-sized, orange-yellow to orange-red, rind medium thick, globose to slightly ellipsoid or ovoid, pulp light orange, juicy, flavour good, seeds 9 to 10, suffers from alternate bearing habit and excessive fruit drop but stands shipment well.

Malta Blood Red

Introduction from the Mediterranean region to Punjab. It has compact growing habit and is gaining popularity in Haryana and western Uttar Pradesh. It is a seeded, mid-season variety and develops good colour upon proper exposure to cold temperatures. Fruits medium-to-large, orange-yellow to blood red, rind leathery, pulp light orange, flavour pleasant, good blend of sweetness and acidity, seeds range from 25 to 30 per fruit.

Mosambi

It is an introduction from Mozambique to central, southern and western India. Fruits are medium-large, slightly oblate to globose or obovoid, rind medium thick, leathery, tight and difficult to peel. Colour marigold to straw-yellow at maturity. Light yellow juice, juicy, flavour insipid, low acidity, moderately seeded 20-25 per fruit. Skin colour remains green or yellowish-green under tropical conditions.

Pineapple

This is an early maturing introduction from Florida and cultivated in the states of Punjab, Haryana and Uttar Pradesh. Fruits medium-sized, spherical to slightly obovate, rind thin, tight, smooth and glossy pulp light orange coloured, juicy, excellent quality with flavour similar to that of pineapple.

Sathgudi

It originated from 'Sathgur' a place in Tamil Nadu and is widely cultivated in southern India. It is widely distributed in districts of Cuddapah, Chittoor and Kurnool in Andhra Pradesh. Fruits are medium-to-large, spherical or globose, skin surface smooth, leathery, bright orange colour when ripe, juicy, flavour good, sweet with mild acidity, seed 12 to 20. It is thought to be a probable introduction from Batavia or Indonesia.

Washington Navel

This variety originated in Bahia, Brazil and is an important cultivar of California. Its special features are crispness of pulp, easy removal of peel and segments, richness of flavour and seedlessness. Widely grown in southern states, *i.e.* Bangalore and Kodagu district of Karnataka and Kodur district of Andhra Pradesh. Fruits medium-to-large, greenish-yellow to pale-orange when immature but attains attractive orange-colour when ripe, spherical to ellipsoid, thick rind, segments 9 to 11, pulp deep orange tender, moderately juicy, flavour excellent, seedless, unfit for processing due to accumulation of bitterness during storage. Navel cultivars are poorly adapted to hot humid tropics.

Shamouti

It has been introduced from Palestine and is considered as a bud sport of 'Bellady'. Fruits are large, oval to ellipsoid, rind thick, colour orange to deep-orange, smooth surface, solid or semi-hollow core, pulp light orange, flesh, tender, juicy, flavour sweet and rich and almost seedless. It is a mid-season variety and highly productive.

Valencia Late

It is an introduced cultivar in European countries and is widely grown in Europe and USA. In India, it is grown in Punjab. Fruits medium-to-large, oblong to spherical, colour deep-golden orange, rind medium thick, smooth, tough, pulp golden-yellow, good blend of acidity and sweetness, excellent flavour and quality, seeds only few, *i.e.* 5-6.

Soh-Niangriang

It is also known as 'Mitha Chakla' and grown extensively in N.E. regions of India. Fruit globular, colour bright yellow, smooth surface, rind medium thick, juicy, seeds 10-25 per fruit.

Mandarins (*C. reticulata* Blanco)

These are loose skimmed oranges, while tangerine, refers to those varieties producing deep orange or scarlet fruits. Edible mandarins are divided in four broad classes *viz.*, Satsuma, King, Willow-leaf and mandarin.

(i) Satsuma Group (C. unshiu M.)

These are popularly grown in Japan for their fineness. Fruits are of excellent quality, seedless and fit for dessert and processing. Some of the popular varieties are Ikeda, Owaru, Mikado, Wase, Silver Hill etc.

(ii) King Group (C. nobilis)

This group is a native of Indo-China and suited for tropical conditions. These are late, seeded and of high quality. Some of the popular types are King (USA) and Kunembo (Japan).

(iii) Willow Leaf Group (C. deliciosa Tenore)

These have originated in the Mediterranean region. The fruit juice has pleasant aromatic flavour, spherical seed and high degree of polyembryony. Ciaculla (Italy) and Willow Leaf (Spain) are the popular cultivars.

(iv) Common Mandarin Group (C. reticulata Blanco)

These are the chief type of mandarins grown world over.

Some of the mandarin cultivars growing in the country have been described below:

Nagpur Santra

It occupies the premier position in Indian market and is also popular as 'Ponkan' in China. It is mostly cultivated in Maharashtra region. Fruits are medium, sub-globose, orange-yellow (cadmium), rind smooth, glossy and thin, segments 10-11, pulp saffron colour, seeds 4 to 11, flavour excellent.

Khasi Mandarin

A commercial cultivar of N.E. hills. Fruits globose-to-oblate, orange-yellow to deep-orange, surface smooth and glossy, short-necked, rind thin, pulp orange coloured, sugar : acidity ratio is good, seeds 9-25.

Darjeeling or Sikkim Orange

Grown widely in Darjeeling (West Bengal) and Sikkim, similar to 'Khasi mandarin' in many respects.

Coorg Mandarin

Commercial cultivar of Karnataka and Kerala. Fruit oblate-to-globose, large, bright orange-yellow, rind thin, pulp golden yellow, flavour excellent, good blend of acidity and sweetness, juicy, seeds 14 to 30.

Kinnow Mandarin

It is a hybrid between King (*C. nobilis*) and Willow Leaf (*C. deliciosa* T.) mandarins developed by H.B. Frost at California in 1915. It was introduced in Punjab and is extremely popular in northern states. Fruit medium, globose, rind thin, peel leathery but removable, surface smooth and glossy, colour yellowish orange-to-deep orange, core semi-hollow, very juicy, flavour rich, seed bold and many.

Satsuma Mandarin

It is the prime cultivar of Japan. The trees can tolerate temperature as low as –10°C. Fruits are medium, oblate to spherical, surface slightly rough, peel orange to orange chrome, rind thin, leathery, core hollow, pulp orange, flavour rich, mainly seedless.

The other popular mandarins grown worldover are King mandarin (USA), Temple, Willow leaf (Mediterranean); Clementine (Algeria); Dancy (USA); Beauty (Australia), Campeona (Argentina and Uruguay), Elendale (Australia), Emperor (Australia); Ponkan (South China), etc.

Lemon (*C. limon* Burrn)

Two types of lemons are found, *i.e.* sweet and acid, of which acid lime is the most extensively grown type. Webber in 1948, classified lemon varieties into 4 group *viz.*, (i) Eureka, (ii) Lisbon, (iii) Anamalouskin and (iv) Sweet lemon group. Later, Hodgson in 1967, classified them into five groups (i) major acid lemon *e.g.*, Eureka, Lisbon, Villafranca, (ii) minor acid lemon *e.g.*, Assam lemon, Elaichi nimbu, (iii) sweet lemon,

Pani Jamir, (iv) Ornamental lemon-variegated prior Lisbon, variegated pink-flesh lemon, (v) Lemon like fruits *e.g.,* Rough lemon, Meyer lemon, Karna Khatta, Galgal.

Some of the lemon cultivars grown in India and elsewhere are briefly described hereunder:

Eureka

It originated as a chance seedlings from Sicilian lemons at California in 1860. In India, it is grown in Punjab and western Uttar Pradesh. Fruits are medium small, lemon yellow, pulp tender greenish yellow, juice abundant, very acidic, quality and flavour excellent, seeds 0-6 per fruit. Bears throughout the year but sensitive to cold.

Lisbon

It is a Portugeese introduction grow commercially in California. Fruit medium, elliptical-to-oblong, seeds few, peel colour deep yellow, rind medium thick, segments 10, core solid. It differs from Eureka as it has more drawn out nipple. Juicy, very acid, quality excellent, seeds few (0-10).

Assam Lemon

It originated in Assam and is widely cultivated in N.E. regions and Andhra Pradesh. Fruits medium large, elliptic to oblong, obovate, base rounded, nipple broad, rind medium thick, firm, pulp greenish yellow, juicy, very acidic segment (9 to 10), flavour good, seedless, prolific bearer.

Italian Lemon

Grown in drier zones of South India and Kodagu district of Kerala. Fruits are round, very juicy, seedless, prolific bearer, peel rich in oil.

Malta Lemon

Cultivated in Andhra Pradesh and Karnataka. Fruits small with thick rind, juicy, seeds 0-20 per fruit, early bearer.

Villafranca

It is an introduction from Florida grown in Punjab and Haryana. Fruits medium sized, colour sunflower yellow, rind thick, pulp fine, juice abundant, very acidic, flavour excellent, seeds 9-15 per fruit.

Elaichi Nimbu

Originated in Assam and is known for its characteristic cardamom flavour. Fruit medium, elliptical to oblong, with short nipple, colour light yellow, rind medium thick, pulp colour pale-greenish yellow, juicy, acidic, seeds few.

Gol Nimbu (*C. jambhiri*)

It is a native of north eastern India grown widely for use as rootstock. In Assam, it is grown as a substitute for lemon. Fruit spherical-to-globose, colour light yellow, surface smooth, rind thin, segments 10, pulp vesicles white or light yellowish white, flavour good, juicy, seeds few, polyembryonic.

Kagzi Kalan

It is a seedless lemon selection made at IARI, New Delhi, is fast gaining popularity in North Indian states.

Lime (*C. aurantifolia* Swingle)

Lime cultivars were grouped into different categories by Hodgson in 1967, *i.e.*

(i) Small-fruited acid lime (*C. aurantifolia*) *e.g.*, West Indian (Mexican, Key), Kagzi, etc.

(ii) Large-fruited acid lime (*C. latifolia*) *e.g.*, Tahiti lime (Persian), Pond, etc.

(iii) Sweet lime (*C. limettoides*) *e.g.*, Indian (Palestine).

Some of the lime cultivars grown in India and abroad are described briefly hereunder:

Kagzi Lime

It is widely cultivated in India, Mexico, Florida, West Indies, Brazil etc. Fruit small, round obovate, rind very thin, surface smooth, tightly adhered, greenish yellow, segment 10-12, pulp greenish yellow, juicy highly acid with distinctive aroma.

Tahiti Lime

Grown in Mexican and West Indian regions. Fruits are large, seedless, oblong, slightly nippled, rind thin, segments 9-10, light greyish yellow, very acid, highly flavoured, almost seedless.

Rangpur Lime (*C. limonia* Osbeck)

It is indigenous to India producing attractive fruits, loose skin, pulp light orange yellow. It is resistant to cold and tristeza.

Sweet Lime (*C. limettoides* Tanaka)

It is similar to Tahiti lime, sweet with little resemblance to typical lime.

Grapefruit (*C. paradisi* Macf.)

It has a West Indian origin and called grapefruit as fruits are borne in a cluster of 3 to 10 or more. Some of cultivars grown in India are described below:

Duncan

Originated in Florida and is the hardiest grapefruit cultivar grown in Punjab and Himachal Pradesh. Fruit light yellow, smooth surface, oblate-to-globose, large size, rind medium, core solid, segments 12-14, flesh light greyish in colour, juicy, excellent flavour, seed 30-40.

Foster

It originated as bud sport of Walters identified by R.B. Foster in California. Fruit is medium large, oblate-to-spherical, rind medium thick, surface smooth, rind contains yellowish pink pigments, pulp colour pink, juicy, flavour good.

Marsh Seedless

It is grown in both North and southern states of India. Fruits medium-to-large, light yellow, rind leathery, juicy, pulp greyish-green, segment 12-14, flavour excellent, seedless.

Thompson Seedless

It is a pink fleshed bud sport of 'Marsh Seedless' identified in Florida and is grown in South India. Fruits medium-sized, rind thin, tough, pulp buff coloured, juicy, flavour good, seeds none, good storage qualities.

Red Blush

It is a buds sport of 'Thompson Seedless' with deep coloured pulp and rind. Rind colour crimson and fruits similar to Thompson, also called Ruby, Red Marsh, Red Seedless.

Pummelo (*C. grandis* Osbeck)

It is a native of Malaysia and Polynesia and commercially grown in China, Thailand, Malaysia. In India, it is grown in N.E. hills, arid central regions.

Nagpur Chakotra

Fruit shape obovate or oblate, yellow, rind thick, segments 13-14, pulp pink and soft, juice scented, seeds 70-159.

Common Pummelo

Fruits medium-to-large, subglobose, spherical, light yellow at maturity, rind thick, surface smooth, segments 12-18, pulp colour pale yellow, moderately juicy, mild acidic, flavour good, seedy.

Besides, some other known cultivar grown in India and other parts of the world are:

1. *Citron* : Bira Jora (Assam); Corsican (France); Ertoz (Israel and USA), Diamonte (Italy and USA).
2. *Sour Orange* : Chakla Tenga (North Eastern Hills).

PLANT PROPAGATION

Citrus trees are propagated by both vegetative means and sexually by seeds. Vegetative methods are preferred because they ensure uniformity in quality and bearing. Seeds of several citrus species are polyembryonic and nuclear seedlings come true-to-type. Mandarins and acid limes are mostly propagated as seedlings. Lemons, citrons, sweet limes are easily propagated by stem cuttings. Air-layering is mostly practiced in pummelo, mandarin, acid lime and seedless lemons. Most of citrus cultivars are propagated by 'T' budding on a suitable rootstock.

The seeds should be obtained afresh and shown in polybags filled with potting mixture. After 6-8 months, the seedlings achieving 8 mm diameter are fit for budding during the season. The bud wood is procured from certified virus-free, indexed stock. Well swollen, unsprouted buds are collected afresh before. 'T' or inverted 'T' budding is generally done at about 20 to 30 cm and after proper bud union, the stock portion is cut in two stages. Buddlings of 1-2 years age are ready for field planting.

Rootstock and Polyembryony

An ideal rootstock should have high degree of polyembryony, graft compatibility, adaptability to wide range of soil conditions and tolerant to biotic and abiotic stresses. No single rootstock possess all the characters together, however, few of the rootstocks commonly employed for citrus propagation are listed below:

Rough Lemon (*C. jambhiri* Lush.)

It is a vigorous rootstock, resistant to gummosis and used in the states of Punjab, Rajasthan, Madhya Pradesh, Maharashtra and Assam. Fruits produced are large, thick skinned, highly acidic and poor in quality. However, bud-union crease with sweet orange has been reported from Punjab and Maharashtra.

Sour Orange (*C. aurantium* L.)

It is most widely used rootstock in the world up to mid of 1990s, however, due to high susceptibility to tristeza, it has lost its usage. It is resistant to several soil borne diseases, high salts, cold and is known to improve fruit quality.

Karna Khatta (*C. karna*)

It has gained popularity in the states of Punjab and Uttar Pradesh for sweet orange and mandarins. It is semi-vigorous, has wide adaptability to different soils and delays fruit maturity.

Rangpur Lime (*C. limonia*)

It is known to possess tolerance for CTV, drought conditions, saline and calcareous soils. Vigorous in nature and gives high fruit yield and quality. Resistant to quick decline and root rot. It is mostly used for budding sweet orange, grapefruit and pummelo.

Sweet Orange (*C. sinensis* Osbeck)

It is not a popular rootstock in India because of poor germination. It is tolerant to quick decline and tristeza. It is best for sandy-loam soils and yield high quality fruits.

Citranges

Citranges are hybrid of sweet orange and trifoliate orange. Of several citranges, 'Carrizo' and 'Troyer' are more popular as rootstock. These are resistant to soil borne diseases, nematodes and tristeza virus, semi-dwarfing and improves fruit quality. Under North Indian conditions, these are mostly used for mandarins and sweet oranges.

Trifoliate Orange (*Poncirus trifoliata* L. Raf.)

It is a popular rootstock for 'Satsuma' mandarin in Japan, China and Australia. Dwarfing, tolerant to tristeza but resistant to soil borne diseases, nematodes and cold. It is a popular rootstock for 'Coorg' and 'Kinnow' mandarin and grapefruits in India.

Cleopatra Mandarin

It is mostly used for 'Coorg' mandarin, sweet orange and grapefruits in India and performs well on saline and heavy soils. It has a marked dwarfing effect. Bears fruits of high quality and increase the yield.

PLANTING AND ORCHARD ESTABLISHMENT

Cultivated citrus genotypes are commonly planted in a square or a rectangular system. Planting density has tremendously increased with the use of dwarfing rootstocks. It is now possible to accommodate high number plants per hectare with the use of dwarfting rootstocks. In square system, the planting density of 4×4 m, 5×5 m, 3×3 m can accommodate 625, 400, 1,111 plants/ha. Similarly, in rectangular system spacings of 3×5 m (667 plants/ha) and 4×6 m (417 plants/ha) are being adopted. High density plantings maximize the yield by improving the productivity (yield per unit area). Ultra high density plantation (>5,000 plant/ha) is popular in Japan. At IARI, New Delhi Kinnow has performed best on Troyer citrange followed by 'Jatti Khatti' and Soh Sarkar with regard to dwarfing effect.

Pits of size $50 \times 50 \times 50$ cm size are dug in summer according to the layout plan. The exposed soil after 30-40 days is mixed with 15-20 kg well rotten FYM and 50 g chloropyriphos (to kill white ants) and filled tightly. The best planting time is beginning of rainy season. After planting, soil should be pressesed firmly and a light irrigaton is given immediatley. Care must be taken that bud or graft point should be at least 10 cm above the soil surface. The irrigation channels/sub-channels basins are also laid out at the planting stage.

INTERCULTURAL OPERATIONS

Training and Pruning

Training of young established plant is essential to get a strong framework. Generally, heading back at 75-100 cm is done after 1st year and in the next season, 3 to 6 strong well-placed secondary branches are allowed to develop to form the framework. Pruning in general is practiced to remove extra/non-balanced branches, water sprouts, growth from rootstock and dead or diseased branches. Light pruning allows good productivity as it permits proper light penetration. In older trees, pruning is generally done to facilitate flowering and fruiting. Apart from above, root pruning is practiced in some parts of the country to induce flowering. Faulty and untimely severe pruning may lead to extensive damage to bearing plants.

Integrated Nutrient Management

For sustainable production of fruits and for proper maintenance of plants and soil health, efficient nutrient management programme must be adopted. Citrus is a nutrient exhaustive crop as plants in the population density of 400 plants/ha can remove about 200 kg N, 50 kg P_2O_5 and 200 kg K_2O/ha. Citrus is grown on a wide range of soil and hence no generalized recommendations can be made for any type. Besides, the dose of nutrients increase with the increase in plant age.

For 'Khasi mandarin', application 300 g N + 250 g P_2O_5 + 300 g K_2O has been found economical. Under Coorg conditions, 600 g N, 200 g P_2O_5, 450 g, K_2O/plant has been recommended for mandarins. Similarly, for lemons, 500 g N, 250 g K_2O was found ideal.

Foliar spray of micronutrients has given beneficial effect on improving the yield and quality of citrus fruits. One to 2 per cent urea, superphosphate and KCl were recommended for declining orchards for their quick revival. Application of 1.0 per cent Zn reduced chlorosis and increased yield. Likewise, 1 per cent Zn + Cu and 0.5 per cent Fe are found beneficial in sweet orange. Combined application of $ZnSO_4$, $MnSO_4$, $MgSO_4$ at 0.5 per cent and 0.25 per cent of $CuSO_4$ on Coorg mandarin reduced the micro-nutrient deficiencies. Foliar application of $ZnSO_4$ and chelated Zn at 0.1 or 0.4 per cent increased the yield in Assam lemon. Light tillage is beneficial to check the weeds that take away major share of applied nutrients. Proper placement of inorganic fertilizers in the region of high root activity and root distribution helps in proper uptake of nutrients. Fertilizer placement at about 1 m diameter and 30 cm deep, allows the maximum uptake of nutrients.

Green manuring of young orchards is essential to enrich the soil with organic matter and improve fertility. Leguminous crops like lentil, *dhaincha*, cowpea, green or black gram are the best green manure crops. Shallow rooted and low spreading crops like pea, cowpea do not interfere with the main crop and hence are also beneficial.

Mulching

Mulches are gradually gaining popularity in present day agriculture for their known merits *viz.*, weed control, reduction in soil temperature, prevention of soil moisture loss, reduction in soil loss, addition of organic matter, etc. Crop residues *viz.*, straw, leaves, dried grasses, husk etc. and synthetic sheets, *i.e.* black or white polythene sheets are now widely used for fruit production. Mulches check weed growth thereby improve yield, maturity and fruit quality. When growing green manure in the orchard, constant check for pests and diseases must be made to keep them under control.

Intercropping

This is not popular and rather never recommended for citrus unlike other fruit crops as it is reported to cause harm to the main crop. Large number of plants show decline. Only in the initial years of establishment, *i.e.* up to 3-4 years, vegetable crops like onion, potato, chillies, pulses, gram, etc. can be grown.

Weed Control

A broad range of weeds compete with citrus plants, namely, *Cyperus rotundus, Cynodon dactylon, Sorghum halpense, Euphorbia khirta, E. microphylla, Convovlulus arvensis, Amaranthus viridis, Paspalum* spp., *Imperata cylindrica, Ageratum* spp. etc. These weeds appear according to the season and occupy the open space and area underneath the plants. Weeds compete for water and nutrients with main crop and harbour pests and diseases. Light hoeing is essential to control the germinated weeds. Chemicals *viz.*, atrazine, simazine (6 kg a.i.) are recommended right from germination to flowering stages. Now-a-days, glyphosate is gaining popularity as a broad spectrum weedicide. Care must be taken to check the problem of spray drift.

FLOWERING, FRUIT SET AND GROWTH

Flowering citrus species bloom during spring and other growing seasons. Members of acid group *viz.*, limes, lemons, citrons, flower throughout the year, while other species exhibit two major flushes under sub-tropical conditions. In North, flowering season starts from February-to-April. In southern parts, where there is no clear winter season, the plants do not exhibit longer blooming season and two seasons are noted. Sathgudi oranges in South India, flowers during December-April and September-December. Similarly, Coorg mandarin flowers during March-April and September-October. In central and western India, oranges flower three times, *i.e.* June, October and February.

Flowering in citrus can be regulated by employing varied practices *viz.*, withholding water, pruning, ringing and girdling, fruit thinning etc. Application of chemicals like CCC (100 ppm), SADH (250 ppm) or BOA (25 ppm) on lemon plants induced flowering. The flower differentiation in sweet orange under North Indian conditions takes place during second to third week of January, *i.e.*, about 30-40 days before flowering.

In general, pollen development is a normal phoenomenon in citrus, except in 'Washington Novel' orange, 'Satsuma' mandarin and 'Bearss' lime in which no viable pollen is produced. In general, self-pollination takes place in species, which produce abundant pollen, yet cross-pollination is not uncommon. Honeybees are chief pollination agents, and cross pollination increases seediness in several cultivars. Both male sterility (*e.g.*, 'Washington Navel' orange and 'Satsum' mandarin) a female (*e.g.*, 'Marsh' grapefruit', 'Eureka' and 'Lisbon' lemons) partial self-incompatibility, and normal male and female fertility conditions have been reported to occur in citrus. As a result of these conditions, different degree of seediness has been found in different species and their cultivars. Hence, on the basis of seediness, citrus cultivars have been categorized as seedless (no seed), commercially seedless (0-10 seeds), and seedy (numerous seeds). Self-incompatibility has been reported in Orlando tangelo, Siamese pummelo, Clementine mandarin, and Italior, Nepali Oblong and Kagzi Kalan lemons. Nagpur santra has been reported to be self-compatible and cross-compatible except with grapefruit. For Pant lemon-1, Nepali Oblong and Italian lemons and for Kagzi Kalan lemon, Pummelo have recommended as a suitable pollinizer for getting adequate fruit-set and profitable yields in commercial plantings.

Application of pacilobutrazol (PP 333) @ 2.5 to 10 g per tree induced flowering and improved fruit set. In citrus, parthenocarpy can be induced by applying GA_1. A spray of 2,4-D (8-10 ppm) markedly increased fruit-set and quality of mandarin and sweet orange. Likewise, application of 2,4-D or 2,4,5-T at 30 and 45 ppm, respectively can reduce excessive fruit drop in citrus.

Heavy flowering coupled with fruiting in any season may lead to poor bearing in the next season. Hence, crop regulation is highly desirable. NAA at 250-600 ppm has been suggested for thinning of fruits in mandarin thereby overcoming alternate bearing. Likewise, MH (1,000-2,000 ppm) and ethrel (300 ppm) have also been found effective for fruit thinning.

Fruits after set grow in a sigmoid fashion on the basis of observations on fruit weight and volume in Kinnow, while sweet orange show a linear growth pattern. There is a continuous increase in fruit diameter up to maturity in Kinnow, while in sweet orange, there is a decline once the winter sets in.

CITRUS DECLINE: CAUSES AND CONTROL

Citrus fruits have a prominent place among popular and expensively growth tropical and subtropical fruits of the world. Citrus fruits are grown in about 100 countries of the world, but Citriculture has attained the status of commercial industry only in about a dozen countries of the world. Various species of citrus are grown in different parts of the world, but by far, oranges are the most common among the citrus fruits, occupying nearly 2/3rd of the total world area under citrus fruits. The leading orange producing countries are the USA, Brazil, South Africa, Japan, China and India. Florida alone produces 90 per cent grapefruits of the world. Italy leads in lemon, Mexico and India are the main producers of acid lime, whereas, Japan grows mainly the mandarins. Israel's agriculture industry has about 80 per cent share of citrus export.

India is the fourth largest producer of citrus fruits in the world. Like many other countries of the world, citrus occupies a place of pride among different fruit crops grown in India, because it is the 3rd largest fruit industry, which occupies about 12 per cent of the total land under fruits in this country. Citrus fruits are rich in vitamin C, fruit sugars, acids, minerals and alkaline salts, which are essential health promoting ingredients in our diet. In addition, various value added products can be made from citrus fruits. Various species of citrus fruits are grown all over the India. Of these, mandarins (*Citrus reticulata* Blanco) are commercially grown in Assam, Coorg (Karnataka), Nagpur (Maharashtra), foothills of Himachal Pradesh, Kerala and some pockets of NEH regions and sweet oranges (*Citrus sinensis* Osbeck) in Punjab, Maharashtra. Kagzi lime (*Citrus aurantifolia*) is cultivated on a large scale in Maharashtra, Tamil Nadu, and Andhra Pradesh and to some extent in Bihar and Gujarat. Other species of citrus, like lemon (*Citrus limon*), grapefruit (*Citrus paradisi*), pummelo (*Citrus grandis*) and sweet lime (*Citrus limettioides*) are grown in different parts, but to a limited extent. Citrus once held a position of prime and pride among fruits cultivated in India. But, due to citrus decline or die back problem, the area and production have shrunk to a considerable extent. The productivity of citrus orchards in India is only 7-8 t/ha, compared to 25-30 t/ha in major citrus growing countries of the world. Various scientists are of the view that there are many reasons for low productivity of citrus, but the malady of citrus decline is the major contributing factor. The available literature suggests that decline is the most devastating malady of citrus, which has resulted in devastation of million of citrus trees all over the world, including India. Citrus decline, also often referred to as 'citrus dieback', is not a specific disease, but is a symptomic expression of many disorders in the plant. The malady has also been named as 'wronching', 'chlorosis', 'decay', 'neglectosis', 'blight', 'declinio' etc., in different countries. However, in India, the terms like 'dieback' and 'decline' are widely used, though; the term *'dieback'* signifies the death of plant from top downwards. Moreover, the term 'dieback' is usually used as a synonym for *'exanthema'* or copper deficiency. Citrus decline is not a new problem, because the first observation on this malady in our country was made in the eighteenth century by Roghoji Bhonsale in the Central Province. In Assam, it was first observed in 1888. However, its seriousness was first recognized by Department of Agriculture, the then Bombay state in 1912. In Punjab, the presence of decline in citrus was reported in 1940. However, the problem became severe only in the early 50's, when sweet orange orchards began to decline rapidly. Considering its importance, the ICAR sanctioned a scheme for survey of this malady in Peninsular India in 1943. Of late, this malady has assumed an alarming position in many parts of the world, wherever citrus is grown.

Symptoms of the Malady

Symptoms of citrus decline usually vary with the cause of the malady. All citrus species and varieties are susceptible to decline, but the mandarins are the most susceptible, followed by sweet oranges. Both budded as well as seedling trees suffer badly from this malady. In this malady, the trees make excellent growth and perform satisfactorily well in the first few years of their orchard life. But, later they begin to deteriorate in health, bear every little crop and become uneconomical. The affected trees usually do not die, but get into a state of decadence and remain impoverished and unproductive for several years. In some cases, the trees may wilt and die relatively rapidly. Such trees, instead of being a source of profit, become a great liability of a grower. In general, the symptoms of citrus decline appear in following four stages.

1. In early stages, the symptoms are restricted only to a few limbs, but eventually whole tree is involved. The leaves may be small with light green interveinal areas, with midrib and lateral veins remaining dark green. The new flush of leaves on affected twigs may be smaller in size.

2. In advanced stages, chlorotic areas gradually increase in some leaves, only basal portion of the midrib is green, while in others, splotches of yellow or pale green are present between the lateral veins.

3. In acute cases, the growth is usually checked and entire canopy of the tree bears short twigs with narrow, small leaves on their lower portions, while distal portions of the twigs are devoid of leaves. The shoots have a tendency to die from the growing points downwards. In a few years, the flushes are reduced and the network of veins becomes prominent.

4. In more severe cases, the young and old leaves may be very pale, the colour of veins gradually fading until only a very pale green midrib.

The cropping and fruit quality of such trees also deteriorate seriously. There is conspicuous reduction in the size of fruits and fruit yield. The quality of fruits is adversely affected. The fruits have rough rind with lack of shine and less juice content. The feeder root system of the trees may get depleted, and roots turn black. Such trees may remain in the state of decadence and die slowly and steadily, which is called as *'slow decline'*. However, in some cases, the trees die within a few days, which is called as 'quick decline'. Either only a few trees in an orchard or entire orchard or a few orchard in a locality may be affected. No orchard, however, has been completely free from this malady. The magnitude of this malady increases with age of trees and after 15-20 years, affected trees become uneconomical.

Causative Factors

The causes of this malady are still unknown. However, scientists world over have come to the conclusion that most of the factors singly or jointly may be responsible for citrus decline. At the same time, it has been concluded that one factor responsible for the cause in one locality may not necessarily be responsible for decline in another locality. In general, following factors have been reported to be associated with the cause of citrus decline in various parts of the world.

A. Edaphic Factors

Scanning of voluminous literature reveals that several scientists have attributed the decline of citrus trees to unfavourable soil conditions as under:

i. Soil Texture

Citrus thrives well in soils, which are deep, loose, well aerated and devoid of impregmented layers of calcium carbonate, which is also governed by soil texture. The relative proportion of coarse and fine particles in the soil plays an important role in the root activity of citrus trees. Many scientists have reported that presence of any impregmented layers of calcium carbonate or clay affects the permeability and aeration of soil, which affects drainage and root growth. Some research workers are of the view that presence of hardpan due to $CaCO_3$ or clay was responsible for citrus decline in Punjab, Maharashtra and Andhra Pradesh, because hard pan/ *kankar* pan in the sub-surface affects the root activity, and if, root activity is poor, the life span of the tree would certainly be less. Further, higher concentration $CaCO_3$ also contributes indirectly to citrus decline, because of decreased availability of micronutrients, like Fe and Zn in the soil. In general, deficiency of micronutrients like Fe, Zn and Mn is quite common in calcareous soils.

ii. Soil Reaction

Soil reaction, as indicated by pH, has a marked influence on soil environment, especially on the availability of plant nutrients. An ideal pH for most species of citrus is considered to be between 5.5 to 7.5, because the soil in this pH range are more fertile and trouble free than those with lower or higher pH. However, scanning of voluminous literature on the subject reveals that excellent citrus is being raised in our country in soils ranging from highly acidic (pH 4.5) to those containing free lime with pH 8.5. For example, pH of citrus soils in Punjab varied from 6.5 to 9.0. However, in general, very high or low pH has definite affects on the availability of micronutrients, like, Zn, Fe and Mn. Similarly, the tolerance range of exchangeable sodium percentage (ESP) for citrus is 2-10. However, in many parts of India, where citrus is grown, this range is usually 20-25. Thus, growing of citrus under very low or high soil pH may contribute to citrus decline.

iii. Improper Drainage

Most of the citrus species are sensitive to excessive moisture conditions and thrive best on well-drained soils. Citrus trees put forth excellent growth in areas where water table is below 1.5 m. However, the problem of citrus decline has been observed in areas where water table has gone up. Under improper drainage, there is poor supply of O_2 to the roots. As a result, the root growth is arrested, leading to less absorption of nutrients. In addition, under poor drainage conditions, soil-borne fungi like *Phytophthora* becomes active, inciting the incidence of various diseases, resulting in the dieback of the trees.

iv. Excessive Salts

Citrus trees are extremely sensitive to excess of salts and should never be planted on land that has more 1,000 ppm concentration of salts. Soil salinity has been reported to be a major cause of citrus decline in India. It has been reported that presence of salts up to 180 cm soil depth would certainly affect tree health. When planted on a soil having higher salts, would lead them to decline.

v. Excessive Free Lime

Presence of excessive lime in calcareous soil, renders P, Fe, Mn and Zn less available to the trees. Surveys conducted in declining citrus orchard soils have revealed that a soil containing more than 5 per cent lime in any horizon up to 180 cm soil depth is quite favourable for initiation of citrus decline.

vi. Disturbed Soil Structure

Cultivation of soil for a number of years may lead to the deterioration of fertility and physical structure. Defective soil structure causes improper air and water relationships, which hinders not only the growth of the plants, but also microbial activity. Continuous ploughing also sometimes results in the development of a layer called as *'plough sole'*, which hinders the growth of trees, resulting in decline.

vii. Low Soil Organic Matter

Presence of adequate amount of soil organic matter improves soil fertility by improving the soil structure. Organic matter also improves the microbial activity by enhancing the population of soil microorganisms. It is a known fact that even under adequate supply of fertilizers, trees do not put forth proper growth in absence of organic matters. Thus, low organic matter content of the soil may lead to citrus decline.

viii. Soil Erosion

Soil erosion in citrus orchards is a common problem in mountainous areas of our country, which washes away the valuable soil. Under such conditions, the trees may decline due to non-availability of adequate soil

B. Varietal Factors

The success of citrus in a particular area depends to a large extent on the type or variety grown. In general, all citrus species are susceptible to citrus decline, but mandarins and sweet oranges have higher susceptibility than grape fruit, pummalo, lemons and limes. Of various varieties of sweet orange, Sathgudi, Mosambi and Blood Red are only commercially grown. Of these, Sathgudi is confined to south India, Mosambi to western India and Blood Red is popular in some parts of north India. Many varieties of sweet orange have been introduced in India. Of these, only Valencia Late, Jaffa, Hamlin and Pineapple have performed well under our conditions. Washington Navel has been a totally failure. Among mandarins, Nagpur, Khasi and Coorg are confined to specific belts, and if grown out of that particular belt, they usually do not perform well and tend to decline. Kinnow mandarin, a hybrid between King and Willow leaf mandarin, has acclimatized very well in semi-arid zones of Punjab, Haryana, Rajasthan and sub-mountainous tracts of Himachal Pradesh (India), but has not at all been successful in southern or western Indian conditions. Hence, review of the varietal situation in our country reveal that certain varieties or species are more prone to decline than others.

C. Rootstock Factors

Rootstock is a vital component of a budded or grafted plant, which plays a vital role in the success or failure of a plant. Rootstocks are known to influence the vigour, precocity and regularity in bearing, productivity, fruit quality, and or longevity of the scion budded or grafted on them. In addition, rootstocks also influence susceptibility of the trees to various biotic and abiotic stresses. Thus, it is evident that for optimum performance of the orchard, selection of proper rootstock is very important.

Table 3.1: Reaction of some Citrus Rootstocks to Biotic and Abiotic Stresses

Sl.No.		Biotic Stresses							Abiotic Stresses	
	Rootstock	Gum-mosis	Root Rot	Tris-teza	Xylo-porosis	Exo-cortis	Citrus Nema-tode	Burro-wing Nema-tode	Salts	Drou-ght
1.	Cleoptra mandarin	T	T	T	T	T	S	S	MT	T
2.	Carizzo citrange	MT	MT	MT	S	S	T	S	HS	HS
3.	Sour orange	T	R	S	T	T	S	S	T	MT
4.	Rangpur lime	T	MT	T	S	S	S	S	R	R
5.	Rough lemon	S	S	T	T	T	S	S	T	T
6.	Trifoliate orange	T	R	T	T	S	T	S	HS	HS
7.	Troyer citrange	T	MT	T	T	S	-	S	HS	HS
8.	Sweet orange	S	HS	S	S	-	S	S	S	S
9.	Grape fruit	T	T	T	T	S	-	S	R	R
10.	Karna Khatta	S	S	-	-	T	MT	-	T	S

In general, selection of an ideal rootstock is a great problem in many fruit crops, but it is of greater importance in citrus. The most commonly and widely used rootstocks for citrus throughout the world are trifoliate orange, sour orange and rough lemon. Of these, sour orange has been most extensively used in the past, but due to susceptibility to tristeza, it has been abandoned in some parts o the world. In India, rough lemon or its stains are mostly use as rootstocks. However, it has shown graft incompatibility with Mosambi in Maharashtra, and Sathgudi in Andhra Pradesh and with Mosambi in Punjab. Thus, in many instances citrus decline has been attributed to graft incompatibility between the stock and the scion. In addition, many problems associated with rootstock may further aggravate the problem of citrus decline. For example, susceptibility of the rootstock to root rots, salinity, water logging, cold or viruses etc., may also lead to decline of scion cultivar budded on it. Reactions of some important citrus rootstocks have been given in Table 3.1.

D. Nutritional Factors

Among various factors responsible for citrus decline, nutrition occupies an important place. Many research investigators are of the view that malnutrition of citrus trees is the most important contributory factor for citrus decline in India. Like other fruit crops, citrus trees also require all major and minor elements in adequate proportion. Any deviation from the recommended dose of manures and fertilizers may result in decline. It has been observed that most of the declining sweet orange trees suffer badly from zinc deficiency. In some cases, lack of copper may also cause the trouble. In hilly areas, deficiency of some micronutrients like zinc, manganese and iron has been observed in declining mandarin trees. In heavy rainfall areas of Coorg (Karnataka) and Palampur (H.P.), where mandarins are grown on sandy or gravely soils under rained conditions, citrus decline may occur due to heavy leaching of nitrogen during rainy season. In hilly areas of Assam and H.P., declining citrus trees had lower level of potassium. In some cases, excess of iron and salinity and high uptake of phosphorus and manganese have also been reported to be associated with citrus decline. Similarly, non-availability of minor elements due to low pH and poor lime status may also be responsible for decline in some species or varieties.

The above discussion reveals that malnutrition may be major factor responsible for citrus decline.

E. Faulty Irrigation Practices

Citrus trees are evergreen and put forth new growth almost regularly. Hence, they require more and frequent water than other fruit crops. However, like other fruit trees, water requirement of the citrus depends upon variety or species, soil type, age of the tree, tree growth, rootstock used, root activity, water table, environmental conditions, irrigation method etc. That is why; it is difficult to give uniform irrigation schedule for all varieties/species grown at different agro-climatic zones of the country. Hence, the recommendations given by the scientist differ greatly.

Various researchers are of the view that both improper amount and frequency of irrigation can lead to decline as it controls growth and vigour of the trees. It has been observed that large scale drying of citrus orchards in our country is largely associated with scarce water sources, frequent drought and lowered water table. In addition, citrus trees are quite sensitive to water stress during their period of active growth; flowering and fruit set, and if, not irrigated during these period, they are liable decline soon. Over irrigation and water logging often cause leaching of some essential elements. Citrus is very sensitive to water logging as it causes root rot. Further, upper salt tolerance limit of citrus is 1,000 ppm. Salty water causes salinity built in orchards leading to decline. Similarly, due to lack of planning and use of conventional methods of irrigation, entire orchard has to be irrigated properly. In one location, water may be in excess, and in other, there may be acute shortage. All these conditions may lead to failure/decline of the citrus orchards in our country.

F. Intercropping

In the newly planted citrus orchards, there is possibility of growing some short season crops during the initial years, because trees occupy relatively very less area in this period of their growth. In most cases, intercropping is advantageous. However, most horticultural workers in our country are of the view that excessive and indiscriminate use of intercrops in citrus orchards has resulted in the decline of citrus. It has been observed that because of economic considerations, orchardists neglect citrus plants initially and they grow intercrops even in the bearing orchards. Deleterious effects of intercropping apparently result from wrong selection of the intercrop (*e.g.* cotton grown in citrus orchards in Punjab), which depletes the fertility of the orchard, exhaustion of mineral nutrients, causes adverse effects on physical properties of soil, causes shading of the orchards, incompatible irrigation, and insects and diseases that attack intercrops. All these effects aggravate as the trees become older, which may lead to decline.

G. Insect-Pests

Citrus trees are attacked by a number of insect-pests, which cause damage either by direct feeding or by acting as vectors for the transmission of viruses. Of various insect-pests only about a dozen (*e.g.*, citrus psylla, citrus black fly, white fly, mealy bugs, leaf miner, aphids, bark eating caterpillar, and trunk borers etc.) are responsible for citrus decline.

Among different insect-pests of citrus, citrus psylla (*Diaphorina citri*) is of utmost importance, which attacks all species of citrus and cause considerable damage by sucking sap from tender plant parts. In addition, it also injects certain toxins to the affected plant parts, which later wither and dry up. The effects of its damage are visible after 2-3 years. It also acts as vector for greening mycoplasma disease, which has been reported to be most common in declining trees. Hence, most of the scientists are of the view that the symptoms of citrus psylla damage are similar to the symptoms of greening.

Citrus leaf minor (*Phyllocnistis citrella*) is a serious pest of citrus foliage. Its larvae make silver shining tunnels in the leaves. As a result, the leaves become distorted and crumpled. In serious attack, defoliation may occur. These tunnels may encourage the incidence of citrus canker, a serious bacterial disease, which later leads to decline of the affected trees.

Both black (*Aleurocanthus woglumi*) and white fly (*Aleurocanthus citri*) are destructive species of fly, which reduce plant vigour by sucking plant sap. These pests also secrete honeydew on which sooty moulds (*Capnodium* spp.) develops, which affects photosynthetic activity of the plants. Some hard and soft scales and mealy bugs (*Pseudococcus filamentotus*) cause damage to citrus plants by sucking large quantity of sap. In addition, these insects inject toxins, which may cause flower, leaf or fruit drop.

Lemon butterfly (*Papilio demoleus*) damages young foliage of plants. Its caterpillars eat away leaves, flowers and fruits and devitalize the plants and ultimately lead to decline.Though, many species of aphid attack citrus, but brown citrus aphid (*Toxoptera citricida*) causes maximum damage by acting as vector for transmission of tristeza virus, which has been considered as the most contributing factor for decline in various parts of the world.

Bark eating caterpillar (*Inderbela quadrinotata*) is very serious in old and neglected orchards of citrus. It girdles the major branches, resulting in tree decline. Similarly, in some instances, trunk borer (*Monohanmus versteegi*) causes severe loss to citrus orchards by girding the trunk, leading them to decline.In addition, there are numerous other insect-pests, which are not directly responsible for citrus decline, but may contribute towards it by debilitating the trees and making them more susceptible to the attack of other pathogens.

H. Nematodes

Nematodes are worm like organisms, too small to be seen with naked eye. Of late, the incidence of nematodes has also been recognized as one of the contributing factors of citrus decline. Several species of nematodes are known to be destructive to citrus. However, two important nematodes, which enjoy worldwide distribution, are (i) citrus nematode (*Tylenchulus semipenetrans*) and, (ii) the burrowing nematode (*Radopholus similis*). The former *i.e.* citrus nematode has been reported to be associated with '*slow decline*' in California by J.R. Hodges in 1912, and the later *i.e.* burrowing nematode with the most destructive '*spreading decline*' in Florida. In addition, some other species, like root knot nematode (*Meloidogyne javanica*) and root lesion nematode (*Pratylenchus prantensis*) also parasitize citrus roots in many parts of the world. These nematodes may affect the citrus tree growth by (i) injuring the root bark, (ii) removing plant nutrients during feeding, (iii) impairing normal growth and functioning of the roots, and (iv) possibly by injecting toxins into the trees. At low population, nematodes may not cause much damage, but severe infestation leads to tree decline.

I. Fungal Diseases

Citrus trees are attacked by various fungal diseases, which contribute towards citrus decline. However, Phytophthora rot, which is also called as *gummosis, root rot,* or *crown rot*, cause maximum damage to citrus. It produces symptoms of decline by (i) rotting of the rootlets, (ii) girdling of the trunks, and (iii) dropping of the blighted leaves. Next to phytophthora rot, scab (*Colletotrichum gloeosporioides*) has been reported to be associated with citrus decline. Besides, different other diseases like, Diplodia gummosis, foot rot, anthracnose, pink disease, *Rhizoctonia* rot etc., also reduce the vitality of the citrus tress, leading them to decline.

J. Viral Diseases

Prominent among the various agencies causing citrus decline are the diseases produced by various viruses. Most of the research workers are of the view that most of the citrus orchards throughout the world have been destroyed by the viral diseases. For example, about 20 million citrus trees were devastated by tristeza virus in Argentina and Brazil between 1930-1940. Though less spectacular, severe losses have also resulted in different parts of the world from infection by the viruses of Xyloporosis, exocortis, porosis, woody gall, strubborn, Satsuma dwarf, bud union crease, favoid and likubin. Among various viral diseases, tristeza is universally present. The presence of all these viruses have been reported from various parts of India, but tristeza causes the maximum loss. In general, virus infected trees do not put on normal healthy growth, become stunted, show die back of twigs and branches, as a result, the affected trees start declining. In the recent past, some new viral diseases like witch's broom, leathery leaf, measles, rubbery wood etc., have also been noticed, which in severe form, cause citrus decline in many species.

K. Bacterial Diseases

Bacteria cause some of the most serious diseases of citrus in the world. These include citrus canker, greening, blast and citrus variegated chlorosis. Of these, canker and greening are present worldwide and are the most destructive bacterial diseases of citrus, whereas, citrus variegated chlorosis (CVC) is the most serious problem in Brazil, and Blast in Australia, Japan and South Africa.

Under Indian conditions, citrus canker is the most serious disease of acid limes, rendering them to decline. Greening, which was previously thought to be a MLO or viral disease, is a serious bacterial disease in many parts of India. It is caused by a bacterium, *Liberobacter asiatiacum*. Greening affected plants bear small leaves with green veins and chlorotic interveinal area. Leaves become mottled and plants are stunted, showing twig die back. At last, plants start declining in growth and fruit production.

L. Weeds

Weeds are unwanted plants, which are responsible for substantial losses in citrus yield, because they complete with growing trees for space, moisture, nutrients and solar energy. In citrus orchards, many monocot (*Cyprus rotundus, Andropogon, Cynodon dactylon* etc.) and dicot weeds (*Euphorbia* sp., *Convolvulus* sp., *Oxalis, Amaranthus* sp. etc.) grow. By way of interference with the citrus trees, they affect the tree health, production and productivity, if not controlled in time. As a result, citrus trees may decline sooner or later.

M. Biochemical Factors

Though, no direct relationship has been established by the scientists between the enzymatic activities and citrus decline, however, some recent studies indicate that the polyphenol contents are increased in the declining citrus trees. Similarly, trizteza infected declining trees have higher peroxidase activity than healthy trees of citrus, which established the fact that declining trees produce phenols in higher proportion to fight against disease/malady, which increases the peroxidase activity.

Management Practices of Reducing Citrus Decline

After scanning the voluminous literature, it becomes evident that the following management practices would be useful to reduce the incidence of decline in the citrus orchards:

A. Use of Resistant Varieties and Rootstocks

One of the best alternatives to avoid citrus decline is to grow decline resistant or partially resistant species or varieties. Though the choice for such species or variety is limited, but we need to introduce new varieties of sweet oranges and mandarins and after their adaptive trials, these may be recommended for commercial cultivation. For example, Honey and Dancy tangerines, Pearl tangelo and Campbell sweet orange should be tried in our country, which may replace existing susceptible citrus types.

Table 3.2: Some Recommended Rootstocks for Various Citrus Scions in different Regions of India

Area	Scion	Recommended Rootstock
Punjab	Malta local	Karna Khatta
	Blood Red	Jatti Khatti
	Mosambi	Jatti Khatti
	Valencia	Jatti Khatti
	Jaffa	Jatti Khatti
	Pineapple	Jatti Khatti
	Santra local	Karna Khatta
	Kinnow mandarin	Jatti Khatti
	Grape fruit	Karna Khatta
Assam	Valencia orange	Kata Jamir
	Khasi mandarin	Soh Myndong
Maharashtra	Ganesh khind	Rangpur lime
	Nucellar Mosambi	Rangpur lime
	Nagpur mandarin	Rangpur lime Jambheri
	Acid lime	Rough lemon
Andhra Pradesh	Sathgudi	Rangpur lime
Delhi	Kinnow	Troyer citrange, Karna Khatta
Tamil Nadu (Periakulam)	Acid lime	Rough lemon
Karnataka (Coorg)	Coorg Mandarin	Rough lemon, Rangpur lime, Cleopatra mandarin
Nagpur	Acid lime	Alemow
Kodur (A.P.)	Sathgudi	Gajanimma
	Acid lime	Gajanimma
Assam	Nagpur mandarin	Rangpur lime, *C. jambhiri*

In a country like India, where different citrus species/varieties are grown in different agro-climates, it is very difficult to recommend a particular rootstock to a particular species or a variety. Similarly, it is equally difficult to develop new rootstock, which is suitable for different varieties or species in all agro-climatic zones. However, growers should use only the recommended rootstock. A lot of work has been carried out in different regions of our country. On the basis of that work, following recommendations has emerged out (Table 3.2).

In addition, many new rootstocks have also been recommended by various research workers, which should also be tried in our country. These rootstocks are alemow (*Citrus macropylla*), *C. volkameriana*, Benton, Swingle, RxT hybrid, Flying Dragon, C-32, C-35, Milan, Miri etc.

Further, the use of salt tolerant rootstocks like, Jatti khatti, Cleopatra, Rangpur lime etc., in salt affected areas, would certainly avoid the problem of citrus decline. Many viral diseases like tristeza, exocortis, xyloporosis and bacterial diseases like, root rot etc. may affect many stionic combinations. To avoid such problems, use of resistant rootstock is recommended as given in Table 3.3.

Table 3.3: Some Rootstocks having Resistance against Diseases

Sl.No.	Disease	Resistant/tolerant rootstock
1	Root rot	Cleopatra, Trifoliate orange
2	Tristeza	Rough lemon, Cleopatra, Rangpur lime
3	Exocortis	Rough lemon, Cleopatra, Rangpur lime
4	Xyloporosis	Rough lemon, Cleopatra, Rangpur lime
5	Greening	Milan, Miri, Vollamar

B. Cultural Practices

i. Site Selection

Citrus requires deep, loose, well-aerated soils having pH between 5.5-7.5. However, in India citrus is being grown on different types of soils with pH up to 9.5. Hence, before starting the citrus orcharding, soil test must be done. Select citrus types as per soil type and if soil does not suit, its management should be done before starting orcharding. For example, application of gypsum has been advocated as a soil amendment in soils having pH more than 8.5. Similarly, heavy organic manuring should be done in light soils. Provision should be made for adequate drainage in heavy and compact soils having hardpan beneath the subsurface.

ii. Indexing

Introduction of new varieties or development of varieties resistant to decline requires much time. Hence, to avoid disappointment, growers should be advised to establish new citrus orchards of healthy and virus free nursery stock. Strict quarantine and bud wood certification programme must be started in all the countries like the one started sometimes ago in the USA, Brazil and other countries. For this purpose, we need to select healthy trees of desired variety/species. The selected trees should be indexed for the known viruses and other bud transmissible diseases before scion material is to be taken for propagation. Nurseryman should also be advised to raise nucellar seedlings of all commercial varieties. Thus, a nation wide programme must be initiated in our country to supply elite/indexed-planting material for raising new orchards.

C. Nutrition

The problem of decline is aggravated if trees are not fed properly. Unfortunately, citrus growers throughout the country do not follow recommendations with regard to application of manures and fertilizers. Though, it is very difficult to make uniform recommendation of manures and fertilizers for different regions, because the requirement is medicated by the variety, species, age of the plant, type of the fertilizer to be used and previous fertilizer use. Hence, growers should get the orchard soil tested every year to know the status of nutrient available in the soil. Afterwards, guidance of a technical expert may be taken for the application of manures and fertilizers. On the basis of research work conducted at different research institutes, the following recommendations have emerged out (Table 3.4).

Table 3.4: Recommended Manurial and Fertilizers Doses for different Citrus Types

Citrus Type	Doses/Full Bearing Tree			
	N (g)	P (g)	K (g)	FYM (kg)
Acid lime	400	200	200	40-50
Kinnow mandarin	400	200	40	50-60
Khasi mandarin	300	250	300	60-70
Nagpur mandarin	600	200	300	70-80
Sweet oranges	600-800	200	250	100

It is not only the quantity, which is important, but it is the time of application of manures and fertilizers, which affects tree's vitality. Farmyard manure should be applied 2-3 months prior to growth initiation *i.e.* by December-January. Generally, the total amount of phosphatic and potasic fertilizers are applied at one time, while nitrogenous fertilizers should be applied 2-3 split doses. First dose of N most likely applied about 15 days before flowering (by mid-February), 2nd dose after fruit set (mid- April) and 3rd dose just after rainy season.

Table 3.5: Chemicals for Combined Nutritional Spray in Citrus

Chemical	Quantity	Chemical	Quantity
Urea	5.0 Kg	Boric acid	0.45 Kg
Lime	4 .0 Kg	Manganese sulphate	0.90 Kg
Zinc sulphate	2.25 Kg	Copper sulphate	1.25 Kg
Magnesium sulphate	0.90 Kg	Water	500 litre
Ferrous sulphate	0.90 Kg		

In addition to major nutrients, citrus suffers badly from micronutrient deficiencies of Zn, Mn, Fe, B etc. Micronutrient deficiency of individual element is difficult to ascertain because the deficiency symptoms of many micronutrients resembles very closely, making it difficult to judge, which of the nutrient is really deficient. In view of this, some research workers have advocated combined nutritional sprays to combat the deficiency of some major and minor nutrients as given in Table 3.5.

The spraying should be done when new flushes have emerged out during February-March and again during September-October.

D. Water Management

Different species or varieties of citrus have different water requirements depending on age of plant, soil and climatic zone. However, in general, following points should be taken into account while irrigating citrus trees:

1. About 60-80 per cent root activity of most of the citrus species is confined to 60 cm topsoil. Thus, more frequent irrigation is required in citrus than the quantity. Neither too much wetting of total root zone nor wetting of a small percentage of root zone in desirable.

2. Avoid moisture stress during the periods of rapid growth, flowering and fruit set.

3. While irrigation, care should be exercised to avoid the contact of water with the tree trunk, because direct contact of water with tree trunk may pre-dispose the tree to root rot or other diseases, rendering tree to decline.

4. Citrus trees are highly sensitive to excessive moisture and water logging. Ensure better aeration near tree trunk by making a heap of soil around tree trunk.

5. Citrus trees are sensitive to salinity. The total salts in irrigation water should not exceed 1,000 ppm.

6. If possible, install drip irrigation system, which has numerous advantages.

E. Intercropping

A great care should be exercised while selecting the right type of intercrops in citrus orchards. Intercropping plays vital role not only in generating more income to the grower, but it is good for better management of the orchard. However, one should be careful in the selection of intercrop. The intercrop should have cultural requirements similar to main crop, should not be too tall or spreading as to shade the tree, should not have higher water requirements and it should not compete with main crop for nutrition. Different experiments have been conducted at different research station in India and it has been recommended that exhaustive crops like wheat, maize, sugarcane, cotton etc., should be avoided as intercrops in citrus orchards. Similarly, intercrops like, tobacco, bhindi, tomato, brinjal, chillies etc., usually favour the incidence of root-knot nematodes in citrus. In general, the following recommendations should be followed:

Intercrops for summer season: Vegetables like tinda, kaddu, karela, onion and pulses such as, mung and mash etc, should be grown

Intercrops for winter season: Vegetables like peas, turnip, cauliflower, cabbage, radish and pulses like gram may be grown easily.

Fruit growers near to cities and big towns should preferably grow vegetables and fodders like *senji* and *barseem*. Fruit growers away from the markets should go for onion, potato, and pulses. Author of this book is of the view that nowadays, strawberry can be a better choice as an intercrop in citrus orchards near to towns and cities. In addition, *phalsa*, papaya and pineapple can be grown as intercrops near cities under proper management.

F. Weed Management

To avoid heavy losses caused by weeds, chemical weed control should be adopted for the complete control of weeds in citrus orchards. Diuron @ 3-4 kg/ha and simazine @ 4-5 kg/ha as pre-emergence spray kills almost all the weeds in citrus orchards. For post-emergence spray, glyphosate @ 4 litre/ha gives complete control of grassy weeds. Dalapon @ 5 kg a.i./ha as pre-emergence and gramoxone @ 2 litre/ha, as post-emergence are equally effective weedicides to kill both monocot and dicot weeds.

G. Management of Insect-Pests

Insects cause direct damage to citrus trees by sucking sap from tender plant parts or eating away some plant parts. Some sucking pests also inject toxins. A few insect-pests act as vector for viruses. All these factors contribute immensely for citrus decline. Hence, their control at right time is mandatory.

Among different insect-pests citrus psylla (*Citrus psylla*) causes maximum damage. It is also known as a vector for greening disease, It can be controlled effectively by spraying systemic insecticides like rogor or metasystox or phosphamidon (0.05 per cent), when its activity is high. It can also be effectively controlled by releasing parasitoid, *Taxarixia radiata*.

Citrus aphid, white fly, soft scales, black fly and leaf miner all are sucking pests, which can easily be controlled by a single or two sprays of any of the systemic insecticides like monocrotophos, dimethoate, phosphamidon at 15 days interval. Citrus aphid ad black fly can also be controlled by predators like *Mallada boninensis* and *Coccinella septumpunctata,* but in that case, spray of insecticides should be avoided.

Citrus trunk borer can be controlled by injecting petrol or celphos tablets (1-2/hole) into the holes and then plugging the holes with mud. In addition, alternate hosts in the vicinity of citrus orchards should also be treated with suitable weedicides or completely destroyed in time to avoid the spread.

Lemon butterfly can be controlled by a spray of malathion (0.05 per cent) or methyl parathion (0.02 per cent). Its young caterpillars are also predated by yellow wasp and preying mantis.

H. Management of Nematodes

Citrus nematode and burrowing nematode are associated with huge losses in citrus. For their control, following integrated approaches should be followed:

1. Add 2-3 kg cake/tree either of neem, groundnut, mahua and mustard.
2. Use resistant rootstock like Carrizo citrange or Trifoliate orange.
3. Use neem cake with fungus, *Paecilomyces likicinus* or *Glomus fasciculatum* for reducing nematode population.
4. Use nematicides like dichlorophenthion (45 kg a.i./ha) or fensulphothion (30 kg a.i./ha) or ethoprophos (40 kg a.i./ha) or phorate (15 kg a.i./ha). Several investigators have reported the effectiveness of these nematicides in controlling various nematodes.

I. Management of Diseases

Various fungal, bacterial and viral diseases attack citrus and thus they have strong relationship with decline. Hence, proper management schedule should be followed for their control in time to avoid the decline of the plants.

(a) Fungal Diseases

Among various diseases, Phytophthora rot is very serious disease. The best ways to avoid its incidence in citrus are:

(i)   Avoid direct prolonged contact of water with tree trunk.

(ii) Ensure proper drainage.

(iii) Avoid excessive irrigation.

(iv) Use resistant rootstocks like, Cleopatra mandarin, Trifoliate orange or Rangpur lime.

(v)   Rear and release *Trichoderma* before the expected attack of Phytophthora rot.

(vi) Spray or drench with metalxyl or fosetyl.

Citrus scab, another serious disease can be controlled by spraying benomyl or carbendazim and powdery mildew by two sprays of karathane (0.1 per cent) at 10 days interval.

(b) Bacterial Diseases

Greening is the most destructive bacterial disease of citrus associated with decline. *Citrus psylla* acts as its vector. Hence, control of vector with some systemic insecticide (metasystox or rogor) also reduces the chances of greening. In addition, resistant rootstock (Milan or Miri etc.) should be used. New orchards should only be established with the indexed planting material.

Canker is another serious bacterial disease of citrus, which can be controlled by spraying 1 per cent Bordeaux mixture or streptocycline sulphate (500 ppm) before monsoon. In addition, population of leaf minor should be kept under control by spraying systemic insecticides.

(c) Viral Diseases

Many viral diseases attack citrus, but a few are serious, which cause heavy losses. The following recommendations should be followed to avoid the incidence of viral diseases:

☆ Use certified bud wood for raising new orchards.

☆ Keep the insect vectors under control by following integrated approach.

☆ Use resistant rootstocks wherever possible.

☆ Try to raise nucellar seedlings.

☆ Never use susceptible varieties or species.

PLANT PROTECTION

A. Major Diseases and their Management

Fungal Diseases

Phytophthora Rot

It is also known by different names like, foot rot, root rot, collar rot, crown rot etc. This disease occurs in almost all the citrus growing areas of the world. In general, citrus trees exposed to poor drainage or to wet soil for an extended period are more prone to this disease. *Phytophthora* spp. including *P. citrophthora*, *P. nicotianae* var. *parasitica* and *P. palmivora* have been found to be commonly associated with citrus species and are considered as one of the major cause of citrus decline. *Phytophthora* infections have been considered as one of the major cause of citrus decline in California, Florida, Cuba, Paraguay, Brazil, Mexico and Trinidad. In India, *P. nicotianae* var. *parasitica* is present in warmer and drier areas of Maharashtra and Andhra Pradesh and is chiefly associated with leaf fall and foot rot symptoms, while *P. palmovora* is prevalent in relatively more humid areas of Maharashtra and Karnataka.

Symptoms

The fungus produces symptoms on different plant parts thus causing root rot, foot rot, collar rot or stump rot on the roots and stem; gummosis and leaf fall on the foliage and brown rot on fruits leading to decline through, (i) rotting of the rootlets, (ii) girdling of the tree trunk and dropping of blighted leaves. The plants usually blossoms heavily and die after fruits mature. The disease manifests itself in the form of large water-soaked patches on the basal portion of the stem near the ground level.

Injury and destruction of small fibrous feeder roots by *Phytophthora* spp. is observed in a few feeder roots. In severe cases infection progresses further, results in complete rotting of relatively small roots and develops bird-eyed spots on big roots. Sloughing of bark is a common phenomenon in roots with lesser rotting. When the tip of main tap root is infected, the root system remains restricted to a shallow depth and such plants show early wilting during drought. There will be no flushing, and after watering these will show excessive flowering followed by poor fruit set which do not develop.

In susceptible rootstocks the root near the collar region gets infected and the infection often extends to the trunk. Due to foot rot, trunk bark is permanently damaged, and the plant vigour is adversely affected leading to slow death. In case of complete girdling of the trunk, plant may be killed suddenly. On the stem dead areas are produced on the bark that remain firm and cankerous and ooze gum. Oozing of gum is more common in varieties like kagzi lime, seedless lime and rough lemon rootstocks.

Leaf blight symptoms are common in the nursery however, in Coorg area in South India severe leaf infections have been observed during monsoon on big trees. Before even the complete leaf has blighted these usually drop off. In due course all the leaves and fruits fall due to infection and only barren branches are visible giving broom like appearance to the tree. Under these conditions especially when the environmental conditions are conducive thousands of trees are infected and killed with in 2-3 years causing 100 per cent loss. Fruit rot is a serious problem in monsoon or if rains occur during late development stages of fruits. Shelling often permits the development of pathogen over dropped pre-mature fruits which also emit characteristic odour.

Management Practices

A number of preventive measures have been recommended to reduce the primary infection of this disease.

1. Improving the drainage to avoid water stagnation in drip circle; by avoiding injuries due to implements and digging in drip circle which decreases root rot. Keep tree trunk free of new shoots or moss growth.

2. Use tolerant rootstocks like, sour orange, Cleopatra mandarin or Rangpur lime for raising new plants.

3. Budding should be done at a higher point (30-40 cm) above the ground.

4. Paint the affected parts (*e.g.* trunk) with Bordeaux paste up to about 50-60 cm. Height.

5. The application of ridomil (0.25 per cent) as spray or soil drench and Phosetyl-Al (0.3 per cent) as spray provide good control of root rot and increase the root volume. Among other non-systemic fungicides, captafol (0.2 per cent), captan (0.2 per cent), chlorothalonil (0.2 per cent) and copper oxychloride (0.3 per cent) have been reported to be effective as foliar spray and soil drench for the control of leaf fall and stump rot, respectively.

6. Release or spray the culture of *Trichoderma harzianum* in the orchards along with organic amendments that support the growth of *Trichoderma* spp.

Die Back or Wither Tip

Wither tip was reported as early as 1900 from Florida, Cuba and West Inidian Islands and now it is serious disease of citrus throughout the world. It is considered as one of the important factors responsible for citrus decline in India. Kagzi lime is the most susceptible followed by mandarin and sweet orange. It affects all aerial parts, including twigs, leaves and fruits. Several fungi have been reported to cause symptoms like twig blight and die-back. *Colletotrichum gloesporioides* is the most commonly reported. The other species like, *Gloeosporium limetticolum* or *G. follicolum* or *Diplodia natalensis* and *Gloeosporium limetticola* are also associated with wither tip of citrus. The fungus survives on dead twigs or tissues of the host. In sweet orange and mandarins, the fungus remains in the stalk left after harvest of the fruit.

Symptoms

Symptoms of anthracnose appear on leaves, young shoots and tender fruits. On leaves, the necrotic spots show acervuli arranged in concentric rings. Dead parts of the twigs assume silvery-grey appearance and the affected buds fail to develop. Shedding of leaves and dieback of twigs characterize symptoms of wither tip. Leaves show light green spots, which turn brown. On dead twigs, black dot like acervuli appear in concentric rings. The stem-end infection of immature fruits, results in fruit drop. In severe cases, branches show dieback and trees die in a few years.

Management Practices

The incidence of anthracnose/wither tip can be minimized by the following practices:

1. Better orchard management practices like proper irrigation, fertilizer application and control of other diseases will also reduce dieback
2. Regular pruning of infected branches reduce the inoculum.
3. Apply Bordeaux paste to the cut ends.
4. Collect and burn the fallen leaves, fruits and destroy them.
5. Spray the affected trees with Bavistin (0.1 per cent) or captafol (0.3 per cent) three times after pruning at 10-12 days interval.

Scab

It is a serious disease of citrus nursery and orchards. It is caused by *Elsinoe fawcetti*. It is more often confused with bacterial canker.

Symptoms

The disease appears in the form of small semi-translucent dots on the leaves, which later become pustules. The lesions mostly appear singly on one side of the leaves. The opposite surface corresponding to warty growth shows circular depressions with a pink to red depression. The lesions may appear on shoots as well. The lesions of fruits have corky projections, which often break in to scabs. As a result of severe infestation, leaves become distorted, wrinkled and infected fruits become hard and often drop pre-maturely.

Management Practices

A number of preventive measures have been recommended to reduce the primary infection of this disease by the Horticulturists and pathologists for the effective control of citrus scab as under:

1. Remove the affected leaves, fruits and twigs and destroy them to reduce the inoculum.

2. Spray Bordeaux mixture (1 per cent) or difolatan (0.2 per cent) or blitox (0.3 per cent) to get effective control of the disease.

Powdery Mildew

Powdery mildew is present in almost all citrus growing regions of the world. In India, the disease is of common occurrence in sub-mountaineous tracts of Coorg, Nilgiris, Palneys, Yercaud, Wynad, Shevroy Hills of South on mandarin oranges and sometimes even on sweet oranges in the plains. It poses a serious problem in nurseries especially in South India. Powdery mildew is caused by *Acrosporium tingitaninum* The fungus survives as dormant mycelium in the buds. Once the new flush matures, powdery mildew disappears. Post monsoon flushes especially during October - November and those after winters in March - April are severely affected by the disease. In mandarins severe powdery mildew infection of vegetative flush during October reduces the development of flowering shoots during March - April which ultimately result in less summer crop.

Symptoms

On infection, whitish powdery growth appears on infected plant parts. Young and tender shoots often get fully covered with white growth of the fungus. The affected leaves get distorted and reduced in size. In severe infection, leaves and fruits fall pre-maturely. Warm and humid weather favours the development of fungus and usually the disease symptoms are more prominent during autumn and spring season.

Management Practices

Following measures have been recommended to reduce the incidence of this disease.

1. Dusting sulphur powder or spraying wettable sulphur or karathane (0.2 per cent or calixin (0.2 per cent) or bavistin (0.1 per cent) brings effective control. Effective control with bitertanol, tridemorph, triadimefon etc fungicides has also been obtained.

2. Rootstock like *Poncirous trifoliata*, its hybrids (Citranges) and Rough lemon 7245 have also been reported to be resistant to powdery mildew.

Pink Disease

The disease was reported in citrus from Phillipines in 1936. The fungus also attack other plants namely, rubber, cacao, tea, cinchona, mango, cashew and jackfruit besides citrus. In India, the disease has been reported on citrus from Assam, Madhya Pradesh and Southern parts. In general, it appears during the fag end of monsoon and infects all above ground parts of citrus trees. The disease is caused by the fungus *Pellicularia salmonicolor (Corticium salmonicolor)*.

Symptoms

The fungus appears on affected plant parts as silvery white leathery growth, which turns to pink colour. The diagnostic symptom is the presence of salmon pink pustules on the infected bark. At first, the fungus is surface restricted, but later it penetrates the host bark and causes necrosis. Wood is also infected which affects the transport in xylem vessels. Once the symptoms appear all around the stem it is girdled and the tree may even be killed. Desiccation, chlorosis, dropping of leaves and finally drying of twigs take place during August-October, when the disease assumes full form.

Management Practices

1. Infected branches should be pruned and cut ends applied Bordeaux paste. Pruned branches should be burnt.
2. Post monsoon spray of Bordeaux mixture (1 per cent) or blitox (0.2 per cent) during monsoon.

Viral Diseases

Tristeza or Quick Decline

Tristeza (a word that describes the sad appearance of a tree) is considered as the most destructive disease of citrus throughout the world. It was first reported from South Africa in 1910, and later from many countries of the world. In India, it has been reported to be present in all citrus growing areas and infects mandarins, sweet oranges, acid lime, and grapefruit with variable severity. Citrus aphids, like *Toxoptera citricida* and *Aphis citricidus* are common vectors of tristeza.

Symptoms of tristeza depend upon scion, rootstock and strain of the virus and hence these differ widely. The infected trees usually show general decline. The leaves lose dark green luster, become dull and chlorotic, and curl lengthwise and upwards. The twigs start drying up from tip downwards. Honeycombing, a fine pitting of the inner face of the bark in susceptible rootstock portion below the bud union, is also a common symptom in grapefruit and acid lime, which are highly scriptable to this disease. Stem pitting is quite common in the susceptible species. In acid lime, another characteristic symptom is vein clearing in young leaves. Where rootstocks are susceptible and the scion tissues allow virus to multiply, the bud union is injured, phloem sieve tubes undergo necrosis and carbohydrate transport from top to the roots is interrupted, resulting in the starvation of roots. Most of the feeder roots are killed, as a result, water cannot reach the top and trees start wilting. Off-season flowering or heavy flowering in the infected trees is observed. Infected trees bear heavily and get exhausted.

Management Practices

Scientists world over largely agree that it is not possible to control the disease once it has infected the trees. However, the following measures have been suggested by them to prevent the spread of this disease in newer plantations:

1. Follow strict quarantine measures at national and international level.
2. Use CTV tolerant rootstocks like, *Poncirus trifoliata*, Cleopatra etc.
3. Follow indexing or shoot tip grafting for raising new planting material.

4. Raise new plantings through apomictic/nucellar seedlings.

5. Keep the population of aphids under control by spraying systemic insecticides like, rogor or metasystox (0.05 per cent).

6. Remove alternate hosts plants like, *Cuscuta reflexa* (dodar) from the vicinity of the citrus orchards.

7. Use cross-protection technique with milder strains of the virus.

Xyloporosis (Cachexia)

Xyloporosis was first noticed in the Philippines in 1954 and later from many other citrus growing countries of the world. It is considered as the second most destructive viral disease of citrus, only next to tristeza in many countries, but in India, it rarely occurs, perhaps because, susceptible rootstocks like, sweet lime and Orlando tangelo are not commercially used. The virus causes the development of peg-like outgrowths from the inner face of the bark that fits into corresponding pit like depressions in the wood. Gum pockets also sometimes develop in the bark. The leaves of the affected plants become smaller, chlorotic and drop-down pre-maturely. The plants become unthrifty and unproductive. In highly susceptible species, there is development of buttonholes. The virus is bud transmissible and no insect vector is recognized.

Management practices

Following measures have been recognized quite useful for the prevention of this disease in citrus orchards:

1. Destroy the affected plants.

2. New plantings should be raised from disease free lines selected through indexing and nucellar selection.

3. Use tolerant rootstocks like, Cleopatra mandarin, Trifoliate orange, rough lemon etc.

Exocortis (Scaly butt)

The name exocortis was first given in 1948 to a 'bark shelling' disorder of Poncirus trifoliata rootstock in California. It occurs in all the citrus growing countries of the world, and in India, it is considered next to tristeza in its destructiveness. The virus is bud mechanically transmitted, but no vector is known. It is carried symptom less in sweet orange, mandarin, grape fruit and rough lemon. The virus causes distinct cracking and bark scaling symptoms in susceptible species like, Trifoliate orange and its hybrids, like citranges. Scaling is characterized by narrow vertical strips of the outer portion of the bark commonly appear on rootstock at bud anion, which gradually extends upwards the union and down to larger roots. Bark becomes dead and dries. The infected plants on susceptible rootstocks are stunted and may die later.

Management Practices

Pomologists feel that it is very difficult to control this disease once it has infected the orchard. However, the following measures are important to prevent its entry to other locality or orchards:

1. Remove and destroy the unproductive and infected trees.
2. Use disease free, indexed planting material for establishing new orchards.
3. Use tolerant rootstocks like, Cleopatoa mandarin, rough lemon etc., for raising new plants.
4. Always disinfect the budding knives, pruning shears etc., before their use in the different operations.

Psorosis

Psorosis is also called as 'scaly bark', and is considered as one of the most destructive viral diseases of the world. Many strains of psorosis occur in different ports of the world, hence symptoms also vary accordingly, but the characteristic symptoms appear on the bark of the trunk, limbs, branches and large twigs. In general, the infected plants show leaf flecking in young leaves, and certain strains cause scaling of bark (Psorosis A and B). The crinkling leaf disease and infectious variegation are mainly the symptoms is lemons. There may be development of oak leaf patterns on young leaves and concave cavities on trunk or twigs. Affected wood is usually impregmented with gum. Limbs often get twisted and irregular in shape. Usually, sour orange, sweet orange and Maxican lime are its important indicator plants.

Management Practices

Following measures have been recommended by the scientists for the prevention of this disease in newer locations/orchards:

1. Remove the affected plants.
2. Avoid mechanical injury to the plants.
3. Raise new nursery stock from indexed plants after the use of indicator plants or from nucellar seedlings.
4. If possible, use resistant rootstocks.
5. Scrap the diseased bark of the plants and disinfest with some chemical.

Woody Gall

Woody gall is also called as 'Vein enation'. Though, it is not a serious disease, but can cause some losses in certain parts of India. The virus mainly infects Rangpur lime, West Indian lime, acid lime, gajanimma, Sathgudi and jambhiri plants in India and results in the development of small swellings on branches, trunks and roots. These swellings increase in size, coalesce to form cauliflower like surface. On leaves, wedge like warts appear on lower surface of the main lateral veins and small indentations appear opposite to these structures on the upper surface of leaves. The disease is transmitted either by grafting or by aphid vector (*Toxoptera citricida, Aphis gossypii* and *Myzus persicae*).

Management Practices

The following measures are absolutely necessary for the prevention of this disease in newer areas:

1. Remove and destroy the affected plants.
2. Use bud-wood from indexed plants.
3. Keep the aphids under control.
4. Use resistant rootstocks.

Leathery Leaf

Leathery leaf appears in some parts of India, where it causes some losses. Aphis gossypii aphid is considered as its vector. The infected plants produce thickened new leaves. The apical bud on the main trunk or branch is killed, resulting in suppression of growth. The leaves drop prematurely. The infected plants present dieback symptoms during the subsequent years. The inoculated Mosambi seedlings exhibit psorosis like symptoms including leaf flecking in spring flush, which persist for about one week. The thickened leaves curl downwards and fall afterwards. Inoculated plants die within 4-5 years of inoculation. It infects almost all citrus species, except Trifoliate orange.

Management Practices

Scientists believe that the under-mentioned measures are required to be positively undertaken by the orchardists to avoid the spread of this viral disease from one orchard to the other:

1. Uproot the infected plants and destroy them.
2. Use disease free propagation material to avoid further spread of the virus.
3. Destroy the alternate host plants like Gomphrena, beans, *Cucumis sativus*.
4. Keep the vector under control by spraying systemic insecticides like, rogor or metasystox (0.05 per cent).

Other minor viral diseases of citrus are Blastomania, Yellow Corky Vein, fovoid, Tumour and bud union crease. Control measures as recommended for other viral diseases would also keep these diseases under control.

Nowadays, a great emphasis has been paid on the production of virus-free plants in citrus. For this, indicator plants play great role. A list of some important indicator plants has been given in Table 3.6.

Bacterial Diseases

Canker

Canker is a widespread disease in all the citrus growing areas of the world which was recognized as a new disease in 1913 in Florida. Earlier it was confused with a fungal disease citrus scab (*Elsinoe fawcetti*). In many countries of the world, severe attack of the disease has resulted in the mass eradication of millions of trees

Table 3.6: Indicator Plants for Viral Indexing in Citrus

Sl.No.	Disease	Indicator Plants	Characteristic Symptoms	Incubation Period
1.	Tristeza	Acid or kagzi lime	Cupping of young leaves, clearing of veins	6-8 weeks or more
2.	Greening	Pineapple and Valencia sweet oranges; grapefruit and Orlando tangelo	Reduction in leaf size and internodal distance, yellowing of leaves with green islands	3 months or more
3.	Psorosis			
	Crinkly leaf	Coorg mandarin, sweet orange, grape fruit, and Rangpur lime	Crinkling and puckering of leaves on one or more branches	6 weeks or more
	Infectious variegation	Lisbon lemon	Vein-flecking, mottling, distortion and variegation of leaves	Not known
	Psorosis A/B	Sweet orange	Vein banding of leaves with chlorotic patches containing greasy spot-like swellings on the under surface, bark-scaling, discolouration of wood and deposition of gum in the xylem	Not known
4.	Exocortis	Rangpur lime and Tifoliate orange	Splitting and scaling of bark	One or more years
5.	Xyloporosis	Orlando tangelo and Palestine sweet lime	Stem pitting with corresponding pegs on inner surface of the bark with gumming	6-36 months
6.	Vein enation	Acid lime and rough lemon	Enation of under surface of veins with corresponding concavities on the upper surface; woody galls on trunks, branches typical mosaic on leaves	4-6 months
6.	Mosaic	Sweet oranges	Typical mosaic on leave	3-4 months
7.	Leaf curl	Sweet oranges and mandarins	Curling of leaves, gum and grooves in the wood, die-back of branches	2-5 months

and nursery stock. The campaign to eradicate citrus canker in southern states of America began in 1915 and the disease was declared to be eradicated from these areas by 1947. In many countries, like Australia, New Zealand, South Africa and the USA, it was completely eradicated, but it again reappeared in early eighties. In India, citrus canker was first reported from Punjab in 1940 and now the disease is known to occur in almost all citrus growing areas of the country. The disease causes heavy losses when the infection occurs at early stages of plant growth. The fruits crack or become malformed as they grow, and the heavily infected ones fall prematurely. Severe foliage infection often causes defoliation, leaving only the bare twigs leading to almost complete loss.

Symptoms

The bacterium infects all the aerial parts of plants and hence the symptoms appear on leaves, twigs, thorns, older branches and fruits as necrotic brown spots with rough, coarse raised surface. Lesions first appear as small, round watery translucent spots (1-2 mm in dia) on lower surface of the leaf and then on upper surface. As the disease advances, spots become white or greyish and give a rough, corky and crater like appearance. The lesions are surrounded by watery yellowish halo. It may cause premature defoliation of the leaves. The cankers are irregular, rough and more prominent on twigs and branches. Cankers on the twigs are elongated and may cause them to break. On larger branches, the cankers are irregular, rough and more prominent. Cankers on fruits are similar to those on leaves, except that the yellow halo is absent and crater like depression in the centre is more pronounced and injury to the fruit is only in the peel and not on juice sacs. The lesions on fruit remain confined to the rind, but sometimes, it causes cracks and fissures on fruit peel. The presence of a large number of lesions on the fruit surface may result in small and misshapen fruits especially when the infection is early.

Causal Organism

The pathogen probably originated in the tropical areas of Asia such as South China, Indonesia and India where citrus species are presumed to have originated. The bacterium causing citrus canker was earlier known as *Xanthomonas campestris*. However, now, it has been reclassified as *Xanthomonas axonopodis* pv. *citri*. It infects all citrus species, but is severe on acid limes. The tunnels caused by leaf miner serve as foci for the development of cankerous symptoms. Three distinct forms (A, B and C) of the bacterium have been reported to exist on the basis of geographical distribution and host range of the pathogen. The citrus canker bacterium known as Asiatic canker or true canker is widely distributed and has a broad host range. The bacterium is gram negative and straight rod measuring 1.5-2.0 x 0.5-0.75 µm having a polar flagellum.

Management Practices

1. Strict quarantine measures are practiced to exclude the pathogen. All efforts must be made to eradicate the canker bacterium from infested areas.
2. Uproot the affected plants and destroy them.

3. Use disease free nursery stock for the establishment of new orchards.

4. Spray Bordeaux mixture (1 per cent) before monsoon.

5. Proper care should be taken for the control the leaf winner. For this, systemic insecticide like rogor or metasystox (0.05 per cent) is quite effective.

6. Streptomycin sulphate (500 ppm) or phytomycin (2,500 ppm) sprayed at fortnightly interval can effectively check the disease. Bordeaux mixture or copper oxychloride, streptocycline and neem cake in combination with pruning during winter, budding stage and after petal fall are quite effective for controlling canker.

7. Some bioagents like, *Pseudomonas fluorescens, Bacillus subtilis* and *Aspergilus terreus* have shown strong antagonistic effect on *X. axonopodisi* pv. *citri*. Though, these agents have not yet been used commercially for the control of citrus canker.

Greening Disease

Citrus greening is a common disease in many citrus growing countries of the world. In India, it is particularly serious in northern states, where it considered more dangerous than tristerza virus. Metabolic alteration induced by greening pathogen in the host has also been reported. Most important being the synthesis of a phenolic fluorescent substance gentisoyl-β-D-glucoside (Greening marker substance) which is absent in normal healthy plants.

Symptoms

The symptoms initially start on the leaves as inter veinal chlorosis which resembles zinc deficiency symptoms. Characteristically, green dots or islands appear against yellow background of leaves. A characteristic feature of greening is that the yellow areas are surrounded on one side by the mid-rib and on the other side by lateral veins. The yellowing expands towards the margins and size of the leaves is reduced drastically. Infected trees are sparsely foliated due to extensive die-back of twigs. Fruits usually become smaller in size and lopsided having oblique columella. The rind surface exposed to sun turns yellow even before the maturity of the fruits. The internodes of the branches are shortened, giving a bushy appearance to the branch. Secondary growth is upright, short with small leaves and multiple buds The diseased plants look stunted; they flower earlier and produce smaller fruits of poor quality.

Causal Organism

Citrus greening was earlier considered as a viral disease for a long time. Later, some pathologists claimed that MLOs are responsible for this disease, but at present, the disease is considered to be caused by a phloem-limited fastidious greening bacterium recently identified as *Candidatus liberobacter asiaticum* and *Libreobacter african* for African citrus greening. Citrus psylla acts as vector for greening disease.

Management Practices

1. Uproot and destroy the infected plants.

2. Raise new plants from certified disease free indexed rootstocks or from nucellar seedlings.

3. Citrus psylla should be kept under control by spraying systemic insecticides like rogor or metasystox (0.05 per cent).

4. Dipping bud sticks for one hour in 500 µg/ml ledermycin or penicillin extract in combination with 500 µg/ml bavistin before budding is also effective.

B. Major Insect-Pests and their Management

Although, about 250 different insect-pests attack citrus, but only a few are of major importance that cause regular heavy losses and require immediate management. Among them, citrus psylla, black and white fly, leaf miner, scale insects, aphids, lemon butter fly, bark eating caterpillar, fruit flies, mites and fruit sucking moths damage citrus considerably and may pose a serious threat and their management strategies to the very existence of citrus industry. A brief description about these insect-pests has been given below:

Citrus Psylla (*Diaphorina citri*)

Citrus psylla is a serious pest of citrus, widely distributed throughout the oriental regions. The pest is active from March-to-November. During the extreme weather conditions, only adults are found in the field.

Damage is caused both by nymphs and adults, which in large numbers, suck the cell sap from terminal shoots, flower buds and leaves. As a result of severe attack, tender terminal shoots, leaves, buds etc., wilt and die. The vitality of plants deteriorates and the plant growth stops. The fruits in the affected plants may be smaller in size with less juice contents. Heavy deblossoming particularly during spring leads to extensive loss of fruit set. Black sooty mould develops on the sweet fluid (honeydew) secreted by the nymphs, lowering photosynthetic activity of the levaves. Severe infestation may result in complete crop failure. Citrus psylla also acts as vector greening, a serious bacterial disease of citrus, responsible for its excessive decline. Sometimes, this pest also injects toxins along with saliva in the plants, as a result, the leaf buds, flower buds and leaves may wilt and die.

Management Practices

1. Trees should be planted in wider spacing (6.0m × 6.0m or 6.0m × 4.5m) to minimize the infestation of this pest.

2. Apply optimum doses of fertilizers in citrus orchards as higher doses of NPK lead to higher infestation of psylla.

3. The period for control measures identified during the flushing period are: i) Ambia (January-February) second fort-night of February and first fort-night of March ii) Mrig (June-August) whole of July, and iii) Hasta (October-December) whole of November.

4. As soon as few adults are seen on the underside of foliage, citrus tree may be sprayed with any of the systematic insecticides such as rogor (0.05 per

cent), metasystoze (0.03 per cent) and phosphamidon (0.03 per cent) to effectively suppress the psylla population. Two or three sprays of these insecticides may be given at 10-15 days interval.

5. Spray of neem (azadirachtin 0.03 per cent) or neem oil (0.5 per cent) is quite effective against this pest.

6. Conserve natural enemies like *Coccinella septumpunctata* by avoiding sprays of harmful insecticides.

Citrus Leaf Miner (*Phyllocnistis citrella*)

Citrus leaf miner is virtually an omnipresent and oligophagous pest appearing wherever citrus is grown. The pest remains active from March to November with its peaks during April-May and September-to-October. The legless larvae feed on the epidermis of tender leaves making serpentine mines, which are characteristically silvery in colour due to entrapped air. It prefers ventral leaf surface than dorsal. The damaged leaves (mined leaves) turn pale, curl up and may dry. The larvae even mine the tender shoots and in rare cases stems and fruits too. The damage slowly spreads to the fresh leaves. The affected plants lose their vigour and there may be a considerable reduction in their yield. Heavily attacked citrus plants can be spotted from a distance by the curled and twisted leaves. In case of nursery plants, it leads to complete loss of rootstock. There are reports that citrus leaf miner infestation encourages the development of citrus canker.

Management Practices

1. Avoid pruning/injury to twigs during active period of growth.

2. Avoid frequent irrigations and split doses of nitrogenous fertilizers, as it favours luxuriant growth, which provides better food for the pest.

3. Collect and destroy infested leaves.

4. Spray fenvalerate (1.0 ml) or cypermethrin (2.0 ml) or monocrotophos/quinalphos (1.25 ml) per litre of water at fortnightly intervals. Follow ET of 30 per cent leaves with mines or 30 per cent infested shoots or 0.74 larvae/leaf.

5. Spray of neem seed extract (2 per cent) @ 1 kg/10 litre is quite effective.

6. Spray nursery plants with fenvalerate 20 EC (5 ml) or cypermethrin 10 EC (10 ml) or decamethrin 2.8 EC (35 ml) or monocrotophos 36 EC (15 ml) per 10 litre of water at fortnightly intervals.

7. Use predators like *Mallada boninensis*.

Lemon Butter Fly (*Papilio demoleus*)

Lemon butterfly (*Papilio demoleus* Linn.) is a major pest of citrus. Among various insect-pests infesting citrus, the damage caused by *P. demoleus* ranks next to citrus leaf miner. Damage is most severe during April-May and August-October. Only caterpillars cause the damage. The larvae, which look like the bird droppings, are voracious feeders and feed on fresh leaf growth and terminal shoots. The larvae are found on the dorsal surface of the leaves, where they feed on leaves by biting and gnawing the leaf lamina from the margin inwards to the mid-rib thereby sketonize the leaf. Later, the mature larvae may even feed on mature leaves and may completely defoliate the entire plant and the heavily attacked plants may not bear any fruit. The pest is devastating in nursery and its defoliation synchronizes with fresh growth of citrus plants.

Management Practices

1. It is the best way to hand pick the caterpillars at various stages of their development and destroy them, because it is very easy to pick up the caterpillars from the young plantations and nursery plants.

2. Yellow wasp, preying mantids and spiders prey upon caterpillars, hence encourage them in citrum orchard.

3. Conserve the natural enemies by avoiding the spray of broad spectrum insecticides.

4. Microbial pesticides like *Bacillus thuringinensis* (Dipel) @ 0.05 per cent against early larval stages and nematode DD-136 strain effectively check the pest population. Neem seed kernel extract (5 per cent) is also effective.

5. In case of severe infestation, spray carbaryl (0.1 per cent), endosulfan (0.1 per cent) at the time of emergence of new flushes.

Citrus White Fly (*Dialeurodes citri*)

Many species of white flies attack citrus, but *Dialeurodes citri* is the most common and destructive which causes considerable damage and is widely distributed in the world. Both nymphs and adults cause the damage, which suck cell sap from the tender leaves thereby reducing the plant vigour. Infested foliage turns pale, curl up and even shed. The nymphs also produce copious quantity of honeydew on which sooty mould develops, covering the foliage with superficial black coating and interfering with photosynthetic activity of the plant. Thus, the growth of the tree is stunted and the affected tree produce few blossoms, most of which are shed. There is poor fruit set, immature fruit drop and also the fruit become insipid.

Management Practices

1. Avoid close planting and water logging in the orchard.

2. In case of localized infestation, the affected shoots can be clipped-off and destroyed.

3. Destroy alternative hosts, particularly the plants of Solanaceae family.

4. The hymenopteran parasitoids like *Encarsia lahorensis, E. citrifolia, Eretmocerus serius* and predators like *Brumus sutularis, Cryptognatha flavescens, Verania cardoni,* lace wings, *Chrysoperla* spp. feed on eggs and nymphs of this pest and must be encouraged in the orchard.

5. Spray monocrotophos or phosalone (0.05 per cent) after egg hatching.

Citrus Blackfly (*Aleurocanthus woglumi*)

Citrus blackfly causes serious damage to citrus plants in central, northern and north-eastern parts of India. Both nymphs and adults cause damage by sucking cell sap from the tender leaves, shoots, flowers and developing fruits. They also secrete voluminous honeydew on which sooty moulds develop, which leads to black layer (called as *Kolshi* in central India) covering entire plants, affecting photosynthetic activity adversely. The plants are devitalized due to excessive de-sapping, which may affect the bearing capacity of the tree. Due to its feeding, organic N content is reduced to less than 2.2 per cent, which is the minimum prerequisite for successful fruit set and thus adversely affect the fruit set. Fruits become insipid in taste.

Management Practices

1. Avoid close planting and water logging in the orchard.
2. Prune and destroy the affected twigs or other plant parts.
3. Avoid excessive irrigation, nitrogen or drought conditions.
4. Destroy alternate hosts, particularly the plants of Solanaceae family.
5. Parasitoids like *Encarsia bennetti*, *E. opulenta* and *Eretmocerus gunturiensis* @ 2,000 adults/tree reduce blackfly population.
6. Field releases of predators like *Mallada boninensis* and *Serangium parcesetosum* @ 25-30 larvae/tree cause about 20 per cent reduction in population of ths pest.
7. Follow spray schedule as suggested for white fly.

Scales (*Aonidiella aurantii*)

More than 50 species of scales have been recorded feeding on citrus. Among them, economically important are *Coccus hesperidus*, *C. viridis*, *Aonidiella aurantii*, *A. citrina*, *A. orientalis* and *Icerya purchasii*. The scales can be classified based on their body structure as armoured scales having hard cover over the body (*Aonidiella* spp.) and soft scales (*Coccus* spp.) having no such separate cover but some may have a hard skin or a protective waxy secretion. Red scale, *A. aurantii* is the most common and thrives best in semi-arid climate. It is polyphagous in nature and feeds on stem, branches, leaves and fruits and suck their sap. In the early stage, the plants lose their vigour and become stunted and gradually dry up. The attacked branches soon loose chlorophyll, turn scurfy and start drying. While feeding, they inject toxic substances into the plant sap, resulting in yellow spots on leaves, twigs or fruits. In case of severe infestation, all the leaves turn pale. The marketing value of scale-infested fruits becomes very low.

Management Practices

1. Prune the affected plant parts and destroy them. Expose the tree centre to sunlight.
2. *Aphytis mytilaspidis* (parasitoid) and *Chilocorus nigrita* (predator) are major natural enemies. Field releases of *A. mytilaspidis* and *C. nigrita* @15 adults/ tree show good activity on different scale insects.

3. *Rodolia cardinalis* can effectively check *I. purcharii*.

4. Fungus *Nectria* sp. keeps purple scale under check in South India.

5. The characteristic sap sucking of scale *i.e.* from parenchyma tissues render the insect insensitive to systematic insecticides.

6. Good control of red scale can be obtained by spraying methyl parathion (0.05 per cent), dichlorvos (0.04 per cent) or chlorpyriphos (0.08 per cent) or monocrotophos (0.04 per cent).

7. Spray of pongamia oil (1 per cent) or NSKE (4 per cent) is also quite effective.

Mealy Bug (*Planoccocus citri* Risso)

P. citri is a serious pest of many fruit trees including citrus. The activity of this pest is higher from April-to-June on grown-up trees. The nymphs and adult females inflict injury by extracting juice from the cells of leaves, tender branches, fruits (at the base near the fruit stalk and in some cases roots as well). Clusters of white formation of bugs are found at the joints of twigs. The affected plants become pale, wither and may die also. The fruits fall-off prematurely and twisted rosetting of young leaves takes place. A high level of decay caused by *Alternaria citri* has been reported in fruits infested with mealy bugs.

Management Practices

1. Orchard sanitation is important to keep this pest below damaging levels.

2. The infested shoots should be pruned and destroyed.

3. The ant colonies should be destroyed by ploughing around tress and by application of quinalphos or carbaryl dust.

4. Predator, *Cryptolaemus montrouzieri* and parasitoid, *Leptomastix dactylopii* are major natural enemies. Field release of *C. montorouzieri* @ 20 adults/ plant is very effective.

5. The pest can be checked by spraying (0.05 per cent) dimethoate or (0.03 per cent) chlorpyriphos.

6. Spraying of dimethoate (150 ml) + kerosene oil (250 ml) in 100 litres of water or carbaryl (10g) + kerosene oil (10 ml) in 10 litres of water checks mealy bugs effectively.

Citrus Aphid (*Toxoptera* spp.)

About 25 aphid species have been reported on citrus around the world. Of these, species found in India includes brown citrus aphid (*Toxoptera citricida* Kirkaldy), black citrus aphid (*T. aurantii*, citrus aphid (*Aphis citricola*), cotton aphid (*Aphis gossypii*), green apple aphid (*A. pomi*) and green peach aphid (*Myzus persicae*). *Toxoptera* spp. are universally present on citrus and occasionally severe outbreaks occur, especially in dry weather following rainy season. *T. aurantii* is a specific pest of citrus and is active from mid-February to mid-April. *M. persicae* and *A. gossypii* are polyphagous in nature and both are active during September-October. *T. citricida* is also found in South India and cause damage during *Ambia* season in February-March.

Both nymphs and adults infest and suck sap from lower surface of the leaves, tender shoots and young fruits, resulting in curling of infested twigs and leaves. Blossoms and newly set fruits are also attacked. Aphids also secrete honeydew on which sooty mould develops. Heavily infested plants may suffer from loss of sap, thereby reducing the yield significantly. Citrus tristiza virus (CTV), a phloem limited virus is transmitted semi-persistently by *T. citricida*. Woody gall virus is also reported to be transmitted by *T. citricola* and *M. persicae*.

Management Practices

1. Predators like coccinellids, chrysopids, syrphids and spiders are found to regulate aphid population and these must be encouraged in citrus orchards.
2. Parasitoid *viz. Lipolexis scutellaris, Aphelinus gossypii* and *Lysephlebus japonicus* are also found to parasitize aphids.
3. Spray dimethoate (0.03 per cent), malathion (0.05 per cent), methyl demeton (0.025 per cent) in the second week of March and again in the first week of September.

Bark Eating Caterpillar (*Indarbela quadrinotata*)

It is mainly the pest of mandarin and sweet orange trees especially in central India, Andhra Pradesh, Punjab and also in North East hilly regions. The caterpillars cause damage during night time by feeding on bark and destroying the translocating tissues of the bark. Their presence can be noticed by the presence of hanging loose mass of the fine pieces of wood and pellets of excreta, mixed with silky adhesive material on the branches and stem of the tree. They eat through the bark into the wood. In severe infestation, the whole tree is girdled; resulting in stunted growth and lastly the tree may die.

Management Practices

1. Cut and destroy the affected branches/trees.
2. Keep the orchard neat and clean.
3. Remove or treat the alternate hosts in the vicinity of orchard with suitable insecticides.
4. Remove the frass from the holes and inject kerosene/petrol or chloroform or carbon disulphide + creosote (2:10) and then plug the holes with mud.
5. Systemic insecticides like monocrotophos (0.02 per cent) should be sprayed once during September-October and again during January-February.
6. Protect natural enemies like *Syrphus* sp. and *Coccinella septempunctata* for keeping the population of caterpillars under control.

Fruit Sucking Moths [*Ophideres* (=*Othreis*) *fullonica* Linn. *O.materna*, and *Achoea janata* L.]

Unlike the other lepidopterous pests, fruit sucking moths are the adults, which cause damage. Moths are nocturnal in habit and about 20-40 per cent ready to harvest fruits are damaged in central India. The adult moths do the damage and the caterpillars are leaf defoliators and are generally found on the other wild host plants. With the help of its strong piercing mouthparts, moth punctures the fruits for sucking juice and makes characteristic pinhole damage in citrus and other fruits. The damaged fruits soon start rotting as the punctured regions are easily infected with bacteria and fungi. The mouth of puncture becomes pale and eventually the whole fruit turns yellow. Infested fruits invariably fall prematurely perhaps due to toxins injected by moths. On pressing, infested fruits give out a jet of juice from the hole. Almost all the fruits are lost, if the attack is severe.

Management Practices

It is very difficult to control fruit sucking moths in the orchards, if their damage has already started. However, the following measures are useful to avoid their attack:

1. Systematic destruction of alternate host plants like *Tinospora cardifolia*, *Cocculus* sp. in the vicinity of orchards is suggested.

2. Use light traps to attract the moths and kill them in water containing kerosene oil.

3. Destroy the fallen fruits, which attract the moths.

4. Bagging of fruits is effective but very laborious and expensive.

5. Creating smoke in the orchards after sun set may keep the pest at bay but this method is too cumbersome and not feasible on a large scale.

6. Spray trees with 1Kg of carbaryl 50 WP in 500 litters of water per acre at the time of maturity of fruits.

7. Dusting the fruits with methyl parathion (2 per cent dust) during evening also checks the fruit drop by 50 per cent.

8. Kill moths with a bait containing gur 1 Kg + vinegar 60g + water 5 litres. Wide mouthed bottles (1 bottle/10trees) containing bait solution should be tied to the plants when the fruits are in unripe conditions.

Citrus Trunk Borer (*Monohammus versteegi*)

It is a serious pest of mandarins in some pats of India and sometimes may cause 40-50 per cent of tree mortality in West Bengal, Sikkim and Assam. Damage is caused by the caterpillars, which on hatching bore into trunk near the base and feed within the pith, making tunnels inside. The feeding holes disrupt the translocation of food material, resulting in the decline or death of the tree. The affected branches gradually dry up and the leaves of these branches soon wither away. Appearance of resinous exudates and saw dust like powder on tree trunks indicate the presence of the pest in the tree. They also feed on leaf lamina along the mid-ribs leaving the leaf margins intact.

Management Practices

1. Prune the severely infected branches and destroy them.

2. Swab the stem up to 2 m height with carbaryl (1 per cent).

3. Collect and destroy the adults during the emergence in May-June.

4. Clear the holes with a flexible wire and then inject petrol or carbon disulphide or dichlorvos into the bored holes and plug them with mud.

5. The natural infection by fungus, *Beauveria bassiana* is also highly effective in killing the borers.

Citrus Mite (*Eutetranychus orientalis* Klien)

About a dozen species of mites are associated with citrus, but *E. orientalis, E. banksi* and *Tetranychus fijiensis* cause considerable damage in different parts of India. Both nymphs and adults abrade the surface and suck the cell contents from the upper surface of the leaves, bark of the tender shoots and fruits. Mite attack causes unpleasant blemishes on fruit surface which initially become silvery and later turns brownish in color. The '*Mangu*' disorder in Andhra Pradesh is developed due to attack of green and red mites. Citrus rust mite cause oranges to be '*rusted*' in mandarins, grape fruit, lemons and limes to show the silvery of rind called '*shark skin*'. Dark brown colouration of about 40-60 per cent of rind area of maturing '*Mrig*' fruits of Nagpur mandarins during November-January is due to rind feeding by rust mites. It is locally known as '*Lalya*' in central India. The population of this mite reaches at its peak during monsoon period and it remains active from April to October.

Management Practices

1. Water stress often aggravates mite problem. Keep fields well irrigated during summer to avoid stress.

2. Use plant products like neem oil or pongamia oil or mahua oil.

3. Spray dimethoate (0.03 per cent) or dicofol (0.05 per cent) or ethion (0.05 per cent) or oxydemeton methyl (0.03 per cent) or wettable sulphur (3g/L) during February-March.

4. Conserve natural enemies like *Agristemus* spp. *Euseiushibisu* and *Amblyseius hibisci*, which are predominant in citrus orchards.

Fruit Fly (*Bactrocera dorsalis*)

Fruit flies are serious pests of citrus fruits especially in cooler subtropical regions. These are highly polyphagous pests. The adult fly punctures the ripening fruit by penetrating its ovipositor and lays eggs inside the rind. On hatching the maggots bore into the ripening fruit and feed on soft pulp. The infested fruits show depressions with dark greenish punctures, get deformed and due to bacterial and fungal activity, fruits rot and fall down.

Management Practices

1. Use bait traps containing 0.1 per cent methyl eugenol and 0.05 per cent malathion.

2. Regularly remove fallen infested fruits and bury them at least 60 cm deep in the soil.

3. Shallow ploughing after harvest exposes and kills the pupating fruitflies, which are mostly present at 4-6 cm deep in the soil.

4. Spray 2.5 ml fenvalerate (0.05 per cent) per litre of water.

Citrus Thrips (*Scirtothrips citri*)

S. citri is mainly a pest of citrus but also attacks *Acacia* sp. The pest prefers dry and hot conditions from April-to-June. *Heliothrips haemorrhaeodalis* also attacks citrus fruits. The nymphs and adults such the sap from different plant tissues. They rasp and suck plant sap from fully developed flowers, leaf buds and also the young and grown-up fruits. The leaves become cup shaped and leathery. Two white parallel lines on either side to leaf midrib or a whitish silvery ring around the fruit neck or whitish irregular patch similar to the spread of spilled over milk on the skin of fruit are characteristic symptoms of thrips damage. Heavy thrips infestation during summer and fall months results in unacceptable scarring and discoloration of the rind with a consequent downgrading of fruits for markets.

Management Practices

1. Vegetation around the orchard should be destroyed or sprayed with insecticides.

2. Two yellow sticky traps (140 × 76 mm consisting of non-fluorescent yellow polyvinyl chloride) should be placed on sunny side of the tree.

3. Spray of endosulfan or dimethoate or monocrotophos (0.05 per cent) is recommended.

4. Coccinelid predators and chalcidoid parasitoids attack citrus thrips, conserve and augment their population.

Nematodes

There are several nematodes known to attack citrus plant. *Tylenchulus semipenetrans* is the main nematode, which leads to citrus decline. Other nematodes are reniform nematode (*Rotylenchulus reniformis*), burrowing nematode (*Radopholus similis*) and lesion nematode (*Pratylenchus coffee*). Nematode infection lead to reduced plant growth and dysfunction of root system thereby impairing all the growth and developmental processes. Soil application of aldicarb @ 6 kg a.i./ha or 1,000 ppm monocrotophos is effective in controlling nematodes but oilcakes *viz.,* neem, karanj, mahua or mustard have been found very effective.

C. Physiological Disorders and their Management

1. Granulation: Causes and control

Granulation is a physiological disorder of the juice sacs of citrus fruits wherein they become comparatively hard, assume a greyish colour and become somewhat enlarged.

It is a serious problem affecting the marketability of fruits thus reducing the profits of the grower. It was first reported from California in 1934. Later it was reported from many citrus growing countries; such as Brazil, South Africa, Egypt, Australia, Vietnam, Japan. Israel, West Indies, India etc. In India, granulation was reported from Delhi and Abohar.

Analogy of Granulation

Bartholomew *et al.* (1937) coined the word 'granulation', to, which they also gave a scientific term 'Selerocystosis'. The disorder is called with different names in different places. In Florida it is called, 'dry end' and in California it is called 'Crystallization'. In Thailand it is named 'Koa Sarn' and in West Indies 'Corkiness'. However, the term crystallization is misleading, of the possible inference that crystals are present, which: is not true. Similarly the term 'dryend' is also misleading because even though no juice was extracted from the affected vesicles, they contained more water per unit dry matter than the' healthy vesicles from the same fruit.

Granulation was often confused with 'blossom end granulation' or 'dry sac' (Xerocystosis) which is also a physiological disorder. Bartholomew *et al.* (1940) reported that xerocystosis occurs primarily in the distal half of the fruit and the juice sacs of xerocystosis affected fruits unlike those of granulated fruits never become enlarged or hardened and never lose all of their colour. Further they lose their water, collapse and become more or less needle-shaped or flattened. Xerocystosis may occur in either granulated or non-granulated fruits and does not usually appear until the latter part of the picking season, thus suggesting that it may be a form of senility.

Physico-chemical Characteristics of Granulated Fruits

Due to granulation, juice sacs become gelatinous and enlarged and the stem end of the fruit will be firmer than that of the normal fruit, but when juice had disintegrated they become less firmer than normal ones. Fruit volume is increased and the juice sacs develop hard thickened cell walls. Cell enlargement of the juice sacs is due to the enlargement of cells, but not due to cell multiplication, while hardening is due to gelatinization of the cell and signification of cell walls. Softening and collapse of the juice sacs is due to progressive disintegration of many 'of the internal cells. Granulation increases the density of the pulp.

Granulated juice vesicles lack in colour. Loss of carotene largely accounted for the lack of colour in granulated vesicles. In addition granulated fruits contain higher percentage of moisture, inorganic matter and pectin, but lower acidity and sugar concentrations than healthy fruits. Granulated juice sacs contain excess of calcium, magnesium, sodium and potassium on dry weight basis. It is also reported that there is 50 per cent reduction in sugars, 70 per cent in acids and doubling of pectin substances in granulated fruits. The soluble carbohydrates and organic acids decrease with increased mineral constituents and pH.

Seed number decreases with increase in intensity of granulation in sweet oranges. In the initial stages there is loss of normal shape and development of wrinkled surface followed by shrivelling of seeds with advanced granulation. On the basis of accumulated information, it is difficult at this stage to answer the question whether granulation affects the seed development or seed development affects the granulation.

Granulated plants are reported to be deficient in zinc as a result of which degeneration of seeds may take place.

Stages of Granulation

Turrell and Bartholomew (1934) described three developmental stages of granulation in the juice vesicles.

- ☆ **Stage I:** A normal juice vesicle, which is slightly enlarged near the stalk end and tapering off towards the free end, becomes harder and enlarged and contains a viscous jelly-like substance in the place of juice. Granulated juice sacs have a higher percentage of moisture and inorganic matter but a lower total acidity and lower concentration of sugars than healthy juice sacs.

- ☆ **Stage II:** Juice vesicles become very much colourless and give granular appearance. They become hardened and greatly enlarged containing jelly-like substance. With the appearance of numerous thick walled cells, a cavity develops in the centre of the vesicles. Juice sacs in this stage contained approximately twice as much total pectin as those in healthy condition.

- ☆ **Stage III:** In this stage, the enlargement of the central cavity is coincident with the softening of the hardened cells, which causes progressive disintegration of the internal cells. Later on the vesicles soften and collapse. Their outer appearance remains granular, owing to few cells containing bubbles.

Factors Affecting Granulation

Although specific causes of granulation are still obscure, yet several factors have been reported to influence granulation:

- ☆ **Humidity:** Humid climate, particularly in coastal districts favours granulation.

- ☆ **Temperature :** Prevalence of high atmospheric temperature in, the spring season has also been attributed to favour the development of granulation. On the other hand, Bartholomew and his associates observed that low temperature augumented the amount of granulation. Frost injury appears not to be associate with granulation.

- ☆ **Light:** The role of light in granulation differs with the age of the tree: Bartholomew and associates (1934) reported that the incidence of granulation is reduced from 35 to 17 per cent in fruits from 14-year-old trees covered with cheese tents. However, in young trees of six year age no significant differences are found in the extent of granulation in covered and uncovered trees.

- ☆ **Tree age:** The incidence of granulation does not differ between young and old trees, but the extent of granulation is considerably higher in young than in old trees. This may be due to the luxuriant growth of the young tree.

☆ **Tree health:** In case of old trees tree health (declining or healthy) has no bearing on granulation. Whereas in case of young trees, the extent of granulation is found to be higher in declining trees than in healthy trees.

☆ **Tree Vigour:** Bartholomew *et al.* (1934) stated that granulation is abundant in fruits produced on old trees which are heavily pruned and produce luxiriant growth. Cultural operations that induce more-vigorous growth like the application of nitrogen and frequent irrigations accentuated the problem.

☆ **Tree aspect:** There are conflicting reports in the literature with regard to the incidence of granulation on different sides of the tree. Bartholomew *et al.* (1934) reported higher incidence of granulation on the northern side and interior or the trees. Some workers have reported that southern side of the tree had higher incidence of granulation than on other sides.

☆ **Use lime sprays:** Lime spray @ 18 to 20 kg in 450 litres of water controls the incidence of granulation. However, it should be used cautiously as the vitality of tree is lost due to lime sprays during desiccating winds.

☆ **Use of growth regulators:** Sprays of 2,4-D at 12 ppm delay the incidence of granulation in Valencia oranges. However, after the initial delay of the onset of granulation, no further effect of the spray is found.

☆ **Nutritional Sprays:** Zinc and copper sprays are effective in checking both the incidence and extent of granulation. The best results are obtained when the two elements are sprayed in combination.

Physiological Fruit Drop

High fruit drop in citrus is primarily due to physiological reasons. This drop is due to auxin deficiency and sometimes it is very severe. Commonly, this drop is also referred to as 'May-June drop' in sub-tropical regions. The developing fruitlets compete for carbohydrate, water, hormones and other metabolites. Excessive high temperature (35-40°C) during fruit development also causes high fruit drop. Application of 2,4-D or 2,4,5-T (10-30 ppm) during fruit development check fruit drop.

Fruit Splitting

This is a common disorder of lemon and acid lime. The splitting starts at the stylar end and progresses towards the pedicelar end. Splitting is basically caused due to factors like soil moisture, stress followed by heavy rains or irrigation, atmospheric temperature and relative humidity. Borax (0.2 per cent) spray can check splitting and timely application of irrigation water must be ensured.

HARVESTING AND YIELD

Citrus fruits are important component of dessert and fruit processing industry. Citrus fruits are typical, non-climacteric fruit and thus lack respiration peak. The fruits are poor in starch content show slow changes in internal quality. The quality of the fruits at harvest is the resultant of several factors *viz.*, cultivar, climate, soil type, rootstock, cultural practices, etc.

Harvesting

Time of harvest in citrus vary with region (tropical and subtropical) and the species. Marketable maturity is generally adjudged with the change in rind colour. Commercially, TSS: acid ratio is the most reliable method and it ranges from 10 to 16:1 depending upon citrus species and flush. 'Khasi' mandarin in Northern-Eastern states is harvested during October-January, while 'Darjeeling' mandarin is harvested during November-December. 'Kinnow' in Punjab is harvested during January-February, 'Nagpur' mandarin during April-July, October-January in Coorg, January-February in Karnataka. 'Malta' orange is harvested during January-February in Punjab. 'Mosambi' is harvested in April-June (1st crop) in Maharashtra while pickings are also made during July and October-December. 'Sathgudi' in Tamil Nadu is maturing during two seasons *i.e.* July (1st crop) and October-November (2nd crop). Grapefruit in northern states ripens from January-March while, in central India, it matures during September to October. Pummelo in N.E. states is available during October-December. Limes and lemons have no fixed harvesting seasons. Hand picking is the most popular method to collect fruits while in some regions, harvesting by shaking of main trunk is also done.

Yield

Plant yield depends on several factors such as the age of plant, soil type, and management practices. Maximum productivity in citrus ranges from 700 to 1,000 fruits per tree and depends upon the cultivar, rootstock and species. Kinnow plant yield 300 to 800 fruits/plant, mandarin gives 500 fruits. Sathgudi yields 500 to 800 fruit/plant, sweet orange 500-600 fruits/plant. Kagzi lime give 1,500 to 2,000 fruits/plant in West Bengal and to 1,000-1,200 fruits in southern India. Lemon yields 600-800 fruits while rough lemon gives 1,000 to 1,500 fruits per plant.

POSTHARVEST MANAGEMENT

Ripening

Citrus fruits are no-climacteric and hence, tree-ripened fruits attain desirable quality and not after harvesting. Hence, chemical ripening in citrus fruits is not practiced. Although, pre-harvest application of ethephon improves rind colour and it does not influence fruit quality attributes and may cause excessive leaf drop. However, in general, few citrus fruits (*e.g.* sweet orange, Nagrpu mandarin) don't develop attractive rind colour. For such, fruits, degreening with the help of suitable chemicals/growth regulates is desirable.

Pre-harvest spray with 10 ppm 2,4-D + 10 ppm GA and 0.2 per cent KCl on maturing sweet orange can be used for greening sweet orange. This treatment delayed the harvest up to 60 to 75 days. On contrast, spray of ethrel at harvest improves the skin colour *i.e.* uniform colour development. Higher concentration must be avoided as it causes large scale defoliation. Postharvest dip of fruits in 500 ppm ethrel improves rind colour significantly.

Grading

After harvesting, the fruits are immediately graded as per the market specifications. This is more stringent in developed and developing nations exporting fresh fruits. In India, eye grading or band graders are chiefly employed. Agricultural Produce Grading and Marketing Act 1937 has formulated specific grades for citrus based on the fruit diameter *viz.*, Extra-special, special, Good, A, B and C. In Andhra Pradesh, grades designated as I, II and III, depending upon fruit size has to be formulated as listed below:

Fruit	Grades (cm)		
	I	*II*	*III*
Sathgudi	6.5	5-6.5	>5.0
Acid lime	4.5	3.5-4.5	>3.5
Lemon	8.0	6.5-8.0	>6.5

Packing

The graded fruits are packed in wooden boxes. Mandarin and sweet oranges are packed for distant markets. Newspaper wrappings and straw are used as packaging material to avoid damage during transit. Lime and lemons are packed in bamboo baskets or gunny bags. Lower grade oranges and santras are packed in bamboo baskets or gunny bags with rice straw as lining material. Wooden baskets with capacity of 30 kg or 18-20 kg are most popularly used.

However, corrugated fibre board paper boxes are most suitable for packing citrus fruits. Such boxes reduce cost of packing, mechanical injury, loss in weight, decay loss and keeps fruit firmed for longer time. Now-a-days most citrus fruits are packed in card board boxes holding 18-20 kg fruits.

Storage

The cold storage of fresh fruits should be done at 3-8°C with RH ranging from 85-90 per cent. Citrus fruits can be successfully stored for 3 to 4 months. Lemon and limes can be stored for a longer time. Shelf life can be prolonged by coating the fruits with wax emulsion as it reduces fruit desiccation and incidence of diseases. Wax coating (12 per cent) along with captan (0.1 per cent) increased shelf life of Banarsi lemon. Similarly, in Mosambi wax emulsion (8 per cent) with 100 ppm 2,4-D enhanced shelf life to 40 days in comparison to control. Kinnow fruits coated with 6 per cent wax enhanced shelf-life considerably.

Processing

Globally large quantities of citrus fruits are processed into value-added products such as juices, squash, RTS, canned segments, concentrate etc. High quality fruits are used for processing, which are free from blemishes. Fruits with TSS > 12 per cent and TSS : total acid ratio ranging from 14 to 16 and possessing good juice colour are preferred. In India, mandarins, sweet orange, grapefruit, pummelo are consumed fresh, however, oranges are mostly processed for variety of products like canned juice, beverages, concentrates etc. Mandarins, such as, Nagpur, Coorg, Assam, Khasi and Kinnow, Sweet orange varieties, such as, Malta, Malta Blood Red, Mosambi and Sathgudi are used for preparation of squash. Some of the processed products made out of citrus fruits are as under.

Squashes and Cordials

These are the most popular beverages containing 25 per cent juice, 40- 45 per cent soluble solids, 1.2-1.5 per cent acidity along with 350 ppm SO_2 or 600 ppm benzoic acid as preservatives. Clear or clarified beverage is called 'cordial' while 'cloudy' mature fruit juice is called squash. Lime juice cordial is the only popular in India and widely produced for domestic market. During peak harvest season, large quantity of semi-preserved juices are stored in big plastic/stainless steel containers and later on during production are blended with different ingredient, *i.e.* citric acid, sugar, colour, preservatives and anti-oxidants before packaging.

Sweetened Juice

Freshly extracted orange or mandarin juice is canned with or without sweetner at 88-90°C.

Ready-to-Serve (RTS) Beverages

Such beverages possess over 10 per cent juice, 10 per cent soluble solids and 0.5 per cent acidity as citric acid along with 70 ppm SO_2 or 120 ppm benzoic acid, colour and may or may not be carbonated. Pasteurization (65.6°C) is usually done for 30 min to increase the storage life.

Juice Concentrates

These are prepared on a limited scale in India. The extracted juice is sieved (0.63-0.88 mm), deaerated and flash pasteurized and concentrated in a vacuum. The concentration of the juice may range from 40 °Brix (short storage) to 65- 70 °Brix for long storage.

Citrus-Barley Water

It is a type of squash, which contains barley starch. In India, lemon-barley is prepared of 30 per cent TSS with 0.25 per cent barley starch.

Citrus Juice Powder

These are ready to reconstitute dried fruit juice powder, free from clumps with Brix : acid ratio ranging from 12:1 to 18:1.

Canned Segments

Canned orange segment is the most popular product in the West and has great potentiality for export. Malta, Coorg, Sathgudi and Nagpur oranges can ideally be used for canning. The segments are filled in can with syrup of 50 °Brix and some flavour

Jam, Jellies and Marmalade

Jam is prepared from ripe or semi-ripe fruits where fruit pulp and sugar are in the ratio of 45:55. Total solids (68.5 per cent) is usually high to maintain a thick consistency. Jelly is a product made from fruit extract, sugar and pectin in a right proportion. The TSS is 66 °Brix and is clear and free from blemishes. Marmalade are made in citrus with suspended matter unlike jelly. It may be clear (jelly marmalade) or opaque (jam marmalade).

Pickles

Both lemon and limes are popularly used for pickle making, which may be spiced, salty or sweet.

Other Products

There are several other products of citrus, *i.e.* essential oils, pectin, cattle feed are often extracted from the bye-products.

CROP IMPROVEMENT

Floral Biology

Citrus plants are evergreen shrub to small trees growing up to 10m. Citrus species bear bisexual solitary flowers or a small cluster of cymes in the axils of leaves. Flower are hermophrodite, regular, white or yellow, hypogynous. A distinct disc present at the base, usually pentamerous. On average, 8-10 stamens are present and may increase to indefinite number, arranged in polyadelphous condition. Filaments are free, bilobed and basifixed. Gynoecium consists of 3 to 5 carpels, fused to form a syncarpous, placentation-axile, ovary trilocular, each locule contains one or more ovules. Fruit is a modified berry called 'hesperidium'. Flowering takes place mostly in spring and other growing seasons. Lime, lemon and citron flowers throughout the year. Both selfing and cross pollination occur in *Citrus* spp. Stigma is receptive up to 6-8 days after anthesis. Honeybees are the major pollinating insects. Self-incompatibility, male and female fertility cause different degrees of seediness. Fruit set in general is poor and high rate of fruit drop is common.

Citrus Breeding

History

The first systematic work on crop improvement was initiated at the University of Florida, USA, in 1893 by Prof. W. T. Swingle and co-workers to evolve disease resistant genotypes. Wide scale freeze injury to several plantations in USA (1884-1895) gave the impetus for breeding freeze or cold-hardy plant types using *Poncirus trifoliata* (trifoliate orange).

Citrus and the related genera have 2n=18 chromosomes and mostly diploids exist. Low frequency of triploids, tetraploids and hexaploids do exist in natural population, which appear as the result of spontaneous mutation or natural hybridization, miototic errors etc. Citrus breeding is not only difficult but also time consuming and as a result, very few varieties exist, which can be called commercial types. The different breeding objectives in citrus are:

(a) Scion Cultivar

1. To evolve compact, vigorous and cultivar with wide adaptability to different climatic conditions.
2. To evolve cultivars resistant to pests and diseases.
3. To evolve early maturing and granulation-free types.
4. To evolve cultivars bearing superior quality fruits, possessing good processing and keeping quality.

(b) Rootstock Genotypes

1. To evolve rootstocks tolerant to insect-pests and diseases.
2. To evolve rootstocks with freeze hardiness and tolerance to salt and heavy metals.
3. To evolve rootstocks with dwarfing effect suitable for high-density plantings and high productivity.
4. Should have wide compatibility with different scion species and be able to give staggered fruiting, *i.e.* maturity for longer availability of fruits.

Citrus species are highly heterozygous in nature and hence hybridization lead to wide complexity in F_1s with large variations. The different breeding problems associated with citrus are summarized below:

(a) The genotypes are highly heterozygous in nature.
(b) Existence of self and cross-incompatibility.
(c) Both male and female sterility exist in different citrus genotypes.
(d) Seedlessness is met in many commercial cultivars, which restrict their use as female parents.
(e) Polyembryony cause difficulty in breeding as zygotic embryo is often overpowered by vigorous nuclear embryos.
(f) Long breeding cycle, *i.e.*, long juvenile phase leads to slow identification of useful types.

Breeding Techniques

Conventional method, *i.e.* controlled crosses by hand and selecting superior types from seedling population is a time consuming process. Generally, for the scion type the female parent is selected with low degree of polyembryony but possessing high degree of combining ability. The freshly extracted seeds from hand-pollinated fruits are planted in field and allowed to grow until flowering and fruiting. In the initial years, the young plants are evaluated for disease and pest resistance/tolerance, freeze-hardiness and overall nature of canopy growth. After 5 to 6 years, when the seedlings come into bearing, they are evaluated for fruit characters. In rootstock breeding, programme, the young seedlings are exposed to different soil borne pathogens *viz.*, Phytophthora, nematodes etc. for screening. The resistant seedlings are then budded with desired scion types, which come into bearing after 4-6 years and are evaluated for compatibility and stionic effects (fruit quality, cold hardiness, yield, disease and pest resistance. The entire process of releasing a desirable hybrid consumes 15 to 20 years.

Methods of Breeding

Citrus breeding is accomplished using the following methods.

Introduction

This is the most popular method in which the germplasms from different sources are introduced, evaluated and released for commercial cultivation. In India, sweet orange cultivars like 'Washington Navel', 'Valencia', 'Hamlin', 'Pineapple', mandarins like 'Kinnow', 'Cleopatra', 'Clemantine', lemons like 'Lisbon', 'Eureka', 'Villafranca', grapefruit cvs. like 'Marsh Seedless', 'Duncan and Foster' were introduced from USA assessed at different institutions and released for commercial plantings. These varieties are also used for hybridization.

Selection

In this method, survey and collection of germplasm is done and these plantings are evaluated to select a superior type. This method is employed for selecting superior types after thorough screening and judging the performance over years. Some of the superior selections made in the country are as under:

1. **Acid lime :** Vikram and Pramalini were selected at MAU, Rahuri. Similarly, Sai Sarbati from MPKV, Rahuri, Jai Debi from TNAU and Tenali from A.P. are some new promising selections Recently, Chakradhar, a thornless and seedless selection of Kagzi lime has been reported.

2. **Mandarin :** Mudkhed Seedless is a selection from Nagpur mandarin.

3. **Lemon :** Several selections *viz.*, Pant Lemon, Baramasi, PAU selection and Kagzi Kalan have been made.

4. **Sweet orange :** Yuvraj Blood Red is a clonal selection from Blood Red orange from Sri Ganganagar, Rajasthan.

Hybridization

Breeding between two different identities is done to incorporate the desirable characters in the resultant off springs. Hybridization work in citrus was initiated in USA with 3 genera *viz.*, Citrus, Poncirus and Fortunella and several hybrids have evolved as both scion and rootstock types (Tables 3.8 and 3.9).

1. **Inter-varietal cross :** Kinnow (King x Willow Leaf mandarin), Kara (Satsuma x King mandarin) and Wilking (King x Willow Leaf mandarin).

2. **Interspecific :** Tangelo (*C reticulata* x *C. paradisi*), Tangor (*C. reticulata* x *C. sinensis*), Lemonime (*C. limon* x *C. aurantifolia*), Lemonage (*C. limon* x *C. sinensis*), Lemondrins (*C. limon* x *C. reticulata*).

Table 3.7: Some Important Selections of Citrus made at different Research Centres in India

Crop	Selection	From which Selected	Where Selected
Orange	Yuvraj Blood Red	Blood Red	MAU, Parbhani
Mandarin	Madhukar Seedless	Nagpur santra	MAU, Parbhani
Acid lime	Pramalini	Acid lime	MAU, Parbhani
	Vikram	Acid lime	MAU, Parbhani
	Jaydevi (PKM-1)	Acid lime	MAU, Parbhani
	Seedless selection	Acid lime	Himachal Pradesh
	Chakradhar lime	Kagzi lime	Periakulum, (T.N.)
	Tenali	Acid lime	Tirupati
	Saisarbati	Acid lime	Rahuri
Lemon	PAU selection	lemon	PAU Ludhiana
	Dabwali baramasi		PAU Ludhiana
	Pant lemon		Pant Nagar
	Kagzi Kalan		Saharnpur

Table 3.8: Some Important Hybrids of Citrus and their Characteristics Developed at USDA

Crop	Hybrids	Parents	Main Characters
Tangelos	Samson,	Tangerine x Grape fruit	Fruit similar in size to sweet orange
	ThoronWilliams	Do	Vigorous, high resistant to gummosis, used as rootstock
	Orlando	Do	Early variety
	Semenole, Minneola	Do	Late varieties
	Clement	Grape fruit x Clementine hybrid	
Tangelolo	Wekiwa	Samson x grape fruit	
Tangor	Hybrid	Satsuma x Sweet orange	
Citranges	Rusk Savage, Cunningham, Rustic	P trifoliata C. sinensis	Used as rootstock
Citrngors	Hybrids	Citranges x sweet orange (back cross)	Non found promising
Citrangequats	Hybrid	Kumquat x P. trifoliata	Non found promising
	Thomasville	Kumquats x citranges	Suitable for cold climate, used as rootstock
Limequats	Eustis, ILkeland, Tavares	Maxican lime x kumquats	Outstanding, hardy, suitable for cold climate
Glencitrangedin	Hybrid	Lillits citrange x Calamondin	Fruits like Mexican lime free from citrange oil, hardy
Hybrid acid citrus fruits	Perrine lime	Mexican lime x Genoa lemon	Immune to frost and wither tip, hardy, lemon like fruit
Mandarin/orange	Robinson	Clementine xOrlando	Large fruit, deep orange red colour
	Osceola	Do	Better fruit colour than Robinson
	Lee	Do	Fruits like orange in shape, sweet in taste
	Nova	Do	Do
	Page	Minneola x Clementine	Fruits like orange, early maturing, deep coloured
	Bower	Clementine x Orlando	Large and deep coloured fruits
Grape fruits	Hybrid	Tample x Nakon pommelo	Mono embryonic

Table 3.9: Some Important Hybrids of Citrus and their Characteristics Developed at the University of California

Crop	Hybrids	Parents	Main Characters
Mandarins	Kara	Satsuma x King	Good fruit size
	Kinnow	King x Willow leaf	Remarkable acidity, good flavour
	Wilking	King x Willow leaf	Remarkable acidity, good flavour
	Kency	King x Dancy	Good fruit size, good flavour
	Honey mandarin	King x Willow leaf	Rich flavour, small fruit size
Tangelos	Pear tangelo	Imperial x Willow leaf	Out standing
Tangors	Mency Tangor	Maltese oval x Dancy	Vigorous, productive, early
	Ruddy	King x Maltese oval	Large, coloured, acid fruits
Lemon	Hybrid	Lemon x Dancy	High, pleasant aroma
	Hybrid	Lemon x King	High, pleasant aroma
	Owarban	Satsuma x lisbon	Seedless, poor fruit appearance
	Orwan Oreman	Lisbon x Blood orange	Fruit and acidity like lemon, aroma like lemon
Tripoid	Thahiti	Not explained by Frost	Seedless
	Bearss limes	—	
	Orangelo	Maltese oval x Marsh seedy	Too acid fruits
Mandarins	Frua	King x Dancy	Good fruit size, tangerine like flavour, ease peeling, few seeds
	Dweet	Mediterranean sweet orange x Dancy	Pear shaped fruit, flavour like Dancy, juice content very high
	Hybrid B 8247	King x Dancy	Seedless, easy to peel, late ripening
	Hybrid	Clementine x Honey	Red rind colour, deep orange flush, good flavour, good fruit size
	Hybrid	Clementine x Maney	
	Hybrid	Temple x Frua	
	Fair child	Clementine x Orlando	Good quality fruit, early ripening
Pomelos	Chandler	Pommelo2240 x 2241	Early ripening, thin rind

3. **Intergeneric :** Citrange (*Poncirus trifoliata* x *C. sinensis*), Citrumelo (*P. trifoliata* x *C. paradisi*), Limequat (*C. aurantifolia* x *Fortunella japonica*), Citrandarin (*P. trifoliata* x *C. reticulata*), Citrudias (*P. trifoliata* x *C. aurantium*), Citrumquat (*P. trifoliata* x *F. japonica*).

Besides, some bigineric and trigeneric hybrids and complex hybrids with distinct rootstock characteristics have also been developed *viz.,* Citrangequate (*Citrange* x *Fortunella margarita*), Citrangedin (*Citrange* x *C. mitis*), Citrangor (*Citrange* x *C. sinensis*), Citrange (*Citrange* x *P. trifoliata*), Procimequat (*F. japonica* x *C. aurantifolia* x *F. hindsii*), Orangiquat (*C. reticulata* x *F. japonica* x *F. margarita*), Alemow (*Papeda* x *Ecitrus*).

In India, PKV Akola has released acid lime hybrids *viz.,* Hybrid-2, Hybrid-4 and N52 tolerant to canker. At CHES, Chethali, use of trifoliate orange as parent has yielded CRH-3, CRH-5, CRH-41 hybrid tolerant to Phytophthora rot and nematodes.

Somatic Mutation and Bud Sport (Limb Sport)

Isolation of novel sports has contributed significantly to the development of citrus. Some of the popular bud ports grapefruit cultivars are Duncan, White Marsh and Walters, which were identified from original seedling grapefruit from Barbados. Thompson is a selection from White Marsh, which yielded Burgundy, Red Blush and Ruby Red Henderson, Ray Ruby and Rio Red are selections from Ruby Red. Flame is a selection from Henderson. Walters, yielded Foster Pink, which further yielded Hudson. Star Ruby is a selection from Hudson.

In mandarin, several sports of Washington Navel has been isolated *viz.,* Atwood, Fisher, Newhall, Marrs (all in USA), Navelina and Navelate (both in Spain) and Leong from Australia. Shamauti is a mid-season limb sport of Beladi from Israel. Similarly, Baianinba is a limb sport of Bahia from Brazil.

Mutations

Somatic mutation randomly take place in nature but artificial mutations can also be undertaken to create variation in heterozygous crop like citrus. Mutation of budwood and seed has been successfully attempted. Seeds are generally more tolerant (LD_{50} = 0.1-0.5 GY) as compared to budwood (LD_{50} = 0.05-0.09 GY). 'Star Ruby' is an example of desirable mutant with deep fruit colour. Both synthetic and acitogenous chimeras are known to appear in citrus. For example, Bizzarria has been reported to be a graft chimera from citron bud grafted on sour orange seedlings. Similarly, two periclinal graft chimeras *i.e.* 'Kobayashi Mikan' and 'Kinkoji Unshu' were initiated when Satsuma was grafted onto Natsu Daidai.

Autogenous periclinal chimeras like Pink Fleshed Thompson and Foster grapefruits were identified from White Marsh and White Walters grapefruit, respectively. 'Sarah' is a pink sport of 'Shamouti' orange.

Biotechnology

Genetic engineering tools and marker technology is gradually becoming the part of modem citrus breeding programme. Isozymes have been utilized for identifying hybrids from embryos generated from nucellus. For early selection of desirable hybrids, marker aided selection (MAS) is fast gaining use in citrus breeding. Hybrid progenies have been categorized using Random Amplified Fragment Polymorphisms (RAPD), Raplon, (RFLP) mini and micro-satellite etc. to locate gene(s) of economic traits. These techniques elucidate the linkages and inheritance pattern of linked characters and early identification of potential hybrids. Efforts are underway to transfer CTV resistance to sour orange rootstock (Carrizo) through genetic transformation. Internodal segments were exposed to *Agrobacterium tumefaciens* with engineered anti-viral coat protein DNA and the transformed cells are allowed to grow on a selective medium containing antibiotics. The shoots, regenerated are allowed to root and transfer to glasshouse for confirming through bioassays.

Somatic hybridization, protoplasmic and cytoplasmic hybrids have been well demonstrated to overcome sexual incompatibility barriers. The fused protoplasts are allowed to regenerate cell wall and then embryoids are produced which finally regenerate into complete plants. Somatic hybrids are 4x, *i.e.* tetraploid and usually employed for rootstock breeding to yield hybrids from distinctly related and non-related species.

4

Guava

Guava (*Psidium guajava* L.) is one of the commercial fruit crops of India, ranking fifth after mango, banana, citrus and papaya in production; however, in terms of area it ranks ninth. Among guava producing countries, India ranks first in area and production followed by China and Thailand. The highest productivity of 15.8 tones/ ha had been recorded in Brazil. Among Indian states, Maharashtra ranks first both in area as well as production of guava, contributing 12.6 per cent of production share alone. The other important guava producing states are M.P., U.P., Bihar, West Bengal, Punjab, Gujarat and Karnataka, which contribute to 9.8, 9.6, 7.3, 6.9, 6.1 and 5.8 per cent towards total guava production, respectively.

It is a medium tree with prolific-bearing habit, which bears more than once a year. It is also known as the 'apple of tropics' in India. Guava has wide adaptability to varied types of soils, climatic and cultural conditions and its trees are quite hardy and are able to give satisfactory return without much care. It is grown in most of tropical and sub-tropical zones of the world. In India, it has been in cultivation since early 17th century.

ORIGIN, HISTORY AND DISTRIBUTION

Cultivated guava is a native to tropical America, where it occurs wild and it was domesticated more than 2,000 years ago. It has been found growing in Mexico and Peru since ancient times. It was spread at an early date throughout the world's tropics by the Spanish and Portuguese. Early Spanish and Portuguese colonizers were quick to carry it from the New World to the East Indies and Guam. It was soon adopted as a crop in Asia and in warm parts of Africa. Egyptians have grown it for a long time and it may have traveled from Egypt to Palestine. It is occasionally seen in Algeria and on the Mediterranean coast of France. Apparently, it did not arrive in Hawaii until the early 1800's. Now, it occurs throughout the Pacific islands. Generally, it is a home fruit tree or planted in small groves, except in India where it is one of the major commercial fruits. In many parts of the world, the guava grows wild and forms extensive thickets–called 'guayabales' in Spanish–and it overruns pastures, fields and roadsides so vigorously in Hawaii, Malaysia, New Caledonia, Fiji, the U.S., Virgin Islands, Puerto Rico, Cuba and southern Florida that it is classed as a noxious weed subject to eradication. Nevertheless, wild guavas have constituted the bulk of the commercial supply. The wide adaptability nature of guava tree helped it to sustain a wide range of environments and soils. It is cultivated in India, Pakistan, Bangladesh, Philippines, Hawaiian Islands, Myanmar and Cuba. It is grown in several states of India but Maharashtra, Madhya Pradesh, Uttar Pradesh, Bihar, West Bengal, Punjab, Karnataka, Andhra Pradesh and Chhattisgarh are the main producers (Table 4.1). In Uttar Pradesh, Allahabad region has the reputation of growing the best quality guavas.

Table 4.1: State-wise Area, Production and Productivity Scenario of Guava

State	Area (000'Ha)	Production (000'MT)	Productivity (MT/Ha)
Maharashtra	36.0	311.0	8.6
Madhya Pradesh	9.7	280.8	29.0
Uttar Pradesh	14.6	241.4	16.5
Bihar	29.4	235.2	8.0
West Bengal	13.6	178.8	13.2
Punjab	7.8	171.0	21.8
Gujarat	10.2	150.7	14.7
Karnataka	7.3	141.6	19.4
Andhra Pradesh	8.5	128.2	15.0
Chhattisgarh	13.4	102.8	7.7
Odisha	14.2	102.1	7.2
Tamil Nadu	7.6	100.8	13.2
Others	32.4	317.9	9.8
Total	204.8	2,462.3	12.0

Source: National Horticulture Board Database-2011.

COMPOSITION AND USES

Composition

Guava is a very rich source of vitamin C and pectin contents. The contents; however, may vary with cultivar, stage of fruit maturity and season. Generally, pink fleshed cultivars are poorer in vitamin C than the white fleshed. Guava is a good source of both thiamine and riboflavin. Among different minerals, calcium and phosphorus are also present in fairly good amount in guava (Table 4.2).

Table 4.2: Nutrient Composition of Ripe Guava Fruit

Component	Per 100 g Edible Portion	Component	Per 100 g Edible Portion
Calories	36-50	Calcium	9.1-17 mg
Moisture	77-86 g	Phosphorus	17.8-30 mg
Crude Fiber	2.8-5.5 g	Iron	0.30-0.70 mg
Protein	0.9-1.0 g	Carotene (Vitamin A)	200-400 I.U.
Fat	0.1-0.5 g	Thiamine	0.046 mg
Ash	0.43-0.7 g	Riboflavin	0.03-0.04 mg
Carbohydrates	9.5-10 g	Niacin	0.6-1.068 mg

Uses

In India, guavas are usually eaten raw out-of-hand, but are preferred seeded and served sliced as dessert or in salads. However, the fruit is cooked and cooking eliminates the strong odour. A standard dessert throughout Latin America and the Spanish-speaking islands of the West Indies is stewed guava shells (*cascos de guayaba*), that is, guava halves with the central seed pulp removed, strained and added to the shells while cooking to enrich the syrup. The canned product is widely sold and the shells can also be quick-frozen. Sometimes guavas are canned whole or cut in half without seed removal. Being rich in pectin, excellent jelly is made from guavas. The fruits can also be utilized for making jam, juice, cheese, fruit butter, canned segments, cordial, nectar and RTS. Green mature guavas can be utilized as a source of pectin, yielding somewhat more and higher quality pectin than ripe fruits.

Guava wood is valued for engravings in India. The roots, bark, leaves and immature fruits, because of their astringency, are commonly employed to halt gastroenteritis, diarrhea and dysentery, throughout the tropics. Crushed leaves are applied on wounds, ulcers and rheumatic places, and leaves are chewed to relieve toothache. The leaf decoction is taken as a remedy for coughs, throat and chest ailments, gargled to relieve oral ulcers and inflamed gums; and also taken as an emmenagogue and vermifuge, and treatment for leucorrhea. It has been effective in halting vomiting and diarrhea in cholera patients. It is also applied on skin diseases. An extract is given in epilepsy and chorea and a tincture is rubbed on the spine of children in convulsions. A combined decoction of leaves and bark is given to expel the placenta after childbirth.

TAXONOMY AND BOTANICAL DESCRIPTION

Guava belongs to family Myrtaceae with gametic chromosome number (n) = 11. It belongs to genus *Psidium*. The genus *Psidium* contains about 150 species. The Brazilian or Guinea guava (*P. guineense*) bears small fruits of poor quality. The common guava (*P. guajava*) has been classified as *P. pyriferum* (pear shaped fruits) and *P. pomiferum* (round/apple shaped fruits) on the basis of fruit shape. The mountain guava (*P. montanum*) is a shrub. The other important species of *Psidium* are *P. friedrichsthalianum* (Chinese guava) and *P. cattleinum* (Cattley guava).

Common guava is a small tree to 33 ft (10 m) high, with spreading branches. The guava is easy to recognize because of its smooth, thin, copper-coloured bark that flakes off, showing the greenish layer beneath; and also because of the attractive, 'bony' aspect of its trunk which may in time attain a diameter of 10 inch (25 cm). Young twigs are quadrangular and downy. The leaves, aromatic when crushed, are evergreen, opposite, short-petioled, oval or oblong-elliptic, somewhat irregular in outline; $2^{3/4}$ to 6 inch (7-15 cm) long, $1^{1/2}$ to 2 inch (3-5 cm) wide, leathery, with conspicuous parallel veins, and more or less downy on the underside. Faintly fragrant, the white flowers, borne singly or in small clusters in the leaf axils, are 1 inch (2.5 cm) wide, with 4 or 5 white petals which are quickly shed, and a prominent tuft of perhaps 250 white stamens tipped with pale-yellow anthers.

The fruit, exuding a strong, sweet, musky odour when ripe, may be round, ovoid, or pear-shaped, 2 to 4 in (5-10 cm) long, with 4 or 5 protruding floral remnants (sepals) at the apex; and thin, light-yellow skin, frequently blushed with pink. Next to the skin, is a layer of somewhat granular flesh, 1/8 to 1/2 in (3-12.5 mm) thick, white, yellowish, light- or dark-pink, or near-red, juicy, acid, sub-acid, or sweet and flavourful. The central pulp, con-colorous or slightly darker in tone, is juicy and normally filled with very hard, yellowish seeds, 1/8 in (3 min) long, though some rare types have soft, chewable seeds. Actual seed counts have ranged from 112 to 535 but some guavas are seedless or nearly so. Botanically, its fruit is called as 'berry'.

IMPORTANT CULTIVARS

Salient features of some important guava cultivars are described hereunder:

Allahabad Safeda

This is the most popular variety of U.P. The plants of this variety are vigorous with broad and compact crown. The plants have upright growth habit with dense foliage and tendency to produce long shoots. The leaves are 9.5-9.8 cm long, 4.8 cm wide and elliptical to oblong in shape. Fruits are medium, round smooth with an average weight of 180 g. The colour of fruits is yellowish-white and flesh is white. The fruits have only few seeds and their keeping quality is good.

L-49

It is a selection made from Allahabad Safeda at Pune by Cheema and Deshmukh in 1927. It is popularly known as Sardar guava. The plants are semi-vigorous with spreading growth habit and flat crown. The leaves are large (about 13 cm long) and oblong in shape. The fruits are large and roundish ovate in shape. The fruits are primrose-yellow in colour with occasional red dots on the skin. The fruits have few seeds and their keeping quality is excellent. It is somewhat hardier than Allahabad Safeda.

Banarasi Surkha

This is one of sweetest guavas, lacking acidity. The plants are medium to tall with spreading growth habit and broad crown. The leaves are long with obtuse base and pointed apex. The fruits are round in shape, yellow in colour having medium keeping quality.

Chittidar

This is a famous variety of Maharashtra. The plants of this variety are vigorous with spreading branches and rounded crown. The leaves are elliptical in shape with rounded base and pointed apex. The fruits are comparatively smaller, round to ovate in shape, straw-yellow in colour and with a few scattered red dots on the skin. The flesh in soft, white and sweet in taste. Keeping quality of the fruits is good.

Baruipur

One of the commercial varieties of West Bengal. Trees are medium to tall, spreading growth and crown broad. Fruits are round in shape and yellowish in colour with white flesh. Keeping quality is medium.

Pear-shaped

Tall and upright tree, with erect growth. Fruits are pear shaped, yellowish in colour, with smooth surface having large dots. Keeping quality is good.

Apple Colour

This variety is mainly cultivated for coloured fruits. The plants are semi-vigorous with spreading growth and broad crown. The leaves are elliptical in shape. The fruits are smaller, spherical in shape, pink in colour with minute dots on the skin surface. Fruits contain few seeds and have good keeping quality.

Red Fleshed

The trees are vigorous with spreading growth habit and vase like open croons. The leaves are large with 9.8-19.04 cm in length, 4.8-5.2 cm in breadth and elliptic oblong in shape. The fruits are roundish ovate in shape, saffron-yellow in colour with few red dots on the surface. The flesh is dawn pink in colour containing few seeds. Keeping quality is poor to medium.

Seedless

The trees are tall having long trunk and upright branches. Leaves are large (14 cm in length and 7 cm breadth) and oblong in shape. The fruits are oblong to globose in shape and straw yellow in colour containing very few seeds. The keeping quality is poor. This is not a commercial cultivar because of shy bearing.

Behat Coconut

The plants are moderately vigorous having flat crown. The leaves are large, twisted and lanceolate in shape. Fruits are borne in bunches, round ovate in shape, aureolin in colour, with rough surface and dots. Flesh is white and sweet. Keeping quality is medium.

Harijha

This variety is known for profuse bearing. The plants are semi-vigorous and sparsely branched. The leaves are lanceolate in shape with round base and acute apex. The fruits are roundish in shape and greenish-yellow in colour. The flesh is white and sweet. Keeping quality is excellent.

Lalit

Lalit was released by the CISH, Lucknow during 1999 for commercial cultivation. Its fruits are medium sized (185 g) with attractive saffron-yellow colour and red blush. Its flesh is firm and pink with good blend of sugar and acid. It gives 24 per cent higher yield than popular variety 'Allahabad Safeda'. This is suitable for both table and processing purposes.

Shweta

It is a selection from the population of Apple Colour guava and was released by CISH, Lucknow for commercial cultivation. This has globose, medium-sized fruits, creamy white exocarp with red blush, snow white pulp, high TSS (12.5–13.2 per cent) content and vitamin C (300 mg/100 g pulp) with good keeping quality.

TRY(G)1

Selection from elite mother plant, origin unknown. It is a off-season variety with shiny greenish yellow fruits having desirable aroma, high TSS (10° brix). It is a drought and sodicity tolerant and high yielder (40.52 kg/tree) variety. It was released by TNAU, Coimbatore for commercial cultivation.

Arka Mridula

Selection from open pollinated seedlings of 'Allahabad Safeda' at IIHR, Benguluru. Round fruits, skin yellow and smooth, white flesh, 12 °brix TSS, soft seeded, good keeping quality, suitable for jelly making.

SOIL AND CLIMATIC REQUIREMENTS

Soil

Guava is a hardy fruit plant and can thrive on all types of soils, ranging from heavy clay to light soils. However, the clay loam, deep, friable and well-drained soils are best. It can be grown in soils having pH up to 9.0, though the optimum range is between 4.5 to 7.5. Under highly waterlogged, saline and alkaline conditions, its cultivation is adversely affected. Maximum root density is confined up to 20 cm of soil depth. Hence, top soil should be rich in organic matter.

Climate

Guava grows successfully in tropical and sub-tropical regions and even under adverse climatic conditions. However, both yield and fruit quality are better in areas with distinct winter. It grows well even at an altitude of 5,000 feet. The areas receiving a rainfall of 40 inch between June-to-September, are considered ideal for its cultivation. The young plants are susceptible to drought and frost. Dry atmosphere at the time of flowering and fruiting are favourable but high temperature during fruit development may cause excessive fruit drop.

PLANT PROPAGATION AND ROOTSTOCKS

Guava can be propagated both by seed and asexual means as discussed hereunder:

(a) Seed Propagation

Usually rootstocks are raised by seeds. Fully mature seeds of current season are colleted from healthy fruits and are sown either in the nursery and polythene bags. The seedlings raised in polythene bags can be transplanted in the nursery when they are 6 to 8 months old. The seeds loose viability within a short period. Hence, these should be sown immediately after extraction. However, treatment of seed with 1 per cent potassium nitrate and para-hydroxybenzoic acid (10^{-3} M) helps m maintaining seed viability to a great extent. Soaking the seeds in cold water or in GA_3 (1,000 ppm) for 24 hours improves seed germination. Soaking of seeds in concentrated sulphuric acid for 2-3 minutes also results in for good germination. Under favourable conditions, the one-year-old seedlings are ready for transplanting or grafting.

(b) Vegetative Propagation

Vegetative propagation methods practiced in guava are described below:

Cutting

Guava cuttings are hard-to-root but these root better under mist if treated with growth regulators. Dipping of semi-hard and hardwood cuttings with NAA (50 ppm) or IAA (100 ppm) encourages rooting. Leafy cuttings give 100 per cent rooting under intermittent mist. Rooting in the cuttings improves under bottom heat chamber. However, cuttings are not used for commercial propagation of guava.

Air Layering/Gootee

It is one of the commercial propagation methods for guava. For air layering, the shoots of previous season's growth with thickness of 1 cm are selected. A ring of bark about 2.5-3.0 cm is removed from the shoots at the base. This ring is covered with wet sphagnum moss and tied with a polyethylene strip. The rooting takes place in 30-40 days. These shoots are detached and planted in the nursery. For better rooting in the layers, application of IBA paste (2,500 ppm in lanolin paste) to the upper portion of the cut is helpful. The best time for air layering is rainy season. It can also be done in spring but gives low success.

Grafting

Inarching has been the most popular method of guava propagation. This method gives more than 95 per cent success. In this method, the required scion variety has to be brought near to the stock. It is always better to grow scion in a pot. Thus, inarching is cumbersome, laborious and time consuming method. Side veneer grafting technique is also a good technique. For side veneer grafting, green wood scion should be used especially for top working. For veneer grafting, the scion must be green, angular and of one month age. It should be defoliated before the veneer grafting operation. About 5 cm long shoot should be used for grafting. Veneer grafting gives 75 per cent success.

Stooling

Stooling was perfected for guava at IARI, New Delhi. It is used for faster multiplication of clonal rootstocks. This method is now commercially used for guava propagation. In this method, the plants are headed back during December-January and allowed to grow new shoots. These shoots (stools) are ringed, treated with IBA (2,500 ppm) made in lanolin paste and earthed up to induce rooting. Profuse rooting occurs within a month. The rooted stools are separated from the mother plant and planted in the nursery during July-August. Rooting is better if the soil in the stool bed is kept moist all the time. 40-50 plants can be prepared from one mother plant per year by performing stooling twice a year.

Budding

Guava can be propagated by budding but it is not commercially done. For budding, selection of proper rootstock is essential. The thickness of rootstock should be 20-25 cm. The budwood should be selected from one-year-old shoots of healthy and desired scion variety. Different techniques like 'T', forkert, patch and chip methods of budding give varying degree of success. However, forkert method is better. The best time for budding is July-August.

Rootstocks

Like other fruits rootstocks exert strong influence on vigour, fruitfulness, fruit quality and resistance against insect-pests and diseases in guava also. The Chinese guava (*P. friedrichsthalianum*) has dwarf growth habit and is resistant to guava wilt and root-knot nematode (*Meloidogyne incognita*). It can be used for commercial planting in areas affected by wilt and nematodes. Allahabad Safeda variety produces fruits with high sugar contents on *P. pumilum*, high vitamin C content on *P. cujavillus* and very high yield on *P. catlleianum*. Pusa Srijan, has been recommended as a dwarfing rootstock for Allahabad Safeda guava by IARI, New Delhi.

PLANTING AND ORCHARD ESTABLISHMENT

The field should be cleared, deeply ploughed, harrowed and leveled. The pits of 1 m × 1 m × 1 m size are dug before monsoon at an appropriate distance and should be kept exposed to sun for 10-15 days to kill the soil-borne pathogens. Pits are filled with top soil, mixed with 25-30 kg of well-decomposed farmyard manure or compost and insecticides like chloropyriphos. In general, guava is planted in a square or a rectangular system in plains and along the contours in hilly regions. The ideal time for planting is monsoon, *i.e.* July-September. However, it can be planted during spring season, *i.e.* February-March, if irrigation facilities are available.

The planting density depends on factors like variety, rootstock and climatic and edaphic conditions. Conventionally, guava is planted in a square system at a distance of 8 m × 8 m. The distance between trees has strong influence on growth, nutrient uptake, yield and fruit quality. Recent trends in guava cultivation is planting at a closer spacing for obtaining higher yield per unit area. However, regular pruning and thinning are considered as pre-requisites for high-density planting of guava. By increasing the population density, the yield increases manifold but fruit size and quality of Allahabad Safeda has been reported to be affected. IARI, New Delhi has recommended the use of dwarfing Pusa Srijan rootstock for Allahabad Safeda, which drastically reduces height and spread of Allahabad Safeda scions. In this case, a planting distance of 10 × 10 feet (1,111 plants/ha) has been recommended. No adverse effect of high-density planting on fruit size and quality of Allahabad Safeda has been observed. On the contrary, the ascorbic acid and TSS contents of fruits were found to be significantly higher than in those from the stooled plants. High-density system produced 2.5 to 3.0 times higher yield in the 8th years after planting. At MPUAT, Udaipur, Ultra high-density orcharding in guava has been standardized.

INTERCULTURAL OPERATIONS

Irrigation

Guava is a hardy fruit plant and requires very less water. However, in the early stages of orchard establishment, plants require frequent irrigation. Later, more frequent irrigation (fortnightly interval) is required from April-to-June, for good growth and fruit yield. Irrigation during winter is beneficial to obtain quality crop. Due to regular growth, flowering and fruiting in South India, guava requires irrigation throughout the year. During winter season, irrigation may be provided at an interval of 20-25 days, while in the summer months, it may be provided at an interval of 10-15 days by the ring method. Drip irrigation has proved to be very beneficial for guava. About 60 per cent of the water used for irrigation is saved. In addition, substantial increase in size and number of fruits is observed.

Mulching

Use of mulch in guava orchards helps in conserving soil moisture, regulating soil temperature and suppressing weed growth. Mulching can be done either with synthetic materials such as black polyethylene sheet or with organic materials like rice husk, rice straw and water hyacinth (*Eichhernia crassipes*), which have been found to reduce the weed population and increased yield in guava cv. Allahabad Safeda, Banarasi and Seedless in comparison to the non-mulched trees. In general, mulching, irrespective of nature, significantly increases the height and fruit yield compared with the non-mulched trees. The better performance of trees with organic mulches might be due to their combination of plant nutrients and organic matter in soil besides other usual roles.

Manuring and Fertilization

Like other fruits, guava also requires adequate mineral nutrition for proper growth, development and production of higher yield of better quality fruits. The amount of manure and fertilizers to be given, however, depends upon the variety, age of plant, soil and climatic conditions, previous fertilizer use and type of the fertilizer to be applied. Hence, it is difficult to give one general fertilizer schedule, but a ten-year-old guava tree should be given about 80 kg of FYM, 1 kg of ammonium sulphate or 800 g of calcium ammonium nitrate, 3kg of super-phosphate and 2 kg of potassium sulphate.

The fertilizers should be applied in two split doses (June and October) when there is sufficient moisture in the soil. 50 per cent of nitrogen and full dose of potash should be applied in June and the remaining N and entire phosphorus in October. Foliar application of nitrogen in the form of urea (4-6 per cent) has been found to be very effective in increasing growth, flowering, yield and quality of guava. The best time for foliar application is January and July. A fertilizer schedule for guava is given in Table 4.3.

Table 4.3: Manurial and Fertilizer Schedule for Guava Plants in India

Age (Years)	FYM (kg)	Nitrogen (g)	Phosphorus (g)	Potash (g)
1-2	10-15	60	30	30
3	20	120	60	60
4	30	180	90	90
5	40	240	120	120
6	50	300	150	150
7 and above	60	360	180	180

Among various micro-nutrients, guava requires zinc and boron, the most. Deficiency of zinc results in interveinal chlorosis and reduction in leaf and fruit size. Spraying zinc sulphate (0.05 per cent) in June-July and August-September helps in correcting zinc deficiency. In zinc sulphate, at least half of the quantity of slaked lime should also be added before spraying. Addition of boric acid (0.5 per cent) in zinc, spray, cures boron deficiency. Spray of 0.4 per cent copper sulphate has been reported to correct Cu deficiency. In general, pre-flowering sprays of the micro-nutrients are better than post-bloom sprays.

Intercropping

In the early years of orchard establishment, inter-spaces can be effectively utilized for growing suitable intercrops. Intercropping inside guava orchard with vegetables and fodder crops can fetch better return to growers; besides, generating more employment. Several intercrops like cauliflower, peas, French bean and *senji* in rabi season; cowpea, clusterbean, black gram and green gram during kharif season and cucurbits during summer can be grown in guava orchards at pre-bearing stage. In Uttar Pradesh, papaya is commonly grown as a filler crop, while pineapple and strawberry could be other possible crops. Under dry land conditions, stylo cover crop can give an additional yield per unit area. In bearing orchards; however, intercropping is not desirable.

Training and Pruning

Training of guava trees has been found to improve yield and fruit quality. The primary objectives of training are to develop single trunk tree with well-spaced scaffold branches, to form a strong framework and for bearing a heavy crop without damaging the branches. The trees should be kept open for better penetration of sunlight leading to more number of shoots and higher yield.

Open system of training is considered as the most ideal. In this system, the plants are headed back and four primary shoots are retained for basic framework and subsequently pruned by cutting $1/3^{rd}$ to half of their length after 3-4 months. After initial framework, the two side shoots are permitted to grow initially and after 3-4 years, subsequent doubling of selected branches is continued.

Guava trees are also trained in delayed open centre or modified leader system, in which, 2-tiers of framework are prepared and the centre is kept open at a height of 140 cm. Four side-shoots are retained in this system and only one-third of their length is pruned. For obtaining the second tier, tree is headed back at a position close to an outward bud. It has been reported that training reduces the canopy area in open centre and delays open centre thereby giving an opportunity to accommodate more number of plants per unit area.

The training system varies from variety-to-variety and region-to-region. For example, in case of varieties with spreading habit (Sardar), primary branches are allowed at least 75 cm above the ground. The guavas are trained in a peculiar fashion in Maharashtra and Kodur (A.P.). In Maharashtra, the branches are bent downward and tied to each other. The dormant buds are forced to grow, which results in higher yield. In Kodur, guavas are also trained as cordons.

In guava, the flowers and fruits are borne on current season's growth. It is therefore necessary to prune its plants annually. A light annual pruning after fruit harvesting is considered necessary for encouraging new vegetative growth.

The length of flowering shoots tends to decrease with delay in time of pruning but increase with the increasing severity of pruning, irrespective of season. An increase in shoot length due to severity of pruning might be due to elimination of growing points which in turn encourages the length of remaining shoots. Pruning by heading back encourages new, long, whip-like shoot growth with sparse flowering compared with cutting at fork. In prunning all the criss-cross, dried, diseased, over-crowded and suckers should be removed annually.

Severe pruning has an adverse effect on productivity of guava. Removal of terminal 15 and 30 cm of branches adversely affects flower production and yield. Pruning at an intensity of 25 per cent in February could regulate fruit yield without affecting fruit quality under high-density planting. Time of flowering in guava can also be altered by suitable pruning. With the increasing pruning intensity, the rainy season yield decreases and one-leaf-pair pruning proved superior to the other pruning treatments for flower-bud initiation and fruit yield of the winter crop.

FLOWERING, FRUIT SET AND FRUIT GROWTH

Flowering

Guava flowers twice a year in northern India. First flowering takes place in April-May, which gives fruits in rainy season. The second flowering takes place in August-September to give fruiting in winter season. The rainy season crop is generally avoided as most of the fruits are infested by fruitfly and the fruits are insipid and of very poor quality. The winter crop is virtually free from fruitfly and the fruits are of high quality. Winter crop is therefore, preferred as it gives very high returns to the farmers. In central-southern India, guava flowers thrice a year, with flowers appearing in October also. In West Bengal, flowering once in April-May and again in September-October has been reported.

The flowers appear either solitary or in cymes of two or three flowers in the axil of leaves of current season's growth. The guava has both terminal as well as lateral flower bearing shoots. The flowering period varies from 25-45 days depending on variety, season and climatic conditions. Anthesis takes place in early morning hours (between 6.30-8.00 a.m.). Guava is a highly cross-pollinated fruit crop and pollination is carried out by honeybees and androna flies.

Crop Regulation

In mild climate of the tropics, guava flowers and fruits throughout the year without any specificity of the period, provided water and temperature do not become limiting factors. In general, rainy and winter season are the major crops of guava. Fruits of rainy season crop are insipid and watery and are of poor quality. Besides, rainy season crop is exposed to fruit fly attack, which renders the fruits unmarketable. Thus, winter season crop is preferred. Considering the problems in rainy season crop, there is a need to regulate guava crop in such, a way that only quality crop is harvested in the winter and rainy season crop is avoided. This is done by *bahar* treatment.

Bahar Treatment

The practice of forcing flowering and fruiting in a desired season is called *bahar* treatment. In this treatment, the plants are forced to take rest during the unwanted *bahar*. It is achieved by different means as discussed below;

i. Withholding Irrigation

In general, irrigation is withheld for 2-3 months during February-March. As soon as the growth of plant ceases and leaves start turning yellow and a result of moisture stress, tree undergoes rest. At the end of May or early June, the soil of the orchard is ploughed, harrowed and manured with FYM to individual tree. Immediately after application of the manure around the basin, the tree is irrigated. The first two irrigations should be given at an interval of three days and the subsequent watering may be given at 10-15 days interval till the monsoon sets in. The plants then resume growth and bloom heavily in August-September producing very high yield of quality fruits in winter.

ii. Root Exposure

Various degree of root pruning has been found suitable for minimizing the rainy season crop. The practice of root exposure and pruning is done in regions with high atmospheric humidity and water table. The procedure involves, exposing the upper roots of the tree, of 1.5 ft radius around the trunk by removing the upper 8 cm soil. The main root system is not disturbed while the fibrous roots on them are removed by a pruning shear. This results in shedding of leaves from the tree. At this stage, the exposed roots are again covered with soil, mixed with manures and immediately irrigated. Root pruning should, however, be avoided as it has adverse effect on plant longevity. In Maharashtra, withholding of water and root exposure is commonly followed to regulate cropping.

iii. Flower Thinning

De-blossoming of rainy season crop is another method for crop regulation. Thinning of flowers and small fruits by hand, twice during April-May at fifteen days interval, has been found quite effective in this regard. However, the technique is costly and cumbersome. The total yield of the plant in a year is also reduced by this method.

iv. Use of Growth Regulators

Manipulation of flowering season by application of growth regulators has been found to be very effective for crop regulation in guava. Growth regulators such as NAD, NAA and 2,4-D have been found very effective in thinning of flowers and manipulating the cropping season in guava. Chemical treatment of NAD at 30 and 50 ppm, NAA at 100 and 125 ppm and 2,4-D at 15 and 30 ppm can be successfully used for thinning of summer flowers. Experiments conducted at Pantnagar have suggested that foliar spray of NAA @ 800 ppm twice in May at 15 days interval has been found effective in minimizing the rainy season crop. First spray should be done when about 50 per cent of flower buds are opened. This method can be applied on a large scale.

vi. Shoot Pruning

The pruning of new shoot (current season) growths in 1st week of May has been found very effective for crop regulation in guava. This method involves the removal of half to $3/4^{th}$ portion of the shoot growth. This pruning automatically removes the flowers and flower buds of spring season flowering and consequently the rainy season crop can be reduced. This method has been found effective and economical without much adverse effect on total yield of the plant in a year.

Fruit Set

The formation of fruit is first noticed after 12 days from flowering. The initial fruit set in guava is very high. About 80-85 per cent flowers set fruits. However, due to severe fruit drop, only 40-55 per cent fruits reach maturity. In cultivar Seedless, the final retention of fruits is only 5-6 per cent. The maximum fruit drop takes place within 15 to 20 days after fruit set. In later stages, drop is very less. Pre-harvest fruit drop is mainly due to the attack of fruit fly. The fruit drop may be due to different physiological and environmental factors. Further, endogenous level of plant hormones such as auxins, gibberrelins and cytokinin also play a vital role in drop and retention of fruits. A spray of GA_3 (15 or 30 ppm) in January can provide better fruit set, retention and yield. Ninety per cent fruit retention can be achieved with GA_3 at 200 ppm. Parthenocarpic fruits in guava can be obtained by the application of GA_3 at 1,000 to 8,000 ppm in lanolin paste.

Fruit Growth

After fruit set, guava fruits grow at a faster rate and show two periods of rapid growth with a period of relatively slow growth in between the periods. During this period there is significant increase in size, flesh colour and quality of fruits. Guava fruits contains lot of seeds but a some varieties contain little seeds (seedless), which is due to autoploidy but not due to parthenocarpic development of ruits.

GUAVA WILT: CAUSES AND CONTROL

Yet, guava is considered as one of the hardy fruit crops, it suffers badly from wilt melody, disease which in certain cases may cause 20 to 100 per cent mortality under different agro-climatic and soil conditions. This disease was first reported in 1935 from Babbakkarpur, Allahabad (U.P.), but yet only limited progress has been made on its causes and management. It has been reported from Kanpur, Jaunpur, Gorakhpur, Banaras, western U.P., Bihar, West Bengal, Punjab, Haryana, Delhi, Odisha, M.P., Gujarat and Maharashtra. It has also been reported from the counties like Brazil, South Africa, Pakistan, Taiwan etc.

Symptoms

In general, two types of guava wilts have been reported to occur in different regions of the world *viz.*, slow wilt and quick wilt.

1. Slow Wilt

In slow wilt, die back of the branches starts. The typical symptoms include the appearance of yellow colouration with slight curling of the leaves on terminal branches. Subsequently, the leaves turn brown and may drop down prematurely. The withering of branches from the top downwards starts. The growing tips starts drying first. As a result, branches may fail to put forth new growth flushes. The affected branches defoliate, giving them a barren appearance. Eventually, the whole tree is affected.

In general, the symptoms of die back appear after raining season *i.e.*, during October-November, when the fruits are small. As the wilt progresses, the fruits of the affected branches remain under developed and become hard and strong. In the later stages, the entire plant is defoliated and eventually dies, but hard, strong, dark fruits remain hanging on the trees. In slow wilt, plants may take 1-3 years for complete wilting/dieing.

2. Quick Wilt

In quick wilt, the trees die immediately after they get infected. However, sometimes there is sequential death of branches, one after another, till the whole tree gets dried/died. It usually happens during rainy season when the trees die within a week or month. Sometimes, it may take even a year. Quit wilt is characterized by yellowing or browning of the leaves at the tip of twigs. Leaves die and the bark of the twigs show splitting. The internal portion of bark turns dark and there may be discolouration in the form of streaks, which may extend into cambium or even deeper. The roots get wilted and develop black steaks, sometimes giving fowl smell.

In general, above 10-years old plants are more prone to wilt disease. Partial wilting is also common in some orchards, in which a few branches wilt in one year and the others in the next year, when full plant dies. The symptoms of wilt in the nurseries and the onwards may be different. Diwedi and Diwedi (1999) have given the following clear-cut distinction between the symptoms in nursery and orchards.

In the Nursery

1. Wilting of seedlings occur in patches.
2. Occurrence of wilting patches is more prominent in the later stages of infection.
3. The apical leaves dry up.
4. Seedling stem first turn light brown and finally to blackish brown in colour.
5. Dried leaves initially remain attached to the stem, but later they shed off.

In the Orchards

1. Rootlets become hard, bearing black streaks and spots.
2. Leaves become yellow, brown and dry, but initially remain intact with the stem.
3. The leaves may drop down, and the branches become completely defoliated.
4. The leaves may completely dry.
5. In advanced stages, fruits dry up and the bark of branches and stem starts rupturing and cracking.
6. The under developed fruits may turn dark brown or black and fall down or remain hanging on the wilted trees.

Causative Factors

Though, guava wilt was reported seven decades back, yet its definite causal factors have not yet been established. However, scanning of the available literature reveal that the following factors have some definite relationship with guava wilt.

1. Role of Varieties and Rootstocks

The commercial guava variety, Allahabad Safeda is highly susceptible to guava wilt. The other cultivars like, Chittidar, Hafsi, Riverside, Stone Acid, Kerela, Behat Coconut have been reported to susceptible to guava wilt. The cultivars like, Lucknow-48 (Sardar), Seedless, Spear Acid, Banarasi, Dholka, Sindh, Nasik, Portugal, Apple Colour and Red Flesh have been reported to be tolerant. However, none of the variety has been reported to immune to guava wilt, though some contradictory reports about resistance/tolerance or susceptibility of different varieties or species have also appeared in the literature. In general, none of the *Pisidium* species like *P. cattleianum*, *P. molle*, *P. guineense*, *P. friedrichsthalianum* show susceptibility to guava wilt. The Chinese guava (*P. friedrichsthalianum*) and strawberry guava (*P. cattleianum*) have been recommended as wilt resistant rootstocks for guava in different countries. The Indian Agricultural Research Institute, New Delhi has, recently released Pusa Srijan, a dwarfing aneuploid (2n+2=24) rootstock for high density planting of Allahabad Safeda cultivar. It has also been reported to show resistance to wilt under field conditions.

2. Fungal Pathogens

There are considerable differences among the researchers about the causal agents (*e.g.*, pathogens) of this disease, because a large number of pathogens have been isolated from the wilt-affected plants by the various workers in different regions of the world. Hence, it is difficult to establish a clear-cut relationship between a particular pathogen and guava wilt. For example, up to 1941, wilt was considered to be caused by *Cephalosporium* sp. in north India. In 1947, *Fusarium* sp. was reported to be associated with wilt, which was later named as *Fusarium oxysporum* f.sp. *psidii*. In addition to *Fusarium oxysporum* f. sp. *psidii*, the other species like, *Macrophomina phaseoli* and *Fusarium solani* have been reported to cause wilt disease in guava in West Bengal. Out of these species, *M. phaseoli* has been found to be confined exclusively to root region and *F. solani* to aerial parts. However, both the pathogens were able to incite wilt in guava singly or in combination.

In Punjab, *Gliocladium vermoesenii*, a saprophytic fungus, has been reported to be associated with guava wilt. From Varanasi, *M. phaseolina* has been reported to incite guava wilt.

More recently, *Gliocladium roseum* has been reported to be most potent pathogen causing wilt in guava by the scientists of CISH, Lucknow, India.

Contradictory reports regarding association of different pathogens with guava wilt have also appeared in the literature from the different countries. For example, *F. oxysporum* and *Colletotrichum gloeosporioides* fungi have been reported from wilted guava trees in Pakistan. In Brazil, *Pseudomonas* sp. and *Erwinia psidii* have been isolated from wilted guava plants. In Taiwan, guava wilt has been reported to be caused by *Myxosporum psidii* fungus. In South Africa, the fungus like, *Septofusidium* sp. or *Acremonium diospyri* have been found to be associated with guava wilt.

From the above information, it is conclusive enough to state that no single pathogen is responsible for the cause of wilt disease in guava. However, the pathogens like, *Fusarium oxysporum* f. sp. *psidii*, *F. solani*, *Macrophomina phaseoli*, *Rhizoctonia bataticola* (*M. phaseolina*), *Cephalosporium* sp., *Acremonium diospyri*, *Colletotrichum gloeosporioides*, *Alternaria* sp. and *Gliocladium roseum* have some definite relationship with the cause of guava wilt.

The fungus penetrates through the thinner tissues of the root and establishes in the conducting tissues. Then, it advances upwards into the conducting tissue of the stem, where it blocks the upward movement of water and other soluble nutrients, perhaps by producing certain substances. As a result, the tree gradually dries up from top downwards. Later, thin as well as thicker walls of the fungal spores are produced on the surface of leader roots, which remain viable for a longer time even under adverse conditions and germinate under favourable conditions to incite infection in the healthy plants. Some fungi like *Fusarium solani* first establishes in the epidermal cells and then spreads into cortical cells and finally in the xylem vessels, whereas *M. phaseoli* first invades the phloem vessels, Certain heat tolerant fungi, like *Macrophomina phaseolina* can tolerate higher temperature and survive in the form sclerotia in summer. Once the infection of wilt causing pathogens occurs and establishes in the orchard, the infection gradually and steadily increases and spreads in all the directions. Most of the pathogens secrete pectin-degrading enzymes, like pectin glycosidase, pectin depolymerase, pectin methyl esterase and cellulase etc., and other toxins, which break the structural integrity of the host cell. Thus, these enzymes appear to be involved in pathogenesis of guava wilt. The incidence of guava wilt is more virulent in areas where the summer temperature goes beyond 40°C in summer. In absence of proper moisture supply, high temperature may start the initiation of wilt disease in summer or onset of rainy seasons, which continues up to September-October. The available literature reveals that the wilting of guava trees starts in the rainy season, with the largest number of plants dying in September-October. The incidence decreases remarkably in November and it becomes negligible with the onset of winter.

3. Insects, Nematodes and Bacteria

Directly, no insect-pest has been reported to cause wilt disease in guava, however, *Coleosterna* beetle is known to cause infection and spread of wilt pathogens. Nematodes like, *Meloidogyne, Helicotylenchus* and *Pratylenchus* have been reported to cause guava wilt in Cuba. A bacterium, *Pseudomonas* sp. has also been reported to be associated with guava wilt.

4. Edaphic Factors

It has been found that guava wilt is more prevalent in alkaline soils with pH 7.5 to 9.0. Most of fungi causing wilt usually don't grow at pH lower than 4.5 and higher than 8.6. Though, some contradictory reports are also available.Soil moisture contents of 60 per cent or more are highly suitable for the growth of fungi causing guava wilt. A few workers are of the view that poor drainage conditions in the orchards favour the occurrence of guava wilt. In contrast, moisture deficit may also be cause guava wilt. The incidence of wilt is more in sandy loam soil than in clay loam soils. Usually, higher proportion of sand in the soil favours the incidence of guava wilt. The incidence of wilt disease starts in the month of June-July, which becomes more dominant in September-October, indicting that high atmospheric humidity may be responsible for guava wilt.

5. Nutritional Imbalance

Malnutrition has definite role with the cause of guava wilt. It has been established by some workers that properly fed guava plants show less incidence of wilt than adequately managed trees. In general, low or high levels of nitrogen and low levels of phosphorus and micronutrients like zinc favour the incidence of guava wilt.

Management Practices for Guava Wilt

Following management practices have been recommended by various workers to reduce the incidence of guava wilt:

Cultural Practices

Following cultural practices have been considered necessary to avoid the incidence of guava wilt:

1. First and the foremost step to reduce the further spread of the disease should be to uproot and burn the infected plants.
2. As far as possible, purchase guava plants from a wilt free nursery.
3. Prune all the infected branches and burn them.
4. While planting, the roots of the seedlings should not be damaged.
5. The tree vigour should be maintained by timely application of manures and fertilizers, irrigations and intercultural operations.
6. Soil may be applied with gypsum or lime @ 1.50-2.00 kg/tree.
7. Green manuring of the orchard may be done at frequent intervals.
8. Apply zinc sulphate and nitrogenous fertilizers judiciously.
9. While planting new orchards, the pits may be drenched with formalin and kept covered for about 3-4 days. The planting in the pits should be done after 2-3 weeks after formalin treatment.
10. CISH, Lucknow has advocated the use of 6 kg neem cake + 2 kg gypsum/plant for the control of guava wilt, which may be used effectively.

Soil solarization is a new technique for controlling soil borne pathogens without the use of any chemical. In this technique, moist soil is mulched with polyethylene sheet to increase the temperature of soil surface considerably. This technique is quite cheaper, having no adverse effect either on soil or plant heath. It has been observed that about 80 per cent population of soil borne pathogens is inhibited by this technique. Hence, it should be used before raising guava nursery or establishing new guava orchards.

Use of Chemicals

On the basis of different experiments conducted by the scientists in different corners of the world, the following recommendations have emerged out:

1. The drenching of soil with 0.2 per cent either benlate or bavistin 4 times a year and spraying twice with metasystox (0.05 per cent) checks the disease to a greater extent.
2. Injection of 0.1 per cent water-soluble 8-quinolinol sulphate to the plants provides protection from wilt for a year. If given to infected plants, it provides partial check.
3. Disinfection of the soil with methane sodium controls the nematodes, which cause guava wilt.

Use of Biocontrol Agents

Biocontrol of soil borne pathogens has been reported by several workers. It has been reported that different species of *Trichoderma viz. T. viride, T. lignorum, T.*

harzianum and *T. piluliferum* inhibit the growth of almost all the soil borne pathogens causing guava wilt. Similarly, *Streptomyces chibaensis* also inhibits the growth of *Fusarium oxysporum* f. sp. *psidii*, a major fungus responsible for guava wilt. The inhibition on growth of these pathogens is mainly due to the production of various mycotoxins or antibiotics, which inhibits the growth of the pathogens. It has been established that *Trichoderma* sp. produces the gliotoxin and viridin mycotoxins, which destroys the pathogens causing guava wilt. Hence, biocontrol of the wilt causing pathogens should be employed by the guava orchardists.

Use of Resistant Varieties and Rootstock

The cultivars, such as Sardar, Banarasi, Dholka, Sindh, Nasik etc., are resistant to guava wilt. Hence, cultivation of these varieties should be encouraged in wilt prone areas. Many *Psidium* species like, *Psidium molle, P. cattleianum, P.cujavillis, P. guaneese, P. friedrichsthalianum* etc., have been reported to be resistant to guava wilt. Similarly, recently, Pusa Srijjan, an aneuploid guava rootstock has been released by the IARI, New Delhi, which is not only dwarfing, but it imparts resistance against wilt under field conditions. Hence, every possible effort should be made to use and encourage the recommended rootstocks for raising commercial orchards.

PLANT PROTECTION

A. Major Diseases and their Management

Amongst the diseases, anthracnose and fruit rot are serious and are described below;

Anthracnose

This disease was first reported from Uttar Pradesh and now it is of common occurrence in all the guava growing regions of India. Anthracnose is a common disease of guava in western districts and *tarai* regions of Uttar Pradesh and some other states, sometimes causing significant losses. The disease appears mostly during the rainy season and results in maximum damage under high humidity and moderate temperature conditions.

Causal Organism

Anthracnose of guava is caused by *Colletotrichum gloeosporioides* (*C. psidii* Curzi.) imperfect stage of *Glomerella cingulata* (Stonem.). Association of *C. acutatum* has also been reported. The disease develops best in ripened fruits at maximum and minimum temperatures of 35 and 10°C at around 90 per cent relative humidity.

Symptoms

The affected plants begin to dry and die from top of the branch, while shoots, leaves and fruits are readily affected. The characteristic symptoms of the disease appear as pin head spots on the unripe fruits during rainy season. These spots gradually enlarge to form sunken, circular, dark brown to black lesions where acervuli appear in concentric rings. The lesions frequently coalesce and the pulp below them becomes soft and brown. Under moist conditions, pink spore masses can be seen within the acervulus. The pathogen also causes rotting of fruits in the storage but the infection possibly takes place on the tree itself. The symptoms are also observed on leaves. Dieback phase of the disease is also evident in which the growing tip and young shoots turn dark brown. The dried twigs develop acervuli during moist weather conditions.

Management Practices

1. The disease can be reduced by the modifications in cultural practices such as keeping the orchard clean and pruning and burning the diseased twigs. Apply Bordeaux paste to the cut portions.

2. Spraying the guava trees with Bordeaux mixture (3:3:50), copper oxychloride, cuprous oxide, Difolatan or Dithane Z-78 has been recommended but the copper formulations cause russetting in fruits in some varieties.

3. Benomyl and triforine sprays are useful for controlling the disease in the field. Postharvest dip of fruits in fungicides like benlate, thiabendazole, aureofungin and mineral oil is quite helpful in reducing storage rot.

Phytophthora Fruit Rot

It is a very serious disease of guava fruits in north India and causes considerable losses in south India as well.

Causal Organism

It is caused by *Phytophthora nicoiana* var. *parasitica*.

Symptoms

The fungus attacks the fruits during monsoon season, with the development of whitish cottony growth on ripening fruits. High relative humidity helps in fast spread of the disease to unaffected fruits on the trees. The affected fruits start rotting, giving a characteristic bad odour. The fruits may drop off the trees.

Management Practices

Growers should adopt the following measures for the control of *Phytophthora* fruit rot in guava:

1. Collect and destroy the affected/fallen fruits.

2. Give two sprays of zineb or captafol (0.2 per cent) at 10 day's interval at the onset of monsoons.

B. Major Insect Pests and their Management

About 50 species of insect-pests have been recorded in guava. However, fruitflies, mealy bugs, scale insects and bark eating caterpillars cause considerable damage as described hereunder:

Fruit Fly (Chaetodacus sp.)

It is the most serious pest of guava. Its attack is so severe on rainy season crop that whole crop is rendered unfit for consumption, causing serious losses. The adult flies lay eggs on fruit surface during monsoon. On hatching, the maggots enter the fruits. Maggots grow inside by feeding on guava pulp. The infested fruits show depression with dark greenish punctures and when cut open, the wriggling maggots can be seen. The following remedial measures can be tried is keep this pest under control: (i) The flies pupate in the soil, so raking the soil and dusting with BHC helps in destroying the pupae; (ii) the infested fruits should be collected and burried/dumped; (iii) use of baits (sugar + malathion) in the ratio of 1: 1 is very effective; (iv) spraying the plants with malathion (0.05 per cent) or dimecron (0.03 per cent) during oviposition period of the insect helps to keep the flies population under control, (v) harvest the fruits at 3/4/maturity stage and (vi) avoid rainy season crop.

Mealy Bug (*Cryptolemus* spp.)

This is another serious pest of guava which causes damage by sucking cell sap from tender leaves, shoots, flowers and fruits. The affected leaves dry up and the fruits. drop off resulting in poor yield. The nymphs and adult bugs crawl on the ground and climb up the tree. The banding of the tree trunk with polyethylene film or Ostico sticky bands is the best method to prevent them climbing up the tree. Further, treatment of soil with Aldrin or Malathion is also effective.

Scale Insects (*Pulvinaria psidii*)

The guava scale insect is a serious pest in south India, Maharashtra, Punjab and Uttar Pradesh. These are green sticky insects found on the leaves, shoots and fruits and cause damage by sucking the sap from tender shoots, young leaves, flowers and developing fruits. These insects also secrete honey dew, which encourages the development of sooty mould. The scale insects can be controlled by spraying crude oil emulsion, Diazinon (0.05 per cent) or Monocrotophos (0.05 per cent).

Bark Eating Caterpillar (*Indarbela* spp.)

Bark eating caterpillar causes damage by feeding on bark under silken gallaries. The infested trees are visible from a distance as the winding silken gallaries full of frass faecal matter are seen on the trunk. For its control, remove the silken gall aries, plug the holes with cotton soaked in diesel/petrol or monocrotophos.

C. Major Disorders and their Management

Fruit drop is a serious disorder in guava resulting in about 45-65 per cent loss due to different physiological and environmental factors. Spraying of GA_3 has been found to be effective in reducing the fruit drop in guava. Bronzing of guava has been observed in places having low soil fertility and low pH. Affected plants show purple to red specks scattered all over the leaves. Under aggravated conditions, total defoliation and fruits characterized with brown coloured patterns on the skin, with reduced yield are noticed. Foliar application of 0.5 per cent diammonium phosphate and zinc sulphate in combination at weekly intervals for two months reduces the bronzing in guava. Pre-flowering sprays with 0.4 per cent boric acid and 0.3 per cent zinc sulphate increase the yield and fruit size. Spraying of copper sulphate at 0.2 to 0.4 per cent also increases the growth and yield of guava.

HARVESTING AND YIELD

Maturity

Seedling guava plants have 5-6 years juvenile phase whereas the grafted or layered plants start bearing after 2-3 years. Change in fruit colour is usually taken as harvesting index. As soon as colour start turning from greenish to yellowish, the fruits should be harvested. The fruits should be harvested immediately after maturity otherwise the ripe fruits can be damaged by birds. Hand picking of fruits at regular intervals is preferred as harvesting by shaking of tree may cause severe damage to fruits and the tree. Rainy season crop should be harvested at 2-3 days interval and winter crop at 4-5 days interval. Guava starts giving economic yields after 8-10 years of planting. Yield, in general, depends on variety, age of tree, season and upkeep of the orchard. In general, grafted plant of 8-10 years age can yield 400 to 800 fruits weighing 80 to 100 kg.

For local consumption, ripe or mature fruits should be used, which are freshly harvested from the tree. However, for distant markets and to avoid the damages during transportation, it is advisable to wrap fruits with perforated polythene bags.

POSTHARVEST MANAGEMENT

Grading and Packaging

After havesting, fruits should be sorted and then graded according to fruit size and colour. After grading, the fruits are packed in baskets made from locally available plant material. Wooden crates or bamboo baskets can be utilized in packaging. For distant markets, wooden or corrugated fiber board boxes are used along with cushioning materials *viz.,* paddy straw, dry grass, guava leaves or rough paper. The fruits may be arranged in circular rows alternated with thin layers of soft hay or straw in the baskets. Good ventilation is necessary to check build up of heat. Guava is a delicate fruit requiring careful handling during harvesting and transportation. Guavas being perishable in nature are immediately sent to the local market after harvesting and only a small quantity is being sent to the distant markets.

Storage and Postharvest Treatments

Guava fruits are highly perishable and must be marketed immediately after harvest. The postharvest life of the fresh fruit is short at ambient temperature. It is possible to keep ripe but firm fruit in good condition for about 4 weeks at 8.3 to 10 °C and 80-90 per cent relative humidity. Several growth regulators have also been tried to enhance the storage life of guava. Fruits treated with maleic hydrazide (1,000 ppm) in addition to 6 per cent waxol treatment, extended the storage life of fruits at room temperature. Dipping of fruits in solution of 1,000 ppm cycocel has been found effective for storage of guavas at room temperature for 14 days. Treatment of fruits with calcium nitrate (1 per cent) has reduced weight loss, respiration rate and disease occurrence. These fruits maintain the edible quality and marketability for about 9 days at ambient temperature.

Trials at Haryana Agricultural University, Hisar, India have shown that weekly spraying with 1.0 per cent potassium sulfate–6 liters per tree–beginning 7 days after fruit set and ending just before harvesting at the pale-green stage, delays yellowing, retains firmness and flavour beyond normal storage life. Food Technologists in India found that bottled guava juice (strained from sliced guavas boiled 35 minutes), preserved with 700 ppm SO_2, lost much ascorbic acid but little pectin when stored for 3 months without refrigeration, and it made perfectly set jelly.

CROP IMPROVEMENT

Most of the guava varieties commercially grown in India, till date, such as Allahabad Safeda, Lucknow-49, Chittidar, Red Fleshed etc., are selections of superior seedlings resulting from sexual propagation. The thrust areas for guava improvement programme are development of guava cultivars with dwarf plant architecture suitable for high density planting; precocity in bearing; good branch angles and spreading tree habit to avoid limb breakage during heavy bearing; branching habit for great number of spurs and higher productivity; fruits of uniform shape, size and colour; fewer and soft seeds; high TSS, pectin and vitamin C content, long storage life and resistance to pests like fruit fly and dreaded disease like *Fusarium* wilt.

Selection

Guava is an open pollinated fruit crop and thus has been used extensively for raising new plantations. At IIHR, Bangalore, 200 open pollinated seedlings of Allahabad Safeda raised from seed collected from Lucknow resulted in a seedling selection *i.e.* Selection 8, which was later released as Akra Mridula. The important characteristics of the selection are dwarf plants, few soft seeds, high TSS and good shelf life. Similarly, from guava plantations around Navalur in Karnataka 16 high performing seedlings were selected from guava Navalur. It is hardy, drought tolerant and resistant to canker. The selected seedlings had good fruit quality and yield.

In Maharastra, 12 strains were collected from Aurangabad and Bhir districts. Of them, ABD3, BHR3 and BHR5 were superior. Likewise, at the NDUAT, Faizabad, 23 strains of guava have been collected. As a result, 3 seedlings of Allahabad Safeda (AS_1, AS_2, AS_3) and 2 of Faridabad selection (FS_1 and FS_2) were found to be promising with respect to fruit quality and yield. Similarly, for developing a better coloured guava, work on selection of seedlings of open pollinated Apple Guava is underway at Allahabad and Central Institute of Sub-Tropical Horticulture, Lucknow. As a result, a number of promising seedlings, with bright red skin and pure white pulp, with soft seeds have been identified. These are CISH-G1, CISH G-2, CISH-G3 and CISH-G4. CISH-G3 and CISH-G4 were later rechristened as Lalit and Shweta, respectively. In case of Lalit, fruits are saffron yellow colour with occasional red blush, medium in size (average weight 185 g), flesh is firm and pink in colour and has good blend of sugar and acid and suitable for both processing and table purpose, while in case of Shweta, fruits are sub-globose with few soft seeds high TSS (14 °brix) and attractive pink blush.

Hybridization

Most of the commercial guava varieties are diploid in nature (2n=22), while triploids or seedless varieties are shy bearing in nature. Hybridization for improvement of guava has been underway at several research institutes. In North India, work at Fruit Research Station, Basti, Uttar Pradesh, crosses of seedless guava, Allahabad Safeda and Lucknow 49, Apple Colour, Red Fleshed, Patillo and Kot Prud were made. None of the 55 hybrids obtained from these crosses were found to be promising.

In order to evolve varieties with less seeds and increased productivity, crosses were made at IARI, New Delhi between Seedless triploid and seeded diploid variety Allahabad Safeda. Of the 73 F_1 hybrids raised, 26 were diploids, 9 trisomics, 5 double trisomics and 13 tetrasomics. Distinct variation in tree growth habit and leaf and fruit characters was observed. Trisomic plants had dwarf growth habit and normal shape and size of fruits with few seeds.

Interest was shown in eighties in aneuploidy breeding both for direct use as scion as well as for dwarfing rootstocks. The imbalance in chromosome numbers in aneuploids imparted ovule sterility resulting in seed reduction in fruits.

In South, intervarietal hybridization work at Fruit Research Station Sangareddy (Andhra Pradesh) resulted in selection of 2 superior hybrids e. g. Safed Jam (Allahabad Safeda x Kohir) and Kohir Safeda (Kohir x Allahabad Safeda) which were released for commercial cultivation particularly in semi-arid tropical areas of Telangana and Rayalseema. The breeding programme at CCSHAU, Hisar resulted into the development of two hybrids *viz.*, Hisar Surkha and Hisar Safeda. At IIHR, Bangalore, hybridization was carried out using Allahabad Safeda, Red Fleshed, Chittidar, Apple Colour Lucknow 49 and Banaras cultivars. Of the 600 F_1 hybrid seedlings, hybrid-1 and hybrid-16-1 were found promising. Their important characters have been given in Table 4.4.

Constraints in Breeding

Most of commercial cultivars in guava are diploids while the seedless varieties are triploids. Producing more triploids will be futile since the fruit shape in triploids is highly irregular and mis-shapen because of differential seed size. Cross incompatibility between several varieties *e.g.*, Lucknow 49, Apple Colour and Behat Coconut has also been noticed.

Inheritance of Characters

Genetical studies conducted in guava have indicated that:

☆ Red pulp colour is dominant to white and this character is governed monogenically.

☆ Bold seeds in guava were found to be dominant over soft seeds and this was also found to be determined monogenically.

☆ A linkage was found between red flesh colour and bold seed size.

Table 4.4: Important Characteristic of Guava Hybrids Developed in India

Name of Hybrid	Parentage	Important Characteristics
Safed Jam	(Allahabad Safeda x Kohir)	Fruit larger than Allahabad Safeda. Good quality and few, soft seeds.
Kohir Safeda	(Kohir x Allahabad Safeda)	Fruit larger, white fleshed. Seeds few, soft.
Arka Amulya (hybrid-1)	(Seedless x Allahabad Safeda)	Heavy yielder, pulp white, seeds soft. TSS high, keeping quality good.
Hybrid-16-1	(Apple colour x Allahabad Safeda)	Fruits attractive, bright red all over. Flesh firm, white, TSS high. Seeds few, soft, good keeping quality.
Hisar Safeda	Allahabad Safeda x Seedless	Upright tree growth, compact crown, round fruits with smooth surface and creamy yellow skin
Hisar Surkha	Apple Colour x Banarasi Surkha	Broad to compact tree crown, round fruits with smooth surface, skin yellow with red dots in low temperature
Hybrid-1	Sardar x Allahabad Safeda	Good yield with soft seeds. Released from BAC, RAU, Sabour
Hybrid-2	Chittidar x Allahabad	Good yield with soft seeds. Released from BAC, RAU, Sabour

5

Apple

Apple is the most important temperate fruit of the world and also of north-western Himalayan region in India. It has long been a staple fresh fruit in temperate parts of the world. It is considered as one of the most delicious fruits, which is liked for its crispness. The old proverb 'an apple a day keeps the doctors away' focuses its importance in human diet as it has high neutraceutical value, primarily because it is a rich source of antioxidants, and has high fibre content. Although, it is a popular table fruit, but its several varieties can also be canned or it can be used for juice, wine, vinegar and cider making.

The apple-growing areas in India do not fall in the temperate zone of the world but the prevailing temperate climate of the region is primarily due to snow covered Himalayan ranges and high altitude, which helps to meet the chilling requirement during winter season extending from mid-December to mid-March.

COMPOSITION AND USES

Composition

Apple is a rich source of anthocyanins, flavonoids and polypehnols. It is considred as a good source of carbohydrates, vitamin C and appreciable amount of thiamine and riboflavin. It is also considered as a good source of fibres and minerals like Fe, P and K (Table 5.1).

Uses

Besides its raw consumption, apples are used to prepare juices, wine (cider), chutney, jam, vinegar etc. Bark and roots of *Malus* sp., contain phloretin, an antibiotic.

Apples can be used to treat both diarrhoea and constipation. Apple is also highly recommended for balancing blood sugar levels. Apples are cooling and anti-inflammatory. Apple tea, usually prepared by infusing minced fruit or peels in hot water, helps uric acid elimination and is helpful as a supportive remedy in the treatment of arthritic and rheumatic conditions as well as rheumatoid kidney and liver disease. Therefore, an apple diet is recommended for gout, constipation, haemorrhoids, bladder and kidney diseases. Apple wood is used as firewood or making fine furniture, tool handles, carving, mallet heads, turned items etc.

Table 5.1: Composition of Ripe Apple Fruit

Component	Per 100 g Edible Portion	Component	Per 100 g Edible Portion
Water (per cent)	85	Riboflavin	1.2
Protein (per cent)	0.2	Niacin	0.6
Fat (per cent)	0.6	Vitamin C	16.0
Carbohydrates (per cent)	14	Calcium	0.9
Crude Fibre (per cent)	1	Phosphorus	1.2
Vitamin A	1.8	Potassium	2.3
Thiamine	2.1	Iron	3.0

TAXONOMY AND BOTANICAL DESCRIPTION

The cultivated apple is usually referred as *Malus pumila* L. (earlier called *M. communis* or *Pyrus malus*), although, the species can be split into *M. x domestica* Borkh (the cultivated apple), *M. pumila* var. *paradisica* (paradise apple) and *M. sylvestris* (wild crab apple). The cultivated apple is derived from natural crosses of *M. baccata*, *M. sylvestris*, *M. floribunda*, *M. micromalus*, *M. prunifolia* and *M. atrosanguina* and that is the reason for depicting cultivated apple as *M. x domestica* Borkh, the 'x' denotes natural cross. It belongs to Rosaceae Family, which also contains several other fruits of temperate region, such as pear, quince, plum, peach, nectarine etc.

Most of the cultivated apple varieties are diploid with chromosome number 2n = 34. Some cultivars are triploids and tetraploids *e.g.* Suntan, Jupiter and Jonagold are triploid and Alpha-68 is a tetraploid variety.

The apple forms a tree that is small and deciduous, reaching 3 to 12 metres (9.8 to 39 ft) tall, with a broad, often densely twiggy crown. The leaves are alternately arranged, simple ovals 5 to 12 cm long and 3–6 centimeters (1.2–2.4 in) broad on a 2 to 5 centimeters (0.79 to 2.0 in) petiole with an acute tip, serrated margin and a slightly downy underside. Blossoms are produced in spring simultaneously with the budding of the leaves. The flowers are white with a pink tinge that gradually fades, five petaled, and 2.5 to 3.5 centimeters (0.98 to 1.4 in) in diameter. The fruit is typically 5 to 9 centimeters (2.0 to 3.5 in) in diameter. Botanically, apple is a pome fruit, which develops from inferior ovary and is derived both from ovary wall and floral tube composed of basal parts of sepals, petals and stamens. This tube is fused with the ovary wall and becomes fleshy. The fleshy thalamus constitutes the edible portion and actual fruit lies within. The center of the fruit contains five carpels arranged in a five-point star, each carpel containing one-to-three seeds, called 'pips'.

ORIGIN, HISTORY AND DISTRIBUTION

The primary center of origin of cultivated apple is thought be the region, which includes the Caucasus, Asia Minor, Soviet Central Asia, and Himalayan India and Pakistan where maximum diversity in wild apple exists. It is believed that apple has been in cultivation in Europe for over 2000 years. Apple cultivation by Greeks and Romans seems to have started from a few centuries BC in the Middle-East and South Eastern Europe as a result of their travels and invasions. The seeds of European cultivars were introduced to North America and domesticated there by early settlers. The traders and missionaries disseminated the apple to other continents. John Chapman popularly known as Johny Appleseed was an American pioneer nurseryman who introduced apple trees to large parts of Ohio, Indiana, and Illinois. Thus, apple seed had a huge contribution in extending cultivation of apple to non-traditional areas. By the middle of nineteenth century, apple had spread throughout the temperate zone of the world. During last four decades a significant increase in apple production was witnessed in mountainous regions of countries such as India where the necessary winter chilling was available. Ambri, a dessert apple variety is considered indigenous to Kashmir (India). All other varieties have been introduced in India. Captain A.A Lee, an Englishman, established the first apple orchard in HP at Bandrole in Kullu valley in 1870. The famous Delicious varieties were introduced by Samuel Nicholas Stokes, a resident of Philadelphia, USA in 1918 at Kotgarh (Shimla). Later, apple spread to different parts of our country. In Nilgiri hills, apple was introduced as early as 1897, but was destroyed by woolly apple aphid.

Europe is considered to be the largest producer of apple. In recent years, China has emerged as leading apple producing country. China contributed nearly 47 per cent (33,266 thousand metric tons) of total world production during 2010-11. This is followed by USA (4,212 thousand metric tons) and India (2,891 thousand metric tons). In India, it is predominantly grown in Jammu and Kashmir, Himachal Pradesh and hills of Uttarakhand, accounting for more than 90 per cent of the total production. Of which, J&K alone contributes 64.1 per cent followed by H.P. (30.9 per cent). Its cultivation has also been extended to Arunachal Pradesh, Sikkim, Nagaland, and Meghalaya in north-eastern region and Nilgiri hills in Tamil Nadu. The agro climatic conditions in these states are not as conducive as in north-western Himalayan region. Early and continuous rains from April onwards do not favour the production of quality fruits besides resulting in high incidence of diseases. The highest productivity of apples had been noted in J&K (13.1 t/ha) followed by H.P. (8.8 t/ha) and Uttarakhand (4.1 t/ha) (Table 5.2).

SOIL AND CLIMATIC REQUIREMENTS

Soil

Apple can grow on a wide variety of soils. However, well drained, deep, fertile, slightly acidic, clay loam soils with pH 6.0-6.5 are considered as the best for profitable apple cultivation. Sites with gentle slope are generally more suitable than too steep areas. Apple does not withstand water logged conditions and therefore, provision for proper drainage should be there in such areas. Soils should be free from any hard sub strata and should be rich in organic matter. As for as possible, windy locations, ridge tops and skylines should be avoided for apple cultivation.

Table 5.2: Statewise Area, Production and Productivity Scenario of Apple in India

State	Area (000'Ha)	Production (000'MT)	Productivity (MT/Ha)
Jammu and Kashmir	141.7	1,852.4	13.1
Himachal Pradesh	101.5	892.1	8.8
Uttarakhand	33.0	1,35.9	4.1
Arunachal Pradesh	12.8	10.0	0.8
Others	0.1	0.2	2.7
Total	289.1	2,890.6	10.0

Source: National Horticulture Board Database-2011.

Climate

For successful cultivation of apple, about 1,000 to 1,500 hours of winter chilling are required to break the dormancy, which can be defined as physiological condition of deciduous fruit trees during which no noticeable growth takes place. This chilling process takes place at a temperature below 7 °C. Snowfall in winters renders winter chilling. But at the same time, an inclement weather particularly, very cold climate ranging from 10 to 20 °C below freezing point may cause extensive damage to plants. Abundant sunshine is necessary for proper growth and colour development in fruits. The sites located in north-eastern aspect at a lower altitude and south-western aspect at a higher altitude is the best for profitable cultivation of apple. Spring frost and hails are the major limiting factors in apple production. Therefore, areas experiencing frequent spring frosts and hails coinciding flowering and fruit set should be avoided. The growing season temperature should be between 21-24 °C for proper growth and fruit maturity. Although, temperature requirement for different stages of flower bud

Table 5.3: Critical Temperature Requirement for different Stages of Apple

Stage of Development	10 per cent Kill (°F)	90 per cent Kill (°F)
Silver tip	15	2
Green tip	18	10
1/2-inch green	23	15
Tight cluster	27	21
First pink	28	24
Full pink	28	25
First bloom	28	25
Full bloom	28	25
Post bloom	28	25

development and post-bloom stages is different (Table 5.3). On the other hand, the greater difference between day and night temperatures favours the colour development. Apple trees are sensitive to low soil moisture supply. Water stress adversely affects fruit set, fruit size and increases June drop. An annual precipitation of 100-125 cm well distributed during growth period is favourable for good productivity. Excessive rains and foggy conditions near fruit maturity result in poor fruit quality with poor colour development and black fungal spots on the fruit surface.

PLANTING AND ORCHARD ESTABLISHEMENT

Like many other fruits, the commercial age of an apple tree varies from 40 to 50 years, depending on the climatic condition and management. Thus, a careful planning of the orchard is of utmost importance before starting a commercial venture. The area selected for commercial apple orchard should preferably be on north-western slopes, which assures sufficient winter chilling, proper soil moisture retention and adequate sunshine. The ideal site should be elevated and free from spring frosts. The site should have a slope 4 to 8 percent. Similarly, the soil of the selected site should be deep and well-drained and should be free from hard pan.

The several methods of planting layout for flat areas are square, rectangular, quincunx and hexagonal systems. In hilly areas, contour or terrace planting is convenient.

Being very flexible in its nature of growth, apple can be raised either as bush trees, half standards or standards. Bush trees are usually smaller trees and are suitable for growing on dwarfing, semi-dwarfing, vigorous, very vigorous and seedling rootstocks; while, half standards and standards are suited for vigorous, very vigorous and seedling rootstocks.

The planting distance in apple orchard depends on variety, rootstock, soil type, sunshine and prevailing temperature and cultural practices, which influence the size of the canopy. The planting distance for different scion varieties and rootstock combinations are given in Table 5.4.

Table 5.4: Planting Distance and Planting Density of Apple on Different Rootstocks

Scion Variety	Rootstock	Tree Vigour	Planting Distance (m)	Planting Density (Trees/ha)
Standard	Seedling	Vigorous	7.5	178
Standard	MM 111, MM 109	Semi-vigorous	6	278
Standard	MM 106, M 7	Semi-vigorous	4.5	494
Standard	M 9	Dwarf	1.5	4444
Spur type	Seedling	Semi-vigorous	5	400
Spur type	MM 111, MM 109	Semi-dwarf	3.5	816
Spur type	MM 106, M 7	Dwarf	3	1111

In general, pits of $1 \times 1 \times 1$ m size are dug and filled at least one month before planting by mixing 40-50 kg FYM or compost, 500 g super phosphate and 50 g insecticidal dust (*e.g.*, metacid dust) at the time of filling of pits. Early planting should be done in dormant season, preferably in December. Care should be taken that the graft union remains at least 25 cm above ground in order to avoid collar rot incidence.

IMPORTANT VARIETIES

The desirable apple variety should be a heavy and regular bearer, the fruits of which should have attractive colour and appearance, very good dessert quality, good handling, storage and processing quality, resistant to pests and diseases and adaptable to the prevailing agro-climatic conditions. Most of the apple varieties have been developed as a result of bud mutations followed by selection. Important commercial varieties of the world are Red Delicious, Golden Delicious, McIntosh, Rome, Beauty, Jonathan, York Imperial, Stayman Winesap, Winesap, Yellow Newton, Cortland, Northern Spy, Gravenstein, Baldwin, Grimes Golden, Black Ben Davis, Wealthy, Cox's Orange Pippin and Rhode Island Greening. According to the time of harvesting, apple varieties can be categoriesed as early, mid season and late (Table 5.5) and on the basis of bearing behaviour, apple varieties have also been categorized as regular and alternate bearers (Table 5.6).

Table 5.5. Recommended Varieties of Apple in different States of India

Season	Himachal Pradesh	Jammu and Kashmir	Uttar Pradesh
Early season	Tydeman's Early Worcestes (P) Michael Molies Delicious Schlomit Starkrimson	Irish Peach Benoni	Early Shanburry (P), Fenny Benoni, Chaubattia Princess
Mid-season	Staring Delicious, Red Delicious, Richared, Vance Delicious, Top Red Lord Lambourne (P), Red Chief, Oregon Super, Red spur, Red Gold (P)	American mother Razakwar Jonathan (P) Cox's Orange Pippin Red Gold (P) Queen's Apple Rome Beauty Scarlet Siberian	Red Delicious Starking Delicious MaIntosh (P), Cortland, Golden Delicious (P)
Late season	Golden Delicious (P), Yellow Newton (P), Winter Banana, Granny Smith (P)	King Pippin American Apirouge Kerry Pippin Lal Ambri, Sunhari Chamure, Golden Delicious (P) Red Delicious, Ambri Baldwin, Yellow Newton (P) White Dotted Red	Rymer Buckingham (P)

P = Pollinizer

Table 5.6: Classification of Apple Varieties on the Basis of Bearing Bahaviour

Regular Bearer	Moderately Alternate Bearer	Strongly Alternate Bearer
Rome Beauty	Delicious group	Baldwin
Stayman Winesap	McIntosh	Oldenburg
Gallic Beauty	Northern Spy	Wealthy
Golden Delicious	Discovery	York Imperial
Idared	Golden Spur	Cox's Orange Pippin
James Grieve	Melrose	Laxton's Superb
Jonagold	Starking	Miller's Seedling
Mutsu	Wagener	Yellow Newton
Tydman's Early Worcester	Winter Banana	Yellow Transparent
Spartan	Granny Smith	Rallies
Bramley's Seedling	Winesap	Newton
Ben Davis		

A number of bud sports of the popular varieties have similar characters as of the parent except the fruit colour of fruit rusting etc. Some of these bud sports having high market values are given below:

Delicious	:	Starking Delicious, Richared, Red King, Hardeman, Rayan Red, Royal Red, Vance Delicious and Top Red
Rome Beauty	:	Gallia Beauty, Cox Red Rome
Stayman Winesap	:	Stamared, C and O Stayman
Jonathan	:	Jonared, Black Jon, Jonagold

Some mutants of the standard varieties have compact tree, shorter internodal length and high spur development potential. The trees of spur type mutants remain $2/3^{rd}$ of the size of standard varieties. Some of such mutants bear better-coloured fruits and have early maturity. The spur type varieties are suitable for areas having problems related to high vegetative growth, poor spur development and fruit colouration. Important spur types of Delicious group are Red Chief, Oregon Spur, Red Spur, Starkrimson and of Golden Delicious are Starkspur Golden Delicious and Golden Spur.

The spur types and colour strains can be categorized into 3 groups according to fruit colour development *viz.*, low, intermediate and high colouring. The important cultivars of the three groups are:

Low colouring	:	*Standard cultivars*: Early Red, Hi Red, Starking (old)
		Spur type: Wood
Intermediate colouring	:	*Standard cultivars*: Red Rich, Harrold, Starking, Huebner, Imperial, Royal Red
		Spur type: Red Spur, Red King, Well Spur

High colouring	:	*Standing cultivars*: Top Red, Red Prince, Chelan Red, Hi Early, Red Queen, Vance Delicious, Clawson Red
		Spur types: Starkrimson, Red Chief, Oregon Spur, Mor Spur, Silver Spur
Low chilling cultivars	:	Tropical Beauty, Michal, Shlomit, Vered, Naomi, Maayan

Besides natural mutations, induced mutation has also been attempted in apple for varietal improvement. Induced mutation of (budwood irradiation with Gamma rays) Early Shanburry resulted in release of Chaubattia Agrim in 1984 from Uttarakhand. Its fruits are small, egg shaped and slightly conical, yellow coloured with slight red tinge.

Scab Resistant Varieties

Some of the scab resistant varieties are Prima, Priscilla, Sirprize, Jonafree, Florina, Macfree, Nova Easy Grow, Coop 12, Coop 13 (Red Free), Nova Mac, Liberty and Freedom. Firdous and Shireen are scab resistant varieties released by SKUAT, Shalimar (Jammu and Kashmir).

Pollinizing Varieties

The most suitable pollinizing cultivars are Tydeman's Early Worcester, Red Gold, Golden Delicious, McIntosh, Lord Lambourne, Winter Banana, Granny Smith, Stark Spur, Golden Spur. A combination of early, mid-season and late-flowering pollinizers provides assured cross-pollination for the main variety. A combination of Tydeman's Early Worcester, Red Gold and Golden Delicious has been recommended in Himachal Pradesh for Starking Delicious plantations. The flowering crabs varieties such as Red Flush, Crimson, Gold, Yellow Drop, Manchurian, Snowdrift, Golden Hornet and *Malus floribunda* are also considered as good pollenizers. Red Free and Liberty varieties are resistant to scab, can also be used as pollenizer.

Apple Hybrids

As a result of extensive work breeding work, following apple hybrids have been released in our country:

SKUAST, Shalimar (J&K)

- ☆ Lal Ambri (Red Delicious x Ambri)
- ☆ Sunehari (Ambri x Golden Delicious)
- ☆ Shalimar-1 (Sunehari x Prima)
- ☆ Akbar (Ambri x Cox's Orange Pippin)
- ☆ Shalimar Apple-2 (Red Delicious x Ambri)

Regional Fruit Research Station, Mashobra, Shimla (H.P.)

- ☆ Ambred (Red Delicious x Ambri)
- ☆ Ambrich (Richard x Ambri)

☆ Ambstarking (Starkign Delicious x Ambri)

☆ Ambroyal (Starking Delicious x Ambri)

Regional Research Station, Chaubattia, Uttarakhand

☆ Chaubattia Anupam (Early Shanburry x Red Delicious)

☆ Chaubattia Princess (Early Shanburry x Red Delicious)

☆ Swarnima (Benoni x Red Delicious)

Some new apple hybrids like Jonagold (Golden Delicious x Jonathan), Fuji (Royal Delicious x Rall's Janet) and Gala (Kidd's Orange x Golden Delicious) have been introduced in our country, which are liked exceptionally well by Indian consumers.

IMPORTANT VARIETIES

The chief characteristics of important cultivars, which are grown in India are briefly described hereunder.

Red Delicious

It is the most popular variety of India. Its fruit is long, conical in shape with protuberance near the calyx, skin yellow with red strips not all over the surface, flesh firm, sweet and juicy, matures in the 3rd week of August. The only disadvantage in the Delicious is that it does not develop superior red colour until fully mature but by that time apple losses its crispness. This has led to search for a better early strain of the Delicious.

Royal Delicious

Its fruit are large, conical, skin yellow covered with red stripes all over the surface, flesh firm, sweet juicy and aromatic, Tree bearing fairly regular, vigorous, upright with narrow crotch angles, fruits mature in the second week of August. It is quite popular in mid hills than Red Delicious because of better colouration. On the higher hills, Starking Delicious develops a dark maroon red when it is fully mature.

Rich-a-Red

It is also an important variety of apple in India. Its fruits is conical in shape, and develops red blush earlier than Delicious but later than Starking Delicious. Fruits are sweet and juicy. It matures in the 3rd week of August. It is more completely coloured than the standard Delicious variety.

Red Gold

Its fruits are round to slightly oblong, skin highly red blushed, waxy and glossy, flesh white with pinkish tint near the surface, firm, flavour mild, fruit matures in the 2nd week of August, forms a relatively small tree and fruit buds are formed freely on shoot terminals and spurs. Regular and heavy cropper, fruit matures in the 3rd week of August. This variety is used as pollinizer for Red Delicious and Starking Delicious.

Golden Delicious

Several consumers likes this variety due its attractive yellow colour, high juice content and high keeping quality. Its fruits are round, conical to oblong, greenish yellow becoming golden yellow with an occasional pale orange blush on the high altitude with temperate dry climate, flesh cream, fine textured, crisp, juicy and of conspicuous pleasant flavour, suitable for processing. Tree is moderately vigorous with wide crotch angles. Fruit matures in the second week of September very useful as polliniser.

Mollies Delicious

It has now become an important variety in mid-hills. Its fruits are large, conical with 50 to 75 per cent and develops red colour. Its sweet blend is similar to Delicious, but it is moderate in quality attributes. It ripens in the 3rd week of July. It is susceptible to powdery mildew.

Ambri

This is the only indigenous variety grown commercially in Kashmir, Originated as chance seedling in Kashmir. Fruit is medium to large in size, and oblong in shape. It has red streaks over a greenish yellow background and is a late season variety. The pulp is white, crisp and sweet. The fruit has long keeping quality (4-5 months under ordinary storage and about 10 months in cold storage).

Vance Delicious

It is a bud mutant of Delicious. Outstanding variety and is in great demand by fruit growers. Fruit is conical skin solid red, colouring atleast two week earlier to Red Delicious and therefore, is sold at premium price. Matures in the second week of August, high yielding.

Top Red

This variety is the bud sport of Shot Well Delicious. The fruits are of long, large size with red streakes; pulp is yellow, sweet and juicy: fruits mature 15 to 20 days before Starking Delicious.

Red Chief

This cultivar is also bud sport of Delicious. Fruits are similar to Delicious with prominent red streaks. The maturity is earlier by 15-20 days: fruits are attractive bright in colour; pulp yellow, soft, sweet and juicy with excellent flavour.

Skyline Supreme Delicious

This variety is a bud mutant of Starking Delicious, Fruit is dark red in colour, matures 10 days earlier than Rich- a-Red. Fruit shape and colour are similar to Delicious, sweet and juicy.

Hardeyman

Fruit of this variety is similar to Starking Delicious in shape and colour, but matures 10-15 days in advance. Tree is vigorous in growth.

Starkrimson Delicious

It is a spur type of Delicious. Trees are dwarf; fruits are oblong, deep red in colour, hard in texture, sweet and juicy. Since the bearing is heavy, fruits are mostly of medium size. This variety is found suitable for cultivation in mid hills.

Oregan Spur

It is a bud support of Delicious, trees are small in stature but with large number of spur bearing branches. Fruits are similar in size and shape as in Delicious cultivar, but mature 10-14 days earlier. Fruit pulp is light yellow, crunchy sweet and a bit hard.

Gold Spur Delicious

It is a bud mutant of Golden Delicious. The shape and colour of the fruit are similar to Golden Delicious but it matures little early. This variety is recommended for all spur types as polliniser.

Red Spur Delicious

It resembles to well coloured Rich-a-Red. Fruits are conical and become ready by the first week of August. Tree has close internodal growth, which may ultimately grow to two-third the size of standard trees.

McIntosh

It is a leading variety of Canada, because of hardiness of the tree, is a mid-season variety which ripens in late July or early August. Fruit is small to medium, round, yellowish green with red blush. The skin is tender, pulp is greenish, juicy and fairly sweet. Tree starts bearing at the age of 4-6 years. Fruit takes 125 to 130 days from full bloom to maturity. Annual bearer, vigorous, strong but is susceptible to scab. Preharvest fruit drop is heavy.

Baldwin

Tree vigorous, large sized fruits, reddish in colour, round to conical in shape, and sour in taste. The fruit matures by the beginning of August and has very good keeping quality. Tree start late bearing and has tendency to bear biennially. Although popular in Kullu but due to poor return it is now quickly replaced by Delicious.

Maharaji

Fruit is large sized, flat in shape with red stripes and white dots. The fruit is sub-acid to slightly sour in taste. Tropical and hence is not suitable as pollinizer. Matures in October.

FLOWERING, POLLINATION AND FRUIT SET

Apple starts bearing at the age of 4-6 years, depending upon the variety, rootstock used, climatic conditions, altitude, and cultural practices. For instance, spur type varieties starts bearing at an early date than standard bearers. Similarly, scion cultivars start bearing much earlier if grafted on a dwarfing rootstock than on seedling rootstock.

Apple bears terminally or on spurs. Fruit bud differentiation takes place in July and flowering in the spring. Some apple varieties don't set fruits unless cross-pollinated with other variety and thus varieties require cross-pollination to ensure a good crop. Almost all apple varieties differ in their degree of self-fruitfulness. For instance, Golden Delicious is considered as a partially self-fruitful, while Delicious group is largely considered as self-unfruitful. Irrespective of the degree of self-fruitfulness, provision for cross-pollination is always desireable in the orchard. Thus, there should be provision for the planting of recommended percentage of pollinisers in the lay out plan of apple orchard for getting commercial crop in future. Pollinizers can be placed in orchard either with 11, 15, 20, 25 or 33 per cent placement plans. Of which, 11, 25 and 33 per cent pollinizers placement plans are most common in India (Figure 5.1). The plan of placing 11 per cent pollinizers can be used with partly self-fruitful apple varieties and in regions where crops are easily set. In this plan, every third tree in every third row is a pollinizer variety. Thus, the plan for 25 per cent pollinizers, requires placement of every alternate tree in every alternate row as pollinizers, while for 33 per cent plan, every third row in the orchard should be a pollinizer variety.

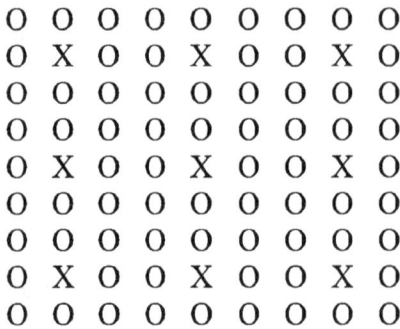

```
O O O O O O O O O          O O O O O O O O O
O X O O X O O X O          O X O O X O O X O
O O O O O O O O O          O O O O O O O O O
O O O O O O O O O          O X O O X O O X O
O X O O X O O X O          O O O O O O O O O
O O O O O O O O O          O X O O X O O X O
O O O O O O O O O          O O O O O O O O O
O X O O X O O X O          O X O O X O O X O
O O O O O O O O O          O O O O O O O O O
```

     11 % Pollinizers   15 % Pollinizers

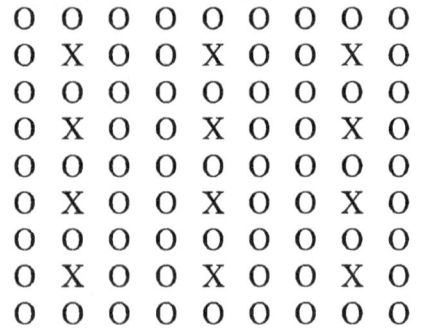

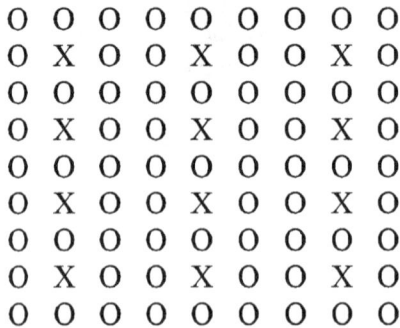

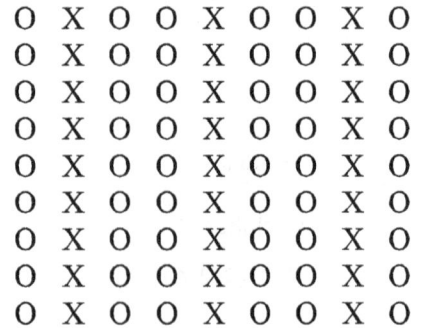

```
O O O O O O O O O          O X O O X O O X O
O X O O X O O X O          O X O O X O O X O
O O O O O O O O O          O X O O X O O X O
O X O O X O O X O          O X O O X O O X O
O O O O O O O O O          O X O O X O O X O
O X O O X O O X O          O X O O X O O X O
O O O O O O O O O          O X O O X O O X O
O X O O X O O X O          O X O O X O O X O
O O O O O O O O O          O X O O X O O X O
```

     25 % Pollinizers   33 % Pollinizers

Figure 5.1: Planting Plans Showing the Placement of Pollinisers in Apple Orchards

This plan reduces the number of trees of commercial varieties and suggested to be used under the conditions where climatic conditions are generally not so favourable for apple cultivation.

Most of the apple cultivars, especially of the Delicious group and its colour and spur mutants are self-infertile despite viable pollens and they require cross-pollination for satisfactory fruit set. Therefore, pollinizing varieties having overlapping flowering period should be planted with the main varieties. Tydeman's Early Worcester, Red Gold and Golden Delicious are good pollinzers, which have early, intermediate and late flowering, respectively covering all the flowering flushes of the Delicious group varieties. For the spur types, Golden Spur and Starkspur Golden can be inter-planted as pollinizers. Some of the other good pollinizers are McIntosh, Lord Lambourne, Ruspippin, Black Ben Davis, Red June, Granny Smith, King Pippin, Winter Banana, Rome Beauty, Fanny, Buckingham, and Jonathan. The flowering crab varieties such as _Malus floribunda_, Red Flesh, Crimson Gold, Yellow Drop, Manchurian, Snow Drift and Golden Hornet are also good pollinizers and must be introduced into the orchard apart from standard pollinizers to ensure pollination. With respect to cross-pollination, all red sports and spur types are considered the same as the parent variety. For example, Yorking is not a pollinizer for York Imperial. Closely related varieties do not pollinate each other. For instance, McIntosh, Early McIntosh, Cortland, and Macoun are closely related but never cross-pollinate each other. Triploid varieties do not pollinate any variety. Otherwise, all varieties with satisfactory pollen are pollinizers to each other, if the bloom periods overlap. Therefore, it is advisable to have at least three pollinizer varieties in all Stayman and Winesap blocks.

A good pollinizer should fulfill the following requirements:

- ☆ It should produce plenty of viable pollen.
- ☆ It should be compatible with the commercial cultivars.
- ☆ Its flowering period should overlap with the flowering period of main/commercial cultivars.
- ☆ It should produce fruits of attractive colour, good size, appearance and quality.

In general, five conditions are necessary to ensure satisfactory cross pollination in apple (i) the blooming period of the pollinizer and main variety must be the same or overlapping, (ii) the pollinizer variety must produce plenty of viable pollen, (iii) the pollinizer variety must be located near the commercial/main variety, (iv) the honeybees and other pollinators must be present in the orchard and be active at bloom, (v) weeds, and plants such as dandelions, mustard, and wild radish, should not be present in the vicinity of apple orchards as such plants attract honeybees, and they don't visit apple blossoms. Thus, improvement in fruit set of apple trees can be appreciably increased by making the provision of beehives in the orchard during flowering. The colony having more numbers of top workers, which inadvertently pollinate the flower by touching stigmas, while collecting nectar, are preferred over side workers or drones, which collect pollen and nectar without touching stigmas. It is recommended to place 6-7 bee hives/ha for effective pollination and to ensure sufficient fruit set in apple.

It is common observation that cold periods during flowering can reduce pollination and subsequent fruit set. Pollen may fail to germinate when temperature is below 41°F, and pollen tube growth is extremely slow below 51°F. Therefore, in some situations, temperatures could be warm enough for bees to fly (65°F), but if the weather turns cold, the pollen tubes may not grow fast enough before the embryo sac deteriorates.

Supplemental Pollination Practices

Sometimes, it is necessary to provide additional pollen in the apple orchard especially when the weather conditions do not favour cross-pollination. Using hive inserts with commercially obtained pollen can increase pollen sources. Inserts are specially constructed to fit in the entrance of hives and are filled on a frequent basis with pollen. The inserts are constructed so that bees are forced to track across the pollen and carry it to the flowers as they forage. It is a common practice in western countries for supplementation of pollination. A second method of increasing pollen is to place bouquets of flowering branches in the orchard in large containers of water. These bouquets should be checked daily and replenished as needed. A third method is to graft selected limbs with a compatible pollinizer branch. The disadvantage of this method is the necessity to clearly mark the limb to prevent it from being pruned out in the winter and prevent harvest crews from mixing the fruit in bins. All these methods should be viewed as supplemental means of increasing pollination. However, the best pollination method is to have an adequate number of pollinizer cultivars and strong and healthy honeybee colonies in the apple orchard.

Practices for Induction of Early Flowering

Several practices can be attempted in apple for inducing early flowering and fruiting, as described briefly hereunder.

Spreading of Branches

Experiments have shown that spreading of apple branches in horizontal position reduces growth and promotes flower bud formation. The bending of branches of young trees at an angle of 45° from the main trunk helps in early flowring. Orientation of apple tree at 45° angle from the soil surface reduces vegetative growth, tree canopy volume and promotes spur formation. Scientific evidences have indicated that bending of branches restricts the movement of carbohydrate and auxins from the upper portion of the limb towards the roots, and accumulated carbohydrates assumed to be favourable to flower-bud formation. Further, orientation of tree towards horizontal position also alters gibberellin translocation and therefore, metabolism and translocation of both gibberellin and auxin may account for the gravimorphic response to flowering in apple.

Pruning

Pruning is considered as most vital cultural practice in apple. It is of vital importance as the tree reaches the cropping phase. Excessive pruning, especially dormant tipping of new shoots will induce excessive vegetative growth and inhibits flower bud formation. If the trees are excessively vigorous, dormant pruning should be limited to cuts that are absolutely required.

Nutrient control

For general vigour and bearing, a precise control on nutrition is very important. Of several nutrients, application of nitrogen at the transitional phase of apple tree is critical. Excessive nitrogen application induces more growth and the tree remains vegetative. The tree should get sufficient nitrogen for optimum growth that would not inhibit flower bud formation. Deficiency of nutrients like zinc, copper and boron is detrimental to flowering in apple.

Ringing and Scoring

Ringing is the removal of a thin strip of bark around the trunk or at the base of main limbs. Whereas, scoring consists of making two or more parallel cuts to the sap wood around the trunk or limb, but no bark is removed. Both ringing and scoring are employed on over vigorous trees to check growth and increase flower bud formation. These practices are not recommended as a general practice. These should not be performed on very young trees, small limbs, on trees low in vigour but should be employed only on those trees that are sufficiently large to produce a crop but, due to excess vigour, have failed to develop fruit buds. Scoring and ringing are performed one to two weeks after full bloom, using sharp knief. In scoring, knief blade should penetrate complejely through the bark to the wood. These practices may however, induce blossom bud formation during the year of treatment, but can cause some trees to show severe shock symptoms. Trees those are not excessively vigorous before treatment, may require several years to recover from the shock effect. It has been observed that scoring can make trees more susceptible to winter injury during the following winter and the scoring and ringing wound on the trunk are also susceptible to severe freeze injury.

Use of Plant Growth Regulators

Plant growth regulators are capable of controlling and influencing tree vigour and cropping. Research reports have demonstrated that application of alar in apple induces flower bud formation during the year of treatment. However, alar should be applied only to those trees, which are sufficiently large to produce a crop, but due to excess vigour have failed to develop blossom buds. Similarly, naphthalene acetic acid (NAA) when used as an additive to paint the pruning cuts, it reduces vigorous shoot growth, which normally arises in the vicinity of large pruning cuts. Paclobutrazol (PP_{333}) can be used to restrict growth and to increase the return bloom. Alar and PP_{333} should be used as late-July spray and should be preferred over ringing and scoring treatments.

ROOTSTOCKS AND PROPAGATION

Rootstock

Rootstock is a vital component of apple production system. In apple, seedlings as well as clonal roostocks are used in commercial production. Traditionally, seedling rootstocks are used in our country. Seedling rootstocks provide strong and well-developed root system with good anchorage particularly in shallow sloppy hilly conditions. However, due to heterozygosity, the seedling rootstocks manifest variable effects on the growth, vigour, juvenile phase and productivity of the scion cultivar. Additionally, collection of seeds from various sources further complicates their likely characteristics. Though, seedling rootstocks can be raised from any apple variety, the most satisfactorily results are obtained from the seeds of Yellow Newton, Wealthy, McIntosh, Rome and Delicious. Apple seeds require after–ripening treatment, characterized by complex enzymatic and biochemical process that certain plant embryos must undergo before they will germinate. The period of after-ripening is usually 70-80 days at temperatures of 4 to 10 °C. A common system used for after-ripening in our country is to bury the seeds in an open container filled with sand, which is kept moist constantly.

In western countries, use of seedling rootstocks has been abandoned and now standard clonal rootstocks such as Malling (M) series and Malling Merton (MM) series rootstocks developed at East Malling Research Station and John Inn's Research Institute, Merton are used. But the history of modern rootstock breeding goes back to as early as mid-1800s, wherein, horticulturists began referring to rootstocks by name. They were called Paradise (or French Paradise) or Doucin (or English Paradise), the former being more dwarfing than the latter. These plants; however, showed much variation in size control. In addition, many new stocks had been introduced inaccurately under these names; undoubtedly viruses and genetic mutations had occurred in the plant material. In the late 1800s, one author described 14 different kinds of Paradise rootstocks. This diversity led researchers at England's East Malling Research Station to gather the selections to determine their trueness to name. The researchers concluded that indeed there were numerous misnamed and mixed collections of plant material. Dr. R. Hatton assigned a Roman numeral to each of 24 selections but did not number them in any order with respect to tree size. Hence, M 9 with a larger number is a smaller tree than M 2. Most of these rootstocks, with the exception of M 9, M 7, M 2, M 8, and M 13, never became commercially important. In succeeding years, some rootstocks were developed from controlled crosses, M 26 and M 27 being the most famous. In 1917, another research station, the John Inns Institute of Merton, England, joined with the East Malling station to begin a breeding program for apple rootstock. The joint efforts of both these institutes resulted in the development of rootstocks resistant to woolly apple aphid, which produced the Malling-Merton series of rootstocks, of which, MM 106 and MM 111 are still used widely. In the late 1960s, researchers began work to remove many of the viruses naturally present in the rootstocks in order to reduce incompatibility problems caused by the viruses. The first rootstock to be partially cleaned up was M 7; it was designated M 7a. Later, viruses were removed from all of the Malling and Malling-Merton series of rootstocks. These were then designated EMLA for the East Malling and Long Ashton research stations in England. While the viruses have been removed, some of the rootstocks' size control was lost. Therefore, the old M 9 produced a smaller sized tree than the 'clean' M 9 EMLA.

In the next few years, several new rootstocks were developed in Europe, of which Budagovsky series need a special mention here. These were designated as either Bud or B, which were developed in the central plains of the Soviet Union for their cold hardiness. Another series of rootstocks were released from Poland, which are called the 'P-series'. Like the Russian series, they are expected to have some cold hardiness. The P-series was developed from crosses between M 9 and common Antonovka. Reportedly, these stocks have good resistance to collar rot.

MM series rootstocks are resistant to woolly apple aphid. Screening the M and MM series rootstocks has resulted in the development of EMLA series virus-free rootstocks. Based on their effect on the scion, clonal rootstocks of apple have been classified in five groups *viz.*, very dwarfing (M 27), dwarfing (M 9, M 26), semi-dwarfing (MM 106, M 7), semi-vigorous (MM 104, MM 109, MM 111) and vigorous (M 16, M 25, MM 106, MM 112). The important features of some of the important rootstocks are presented in Table 5.7.

The performance of apple trees on clonal rootstocks has not been satisfactory in India, because of the following reasons.

☆ Poor anchorage of the rootstocks due to lower fertility of the soil.

☆ Slopy lands.

☆ Inadequate irrigation facilities.

☆ High suckering of rootstocks.

Considering these points in mind. Pomologists have advocated the use of crab apple seedlings as rootstock in apple.

Table 5.7: Clonal Rootstocks of Apple

Category	Rootstock	Characteristics
Very-dwarfing	M 27	Very samll size, poor anchorage
Dwarfing	M 9, M 26	Short juvenile phase, weak anchorage, suitable for high-density planting in flat and irrigated areas only
Semi-dwarfing	M 4, M7 and MM 106	Suitable for high-density plating and well-drained soils; resistant to wooly apple aphid but susceptible to collar rot
Semi-vigorous	MM 111 MM 109	Tree size is 70 per cent of standard, drought tolerant and resistant to wooly apple aphid
Vigorous	Merton 793 M 16, M 25 M 104	Wooly apple aphid and collar-rot resistant, early-fruiting, recommended for Kumaon hills of Uttar Pradesh

An Ideal Apple Rootstock

Although, a lot of success has been achieved in the development of several rootstocks at different research stations of the world; however, scientists still feel that there is no ideal apple rootstock in the market. An ideal apple rootstock should have the following characteristics:

1. It should make a successful union with the scion cultivar.
2. It should provide good anchorage to the tree.

3. It should be easily propagated.

4. It should be easily available.

5. It should be dwarf enough to control tree size of the scion cultivar.

6. It should have the ability to grow on a variety of soils.

7. It should be resistant to biotic and abiotic stresses.

Propagation

Mound layering or hardwood cuttings are commercially used to propagate the clonal rootstocks. The commercial apple varieties are grafted on the rootstocks by tongue grafting during February-March. This gives more than 90 per cent graft success. Chip budding is also quite successful during August-September, which gives smooth and strong graft union.

Owing to presence of viruses in rootstocks and gradual spread of dreaded viral diseases such as Apple Proliferation and Apple Stem Grooving Virus through use of infected planting materials, there is an urgent need for the production of disease free planting materials. Keeping in view the limitations of conventional propagation methods in handling, such infected planting material; the use of micropropagation seems to be an apt strategy for multiplication of superior quality virus-free plants. In India, at Micropropagation Technology Park (MTP), and Dr. Y.S. Parmar University of Horticulture and Technology, Solan, success had been achieved in obtaining consistent multiplication of apple cultivars and three rootstocks *viz.*, MM 106, MM 111 and M 7 by tissue culture technique.

INTERCULTURAL OPERATIONS

Training and Pruning

Plants are trained according to growth habit of the varieties and the vigour of the rootstock. Training helps to establish a strong framework of scaffold limbs capable of supporting heavy yield of high quality fruits; regulate annual succession of crops; admit plenty of light and air and expose maximum leaf surface to the sun; direct the growth of the trees so that various cultural operations like spraying, thinning and harvesting become economical; protect the trees from sun burn and wind damage; secure a balanced distribution of fruit bearing, parts of the main limbs of the tree and promote early production.

A. Training of Standard Trees

Standard trees are trained to 'modified leader'. In this system a leader develops on the young tree until it reaches the height of 2 to 3 m when the growth is restricted. Laterals are selected to ascend in a spiral fashion up the central leader and are cut back until the proper number and distribution of branches have been obtained. An ideal tree can be developed as follows:

1. Apple plant is pruned to 60-90 cm above ground immediately after being planted.

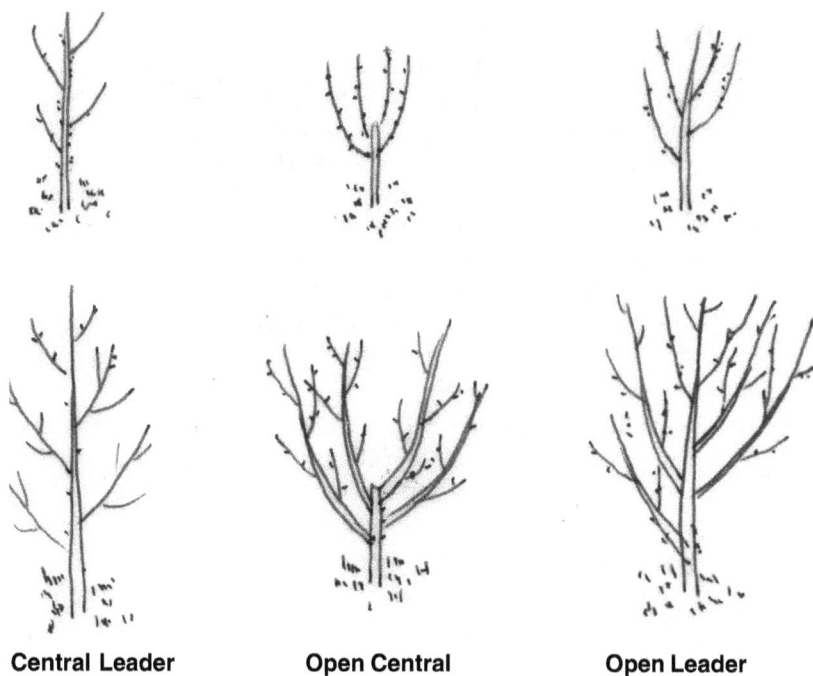

Central Leader   **Open Central**   **Open Leader**

Figure 5.2: Some Commonly Used Training Systems in Apple

2. **First year-summer pinching:** Well spaced shoots as scaffold limbs are selected with the lowest at 30 to 60 cm above the ground and the others are spaced vertically at 10 to 15 cm apart. The growth of these scaffold limbs may be encouraged by depressing the growth of other shoots by pinching the top 3 to 5 cm of other shoots. It should usually be done from mid-April to mid-May or when the shoots are 10 to 20 cm long. However, pinching shoots on weak trees is not recommended.

3. **First year dormant pruning:** This is done in winter following the first summer's growth. Three scaffold branches are selected and these are spaced about equally around the trunk preferably with a vertical spacing of 10 to 20 cm. The primaries are headed back to achieve a balance in growth among them.

Three well spaced primary branches cannot always be obtained in the first year. If two good limbs are available, they may be headed back to 45 cm from the trunk to suppress their development until a third scaffold can be selected at the end of second growing season.

With an occasional tree, it will be necessary to cut laterals back to one bud and defer scaffold selection until the next year. Such trees will generally grow vigorously and will benefit by summer pinching during the second growing season.

4. **Second year dormant pruning:** Five to seven secondary branches per tree usually two on each primary scaffold, are selected in the second dormant season. These secondaries are directed partially outward. New growth from such branches will fill in laterally, giving the tree a well shaped framework. Secondary limbs should be left full length and not headed back until it is necessary to maintain balance with the primaries or other secondaries. Any forked tips may be thinned out to single terminal. If a primary scaffold branch is over-vigorous it can he retarded by heading back to balance with the rest of the tree. Upright growing shoots should be removed. Most trees will produce short laterals and spurs during the second year and these should not be immediately removed as these provide shade and some produce early fruits.

5. **Third year dormant pruning:** Third year training consists of thinning out of unwanted branches and cutting others to desirable side limbs. Secondary -branches often form fruit spurs during the third growing season, but on young trees, fruit bud may be formed at the tip of unpruned shoots.

6. **Fourth year dormant pruning:** By the fourth year, tree training is largely completed. From here on, pruning should be done in such a manner as to permit the best development of the crop. The tree should be thinned out and the remaining branches cut to side limbs as is done in the third dormant pruning. A moderate number of small shoots should be left in the centre of the tree for fruit bearing wood. Adequate thinning out of branches and fruiting wood will maintain a well distributed bearing area throughout the tree and reduce excessive spread and breaking of branches. By about fifth summer, after planting, most apple varieties produce enough crop to be considered a bearing orchard.

B. Training of Dwarf Trees

Dwarf trees are trained to Spindle bush, Dwarf Pyramids, Palmettes and Cordon. Of all these systems, spindle bush is most suitable for adoption according to the prevailing soil and weather conditions in India.

1. One-year-old apple plants are cut back to a bud about 60 cm above the ground.

2. **First summer pruning:** During the first summer, 2 to 3 laterals or feathers, 30 cm from the base, may be used to form the main scaffold branches. Such feathers should have a wide angle with the main stem and be well spaced around it. Tying down of these feathers during the first summer is not usually needed, if the growth is moderate-to-weak. However, vigorous growing laterals are tied down in August. If they are tied down before the extension growth has ceased, the tip will turn upwards and need tying again. If tied too late, the main purpose of tying will be missed and the branches will not be fixed in position. The ties can be removed after leaf fall or left till the following summer. The shoots should be tied so that they are straight, not arched, otherwise strong vertical shoots will arise from the top of the arch. Tying is done by putting a loop around the shoot and fastening the other end to pegs pushed into the ground.

3. **First winter pruning:** During the first winter pruning, 2 to 3 well spaced laterals should be retained and unwanted branches arising on the main stem should be removed. The scaffold branches may be cut back by one third to half of their length to a outward growing' bud, if the growth is weak. Otherwise, the main branches should not be pruned. The leader should also be cut back to a more or less erect but weaker lateral. Any vigorous upright growing shoot arising directly below the shoot formingthe leader and competing with it should be cut. If there is no suitable lateral to serve as a replacement leader, the extension growth of the original leader should be cut back by about one third to one half, depending upon the growth.

4. **Second summer growth:** Spindle bushes may carry a little fruit crop during the third summer. During August of the second year, suitable laterals are tied down to form branches.

5. **Second winter pruning:** To check excessive vigour of the central leader, it may again be cut back to the next suitably weaker growing lateral, which is tied and trained to take the place of the central leader. Delaying pruning until late winter or early spring will also help in checking vigour.

6. **Subsequent growth and pruning:** Branches are allowed to grow from the central leader at regular intervals, choosing wide angled shoots. The higher placed branches must be kept shorter than the lower ones to allow sufficient light to reach the lower parts of the tree. In order to keep sufficient wood in the spindle bush while it is being built up, a small surplus of wide angled branches are retained, which may be cut when they require more space.

The main branches should be so trained and spaced that there is plenty of room for fru1ting laterals and for the same reason they should not be allowed to fork.

Precautions

For the development of strong scaffold limbs, it is essential that the lateral branches are spread to about a 45° angle to the trunk. If the central leader becomes too dominant, the branches remain weaker and the tree remains narrow. On the other hand, if it is less dominant, the branches become thicker and the tree wider. The balance may be influenced by the angle of the branches; if they are inclined downwards too much, growth of the top of the tree will be too weak. If the growth of the leader in comparison with the lower branches is too vigorous, it is best to prune the central stem back to a weaker lateral and train it as a replacement.

C. Pruning of Standard Trees

The primary consideration in pruning bearing trees is to maintain a proper balance between vegetative growth and fruit production. Excessive growth by the young tree is usually produced at the expense of fruit production while overbearing is accompanied by less growth and if continued, it results in loss of vigour. An ideal mature tree, should produce at least 25 cm of new terminal growth each year. Thus, the tree increases and maintains its fruiting area while producing satisfactory annual crops. The trees receiving regular pruning require light thinning and cutting back of the terminal branches to laterals. This prevents the tree from getting too high for economical spraying and picking operations.

The centre of the tree should be kept fairly open in varieties for which fruit colour is an important factor. If this has been-neglected for several years, it is best to thin out surplus branches over a period of two or more seasons. Heavy removal of branches can upset the balance of the tree and expose scaffold branches to sunburn damage. Heavy pruning will also encourage vegetative growth. Old bearing trees producing small fruits will benefit from heavy pruning, which stimulates new shoots and fruiting spurs.

Precautions

1. Whenever a limb larger than 3 cm in diameter is removed, the cut should be made as close as possible to the branch from which the limb arises so as not to leave a stub.

2. Large pruning wounds should-be protected with some covering in order to exclude rot causing fungi. Bordeaux paste is a good temporary covering, but Bordeaux paint provides a good permanent covering.

3. In one and two year old shoots, cutting back to promote the growth of the side shoots and in three year and older shoots, pruning should be shifted to thinning cuts to reduce the number and length of branches and to promote fruiting.

Steps in Pruning

1. Start at the top of the tree and work downwards.

2. Cut upward growing limbs back to strong laterals.

3. Remove the crowding branches and thin out the remaining limbs, leaving the vigorous fruiting wood well spaced along the length of the limbs.

4. Remove dead, broken and diseased wood as dead wood harbours and disseminates diseases such as bitter rot and black rot (frog eye).

5. Remove large limbs if growing parallel and crowding other limbs. However, while making large cuts, direct sun light should not sunscald remaining exposed branches.

6. Remove all water sprouts except an occasional one, which may be needed to fill a vacant space in a tree.

7. While removing a thick side branch, the first cut should be made on the underside of the limb where the weakest wood occurs.

8. Thin out badly placed interfering branches growing upward from the top of the limbs.

9. Thin out the poorer wood, which arises from either side of the limb.

10. Divert branches to open areas by pruning back to desirable laterals.

D. Pruning of Dwarf Trees

Spindle Bush

1. Once the central leader has reached its allotted height of 2.5 m, the extension growth should be cut back each year to a weaker side branch.

2. Strong growing shoots and branches towards the top of the tree should be removed completely.

3. Renewal pruning of the lower fruiting branches should be carried out each year for maintaining the vegetative growth and fruit quality in the lower parts of the tree.

4. Lower branches should not be heavily shaded by higher branches and some of them must be completely removed.

5. In cutting back branches to the main stem a stub should be left so that regrowth of the moderately vigorous fruitful bud is encouraged.

6. Avoid removing too many branches which may reduce potential fruit yield. If the main branches lose vigour, these can be stimulated by pruning. Worn out wood of heavy cropping varieties should be removed periodically and bending or typing down in strong growing varieties may be necessary.

Nutrition Management

In India, apple is cultivated on the slopes and its nutritional requirements are different to plantations of Europe and North America where apple is cultivated in the plains. However, in general, apple has high nutritional requirement as it removes heavy amount of nutrients from soil. Application of manures and fertilizers should be done from the beginning itself. However, doses of manures and fertilizers depend on several factors, like soil fertility, soil type, age of the tree, climatic conditions, previous fertilizer use and cultural practices. Thus, the fertilizer schedule should be based on soil and leaf analysis. The optimal leaf nutrient levels in apple are: N = 2.42-2.65 per cent, P = 0.17-0.2 per cent, K = 1.34-1.71 per cent, Ca = 1.29-1.47 per cent, Mg = 0.41-0.62 per cent, Zn = 28.5-44.9 ppm, Mn = 73.1-91.1 ppm, Fe = 350-482 ppm, Cu = 17.2-24.3 ppm. In general, 10 kg FYM, 70 g N, 35 g P_2O_5 and 70 g K_2O should be applied to one-year-old plants. The dose should be increased in the same proportion till the plant gains the age of 10 years when the doses are stabilized. A mature tree requires 100 kg FYM, 700 g N (2.8 kg CAN), 350 g P_2O_5 (2.0 kg single superphosphate) and 700 g K_2O (1170 g muriate of potash) in an 'on' year. In the 'off' year, the fertilizer doses should be reduced to 500 g N, 250 g P_2O_5 and 400 g K_2O. The fertilizer schedule recommended for 3 northern India states for apple is given in Table 5.8.

NPK fertilizers should be broadcasted in the tree basin, 30 cm away from trunk to tree dripline and mixed well in soil. FYM, P_2O_5 and K_2O should be applied during the winter before snowfall at the time of basin preparation whereas nitrogenous fertilizers should be applied one month before bud break. If irrigation facilities are available, nitrogen should be applied in two split doses, first half should be given 2-3 weeks before flowering and the second half one month later. Apple trees also respond to foliar application of nitrogen in the form of urea (0.5 per cent), which can be sprayed twice after fruit set at one month interval.

Table 5.8: Recommended Fertilizer Schedule for Apple in North Indian States

State	Age of Tree	FYM (kg/tree)	Nutrient Dose (g/tree)		
			N	P_2O_5	K_2O
Himachal Pradesh	1-3	10-30	70-210	35-105	70-210
	4-6	40-60	280-420	140-210	280-420
	7-9	70-90	490-630	245-315	490-630
	10 and above	100	700	350	700
Jammu & Kashmir	15 and above	50	600	130	750
Uttrakhand	15 and above	50	500	100	400

In some apple growing areas, zinc, magnesium, iron, copper, boron, manganese and calcium deficiencies are common, which can be determined by leaf analysis. In case of deficiencies, the following spray schedule should be followed (Table 5.9).

Table 5.9: Correction of Nutrient Deficiencies through Foliar Sprays

Element	Salt	Concentration (per cent)	Time of Spray	No. of Sprays
Calcium	Calcium sulphate	0.5	June-July	2, at 15 days interval
Magnesium	Magnesium sulphate	1-2	June-July	1-2 at 10-15 days interval
Zinc	Zinc sulphate	0.5	May-June	1-2 at 5 days interval
Boron	Boric acid	0.1	June	1-2 at 5 days interval
Copper	Copper sulphate	0.3	May-June	1-2 at 10-15 days interval
Iron	Ferrous sulphate	0.5	May-June	1-2 at 15 days interval
Manganese	Manganese sulphate	0.4	June	1-2 at weekly interval

Foliar sprays are usually given when new growth flush has emerged and leaaves are fully expanded. In the spray solution, stickers like Tween-20, Triton-x-100 etc. should be added.

Orchard Floor Management

Growth, productivity and quality of apple are also governed by orchard soil management. Suitability of soil management system largely depends on type of soil, topography, rainfall, planting system etc. However, an ideal orchard soil management programme should conserve soil moisture, prevent soil erosion, control weeds, add organic matter to the soil, and provide proper aeration to the root growth. For this, mulching or sod culture is usually followed in India. Mulching with straw, hay, sawdust, oak leaves or other organic matter increases the humus content of the soil and its moisture holding capacity. Various plastic and polythene mulches are also used. Mulching followed by herbicidal application has been the most effective for floor vegetation management and soil moisture conservation. Oak leaves and hay mulch has been found beneficial. Black alkathene mulch in cooler conditions is the most effective in weed control, moisture conservation, reduction in fruit drop and improvement in size, colour and quality of fruits. In warmer conditions, black alkathene mulch has adverse effect on root growth due to rise in soil temperature. Some of the effective floor management methods in apple are given in Table 5.10.

Table 5.10: Recommended Thickness of Mulching Materials for Apple Orchards in India

Mulch Material	Method of Application
Grass	10 cm thick in tree basin after spring rains
Grass plus herbicide	10 cm thick mulch followed by glyphosate at 0.8 kg/ha
Composted conifer leaves	10 cm thick mulch
Composted conifer leaves plus herbicides	10 cm thick mulch followed by glyphosate at 0.8 kg/ha
Black alkathene	1.1 m wide, 100 gauge thick spread over basin

Weed Management

Gramaxone (1,000 ppm) or mixture of 2,4,5 T (100 ppm) and gramaxone (500 ppm) is effective in controlling shrubby weeds in apple orchards. Diuron @ 2 kg/ha, Tok E-25 @ 4 litre/ha and trafazine at 4 kg/ha can be used to check weed growth in apple nurseries. Important herbicides used for weed control in apple orchards and nursery are given in Table 5.11.

Table 5.11: Recommended Weedicides for Weed Control in Apple Nurseries and Orchards

Apple nursery	Herbicide	Dose (kg/ha)	Time of application
	Diuron	0.8-4	Pre-emergence
	Nitrofen	0.5-1	Pre-emergence
	Simazine	0.2-4	Pre-emergence
	Terbacil	1-3	Pre-emergence
Apple orchard	Atrazine, Diuron, Simazine	2-6	Pre- and post-emergence
	Dalapon	4-10	Post-emergence
	Lasso	2.5	Post-emergence
	Paraquat	0.75-1.0	Post-emergence
	Terbacil	1-3	Pre- and post-emergence
	Tok E-25	1-2	Pre-emergence
	2,4-D	1-2	Post-emergence
	2,4-D (Na salt)	4	Post-emergence
	Atrazine or diuron + glyphosate	4+0.8	Pre and post-emergence

Role of Growth Regulators in Apple Production

Plant growth regulators play an important role in modern apple production by means of regulating crop and induction of branching in nursery. Crop regulation in apple can be done through flower and fruit thinning, control of pre-harvest fruit drop, improvement in photosynthetic efficiency, fruit set, fruit coloration and quality by the use of plant growth regulators and agrochemicals. Different uses of PGRs in apple production have been discussed briefly hereunder:

(a) Fruit Thinning

The judicious thinning of fruits at proper stage of fruit development can regulate cropping and improve fruit size and quality considerably. Hand thinning is cumbersome, hence plant growth regulators, dinitro compounds and insecticides are used for fruit thinning in apple. Carbaryl (1,500 ppm) applied 3 weeks after petal fall induces 60 per cent fruit thinning in Red Delicious apple. 2,4,5-T can induce 35-40 per cent fruit thinning. In Golden Delicious, application of NAA (10 ppm) and carbaryl (750-1,000 ppm) at petal fall is effective for optimal fruit thinning. Some of the fruit thinning chemicals and their doses are given in Table 5.12.

Table 5.12: Some Recommended Growth Regulators and Chemicals for Fruit Thinning in Apple

Growth Regulator/Chemical	Dose (ppm)	Stage
NAA	10-15	Full bloom to 4 weeks after petal fall
NAA	20-100	Petal fall
2,4-D	2-10	Full bloom to petal fall
2,4,5-T	2-2.5	Full bloom to petal fall
Carbaryl/Sevin	1,000-2,000	Petal fall to 4 weeks after petal fall
DNOC	1,000-2,000	Full bloom

(b) Fruit Drop and its Control

Most of the commercial varieties of apple have 3 waves of fruit drop: (i) early drop, (ii) June drop, and (iii) pre-harvest drop. The early drop is considered natural and is due to lack of pollination and fruit competition. June drop is caused by moisture stress and environmental conditions. These two drops neither cause substantial economical losses nor can be effectively controlled. However, the pre-harvest drop results in serious economic losses as full-grown marketable fruits fall before harvest. This is caused due to hormonal imbalance, especially reduction in the levels of auxins. Pre-harvest fruit drop is really a problem in early ripening cultivars like Tydeman's Early Worcester, Red Gold and Pippins, in which, it may range from 40 to 60 per cent of the total crop load. The mid-season cultivars like Delicious group and Golden Delicious are also not free from this malady and suffer about 15-20 per cent loss. Application of NAA (10 ppm) before the expected time of fruit drop or 20-25 days before harvest can check the fruit drop effectively. Now-a-days, 'ReTain' has become popular among the growers for controlling fruit drop in apple in western countries. The active ingredient of 'ReTain' is aminoethoxyvinylglycine (AVG), an ethylene inhibitor, which binds irreversibly with a key enzyme; thereby preventing the ethylene precursor from binding and ultimately blocking the production of ethylene. Natural ripening processes such as stem loosening, fruit flesh softening, starch disappearance, and red colour formation are slowed down by 'ReTain'. In order to make AVG to be more effective, it must be applied well in advance of the climacteric rise in ethylene production that signals the onset of fruit maturity. It is recommended to be applied 4 weeks before anticipated harvest.

(c) Colour Improvement and Enhancement of Ripening

In Delicious varieties of apple, colour development is generally poor in marginal warmer areas and thereby apples of such areas fetch poor price in the market. Application of 1,000 ppm 2-chloroethyl phosphonic acid (ethrel/CEAP/ethephon), an ethylene releasing hormone, about 10 days before harvest improves fruit colour substantially but impairs shelf life. Since this chemical accelerates fruit abscission, NAA (10 ppm) should be added to check fruit drop. In very high cool elevations, fruits develop intense red colour even before they are mature, which also do not fetch optimal price in the market. An application of ethephon (500 ppm) + NAA (10 ppm) aboud 3 weeks before harvesting can enhance maturity of apples by 7-10 days.

(d) Increasing Branching

A growth regulator composed of cytokinins and gibberellic acid (BA+GA) such as Promalin, Perlan, or Typy can be used to stimulate additional branches in young trees. Foliar applications of these growth regulators should be done when new shoot growth in the plants is approximately 1 to 3 inches long, *i.e.,* approximately 2 to 4 weeks after bloom. BA+GA may also be mixed with latex paint and applied directly to buds. Application of such compounds should be made in the spring when terminal buds begin to swell, but before shoots emerge and it should not be applied after buds break. Applications made after bud break, may cause injury to tender shoot tips and fail to promote shoot growth from that point. The application rate of these compounds is 5,000 to 7,500 ppm (0.2 to 0.33 pt/pt of latex paint). Ensure addition of a buffering agent or a nonionic wetting agent to the latex paint @0.5 to 1.0 percent (0.1 to 0.15 oz/ pt of paint) before adding BA+GA. The wetting agent improves the dispersion of BA+GA in the latex paint; it also improves wetting and absorption through the waxy layer of the tree bark.

(e) Shoot Growth Suppression

Apogee®, a plant growth regulator for vigour control in apples, reduces the length of shoot growth considerably. Apogee® is also labeled for the control or reduction of fire blight in apples in western countries. Shoot growth suppression by Apogee® is very consistent when the first application is properly timed and where a sufficient dose is applied during the active growth season. Apogee® acts to retard shoot growth by blocking the production of gibberellic acid (GA). By decreasing the level of GA in the plant, Apogee® inhibits the shoots' ability to elongate, thereby resulting in shorter shoots. Since there always remains some residual GA in the plant, it usually takes about 10 days for shoot extension growth to slow down.

SPECIAL PROBLEMS IN APPLE CULTIVATION

A. Heavy Fruit Set

It is a common place observation that under favourable conditions, most of the apple varieties tend to bear heavily than the actual capacity. During an 'on' year, excessive fruiting is an exhaustive process and results in smaller, low quality and unmarketable fruits and limb breakage. Therefore, the chief goal is to permit as large a crop on the tree to mature as possible and yet conserve sufficient nutrients and carbohydrates for good shoot and spur growth, leaf development and flower bud formation for the next year. During the heavy cropping year, removal of the excess blossom or fruitlets becomes absolutely important to ensure satisfactory development of colour, shape and size of the apples remaining on the trees and to encourage flower bud formation for the following year. This may be accomplished by chemicals or hand thinning of blossoms or fruits.

(a) Chemical Thinning

To regulate the production of marketable fruits and flowering in apple trees, chemical thinning of blossoms and young fruits is useful. The effect of fruit thinning on fruit size is probably related to leaf/fruit ratio. As this ratio is reduced below 30 : 1, fruit size is also reduced. The time at which fruit thinning is done is as important to fruit size as the amount of fruit thinning. Thinning in apple should be done within 30 days after full bloom to improve flower bud production and fruit size. This is important because the period of cell division in apple is brief, which ends approximately 20 days after full bloom. Thinning apples during this period stimulates cell division within the remaining fruits. This early period is also of critical importance for floral initiation, which will partially determine yield in the following season.

Chemical fruit thinning agents are influenced by many factors like climate, cultivar and internal tree factors etc. The most commonly used thinning agents are NAA, and carbaryl (napthyl-N-methyl carbomate), which reduce fruit set effectively in many apple cultivars. The application of 10 ppm NAA, 7 to 15 days after petal fall or when average fruit length is about 15 mm, is most effective for fruit thinning. Carbaryl (Sevin) at the rate of 750 ppm, applied 7 to 10 days after petal fall is also effective for fruit thinning of Red Gold cultivar. Napthalene aceta-amide (NAAm) at a concentration of 35-50 ppm is quite effective in McIntosh, Cortland and Ben Devis apple cultivars. Ethrel (2-chloroethyl phosphonic acid) has been found effective in fruit thinning of Golden Delicious.

(b) Hand thinning

During 'on' year, hand thinning of fruits may be employed to improve fruit size and quality of the remaining fruits. However, hand thinning should not be done until the natural fruit drop has been completed or when it becomes apparent, which apple fruits are going to drop naturally and which will be retained on the tree until maturity or near maturity. Although, it is difficult to give a specific guideline about thinning of fruits in apple, yet it is always desirable to give the determining factors for a proper load of fruits to be left on the tree after thinning. A proper balance between fruit to leaf ratio will ensure the production of high quality fruit.

1. Age, vigour and variety of the tree.
2. Size of fruit at thinning time and size of the fruit desired at the harvest time.
3. Condition of the fruit, *i.e.* shape and freedom from blemishes when thinning, it is desirable to space the fruit as required, leaving no more than one fruit per spur.

While removing the fruit, it is advisable to hold the stem between the thumb and forefinger and push the fruit from the stem with the other finger. Such method of hand thinning will avoid damage to the spur and other fruits. Early hand thinning is always desirable as it will be more effective to the control to flower bud formation for the next year's crop, since flower buds are initiated during four-to-six weeks period following full bloom.

B. Problem of Pollination and Fruit Set

Many apple varieties set very little fruits with their own pollen due to self-incompatibility, which are called self unfruitful. Some are partially self fruitful, whereas others are self fruitful. Even self fruitful varieties will produce larger and regular crops when cross pollinated. In Delicious group, the flower structure is such that the honeybee can extract nector without crawling over the anther and cannot transfer its pollen to the stigma, resulting in poor fruit set, unless interplanted with some other compatible varieties. Short duration of stigma receptivity is also bound to have adverse effect on fruit set. Knowledge of flowering period and value of varieties as cross pollinizer is necessary for fruit grower. Triploid varieties such as Baldwin, Staymari's Winesap are unsuitable as pollinizer as these produce non-viable pollen. The varieties classified according to flowering sequence and value as cross pollinizer are listed as under:

(a) **Early flowering:** Mclntosh, Black Ben Davis.

(b) **Mid-season flowering:** Golden Delicious, Tydeman's Early Worcestor, Winter Banana, King of Pippin, Yellow Newton, Rus Pippin, Lord Aambourne, Red Gold.

(c) **Late Flowering :** Rome Beauty, Granny Smith, Worcestor Pearman.

The varieties such as Jonathan, Golden Delicious, Yellow Transparent, Ben Davis, Wagener, Mcintosh, Grime Golden, Wealthy, York Imperial and Rome Beauty have been reported to be a very good pollinizer. In Himachal Pradesh, Tydeman's Early Worcester, Red Gold and Golden Delicions are most desirable pollinizers. Under normal conditions, 25 per cent pollinisers are enough, however, in areas where the unfavorable climatic conditions coincide with the time of bloom, atleast 33 per cent of three polliniser in equal ratio is desirable. Top working of a branch of commercial cultivar with polliniser, providing two beehives per acre of orchard and placing of flower bouquets of the pollinisers of Delicious cultivar trees also facilitate better pollination.

C. Pre-harvest Fruit Drop

In apple, there are three distinct waves of fruit drop, (i) early fruit drop resulting from un-pollinated or unfertilized blossoms, (*ii*) June drop, and (iii) Pre-harvest drop. Of these, pre-harvest drop is of great economic importance, which can cause serious crop loss.

The problem of pre-harvest drop is more severe in early cultivars such as Mcintosh, Tydeman's Early Worscestor where 40-60 per cent of the fruit drops. In mid-season cultivars *viz.* Red Delicious, Royal Delicious and Golden Delicious, the pre-harvest drop is about 15-20 per cent.

A definite relationship between the auxin content of the seeds and the abscission of fruits during various stages of development has been established. High amount of auxins appear in the seed 30 days after fruit set, which coincide with the formation of endosperm and cessation of post blossom drop. Later, a reduction in hormonal content in seed is observed approximately upto 75 days after petal fall, which rises again when the embryo attains the full size. In the final stage of the fruit growth, a rapid decline in seed hormone content is correlated to degeneration of endosperm causing pre-harvest fruit drop. These investigations have led to believe that high concentrations of auxins supplied exogenously may inhibit fruit drop. It has been observed that spraying of NAA (Napthalene acetic acid) three weeks before harvest reduced pre-harvest fruit drop. Chemicals such as 2, 4, 5-T (20 ppm), 2, 4, 5-TCPA (15 ppm) are also quite effective in controlling fruit drop.

D. Improper Colour Development

Poor colour development is one of the major constrains in the marketing of apple fruit. Poor colour development is the major problem of the orchards in the mid-hills and warmer areas. Poorly coloured fruits fetch low price in the market. At higher elevation, the maturity is delayed and farmers fetch poor price of their product due to glut at late arrival and deep colour.

Ethylene generating chemical ethrel (2-chloroethyl phosphoinc acid) @ 500-1200 ppm enhances fruit maturity and intensity of surface colour in apple. Lower rate may ripen fruit in 10 to 14 days and higher dose in 5-7 days. The spray should be done a week prior to expected harvest time. Ethephon causes rapid fruit drop. So it should be applied with NAA, 20 ppm or 2, 4, 5-7P 20 ppm. Ethephon sprayed fruit should not be left on the tree beyond 14 days of application to avoid over ripening. Ethephon treated fruit should be marketed for fresh eating as they have poor storage life. In several countries, fruit bagging has been recommended for proper colour development. In fruit bagging, fruits are covered with bags about 30-40 days before harvesting with bagrs. Such bags should be removed about 3-5 days before harvesting. Fruit bagging helps to induce good colour. In addition, bagged fruits are free from abrasions and chemical residues.

E. Alternate Bearing

In apple, some varieties tend to bear heavily in one year, and bear little or no crop in the following year. This behaviour has been referred to as alternate or binnieal bearing bahaviour. Scientists assume that due to heavy bearing in one year, crop exhausts the tree and thus tree fails to bear satisfactorily in the next year. Alternatively, poor fruit set caused by adverse climatic conditions at the time of flowering, lack of pollination and fruiting the following years and thus the cycle of 'on' and 'off' years are triggered. There exists cultivar differences in apple in respect to regularity of bearing, which can be categorized as regular (Rome Beauty, Golden Delicious, Jonagold, Tydeman's Early Worcestor, Stayman Winesap), moderate alternate (Delicious, McIntosh, Golden Spur, Starking) and strongly alternate bearers (Baldwin, Wealthy, Laxtons's superb, Miller's seedling, York Imperial) (Table 5,6).

It is difficult to outline the precise sequence of disturbances in the internal system participating in alternation and its perpetuating mechanism. In the year of a heavy crop (on year), the demand for the carbohydrate supply is such that few flower buds are formed for the next year. During the following year, when little or no crop is borne, the carbohydrate reserve in the tree builds up and an excessive number of fruit buds tend to form. Seeds of growing fruits in apple may also exert inhibiting effect on flower bud production. Seeds usually increase the number of persisting fruits and reduce self thinning. Both effects are due to enhanced production of growth regulators and induce strong sink activity.

There are several reports to suggest that seeds in the growing fruits have inhibitory effect on flower initiation in apple. Gibberellins produced by seeds cause inhibition of flower bud formation upon diffusion resulting into the 'off' year. In an alternate bearing cultivar, diffusible gibberellins have an early peak two weeks after bloom. But this peak is suppressed by the application of fruit thinning chemicals. In regular bearing cultivars, such peak occurs only 5 weeks after bloom, apparently too late to inhibit flower bud formation.

Control of Alternate Bearing

While alternation is still a very important problem in crops like mango, avocado, pecan, pistachio, but can be partially overcome in apple. All the control methods are aimed to reduce excess crop in a year or to increase bloom or set in the next year, thus achieving a delicate balance, which is essential for preventing alternation. Major methods suggested to overcome alternate bearing in apple are as follows:

(a) Use of regular or fairly regular bearing cultivars.

(b) Reduction of flower production in the 'on' year following 'off' year or flower bud development during 'off' year.

(c) Thinning of fruits/flowers during an early stage in 'on' year.

(d) Early harvest of heavy crop.

(e) Balancing of reproductive and vegetative growth through pruning and dwarfing rootstocks.

(f) Use of paclobutrazole (PP_{333}) @ 5 g a.i./tree.

The practical way of the control of alternate bearing in apples and other fruits is the use of regular bearing especially self compatible cultivars and the use of dwarfing or semi-vigorous rootstocks with response to regularity of bearing. Heavy pruning in the winter following 'off' year and mild pruning after the 'on' year will help in maintaining the optimal reproductive growth (spurs) year-after-year. The fruit or flower thinning at early stages during 'on' years and improvement of fruit set with use of plant bio-regulators during 'off year' can also avert the acute alternate bearing phenomenon. Sometimes, the summer pruning is also advantageous in retarding vegetative growth and enhancing flower bud differentiation in 'on' year.

REPLANT PROBLEM: CAUSES AND MANAGEMENT

Like many countries of the world, old apple trees in India have either outlived their economic bearing life or declined due to the adverse effects of non-curable insect-pests or diseases and/or natural calamities. Moreover, many growers want to introduce newly improved and highly productive varieties in place of their old apple orchards. However, due to scarcity of land, farmers are compelled to plant new apple plants on the old apple sites. However, it has been observed that apple orchards so planted have poor growth and productivity or such orchards do not bear good crop of quality fruits. This problem has been called as 'replant problem'. This situation is widespread throughout the world. Depending on the causal factors, it has been called as 'apple replant disease' (ARD) or 'apple replant problem' (ARP).

Causes of the Problem

There may be several reasons for poor growth of young apple trees but in this case, only soil related causes including bitoic (harmful microorganisms) and abiotic (nutritional deficiency or excess, soil pH etc.) factors have been discussed below:

A. Biotic Factors

Various types of harmful microorganisms present in soil may become the cause of apple replant problem. When we uproot the old apple tree, several microorganisms remain adhered to left-over portion of the trees, which cause great losses to new planting. Of these microorganisms, species of *Phytophthora*, and *Pythium* are mainly associated with apple replant problem. For example, *Phytophthora sylvaticum* has been identified as causal organism of apple replant problem in England and *P. irregulare* in Canada. In Hungary, *Rossilinia nectrix* and *R. hypogaea* have been reported with apple replant problem. In addition, the bacteria like *Bacillus subtilis* and *Pseudomonas florescence* have been identified as causal organism of apple replant problem in Canada and Sweden, respectively.

Some nematodes are also associated with replant disease in apple. A root lesion nematode (*Pratylechus penetrans*) is the most important pathogen. Other parasitic nematodes like dagger nematodes (*Xiphinema* spp.), ring nematode (*Macroposthoma* spp.), and root-knot nematodes (*Meloidegyne* spp.), which exist as ecto-parasites and feed on the surface of tiny roots, thereby causing excessive damage to the newly established plants in the old locations.

B. Abiotic Factors

Imbalanced soil nutrients (particularly N, P, K) can cause physiological stress that leads to replant problem. It has been demonstrated that growth of apple trees in soil from old apple orchards was generally increased by pre-planting treatment with mono ammonium phosphate, indicating that young apple trees need P for increased growth. Not only deficiency but the excessive fertilization prior to and after replanting can also become a part of the failure of replanted orchards. The high N doses in the form of ammonium nitrate drastically increase soil acidity and subsequently increase concentration of Al and Mn elements in the root zone, which become toxic and cause replant problem. Besides, heavy metals such as arsenic derived from pesticides, have also been implicated in apple replant problem. Soils with low as well as high pH are more conducive to replant problem than the neutral ones. This is mainly related to the non-availability of vital nutrients to the newly planted trees. The adverse effects of poor soil structure, drainage, aeration and orchard tillage on root growth can also be the important factors in the development of replant problem. Though the role of phyto-toxins has been well established in peach replant problem, these are also suspected to be involved in apple.

Management Strategies for Replant Problem

Based on the causal factor(s), different management practices have been recommended for the management of replant problem in apple. The recommendations include various cultural, chemical and biological methods *viz.*, site selection, liming to correct soil pH, intercropping, drainage improvement, fertilizer applications, soil sterilization and use of biocontrol agents as discussed briefly hereunder:

Liming

Liming has been found to be quite effective in soils in which *Dematophora* and *Phytophthora* are the causal organisms of apple replant problem. Deacidification of the soil also helps in elimination of replant problem. The elements like N and P have been reported to suppress the growth of replant disease caused by fungi and bacteria, and subsequently to promote the growth of bacterial antagonists to these causal organisms.

Intercropping

Intercropping with herbaceous crops and growing mustard, radish etc., for 2-3 years before planting new apple trees is highly useful. Similarly, growing of marigold (*Tagetes erecta*) reduces the population of nematodes and fungi in replant soil.

Soil Sterilization

Soil sterilization by solarization, fumigation, steaming etc., helps in reducing the population of harmful pathogens in replant soil. In general, pre-plant treatments are more beneficial than post-plant treatments. Of these treatments, although soil fumigation is comparatively expensive, hazardous and even cumbersome, yet it has proved to be most effective and is still a common practice. Fumigation greatly improves the tree growth in replant sites.

Use of Biocides

Biocides like dazomet, formalin, mancozeb, metham sodium, methyl bromide, and choloropicrin can be used in soil. Experiments have shown that soil treated with any of these biocides improved the growth of new plantings by 20-50 per cent in replant soil.

Biocontrol Measures

Some attempts have also been made to develop biological alternative approaches for the control of apple replant problem. For example, inoculation of apple seedlings with AM fungi (*Glomus* sp.) or with *Agrobacterium radiobacter*, increased tree growth by improving the availability of P and protecting the roots from infections, respectively.

Miscellaneous Approaches

In addition, we should take atmost care while selecting the site. In the field, there should be proper provision of drainage.

PLANT PROTECTION

A. Major Diseases and their Management

More than hundred diseases caused by fungi, bacteria, mycoplasma like organisms and viruses have recorded in apple. Major symptoms and control measures of important diseases of apple are given below:

Fungal Diseases

Apple Scab

This disease is caused by a fungus, *Venturia inaequalis,* which occurs throughout the apple growing belts in India. The first severe epidemic of apple scab in India occurred in 1973 in Jammu and Kashmir. The disease was reported in Himachal Pradesh in 1977 and in 1982-83 severe epidemic occurred.

The symptoms of scab appear as light-brown or olive-green spots on leaves and fruits, which soon turn black. The leaves may fall pre-maturely and fruits show scabby, knotty and misshapen appearance. Mild temperature and rainy weather are most favourable for the spread of scab. The recommended spay schedule for the control apple scab is given in Table 5.13.

Table 5.13: Recommended Spray Schedule for the Control Apple Scab in Himachal Pradesh

Stage	Fungicide (per/100 litre water)
Green tip	Dodine (100 g), captan (300 g) or ziram (300 ml)
Pink bud	Benomyl (50 ml), carbendazim (50 g) or thiophanate methyl (50 g)
Petal fall	Fenarimol (40 ml), hexaconazole (30 ml) or penconazol (50 ml)
Pea stage	Mancozeb (300 g), dodine (75 g) or zineb (300 g)
Fruit development (20 days after pea stage)	Benomyl (250 g), carbendazim/thiophanate methyl [20 (50 g] + mencozeb (250 g) or dithianon (50 g)
Fruit development (after 20 days of previous)	Mancozeb (300 g), captan (300 g), carbendazim (50 g), mancozeb (300 g), zineb (300 g) or propineb (300 g)
Pre-harvest (20-25 days before harvest)	Mancozeb (300 g), captan (300 g) or ziram (300 g)
Before leaf fall	Urea (5 kg)

This spray schedule should be used only when disease has in epidemic form. However, the spray can be reduced according the disease incidence and prevailing climatic conditions favouring disease spread.

A sound forecasting system has been developed for the prediction of apple scab based on weather conditions particularly atmospheric temperature and duration of leaf wetness conditions. Usually 17 to 23°C temperature and leaf wetness for 9 hours can induce the disease and the symptoms are expressed in 9 days if the disease inoculum is present. Such systems have been installed at Mashobra, Kotgrah, Kotkhai (H.P.).

Powdery Mildew

Powdery mildew is the major disease of apple and is found in all apple growing countries of the world. The disease is highly destructive in the nurseries where the young leaves are killed leading to dieback symptoms and death of seedlings, however, some of the cultivars are also susceptible to this disease at bearing stage in the field. Powdery mildew is caused by fungus *Podosphaera leucotricha*, which infects young leaves, blossom and fruit. Infection takes place when the temperature is around 20-22°C, sky is clear by day and dew occurs at night (RH>70 per cent). Young spur buds are only liable to infection during the early stages in their development, until the early fruitlet stage.

Symptoms

On the freshly infected leaves, the mildewed area appears on the lower surface as indistinct white patch. Later, it is more readily visible and may result in curling of leaves. Eventually, these patches may turn pale purple. The disease symptoms are observed immediately after bud-burst when all the freshly produced tissues are completely mildewed and resulting flowers and leaves appear as white rosette. Infected lateral buds may produce completely mildewed side shoots in the same manner. From the mildewed surfaces, numerous spores are released, which infect young unexpanded leaves, young buds, lateral buds on new shoots and growing points.

Management Practices

Following measures should be adopted to control this disease effectively in the fields:

1. Winter and spring pruning by removing the top few buds up to 1-2 inches. At the pink bud stage, the diseased blossom and leaves should be pruned with scissors or pinched off by hand.

2. Sulphur based preparations provide cheap and best control of mildew but some varieties are 'sulphur shy' and are injured when sprayed with lime sulphur. Four sprays of carbendazim (0.05 per cent), binapacryl (0.1 per cent), or morocide (0.1 per cent) given at dormant, bud swell, petal fall and fruit development stages are effective in reducing the disease. EBI fungicides such as tetraconazole(1.0l/ha), triadimenol (0.18 l/ha), hexaconazole (0.28 and 0.34 l/ha) and bitertanol (0.4 l/ha)sprayed at 15 days interval have been reported to control powdery mildew.

3. Delicious group of apple varieties is highly susceptible to this disease, however, crab apple cultivars *viz.* Renetka, Purpurnaya, Yantarka, Altarkya, Dolgoe and Saynets Karchenio and two apple cultivars Maharaja Chunth, Golden Chinese, *Malus zumi* and *M. baccata* -Shillong are resistant to powdery mildew. Recently, developed variety, Co-op29 has resistance against scab, cedar apple rust, fire blight and is moderately resistant to powdery mildew.

4. Coccinellid *Halyzia hauseri* feeds on the mycelium of *Podosphaera leucotricha* and provide 95 per cent control of powdery mildew in China within 20 days. *Ampelomyces* a common mycoparasite of powdery mildews over-winters as resting hyphae in the dried powdery mildew mycelia covering the shoots and in the parasitized ascomata of *P. leucotricha* on the bark and the scales of the buds. Soon after the bud burst both the fungi start their life cycle but *Ampelomyces* slowly follows the spread of its myco-host on infected apple leaves.

Collar Rot

Collar rot of apple is also known as common rot, basal rot or foot rot, based on plant part infected by the fungus. The disease has been reported from Canada, Newzealand, UK, Germany, Holland, Belgium and Korea. The earliest record of apple bark rot attributed to *Phytophthora omnivora*. It also attacks pear, peach, plum, apricot, almond and cherry. Saturated soil moisture coupled with high soil temperature of 20-25°C is favourable for its spread.

Symptoms

Collar rot appears as cankers on trunk at or near the soil line. At first, there is production of water-soaked patches on the trunk, which enlarges in all areas. The bark in the diseased plant parts is discoloured and turns into a moist rot in which necrotic tissues eventually turn dark brown and develop an alcoholic odour. In advanced stages, more areas are involved and the rotting is extended even to the main branches and roots. On roots, the infection always starts from the crown. On the above ground parts, the leading edge of the lesion is irregularly mottled and attains olive green, orange or brown colour, which merges imperceptibly into the healthy tissues. In the first few years, there may not be any effect on fruit yield, but later, trunk rot may cause even death of the trees. The fungus may cause fruit rot as well.

Management Practices

A number of preventive measures have been recommended to reduce the primary infection of this disease for its effective control as under:

1. Elevating susceptible scions out of the soil splash zone by high grafting at 40-70 cm above soil level.
2. Planting at properly drained sites.
3. Planting grasses is useful in controlling collar rot. The use of non-host crops as green manure in apple orchards has also been suggested.
4. Removal of soil from the tree base to reduce infection where the scion is at or below soil level.
5. Removal of crop refuse, weeds, long grasses and fallen fruits prevent the accumulation of inoculum.
6. Avoidance of injuries to stem and pre-plant dipping of seedlings in fungicides is useful

7. Injection of effective fungicides such as metalaxyl and fosetyl-Al into the soil near the crown region is very effective.

8. Copper fungicides can also be applied as paint or paste on the affected parts for treating wounds. The diseased portion is exposed by removing the soil at the crown region and is scarified upto healthy tissue, disinfected with methylated spirit or mercuric chloride and after its evaporation, Bordeaux paint, copper paint, metalaxyl paint or Chaubattia paste is applied.

9. The rootstocks such as M2, M4, M9, and M26, are resistant to collar rot and must be used.

10. *Enterobacter aerogenes*, a potential bacterial antagonist, has been isolated from orchard soil, which inhibits *P. cactorum* population.

White Root Rot

The disease was noticed for the first time on apple in 1890 in Norwich and later in 1913 from Canterbury. In 1925, *Rosellinia necatrix*, the perfect stage of *Dematophora necatrix* was identified as the cause of the death of apple trees. Since then, the disease has been reported from different parts of the world causing huge losses. The fungus has a very wide host range, comprising of fruit and forest trees, vegetables, cereals, ornamentals and other plants of economic importance. In India, the disease was recorded for the first time in Uttar Pradesh hills and later from Himachal Pradesh. It assumes serious concern in India under waterlogged areas and in black acidic soils. The disease is caused by *Rosellinia necatrix* (*Dematophora necatrix*). The disease is favoured by high soil moisture and temperature (20-25°C) in acid soils at pH 6.1 to 6.5. The pathogen spreads very fast along drip irrigation lines. In ill-drained soil, the disease develops very fast and mycelium spreads from infected to healthy roots, through soil particles.

Symptoms

In the initial stages, rooting of bark and yellowing of the foliage are the important visible symptoms. The rotting starts from the fine roots, which further extends to the upper parts ultimately reaching the tree trunk. After a few days of infection, the roots are covered with white cottony mycelial growth of the fungus, which may extend to adjacent areas. Subsequently, the mycelium on the roots turns greenish grey or black and aggregate to form thick hyphal strands. In advance stages, the roots are completely devoured and diseased seedlings or trees are easily uprooted from the soil. The diseased roots are sometime found covered with small sclerotia on the surface, which are rarely observed under Indian conditions. The infected trees give a sickly look and have are premature chlorosis of leaves. Rapid spread and development of pathogen may result in bronzing of foliage and premature defoliation. These symptoms are similar to those produced by other soil- borne pests and pathogens and thus under ground diagnosis of the disease is required for confirmation. The trees may get wilted and die after sometimes.

Management Practices

1. Trimming off of diseased roots in the early stages of infection and application of disinfectant paste.

2. Digging trenches to isolate the infected tree and soil turning to starve out the pathogen.

3. The disease can be controlled by soil solarization particularly in the early stages of infection and it has long term effect in controlling the pathogen in naturally infested soils.The pathogen is sensitive to temperatures higher than 40°C.

4. Soil amendment with neem cake and deodar needles has also been reported to reduce the incidence of the disease. Green amendments such as soybean, black gram, mung bean along with VAM *i.e. Glomus heterospoum* and *Glomus* sp. helps in reducing the disease and in improving plant health.

5. Poorly-drained soils should be improved by following central drainage system.

6. For application of fungicides, first the soil moisture of tree basin is brought to 30-40 per cent level and then deep holes (15-20cm) are made in the basin at 30cm apart with a crowbar. Subsequently, carbendazim (0.1 per cent) or any other effective chemical solution is poured through these holes. The treatment is done 3-4 times every year during the rainy season for suffering trees. Drenching of ailing trees with a combination of aureofungin (0.02 per cent) + copper sulphate (0.02 per cent) improves the disease control and increase the yield and growth of trees.

7. In nursery, diseased plants should be dipped in carbendazim suspension (0.05-0.1 per cent).

8. Some of the rootstocks *viz.* MM109, M16, MM104, are less susceptible to *D. necatrix* and must be used.

9. The antagonists such as *Trichoderma viride*, *T. harzianum* and *Enterobacter aerogenes* have been found to protect the plants from *D. necatrix* and improve plant health and prevent the appearance of the root rot symptoms .

Sooty Blotch and Fly Speck

Both diseases, are economically, especially on late-maturing cultivars and on scab-resistant cultivars that are grown without fungicides. Sooty blotch and fly speck diseases can cause serious losses to pome fruits in certain localities. These two diseases were first recorded in 1940. These are separate diseases but normally both are present on the same fruit. They cause only surface blemishes that detract from the appearance of the fruit. Sooty blotch will shorten the storage life of fruit due to increased water loss. Sooty blotch is caused by *Gloeodes pomigena* and fly speck by *Schizothyrium pomi*. Both these diseases appear annually in late summer with the onset of rains and foggy weather and continue to increase the severity with high atmospheric humidity. Both fungi overwinter on the twigs of numerous wild hosts as well as apple and pear. The spores are windblown throughout the orchard. Cool temperatures and rains or high humidity are ideal for infection. Infection can occur anytime after petal fall but is most prevalent in mid to late summer. Disease outbreaks are favored by extended periods of above normal summer temperatures combined with frequent rainfall and high humidity.

Symptoms

Sooty blotch appears as sooty areas on the fruit surface. Their outline is not definite and the smudge can be removed by rubbing vigorously. The sooty areas are olive-green to black in color. Flyspeck looks like true flyspecks in groups of 10 to 50 or more. These diseases usually appear on fruit late in the season.

Management Practices

Adopt the following measures for the control of these diseases effectively:

1. Dormant and summer pruning that opens up the tree canopy and facilitates air movement and the drying of fruit after a rain period will help in the control of these diseases.
2. Thinning to separate the fruit clusters will also help prevent sooty blotch and flyspeck. No cultivar is resistant to these diseases.
3. Give at least two sprays of captafol (0.15 per cent) or mancozeb (0.3 per cent) at 15 days interval so as finish the last spray 20 days before fruit harvesting.
4. Spray schedule adopted for apple scab also provides effective control of these diseases as well.

Pre-Mature Leaf Fall

This is called as Marssonina blotch, as it is caused by a fungus, *Marssonina coronaria*. This disease is now considered as more serious than scab in some parts of our country as it is more devastating than scab. As the name indicates, it is responsible for the excessive fall of leaves prematurely and untimely. Initially, dark green irregular areas develop on the upper surface of mature leaves, which later turn to brown spots usually 5-10 mm in diameter. The infected leaves turn yellow and fall in the mid-season. Leaf fall is so severe that only fruits are seen hanging on the defoliated branches. Symptoms on the fruits appear as brown spots of 4-6 mm in diameter, which become depressed and dark brown and almost black near harvest. Adopting spray schedule given in Table 5.14 can control it.

Bacterial Diseases

Fire Blight

The name 'fire blight' was first used in 1817 to describe the sudden browning of leaves as if they have passed through a hot flame, causing a morbid matter to exude from the pores of the bark. *Erwinia amylovora* bacterium causes this disease. In countries like India, where the occurrence of this disease has not been confirmed, strict quarantine measures have been enacted to prevent its entry.

Table 5.14: Recommended Spray Schedule for Controlling Premature Leaf Fall in Apple

Stage	Fungicide and its Quantity (per 100 litre of water)
Pink bud	Mancozeb (300 g)
Petal fall	Cabendazim or thiophenate methyl (50 g)
Fruit set (pea stage)	Dodine (75 g)
Fruit development-I (walnut size)	Mancozeb (300 g)
Fruit development-II (Half grown stage)	Cabendazim or thiophenate methyl (50 g)
Preharvest spray (20-25 days before harvest)	Mancozeb (300g)
Pre-leaf fall	Urea (5 per cent)

Symptoms

Infection occurs on flowers, twigs, fruits, branches and the trunk. Usually the first symptom appears on flowers, which is evident in early spring immediately after bud burst. Blossoms first appear water-soaked, then shrivel, wilt and turn brownish to black. The blight progresses into the peduncle and during warm humid weather bacterial ooze droplets exude from the peduncle. Young fruitlets often become infected and both the diseased blossoms and fruitlets may fall or remain attached to the tree. The flowers shrivel and fall or hang on the trees. After the blossoms, the succulent twigs or shoots and water sprouts or suckers are the next most susceptible parts of the plant. In a few days, infection can move 15-30 cm or more into the twig. Infected shoots bark and leaves usually appear light to dark brown. Blighted twigs and water sprouts often form a can like or shepherd's crook at their tips, a characteristic symptom of the disease. The affected parts become water-soaked from which white to clear droplets ooze out. The bacterial ooze also dries out. The infection progresses on to leaves. The leaves curl, shrivel, and usually hang down like flowers. Leaves may become infected after bacteria enter directly through stomata, trichomes, and hydathodes but more frequently through wounds. If infection occurs in the blade, necrosis develops. This part of the leaf may dry out, but infection spreads through the secondary veins into the midrib, then into the petiole and the stem. Terminal twigs and suckers are usually infected directly and wilt from the tip downward. The tip of the twig is hooked, and the leaves turn black and cling to the twig.

Current season's growth also gets infected and wilts from tip downwards. The infection slowly reaches the branches and trunk, and kills most of the parts slowly and steadily. These infections form the canker on branches and stem, which girdles them. If the branches are not girdled completely, the cankers remain dormant as sunken cracked margins. Bacteria invade the fruits through pedicels or lenticels or wounds or from infected spur into the fruit. A sticky, milky to amber coloured fluid collects at the core and sometimes oozes from the lenticels. Developing fruits become water-soaked, turn brown, shrivel, get mummified and hang on trees for months together. Finally, they turn dark-brown to reddish and fall down the tree. The disease symptoms may advance downward from the blossoms, shoots or fruit through the larger twigs and branches causing stem cankers. The cankers may be water-soaked, slightly sunken, varying in size and surrounded by irregular cracks in the bark. Active fire blight cankers have a dark, appearance. Cankers formed early in the season, especially the small ones, usually are surrounded by a callus. They may girdle entire limbs and thus kill that part of the limb above the girdle. Characteristic reddish-brown streaks are common in the sapwood.

Fire blight cankers at the collar region are usually referred as 'collar blight'. It may spread from the collar into the roots or sometimes from the roots into the collar. Bark in the roots is killed in the same manner as that on the trunk. Invasion of the crown and roots may occur through washings of bacteria from infected twigs and fruit down the trunk into the soil, and downward internal translocation of pathogen from infected plant parts to the roots. Such symptoms are common in apple rootstocks, used for dwarfing purposes.

Management Practices

Fire blight is a threatening disease, which can't be easily controlled by a few simple measures. Hence, the following measures should be followed in an integrated manner to achieve the desirable results:

1. Strict quarantine measures should be adopted during the import of fruits, seed, bud-wood and other plant parts.
2. Grow resistant varieties, whenever possible. Some of the apple varieties, which have resistance to fire blight are Red Delicious, Northwest Greening, Cox's Orange Pippin, Stayman, Winesap, Britemac, Carroll, Primegold, Pricilla, Quinte, Splendour and Viking.
3. Ensure better drainage in the orchards.
4. Sprinkler system of irrigation should be avoided in vicinities having problem of fire blight.
5. Excessive use of nitrogenous fertilizes should be avoided.
6. Pruning should not be heavy, because it favours vigorous growth, which is susceptible to fire blight attack.
7. Spray streptomycin (100 ppm) or oxytetracycline (100 ppm) during flowering. It should be repeated after 4-5 days. Slaked lime (20g/l) always gives better results and reduces population of _E. amylovora_ and provides complete protection from the disease. Application of prohexadione-Ca can reduce fire blight infection on shoot and blossoms.
8. Suppression of infection at primary infection stage by antagonists such as _Pantoea agglomerans_ has been reported.

Crown Gall and Hairy Root

Crown gall was first discovered in 1870 in the USA, but bacterium as its causal agent, was reported in Italy during 1907. It is found in almost all the countries on apple, pear, apricot, grape, peach, cherry, plum and almond.

Symptoms

Initially crown gall symptoms appear as small, round, soft overgrowths on the stem and roots, particularly near the soil line. As the galls or tumors start increasing in size, the surface becomes convoluted having dark brown outer tissues due to the death of peripheral cells. Anatomically, a gall is a sphere of disorganized vascular and parenchymatous tissues, having soft and spongy-to-hard texture, depending upon the amount of vascular tissues they contain. The tumor may surround the stem or root, or may be connected to the host through a narrow neck of tissues. Tumors on pome and stone fruits become woody and hard, appearing knobby and reaching up to 30 cm in diameter. Certain tumors on herbaceous host plants and vines are spongy and may become detached from the plant. Few tumors may rot during fall and winter and reappear during spring.

In addition to forming tumors, affected trees may become stunted and produce small, chlorotic leaves. Such plants are more prone to frost injury. Number of tumors on root and stem vary; they may be continuous or in bunches. Tumors on vines and trees may also appear upto 2 m from the ground as well as on branches of trees, on petioles and leaf veins. Small galls require careful diagnosis, because they are easily confused with excessive callus growth at wound sites or with galls induced by nematodes and insects. Hairy root symptoms on young apple trees are characterised by an extensive proliferation of adventitious roots, singly or in clusters, and wounds at the base of stem, crown and roots. Crown gall and hairy root symptoms can occur on the same tree. In some cases, tumors develop first, and roots emerge from the tumor surface. The crown gall is caused by a gram-negative soil bacterium, *Agrobacterium tumefaciens*, belonging to Rhizobiaceae family. *Agrobacterium rhizogenes* is the causal organism of hairy root.

Management Practices

1. Good sanitation and cultural practices are important for reducing the disease, which includes discarding all symptomatic planting stock, budding rather than grafting, choosing a rootstock with low susceptibility, and adopting management practices that minimize wounding.

2. Susceptible nursery stock may not be planted in areas known to be infested with *A. tumefaciens*. Such infested fields should be planted with monocots for a few years before they are planted with nursery stock.

3. Root damaging insects in nurseries should also be controlled to reduce disease incidence.

4. Nearly cent per cent control of crown gall is achieved by dipping seeds or root system of nursery seedlings or rootstocks upto crown region in *Agrobacterium radiobacter* strain K84 suspension.

B. Major Insect-Pests and their Control

More than 100 insect-pests attack apple but about a dozen cause economic losses. The most important insect-pests of apple found in India and their control measures are given below:

San Jose Scale (*Quadraspidiotus perniciosus*)

It is the most serious insect-pest of apple, which attacks pear and many other stone fruits and is thus widespread in prevalence. These scales are very small in size and grayish in colour and cover the entire surface of tree bark. Heavily infested trees appear as if covered with wood ash. Both young and adult scales cause damage by sucking the cell sap. Application of 2 per cent miscible oil or 5 per cent summer oil during February-March efficiently controls the pest. The summer oil formulations, like Orchex 796, Caltex 1 POL Summer oil, etc., can be applied at the rate of 1 per cent at petal fall stage. The natural predators of the pest such as *Aphytis procila*, *Chilocorus bijugus*, *Pharoscymnus flexibilis* and *Coccinella septempunctata* and endoparasitoid, *Encarsia perniciosi* have been found to contain scale population.

Woolly Apple Aphid (*Eriosoma lanigerum*)

This is a serious pest of apple. The insect has been found to migrate from root to shoot and *vice versa* throughout the year. The adults and nymphs cause damage by sucking cell sap from the young shoots, cut ends and cracks of the main limbs, trunk and roots, as a result, there is formation of galls on roots, stem and shoot and thereby, stunts the growth. Soil application of phorate or carbofuran granules during May and October/November checks its incidence and spread. The foliar spray of chlorpyriphos (0.02 per cent), fenitrothion (0.05 per cent), dimethoate (0.03 per cent) or phosphamidon (0.03 per cent) also controls the pest effectively. Malling Merton rootstocks like MM 106, MM 109, MM 111 and Merton 793 are resistant to this pest. Use of an endoparasite, *Aphelinus mali* is quite promising in controlling this pest.

Root Borer (*Doresthenes hugelii*)

It is a more destructive pest in sandy-loam soils in hilly areas up to a height of 2150 m. The grubs either bore or girdle the roots. Consequently roots get severed and branches wither and die. However, damage symptoms are often observed when substantial loss has been caused. The chemical control for the pest includes drenching of basins of trees with chlorpyriphos or triazophos (0.04) during September, besides trapping and killing the adults during June-to-September.

Blossom Thrips (*Thrips flavus*)

Several species of thrips attack flowers of apples. Hot and dry weather conditions are more conductive for quick build up of its population causing extensive damage. Young and adults feed within the flower buds and cause extensive damage to blossoms by sucking cell sap. Heavily infested flowers fail to open and fall down. Chlorpyriphos (0.04 per cent) or fenitrothion (0.05 per cent) at pink bud stage is recommended to control the pest if population exceeds 15 per flower bud.

Codling Moth

Presently this dreadful pest is confined to Leh and Kargil region of Jammu and Kashmir. The larvae of the pest enters the fruit from any point on surface and tunnels upon seeds and cause excessive damage in core region. The pest is found both on apple and pear. The control strategy for this problematic pest includes mass pheromone trapping (25 traps/ha), collection and destruction of over-wintered pupae in cocoons during April-to-June, deep burying of fallen fruits during August in addition to 2 sprays of DDVP or phosphorus (0.04 per cent) during June-July at an interval of 2-3 weeks. Use of egg parasitoids of *Trichogramma* spp. has shown great potential against this dreadful pest.

European Red Mite

It is spreading in different apple growing belts on an alarming rate in Himachal Pradesh. The pest extracts vital plant juice and chlorophyll from the leaves causing discolouration and premature defoliation and ultimately affecting the fruit yield. The sprays of miscible oil at late-dormant stage provide effective control of mite at the egg stage. Spray of dicofol (0.05 per cent) followed by malathion (0.05 per cent) can also provide some control. The reduction of mite population has also been observed by the predators like coccinellids, predatory mites and anthocorid bugs.

Sporadic incidence of other pests such as defoliating beetles, apple leaf rollers, apple fruit moth, tent hairy caterpillar and shoot borers has also been noticed in certain apple growing areas.

C. Physiological Disorders and their Management

Physiological disorders are abnormalities of the fruit that are not associated with diseases or insect-pests. They can appear during the growing season or after harvest when the fruits are stored, and affect the appearance and usability of the fruit. In apples, such disorders include cork spot, bitter pit, Jonathan spot, water core, internal breakdown, and storage scald.

Bitter Pit

Bitter pit develops late in the growing season or in storage. Its symptoms appear as small, dry, brown pockets usually spherical in shape below the peel and also in the cortex. Cultivars like Northen Spy and Delicious group is more prone to bitter pit. Early picked apples are more prone to bitter pit. In addition, unbalanced or high nitrogen promotes its incidence. However, calcium deficiency is the primary cause of bitter pit in apples. Spraying calcium chloride (0.4 per cent) or dipping apples in calcium chloride (2 per cent) is effective in reducing bitter pit. Boron and zinc sprays increase the Ca content of fruit and reduce the incidence of bitter pit effectively. Similarly, application of adequate amount of 'N' and 'K' is equally important. Harvesting should not be too early or late.

Cork Spot

Cork spot occurs during the growing season, and is characterized by localized green-to-brown sunken spots on the fruit or in the pulp. The spots in the pulp, which may or may not develop just under the skin, are brown, corky, dead tissue, and are more prominent towards the calyx half of the fruit. The spots in the flesh have a bitter taste. Under severe conditions, fruit cracking can occur.

Jonathan Spot

Jonathan spot occurs late in the growing season or in storage. It is characterized as small, brown-to-black spots on the peel that may or may not be sunken. The spots develop most frequently on the sun-exposed side of the fruit, and are often associated with the lenticels (dots on the skin's surface). The flesh under the spot has a water-soaked appearance, but does not develop an off-flavour.

Water Core

Water core occurs before harvest, and is characterized as translucent, water-soaked areas in the pulp. These water-soaked areas commonly develop near the core, but in severe cases, can radiate out to the peel. Fruits with water core are edible, and may have a sweeter flavour, but their storage life is reduced. Mild cases of water core can disappear in storage.

Sun Burn

Sun burn to apple fruits can occur in all apple growing regions but is especially a problem where temperatures are high and skies are clear. Found more frequently on fruits on the southwest quadrant of the tree. Heavy crops that cause branches to bend over mid-season can increase sunburn incidence as a sudden exposure to heat and sun promotes sunburn development. Water stress can increase the incidence of sunburn. Granny Smith and other light-skinned apple varieties are more sensitive to sunburn. Also, sensitivity may be associated with low calcium concentration in the fruit.

Initial symptoms are white, tan or yellow patches found on the sun-exposed side of the fruit. With severe skin damage, injured areas can turn dark-brown on the tree. White, tan or yellow patches often turn brown within a few weeks in cold storage (sometimes called sunburn scald but not controlled by antioxidants, such as diphenylamine, as is storage scald). During unusual hot weather following cool or moderate conditions, and especially when accompanied by water stress, injury to the skin and flesh can occur. Injured cortex tissue is brown and firm and may become spongy and sunken. Fruit exposed to the sun after removal from the tree, either on the orchard floor or in field bins, can develop severe sunburn. It can be reduced by the following management practices:

1. Avoid sudden exposure of fruit to intense heat and solar radiation.
2. Adopt proper tree training and pruning to avoid excessive sunburn.
3. Avoid water stress in cropped orchard trees to reduce heat stress.
4. Careful sorting to remove affected fruit upon packing is the only solution once the injury has occurred.

Internal Breakdown

Internal breakdown is a storage disorder that is characterized by a browning of the flesh and eventual softening and disintegration of the fruit. All fruit will eventually develop internal breakdown. The cultivar and storage conditions determine the onset of normal internal breakdown. However, its premature development can be a serious problem.

Storage Scald

Storage scald is a disorder that results in brown discoloration of the skin during storage. It is usually only a problem when fruits are stored for a long period. Fruit that develop storage scald are edible, but not very attractive.

Management of Physiological Disorders

Although, low levels of calcium have been associated with these physiological disorders, calcium sprays alone will not eliminate the problem. However, it is necessary to integrate orchard cultural practices to reduce the incidence of these disorders. The effective cultural practices are:

1. Establish apple orchards on well-drained sites/locations.
2. Irrigate the orchards regularly during periods of drought.
3. Control nitrogen fertilization to avoid excessive vegetative growth and over-sized fruit.
4. Prune the tree moderately to maintain a proper balance between vegetative growth and fruiting.
5. Promote normal crop yields from year-to-year by proper pollination management and thinning of fruits to avoid biennial bearing.
6. Harvest the fruits at the proper maturity. Apples harvested too early are more prone to storage scald, and bitter pit. Late harvested apples are prone to Jonathan spot, water core, and internal breakdown.
7. Spray trees with calcium chloride in years when unfavourable soil moisture conditions develop in the spring.
8. Dip the harvested fruits in calcium chloride (0.05 per cent) solution before storage.

HARVESTING AND YIELD

Apples should be harvested at proper stage of maturity as immature fruits develop poor flavour, have poor quality and usually shrink during storage. Over-mature fruits are also poor in quality and are more prone to storage disorders. Several postharvest management techniques for apple have been standardized in India to reduce postharvest losses and extend the shelf and storage life.

POSTHARVEST MANAGEMENT

Maturity Indices

Apple fruit undergoes several changes in physical, chemical and physiological characters after which it becomes ready for harvesting. The time required for maturity depends on cultivar and climatic conditions. However, the degree of maturity for harvesting is of prime importance because the concerted efforts put in by the growers for producing quality fruits are wasted if the fruits are not harvested at appropriate stage of maturity. Several maturity indices have been standardized for harvesting apple at a right stage of maturity. However, in practice, more than one methods are used as no index used alone is reliable measure of readiness of fruits for harvest. Each method has its own advantages and limitations. Different maturity indices used in apple are as under:

1. Days from Full Bloom to Harvest

The number of days from full bloom (DFFB) to harvest is the most reliable maturity index for harvesting apple at a right stage of maturity. The number of days almost remain constant under a wide range of climatic and cultural conditions. The dates of full bloom should be recorded when about 60 per cent of the flowers of the northern side of tree are in bloom. This period varies from 90 ± 4 days (Tydeman's Early Worcesster) to 180 ± 5 days (Granny Smith) in apple depending on cultivar, climatic conditions and the elevation (Table 5.15).

Table 5.15: Some Recommended Maturity Indices for Apple

Cultivar	DFFB (Days from Full Bloom)	Firmness (kg/sq. cm)	TSS (per cent)
Tydeman's Early Worcester	90±4	7.8 + 0.15	12.0-13.0
Starkrimson	103±3	8.2 + 0.20	11.0-13.5
Lod Lambourne	103±3	8.2 + 0.20	12.5-13.5
Royal Delicious	120±5	8.2 + 0.40	13.0-15.0
Red Gold	122±3	8.3 + 0.20	12.0-15.0
Rich-a-red	128±3	8.6 + 0.25	12.0-13.0
Red Delicious	134±5	8.4 + 0.40	11.0-14.0
McIntosh	135±6	6.8 +0.25	11.5-13.5
Golden Delicious	148±6	8.4 + 0.40	12.0-14.5
Granny Smith	180±5	8.7 + 0.30	11.5-13.0

2. Calendar Dates

Based on the experience of the orchardist over the years, optimum maturity dates can be specified for harvesting the crop in the following years. Orchardists usually start harvesting their crop on a particular date. It is one of the most commonly used maturity index and is reasonably accurate. For example, orchardists start harvesting 'Royal Delicious' after 15th August in Kullu Valley and by 25-30th August in Shimla hills and by 2nd week of September in high hills (Table 5.16).

Table 5.16: Calendar Dates for Harvesting Apples in India

Cultivar	Calendar Dates	Place
Early Shanburry	4th week of June to 1st week of July	Chaubatia (UK)
Golden Delicious	28th August	Mashobra (HP)
Red Delicious	25-30th August	-do-
Rich-e-Red	5-10th September	Rohru (HP)
Royal Delicious	15th August	Kullu (HP)
	4th week of August	Chaubatia (UK)
	20-25th September	Kinnaur (HP)

3. Peel Colour

There is increase in red colour in most of the apple cultivars as the fruits approach maturity. Although, colour is not considered as an index of enough practical value for several apple cultivars, yet it is recommended that when fruits attain 70-85 per cent of red colour on their peel, they are ready for harvesting. However, in practice, orchardists start harvesting apples when they attain 50 per cent or more red colour.

4. Fruit Firmness

Fruit firmness declines gradually as apples mature. It is considered as one of the most reliable maturity indices for harvesting apples at a right stage of maturity. Fruit firmness varies from variety-to-variety. However, fruit firmness of 16-18 ibs/square inch is considered optimum for all the varieties. It is also used as an indicator for determining their ultimate destination (Tables 5.15 and 5.17).

Tables 5.17: Recommended Firmness of Apples in Relation to Destination

Destination	Firmness (lbs/2.5 cm²)
Local markets	13.0 – 14.0
Short distance markets	14.5 – 15.5
Long distance market	18.0 – 20.0
For export	18.0 – 22.0

5. Total Soluble Solids

TSS is an important chemical method for assessing maturity of apples. TSS depends on several factors including the variety, elevation and climatic conditions because of great variation in TSS values of apple cultivars. For instance, the TSS of Red, Royal, Richared and Golden Delicious apple cultivars in orchards at 1,500-2,000 m above mean sea level were 13.3, 15.01, 14.03 and 12.1 per cent, respectively whereas these contents were 14.1, 14.4, 15.0 and 11.5 per cent, respectively at 2,250-2,750 m above mean sea level. Hence, some researchers don't consider TSS as a reliable maturity index. However, TSS of 11-14.5 per cent is considered best for harvesting several varieties at appropriate maturity (Table 5.15).

6. Titratable Acidity

A few researchers consider acidity as a good maturity index for harvesting apples. The acidity is greatly influenced by several factors such as variety, elevation and locality. For example, the acid content of Delicious group of apples grown at different elevation varies from 0.20 – 0.41.

7. Starch Content

Now-a-days, starch pattern index (SPI) is used as an important maturity index. For judging maturity, starch-iodine test is performed to correlate the correct picking time of apples. A starch index of 4.5/10 for Delicious and 4.0/10 for Golden Delicious has been recommended for harvesting apples at a right stage of maturity.

8. Specific Gravity

A specific gravity between 1.03 to 1.06 has been recommended for harvesting apples at full maturity. Although, a few researchers has reported this parameter to a unsuitable index for determining optimum maturity in apple.

9. Respiration Rate

It has been recommended that apple should be harvested with the onset of respiratory climacteric.

Grading

Proper grading of the produce is one of the most important operations to bring uniformity, avoid wastage, reduce marketable cost, fetch better returns and to establish good trade in the market. The efficiency of grading depends upon careful handling of fruits during harvesting, sorting and grading.

Apples are usually harvested manually. Picking is done by lifting the fruits from bottom followed by gentle twisting with fore fingers. Apples during this operation should be handled carefully to prevent bruising injury. The harvested apples should be put in baskets lined with gunny cloth bags or polyethylene sheets and placed into krates and carried further to packing and grading sheds or rooms. In general, apples are graded into different grades depending upon their soundness, visual appearance, typical varietal characteristics etc., followed by size grading.

A. Quality Grading

Depending upon the shape, colour, soundness, freedom from defects like injuries, blemishes, disease spots, bruises etc., apples are commonly graded into 3 recognized quality grades, *viz.*'A', 'B' and 'C'.

Grade A

Apples of 'A' grade have the best shape and colour of the variety. They are sufficiently mature, free from blemishes, injuries or bruises.

Grade B

Apples with slightly abnormal shape, 50 per cent colour characteristics of the variety are graded into 'B' grade.

Grade C

Apples, which are not graded into 'A' or 'B' grades and can't fetch good price in the markets are classified as 'C' grade apples. Such fruits are mainly used for processing purposes.

B. Size Grading

In apple, the following proportion of fruit size of different grades is found on a tree; (i) Large, super and extra large sized apples: 20-24 per cent (ii) Medium-sized apples: 35–40 per cent (iii) Small-sized apples: 15-20 per cent, and (iv) Extra small apples, pittoo and culls: 20-25 per cent.

Table 5.18: Conventional Method of Size Grading in Apple

Grade	Minimum Fruit Diameter (+2.5mm)	Grade	Minimum Fruit Diameter (+2.5mm)
Super Large	85	Extra Large	80
Large	75	Medium	70
Small	65	Extra Small	60
Pittoo	55		

Therefore, quality grading should always be followed by the size grading. Each quality grade is separated into different size grades before packing. In India, apples are usually graded into 7 size grades with the conventional method of measurement with finger and thumb by the orchardists (Table 5.18).

Methods of Grading

Manual size grading consists of holding the fruit in left hand at the broadest point and placing the fingers of the right hand in between the uncovered space. The size grade is then determined by referring to Table 5.18, which depends upon the number of fingers that can be accommodated in the uncovered space.

The method of size grading has been further improved by the introduction of hand-grader. This operation consists of sliding or passing the fruit gently through the hand-grader from the broader side and taking out the fruit, when it reaches the appropriate size grade However, manual methods of grading and packing are both time-consuming as well as labour intensive. Grading, and packing alone add about 15 per cent of the total cost of standard wooden box. These operations can, however be, simplified by using mechanical graders.

Mechanical grading coupled with visual screening for colour, diseased spots and uniform shape improves the overall efficiency of the whole operation. Besides, it also lowers down the labour cost in packing. In Himachal Pradesh, modem mechanical grading of apples was introduced by the HPMC (Horticultural Produce Marketing and Processing Corporation Limited) in the early 1980's with the establishment of 11 packing and grading houses in apple growing areas. Some local growers avail this facility of grading their produce through mechanical graders after paying some nominal charges. Since mechanical grading is more efficient in time, labour and uniform sizing, commercialization of this facility in all apple growing regions of the country will possibly bring uniformity in all Indian apples, especially for export purposes. However, keeping in view the high initial investment, such grading and packing houses need to be installed under government, private or cooperative sectors, where growers can, in turn, pay the reasonable charges for this facility. Now some orchardists have established their own grading units.

Packaging

Packaging is one of the major factors influencing quality of the produce during transportation, storage and marketing. Proper fruit packaging not only reduces bruising and compression injuries but also checks moisture loss, microbial spoilage and pilferage. It also helps in maintaining the freshness and eye appeal of the fresh apple. Thus, packaging is a coordinated system of preparing apples for transport, distribution, storage, retailing and end use. Different types of packing materials used in apple are:

Gunny Bags

Packing of low-grade culled apples in gunny bags is an age-old practice. Such fruits are meant for processing purpose. Before filling apples in gunny bags, a layer of pine needles is placed at the bottom to serve as a cushioning material. After packing, the bags are shaken to settle fruits, padded at top with cushioning material, and then sewed tightly to prevent vibration bruises during transport. However, this method is not recommended for packing quality apples for fresh consumption, since packing in gunny bags give rise to high incidence of bruising injuries and rotting.

Wooden Boxes

Substantial quantity of apple is conventionally packed in wooden boxes. Generally, these boxes are 30, 28 and 25 cm in height for packing large, medium and small-sized apples, respectively. The fruits are packed by lining the inside of box with newspaper sheets and keeping margins for overhanging the flaps. The wrapped fruits are initially padded with wood wool/pine needles at the bottom of box and later in between well arranged layers. The top layer of fruits is covered with paper by bringing together the over-hanging flaps followed by nailing. For packing, wrapped fruits in wooden boxes, the arrangement for different grades of apples is given in Table 5.19.

Table 5.19: Arrangement of Different Grades of Apples in Wooden Boxes

Grade	Size of Box (cm)	Size of Wrapping Material (cm)	No. of Fruits/Box	No. of Layers
Super Large	45.7 × 30.5 × 27.9	27.9x27.9	54-57	3
Extra Large	45.7 × 30.5 × 25.4	26.7x26.7	60-63	3
Large	45.7 × 30.5 × 30.5	25.4x25.4	96	4
Medium	45.7 × 30.5 × 27.9	24.1x24.1	112	4
Small	45.7 × 30.5 × 25.4	22.8x22.8	128-132	4
Extra Small	45.7 × 30.5 × 25.4	21.5x21.5	160	5
Pittoo	45.7 × 30.5 × 25.4	Not wrapped	Variable	Loose

However, packing apples in wooden boxes results in high incidence of bruising injury and more loss in fruit weight. The loss in fruit weight is due to its water absorption by the timber and padding material.

Corrugated Fibre Board Cartons

Corrugated Fibre Board (CFB) boxes are made of fluted paper, which is pasted on flat paper sheets. Such boxes are also prepared from bagasse and paddy husk. Similarly, fruit packing trays, can also be made from non-woody materials, thus leading to substantial saving in woodltimber. In India, tray-packed telescopic cartons (CFB) owe their introduction in apple trade during 1980's. These cartons are suited to withstand various transportation hazards besides causing minimum weight loss and bruising damage to fruits. For packing in tray packed telescopic CFB containers, apples should be arranged in layers (Table 5.20).

Table 5.20: Arrangement of different Grades of Apple in CFB Cartons

Grade	No. of Layers	No. of Fruits/Layer	No. of Fruits/Box	No. of Trays/Box
Super Large	4	18	72	5
Extra Large	4	20	80	5
Large	5	20	100	6
Medium	5	25	125	6
Small	5	30	150	6
Extra Small	5	35	175	6
Pittoo	Loose		Loose	

Plastic Crates

Plastic crates (collapsible and non-collapsible) have also been recommended and introduced to some extent in apple trade in Himachal Pradesh; as field boxes for collection of fruits, for stacking of fruit in cold stores; carriage to nearby markets; and to supply fruits to processing units. However, impact bruising of fruits packed at the bottom of these crates has been noticed by many orchardists. Bottom padding with suitable material like thin foam sheet can possibly overcome this problem.

Wrapping Material

Newsprint paper, tissue paper, polyethylene liner or bags, paper moulded trays are commonly used wrapping material in apple.

Storage of Apples

Different methods of storing apples and some pre-storage treatment have been briefly discussed hereunder.

Pre-cooling

Quick removal of field heat from the fruits is a pre-requisite for an effective storage of apples. Pre-cooling is particularly essential for Delicious apples, which otherwise show poor shelf life during August and September when temperature is high. Ambient temperatures in apple-producing areas normally range between 15 to 25°C during August-September, while in Delhi (a major terminal market), it ranges from 40 to 44°C. Thus, rapid cooling of fruits in the production areas is a key factor for their successful marketing or for their cold storage. However, mechanical pre-cooling to 7°C requires huge capital investment for building, machinery and energy distribution. Alternative means of pre-cooling fruits under Indian conditions are: picking fruit in early morning, evaporative cooling, and/or watering down and forced air over stacked fruits etc. Keeping fruits over night near a tree basin is an another practical way to remove their field heat.

Low Temperature Storage

The recommended storage temperature for each cultivar is the temperature, which is most effective in retarding ripening and growth of decay producing organisms without causing freezing injury. For most apple cultivars, optimum storage temperature is –1 to 0°C with 90-95 per cent relative humidity. The apples held at –1.1°C require 25 per cent longer time to ripen than at 0°C. At 4.4°C, the extent of ripening is double than at 0°C, while at 15.5°C, it is 3 times more than at 4.4°C. Therefore, Delicious apples show higher storage life at –1°C than at 0°C. However, somewhat higher storage temperatures than –1 to 0°C are recommended for some cultivars because of their susceptibility to disorder induced by low temperature. Jonathan apples in some areas often develop soft scald in regular cold storage at 0°C. Therefore, they should be stored at 2°C. Mcintosh apples often develop brown core during extended storage at 0°C, as such, they are stored at 2- 3°C. Fruit ripening is much faster at warmer temperatures, with reduced storage life. Decay and other disorders such as Jonathan spot and bitter pit may be worse at storage temperatures higher than –1 to 0°C. On tonnage basis, apples are stored more than any other fruit and their average storage life is also longer. To avoid gluts and regulate supply in consumer markets, setting up of cold storage facilities in the apple producing areas, is therefore, envisaged

Air-Cooled Stores

Air-cooled stores are used to keep apples on the farms, in areas experiencing low night and high day temperatures. The temperature in such on-farm stores is brought down by opening air inlets/vents near the floor and the exhaust openings near the ceiling across the opposite walls to admit cool air at night. The vents and inlets are closed when outside temperature rises above the storage room temperature. Efficiency of these stores can be improved further by using evaporative coolers in areas, which experience high temperature and low humidity. Stacking of apples inside the store must allow the free movement of air. Humidification of incoming air can also be used to restrict water loss.

Evaporative Coolers

Owing to energy shortage and its rising cost, evaporative (dessert) coolers are becoming more popular as a substitute to air conditioners to lowerdown the room temperature in dry and hot climate. Dessert coolers consume only one-fifth energy per unit compared to refrigerated cooling, besides creating desirable humidity (around 80-90 per cent), which is difficult to obtain in refrigerated cooling system. A simple cooler using a forced draft air (by an exhaust fan) through wet porous walls has been successfully used for cooling and short term holding of grapes and citrus fruits with negligible wilting and weight loss. Such facilities are required to be created for apples near market centres in plains having dry and hot climate.

Zero Energy Cool Chambers

An on-farm, low cost, low energy, cooling chamber has been developed at the IARI, New Delhi, by using locally available materials. The double-walled chamber constructed by bricks, sand and bamboo cover is kept soaked with water. The evaporative cooling effect reduces the inside temperature by about as 17-18°C and maintains a relative humidity of more than 90 per cent during peak summer. A large number of fruits and vegetables have been evaluated for their suitability for storage in this chamber with encouraging results. It has good scope for use in apple storage.

Controlled Atmosphere Storage

Controlled atmosphere (CA) storage at 3.3°C in 2-5 per cent CO_2 and 3 per cent O_2 prolongs the storage life of apples cultivars such as Mcintosh, Newton, and Cortland, which do not keep well at $-1.1°C$ to $0°C$. However, it can also be used for Royal Delicious, Golden Delicious, Rome Beauty and Stayman, Winesap, Early Worcestor Ethylene plays an important role in fruit ripening and high CO_2 and low O_2 sometimes acts as a deterrent in ethylene action, thus delays the onset of endogenous ethylene production by apples. Although, there is no evidence to conclude that CA storage increases storage life by retarding ethylene production, but it does slow down the metabolic activity of fruits. The lower concentration of O_2 limits oxidation process of respiration. Thus, modifying the cold storage atmosphere to 2-3 per cent CO_2 and 2-3 per cent O_2 slows down the rate of respiration of apples, thereby, extending its storage and post-storage life. Some recent developments in CA storage of apples include use of ultra low oxygen CA (1.0-1.5 per cent O_2) rapid establishment of CA and ultra low O_2 and CO_2 (1.0-1.5 per cent O_2 and 1.0 to 1.5 per cent CO_2) CA- systems, in conjunction with C_2H_4 removal or without its removal.

Sub-Atmospheric Storage

The storage of apples under reduced pressure conditions improves their quality. Under hypobaric storage, apples retain characteristic flavour and texture. The apples held under low pressure storage ripen much slower than CA stored ones. Similarly, storing apples at 500-600 mm Hg pressure is also effective to reduce their weight loss and delay the ripening for about 5 months.

Removal of Ethylene during Storage

Ethylene is the principal ripening harmone in climacteric fruits like apple. The control of ethylene biosynthesis and its action are, therefore, key factors to check ripening of fruits and increasing their shelf-life. Use of chemical methods to alter the undesirable effects of ethylene require less sophisticated equipment and are less expensive. Of the chemicals, only potassium permanganate ($KMnO_4$), a strong oxidizing agent, has shown promise for its commercial application. Red Delicious apples packed along with sachets of 'Purafil' or $KMnO_4$ in CFB cartons showed delayed ethylene evolution up to 90 days of storage at 0°C. Potassium permanganate ($KMnO_4$) presumably absorbs and oxidizes the ethylene produced by the fruits to ethylene glycol, which, in turn, keeps the concentration of ethylene lower in vicinity of the fruits and thus, partly delays their ripening process and helps in extending their storage life.

Ethylene antagonistic properties have also been demonstrated in silver nitrate, 2,4- DNP, cycloheximide (CHI), benzothiodiazole and sodium benzoate. Pre-harvest spraying of benzothiodiazole (200 ppm), silver nitrate (100 ppm), 2,4-DNP (200 ppm) and sodium benzoate (500 ppm) were effective in enhancing the overall quality of Starking Delicious apples both at harvest and after storage.

POSTHARVEST DISEASES AND THEIR MANAGEMENT

In apple, about 25-50 per cent losses occur worldwide. Most of these losses are caused due to postharvest diseases. Postharvest diseases in apple have been categorized into two groups *viz.* in the first group, latent infections in the orchard, which are initiated late in the summer. In such infections, the infection on the fruit may not be visible but develop quickly after harvesting. Heavy rains in apple growing areas favour the development of such infections. The second group consists of those pathogens, which enter through wounds after harvesting (*Penicillium, Aspergillus* and *Rhizops* sp.). Most of the postharvest problems are caused by the pathogens that invade apples after harvesting. The major postharvest diseases of apple are as under:

Alternaria Rot

This disease is caused by *Alternaria alternata*. Its typical symptoms are the development of brown-to-blackish, slightly sunken and circular lesions. The lesions gradually enlarge with ivory black centre surrounded by a deep-brown zone. The rotting is dry and rather shallow, rarely exceeding 10.8 cm in diameter and 1cm deep. Dark grey mould may colonize the core of fruits and proceed to rot the surrounding tissues.

Aspergillus Rot

Aspergillus rot is one of the important postharvest diseases of apple. It is caused by *Aspergillus niger* and appears mostly in injured fruits or those stored under warm conditions. The symptoms appear as small, water-soaked, circular and brown spots, which later increase in size and turn black.

Bitter Rot

It is caused by *Glomerella cingulata* and cause heavy looses under warm wet conditions. Brown and dark-brown spots of varying sizes from specks-to-large lesions, covering the whole side of fruits are initial symptoms. The partially-decayed apple has a bitter taste.

Blue Mould Rot

Blue mould rot is the most destructive postharvest disease of apple. It is caused by *Penicillium expansum*. Its initial symptoms are soft, watery spots, which undergo rapid enlargement between 20-25°C. There is a distinct margin between soft rotten and firm healthy tissues. *P. expansum* is a normal pathogen, although it may also enter through lenticels. Bruised, injured, over-mature or fruits stored for a long time are more susceptible for its attack.

Grey Mould Rot

It is also a serious postharvest disease of apple caused by *Botrytis cinerea*, and causes heavy losses. Its symptoms include development of small circular greyish spots on the fruit, which enlarge slowly, covering the greater part of the fruit. It is difficult to sell the affected fruit.

Brown Rot

It is caused by *Monilinia fructigena* and *M. laxa*. Appearance of small circular, brown spots are its initial symptoms. The pathogen enters through lenticels, stalk end, styler end and wounds.

Management of Storage Diseases

Cultural Practices

To reduce primary inoculum of postharvest pathogens, strict orchard sanitation and maintenance of proper tree vigour are recommended. Prevention of mechanical and insect injuries to fruits in the orchard help in reducing fruit decay during storage. The incidence of bitter rot may be reduced by pruning dead wood and removing mummified fruits from the orchard. Similarly, to reduce losses from Mucor rot, it is advisable to collect and destroy fallen fruits before they are colonized by the fungus. Such a measure helps to minimize the build-up of inoculum in the soil.

In irrigated orchard, trickle irrigation is preferable to sprinklers to reduce losses from bitter rot. In areas where Phytophthora rot is serious, use of dwarf rootstocks and pruning systems, which encourage bearing of fruits near the ground should be avoided. Control of grey mould to some extent may be obtained by keeping the area under tree canopy free from grasses and weeds, thereby permitting better air circulation and reducing dampness and other conditions favourable for disease spread.

Pre-harvest Application of Chemicals

Sometimes orchardists overlook that postharvest diseases are related to field problems. Occurrence of postharvest diseases has a close link with production in the field. Thus, the first attempt to reduce postharvest losses is by preventing infection in the orchard. Spraying of fungicide at blossom time is recommended to control Altenaria rot, whereas spraying insecticides protects fruits from insect attack, and reducing the incidence of primary. Infection of Rhizopus. To control bitter rot, spraying trees with fungicides a number if times during growing season is recommended. Pre-harvest spraying with carbendazim and thiabendazole prevents postharvest blue mould rot of apples stored both at ambient temperature and in air-cooled storage. Calcium sprays for the control of bitter pit also confer resistance to *Penicillium expansum* during storage. Pre-harvest sprays of captafol and carbendazim enhance keeping quality of apples by protecting them from various storage rots like blue mould, brown rot, pink rot, black rot, chocolate brown rot. In areas where Phytophthora rot is serious and the soil under the trees is kept bare, it may be drenched with an appropriate fungicide.

Spraying of fungicides may be done according to a protective and curative spray schedule to control apple scab. Orchard inspections coupled with a reliable forecasting service are of assistance in timing the spray correctly. Apple scab and certain other rots were effectively controlled by 2 pre-harvest sprays of dithlanon and mancozeb. Similarly, pre-harvest sprays of bitertanol and captan are most effective in controlling storage scab. It can also be beneficial to apply urea after harvesting but before leaf fall. Leaves from urea sprayed trees decompose rapidly after falling from tree. This results in depletion of food source available to *V. inaequalis*, reducing the initial inoculum available for primary infection in spring season.

Pre-harvest spraying of captan, captafol, benomyl and carbendazim has been found ideal to control flyspeck and sooty blotch. Application of fungicides in the apple scab spray schedule also provide effective control to flyspeck and sooty blotch.

Harvesting

To ensure sufficient time for distant marketing, apples should be harvested before they ripen completely, *e.g.* before the rise in respiration. Temperature of the harvested fruits should be lowered to suppress the respiration. In apple, days from full bloom has been used as main criterion for harvesting, besides fruit colour, flesh firmness, starch content and rate of ethylene production.

Avoiding harvesting of fruits in wet weather reduces losses from Mucor rot. During wet season, particular care should be taken to remove all harvested fruits from the orchard immediately after picking to reduce Phytophthora rot. If harvesting is done in summer, fruits may be protected from heat by keeping them in shade.

Handling and Sorting of Fruits

Fruit should be handled carefully through all stages of harvesting, packing, storage and marketing. These may get bruises and punctures, which may be responsible for serious losses in storage. Some practices such as washing, trimming and curing also reduce the severity of infection. Boxes and bins to collect apples in the orchard should be cleaned periodically. Packing houses and surroundings should be kept free from cull and decaying fruits.

Chief means of avoiding blue mould rot involves careful handling of fruits and strict sanitation in packing houses. Rotting fruits should be destroyed before they become a source of *P. expansum*. Washing water should be frequently changed and disinfected. Careful handling and prevention of mechanical injuries help to reduce brown rot, Mucor rot, pink mould and Altemaria rot in apples

Medium-sized fruits are stored well for a longer period, while large-sized are prone to a number of infections and disorders. Storage of large-sized fruits for a longer period should be avoided.

Ripening fruit is a source of ethylene production. If some ripening fruits are present in a lot or box, ethylene produced from these may trigger ripening of other fruits. Similarly, ethylene from wounded fruits may be sufficient to trigger ripening of unwounded fruits. Wounded fruits are attacked by a number of postharvest pathogens and large quantities of ethylene are evolved from these rotting fruits. Besides rotting fruits, many pathogens also produce ethylene. Such wounded and rotting fruits should be removed from healthy ones. The production of ethylene by rotting fruits is perhaps the basis of traditional saying 'one bad apple spoils the lot'.

Postharvest Chemical Treatments

Postharvest chemical treatment is an effective and one of the important means of controlling apple rots during storage. The fruits are dipped for one-to-several minutes in a fungicidal suspension before packing. Dip treatment with fungicides is effective to control Alternaria rot. Dip treatment with iprodione, thioacetamid, aureofungin and malic hydrazide provide effective control to Altemaria rot. Dipping apple fruits in carbendazim and thiabendazole control blue mould rot effectively.

The effectiveness of postharvest dipping in fungicides or drenching can be increased if applied in conjunction with calcium treatment. The time of application is critical and its delay for few hours can result in huge losses. Use of the same or closely related chemicals, before and after harvesting, is not advised because it can lead to the rapid development of resistant strains of *P.expansum*. It is preferable to use a range of fungicides with different modes of action. There are reports of increased losses due to Alternaria rot following the use of systemic fungicides to control blue mould rot.

Botryodiplodla rot, pink rot and Aspergillus rot can be controlled by postharvest dipping treatment with dithianon, carbendazim and thiabendazole. Thiabendazole and benomyl provide effective control to Botryosphaeria rot, bitter rot and Rhizopus rot. Aromatic compounds such as dichloronitroaniline, sodium orthophenylphenate and biphenyl are quite effective in controlling pink mould rot, brown rot, bitter rot, blue mould-rot and Rhlzopus rot. Treatment with a mixture of 2-aminothiozole and urea give better protection against Rhizopus rot. Similary, postharvest dipping treatment of apples in bitertanol give cent percent control to storage scab up to 90 days, followed by carbendazim and captan. Postharvest dipping treatments with growth regulators also reduced incidence of certain diseases of apple during storage.

Fumigation

Relatively very few fumigation treatments have been developed to control postharvest diseases of apple. Sulphur dioxide fumigation by means of sodium metabisulphite eliminated *Trichothecium roseum* and checked *Glomerella cingulata* and *P. expansum*, effectively. Similarly, sulphur dioxide was better than ammonia to control brown rot.

Chamically-Impregnated Wrappers

Various antifungal chemicals have been used as impregnating fruit wrappers· and box liners. Potassium iodide wraps provided effective control to *Gliocladium roseum* on apples. Wrappers impregnated with dichioro-nitroanaline, diphenylamine and sodium orthophenylphenate provided some protection against *T. roseum, M. laxa, G. cingulata, P. expansum* and *R. stolonifer*. Keeping fruits in newspaper and tissue wrappers impregnated with carbendazlm and thiabendazole fungicides checked blue mould rot when stored in cold storage.

Fruit Peel Coatings

Peel coatings can improve the keeping quality of apples by decreasing the water loss and retarding ripening and rotting by various pathogens. Coating is generally done with oils, waxes and colloidal solulions of carboxymethyl cellulose. Apples coated with mustard 011, paraffin and castor oil checked infection of *T. roseum, M. laxa, G. cingulata, P. expansum* and *R. stolonifer*. Peel coating with neem oil completely checked blue mould-rot.

Irradiation

Irradiation has been tried as a substitute to chemical control to prevent infection in apples. Ionising irradiation may delay ripening and thereby reduce microbial postharvest spoilage. But injurious reactions of irradiation on living tissues have been largely ignored. Germination of *A. niger* conidia was reduced considerably by gamma irradiation at 100 Krad. Apples with *A. nigar* and exposed to 150-200 Krad could be stored for 5-7 days at 20°C without any rotting.

Hot Water Treatment

Hot water treatment may kill the incipient or other infections in fruits. But this treatment may alter the physiology of fruits sometimes making it more vulnerable to fungal spoilage. Treating apples in hot water-bath immediately after harvesting reduces blue mould rot. If fruits are stored at ambient temperature, hot-water treatment can inactivate deep seated infections of *T. roseum*. Treating apples in hot water at 50°C for 5 minutes significantly controlled various postharvest rots caused by *T. roseum, M. laxa, G. cingulata, P. expansum* and *R. stolonifer*.

Refrigerated Storage

Lack of refrigeration facilities in our country is mainly responsible for huge postharvest losses in apple. The development of various postharvest pathogens become very slow at a temperature below 5°C. Refrigeration storage and transportation of apples is so critical that all other methods are sometimes described as supplements to refrigeration. Low temperature not only slows down the development of postharvest pathogens but it also maximizes the physiological postharvest life. To increase the effect of refrigeration, field heat of the fruit is removed immediately after harvesting by forced air cooling or hydrocooling and their temperature is lowered to near 0°C. For longer and best storage, apples must be kept in storage while they are in pre-climacteic stage. Depending upon varieties, apples have a maximum storage life of 2-12 months or more. Storage temperature varies from variety-to-variety. If not chilling sensitive, they are usually stored at –1 to 0°C. Some apple varieties, which are chilling sensitive, must be stored at 3-5°C. Relative humidity of the storage should be 90-95 per cent. Storage should have adequate ventilation to prevent ethylene accumulation.

Refrigerated storage slows down the rate of brown rot in apple. It is checked more readily in incipient and very earty stages than after it becomes well-established in fruits. Normal cold storage prevents development of pink rot and blue mould. Refrigeration can arrest Rhizopus rot and grey mould completely if fruits are stored below 4°C. The *Rhizpous stolonifer* neither grows nor its spores germinate below 7.5°C. Warmer temperature favours the development of Alternaria rot and black rot, therefore, keeping apples at temperature of 0-4°C when moving them from storage to the consumer, helps in preventing their spoilage.

Modified or Controlled Atmosphere Storage

Apples of some cultivars do not tolerate storage temperature of –1 to 0°C, so they must be stored at higher temperature. The storage temperature of 3-4°C results in fast ripening of fruits. Controlled atmosphere (CA) storage at 3.5°C in 2.5 per cent oxygen and 5.0 per cent carbon dioxide solves this problem of fast ripening. Thus, apples can be stored for a year. Fruits stored in conditions low in oxygen, high in carbon dioxide or both is known as CA storage. Keeping fruits in these storage conditions result in good physiological conditions of the fruits, which helps in retaining considerable disease resistance to various postharvest pathogens. Such storage conditions also suppress respiration rate of thie fruits, which ultimately reduces diseases in storage and transportation by suppressing fungal growth.

In CA storage, level of oxygen is commonly maintained at about 2-2.5 per cent. This level of oxygen is the lowest, which can be maintained efficiently with the presently available atmosphere control methods usually available in storages. Excessively low levels of oxygen result in anaerobic respiration by which fruits develop off flavours. For example, the growth of *B. cinerea* in 2 per cent oxygen atmosphere is reduced only 15 per cent below the growth in normal air of 21 per cent oxygen. The growth is checked markedly If oxygen in CA storage is reduced to 1 per cent but this level has deleterious effect on apples.

Air contains about 0.03 per cent carbon dioxide. If fruits are stored in an atmosphere where carbon dioxide is elevated to about 5 per cent, the respiration in fruits is suppressed. If concentration of carbon dioxide is excessive, fruits are injured and off flavour develops. Apple cultivars differ in their susceptibility to elevated carbon dioxide.

Moderate suppression of fungal rots in low oxygen and high carbon dioxide atmosphere suggested the use of fungistatic gas carbon monoxide. Addition of 10 per cent carbon monoxide in an air, reduce growth of postharvest pathogens moderately. However, suppression is greater when 10 per cent carbon monoxide is added to an atmosphere low in oxygen (2-2.5 per cent). If added in an atmosphere containing 2.2 per cent oxygen and 5 per cent carbon dioxide, suppression is always higher because of additive effect of carbon monoxide and carbon dioxide. Suppression of various fungal rots by carbon monoxide is still greater at a lower temperature (5°C) than at higher (12°C).

Sub-Atmospheric Storage

Storage and transportation under low oxygen or hypobaric atmospheres also reduces postharvest diseases of apples. Pressure is reduced by vacuum pumps and then flow of air is regulated. When pressure is lowered to 100 mm mercury, the amount of available oxygen is lowered from 21 per cent at normal atmospheric pressure to 2.8 per cent. At this concentration of oxygen, growth of fungi is suppressed. The sub-atmospheric storage delays ripening, softness and fruit deterioration. Storage of apples at 0.26 atmosphere pressure has been found satisfactory.

Biological Control

For biological control, naturally occurring microbial antagonists on fruit surface can be promoted and managed or antagonists can be introduced artificially. Naturally occurring antagonists on the surface of apple have been multiplied and used effectively as biocontrol agents. There are a number of artificially introduced antagonists to control different postharvest diseases. For example, *Pseudomonas cepacia, P. syringae, Acreomonium breve* and *Debaryomyces hansenii* as biocontrol agents have been effectively used to control pink mould, Mucor rot and grey mould. Biocontrol of postharvest diseases of apples appear to hold greater promise but the hindrance is rather social and economical.

Integrated Post Harvest Disease Management

The effectiveness of chemical treatments is greatly improved by combining it with careful sorting of fruits, refrigerated storage, careful handling, expeditious transportation and harvesting at a proper fruit maturity. Combination of chemical treatment with storage in cold room extend the shelf life of apples by checking a number of postharvest diseases. For example, two days old infection of *P. expansum* was completely checked by the combined application of 150 Krad irradiation and benomyl dip treatment of hot water at 50°C for 10 minutes. The thiabendazole, benomyl and carbendazim do not prevent the development of storage problems caused by Altemaria, Mucor and Phytophthora. However, a mixture of carbendazim and prochloraz give good control to Altemaria and Pencillium rots on apples stored at 0°C. Hot water treatment combined with carbendazim is recommended to control brown rot. Wax, sometimes containing a fungicide, may be applied immediately prior to packaging of apples for reducing losses caused by several postharvest diseases.

CROP IMPROVEMENT

Objectives of Improvement

Apple is grown as a composite tree consisting of a rootstock, scion and occasionally interstem. Thus, genetic improvement must involve both rootstock and scion. The scion breeding objectives are to evolve varieties, red in colour with early maturity, high yield, superior dessert and storage quality and disease (emphasis on scab) and pest resistance. Besides, a new wave of clonal rootstocks capable of surviving under a wide range of environmental conditions, inducing precocity, enhancing productivity and fruit quality in scion required to be bred.

Problems in Breeding

Sterility

There are two main causes of unfruitfulness in the apple: sterility and incompatibility. The sterility can be brought about by the failure of any of the processes concerned with the development of pollen, embryo-sac, embryo, and endosperm.

Self-Incompatibility

Sexual incompatibility that is due to the failure of the pollen to grow down the style and bring about fertilization is widespread in apple. Self-incompatibility is particularly common, although cases of cross-incompatibility are also known. Apples have a gametophytic incompatibility system whereby the pollen tube growth is arrested in the style. Almost all apple cultivars are self-incompatible to some extent, and some are completely so. Even those that appear to be self-compatible, set more fruits with higher seed content when pollinated with a cross-compatible cultivar. Triploid cultivars behave in much the same way as diploids, varying in the degree of self-incompatibility and producing an increase in fruit set when pollinated with diploids, but varying considerably when crossed with other triploids. Natural tetraploids are also very variable, some being possibly self-fertile and others only partly so because gametophytic incompatibility is typically voided in diploid gametes. A large number of apple cultivars have been S-genotyped. Triploid cultivars show cross-incompatibility when pollinated by any of the diploid cultivars of the same group. Triploid cultivars are themselves bad pollinators, producing merely sterile pollen grains.

Apomixis

Apple in some cases produces seeds in the normal fashion, actually reproduces from unfertilized eggs, from diploid cells of the nucellus, or from some cell of the megagametophyte called 'apomixis'. These give rise to two different types of seedlings *i.e.*, zygotic and nucellar. Nucellar seedlings generally are homozygous and represent excellent experimental material for cytological and genetic studies. The apomictic species that have been reported as polyploids are: *Malus sikkimensis* is a triploid; *M. coronaria, M. hupehensis, M. lancefolia* Rehd, *M. platycarpa, and M. toringoides* (Relid.) Hughes are known to occur as both triploid and tetraploid forms; *M. sargentii* is tetraploid, and *M. sieboldii* is known in diploid, triploid, tetraploid, and pentaploid forms. It is probable that the so-called diploid forms of *M. sieboldii* are sexual hybrids with other species.

Parthenocarpy

The apple usually has 10 ovules but it is not necessary for them all to be fertilized and develop into seeds for a fruit to be produced. Often a single seed is sufficient for the development of the fruit; thus fruitfulness may still be maintained even when a high degree of generational sterility is present. It is even possible in some cultivars and in certain conditions for fruit to develop parthenocarpically without fertilization. Fruits that arise in this way vary according to the cultivar. In some, they are small and often misshapen, and in others, they are normal in size and appearance. These fruits tend to ripen earlier and do not keep as well in storage as seeded fruits.

The use of cultivars that produce parthenocarpic fruits consistently has been recommended because of their ability to produce good yields in years when flowers are damaged by late-spring frosts or when conditions are unfavorable for pollination. Some quite heavy-cropping cultivars have been described that produce parthenocarpic fruit, but none seems to have been grown to any extent.

Inbreeding

Inbreeding in apples has generally been considered to be an unsuitable method of breeding because of the expected loss of vigor. Vigor could be restored by intercrossing inbreds but this is an unpromising breeding approach because heterozygous individuals obtained by intercrossing two heterozygous parents, which can be clonally propagated. Inbreeding is increased when parents have common ancestors. Thus, it is important when making crosses between apparently widely different cultivars to consider their pedigrees. Inbreeding may result from intercrossing new cultivars because so many of them are based on 'Golden Delicious' or 'Delicious'.

Inheritance Pattern

Characters Controlled by Single Genes

All the important characters, such as, fruit shape, size, and color, have been shown to be inherited quantitatively. In those days, this was not easy to understand, and Mendelian genetics was thought to have little to offer the fruit breeder. Since then, a number of single gene characters have been found, particularly in relation to disease resistance, but also other characters such as albinism, apetaly, lethals, and columnar growth.

Characters Controlled by Polygenes

In polygenic traits, a narrow-sense heritability is usually estimated. The value of heritability differs greatly according to the characters. Although, it is probable that characters with low heritability are most closely related to fruit quality, for which populations have long been selected, and characters with high heritability hardly influence palatability, the characters with low heritability are, rather, those whose phenotypic values are easily influenced by environmental circumstances. For example, the heritability for texture, crispness, and sweetness were generally low (0.1-0.3), whereas those for appearance, such as fruit size, amount of overcolour, and attractiveness, were moderate to high (0.3-1.0).

Inheritance Pattern for Specific Character

☆ *Chilling requirement*: Controlled by quantitative genes, although the low chill requirement of 'Anna' is controlled by one major gene along with several minor genes but budbreak number is highly heritable .

☆ *Cold hardiness*: Controlled by quantitative genes, mainly additive genes.

☆ *Growth rate*: Controlled by quantitative genes.

☆ *Vigour*: Controlled by quantitative genes, mainly additive genes

☆ *Season of flowering*: Controlled by quantitative genes, mainly additive genes.

☆ *Harvest date*: Controlled by quantitative genes.

☆ *Flower number*: Controlled by quantitative genes.

☆ *Fruit number*: Controlled by quantitative genes.

☆ *Juvenile period*: Controlled by quantitative genes.

☆ *Incompatibility*: Numerous S-allele exist, are semi-compatible combinations.

☆ *Fruit acidity*: Character is controlled by quantitative genes.

☆ *Fruit firmness*: Controlled by quantitative genes.

☆ *Fruit skin colour*: Inherited quantitatively, but may be regulated by a few major genes; one possibility is that three dominant, major genes produce colour.

☆ *Fruit shape*: Controlled by quantitative genes with a low genotype by environment interaction .

☆ *Fruit size/weight*: Controlled by quantitative genes, highly heritable.

☆ *Storage disorders*: High heritability for soft scald and superficial scald; moderate heritability to water core; low heritability for external pit, internal pit, brown heart, breakdown and chilling injury.

☆ *Sugar content*: Controlled by quantitative genes.

☆ *Texture:* Controlled by quantitative genes.

☆ *Fire blight (Erwinia amylovora)*: Immunity reported in some *Malus* species.

☆ *Alternaria blotch (Alternaria mali)*: Resistance is controlled by a single dominant gene.

☆ *Apple blotch (Phyllosticta solitaria)*: The susceptibility is regulated by two dominant genes.

☆ *Apple canker (Nectria galligena):* Highly resistant cider apples and rootstocks have been identified.

☆ *Apple scab (Venturia inaequalis):* It is controlled by both quantitative and qualitative genes, major source is a dominant gene.

☆ *Powdery mildew (Podospharaera leucotricha)*: Controlled by several dominant genes and polygenes; resistance may be enhanced by polygenes.

☆ *Wooly apple aphid (Eriosoma lanigerum)*: Controlled by a single dominant gene (Er); 'Northern Spy' has high level of resistance, along with several other cultivar.

Heritability

The heritability of a trait within a population is the proportion of observable differences in a trait between individuals within a population that is due to genetic differences. Factors including genetics, environment and random chance can all contribute to the variation between individuals in their observable characteristics (in their "phenotypes"). Heritability thus analyzes the relative contributions of differences in genetic and non-genetic factors to the total phenotypic variance in a population.

Breeding Methods

Selection

In apple, lot of emphasis had been laid on introduction and selection during the last few decades. Accordingly, several spur type varieties _e.g._ Red Spur Delicious, Golden Spur Delicious; colour sports _e.g._ Royal Red, Vance Delicious; low chilling varieties e. g. Vared, Michael, Tropical Beauty; scab resistant varieties e. g. Prima, Priscilla, Liberty; and early maturing varieties _e.g._ Yandik- Ovskoe and Papisovka Canniaga have been introduced through NBPGR in apple growing states of India. Of the scab resistant varieties, Co-op 12 has shown better performance as an early sub-acidic variety whereas Florina holds promise as a coloured sweet. However, none of these compare favourably well with the popular Delicious and its commercial sports. It is, therefore, logical that some of these traits are incorporated in our commercial varieties. At Srinagar, some promising selection _viz._, Maharaji Mutant, New Ambri, Regular Ambri and Velayati Ambri were made and are under evaluation. From TNAU, Tamil Nadu, Kodaikanal 1 was released in 1987. It is a selection from Parlin's Beauty. The tree is medium sized with 22 t/ha yields. It is a mid-season variety.

Hybridization

Apple improvement work has been going on at Regional Fruit Research Station, Mashobra, Himachal Pradesh, Fruit Research station, Shalimar, Jammu and Kashmir and Directorate of Horticulture and Food Processing, Chaubattia, Uttarakhand resulting in release of a number of varieties.

Hybridization work was initiated in Kashmir in 1956 with a view to combine high dessert quality of Delicious group with good keeping quality of Ambri. Two hybrids Lal Ambri and Sunehri were released. Work on similar lines was started in Himachal Pradesh in 1960. As a result, three promising hybrids namely Ambred, Ambstarking and Ambrich were selected. Subsequently, hybrid Ambroyal was also selected. Work on apple improvement was started at Chaubattia in 1970 with a view to develop early maturing cultivars with good dessert quality. Two promising hybrids Chaubattia Princess and Chaubattia Anupam were evolved. Induced mutation of (budwood irradiation with Gamma rays) Early Shanburry resulted in release of Chaubattia Agrim in 1984. Fruits are small, egg-shaped and slightly conical. Yellow coloured with slight red tinge. Flesh is dirty yellowish white coloured. Sour to sweet in taste. Matures in last week of June. Some crosses were also made to develop scab resistant cultivars and 2 outstanding hybrids Firdaus and Shireen have been released for adoption. Hybrids developed with their parentage and important characteristics are given in Table 5.21.

Low Chilling Cultivars

Sufficient attempts have not been made to introduce low-chilling apple cultivars in North Indian plains and in South India. From early introductions, cultivars Tropical Beauty and Parlin Beauty were found to be best with respect to productivity and fruit quality. Some cultivars of sub-tropical apple _e.g._, Maayan, Michael, Schlomit and Vared were introduced from Israel and are being evaluated.

Table 5.21: Important Apple Hybrids Developed in India

Hybrids	Parentage	Important Characteristics
Jammu and Kashmir		
Lal Ambri	Red Delicious x Ambri	Ambri with red colour.
Sunehri	Ambri x Golden Delicious	Ambri with colour of Golden Delicious.
Shireen	Lord Lambourne x R-12740-7A	Average fruit yield 50-60 kg tree^{-1}. Fruit small to medium in size, sweet, juicy having good flavour and resistance to scab.
Firdous	Golden Delicious x Rome Beautyx Prima	Yield 50-60 kg tree^{-1} (12-15 mt/ha). Fruit medium in size, sweet with slight acidic blend, crisp, juicy having resistance to scab and moderate resistance to Alternaria and San jose scale.
Akbar	Ambri x Cox's Orange Pippin	Average fruit yield 245 kg tree^{-1} at 16-20 years of age, fruit medium to large in size, red coloured, matures in about 157 days after full bloom.
Shalimar Apple-1	Sunehari x Prima	Resistant to scab, yields about 95 kg/tree (23.75 t ha^{-1}). The variety belongs to mid season group and has reddish pink, small to medium sized, crisp, juicy and sweet fruits.
Shalimar Apple-2	Red Delicious x Ambri	The variety has average fruit yield of 106 kg/tree (26.50 t ha^{-1}) at 25 years on seedling rootstock. Moderately tolerant to scab and Alternaria leaf spot, the fruits are roundish, red mottled, juicy, crisp and sweet. The fruit has a long shelf-life.
CITH Lodh apple-1	Bud sport of Red Delicious	Yield and fruit quality with improved overcolour.
Himachal Pradesh		
Ambred	Red Delicious x Ambri 57	Keeping quality good; low incidence of powdery mildew, sooty blotch and scab.
Ambstarking	Starking Delicious x Ambri 81	Tolerant to scab.
Ambroyal	Starking Delicious x Ambri 84	Semi-dwarf tree, semi spur type; good dessert quality.
Ambrich	Richared x Ambri 15	Semi dwarf tree, semi spur type; good dessert quality; tolerant to scab.
Uttarakhand		
Chaubattia Princess	Red Delicious x Early Shanburry	Early (ripens in last week of June), fruit with deep red streaks on pale background, very sweet and good keeping quality.
Chaubattia Anupam	Red Delicious x Early Shanburry	Ripens in second week of July; fruit skin with shining red streaks with red blush on pale background.
Chabattia Swarnima	Benoni x Red Delicious	Vigorous, spreding type, highly productive and regular. Fruit conical, medium size, white flesh.

6

Papaya

Papaya (*Carica papaya*) is the widely accepted vernacular name for one of the most popular sweet, tropical tree fruits. The common names are papaya, papaw or paw paw (Australia), mamao (Brazil) and tree melon. Papaya is one of the important quick-growing fruit crops, which is rich in several vitamins and minerals. The other related species are babaco (*Carica pentagona*), mountain papaya (*C. pubescens*) and chamburo (*C. stipulata*).

ORIGIN, HISTORY AND DISTRIBUTION

The papaya is a native of tropical America and was introduced to India in the 16th century. Cultivation of papaya began in Central America (no old records, unfortunately), but quickly after the discovery of the New World, seeds of papaya were carried to Old World tropical areas. This revered tropical fruit was reputably called 'the fruit of the angels' by Christopher Columbus. In the 20th century, papayas were brought to the United States and have been cultivated in Hawaii, the major U.S. producer since the 1920s. Today, the largest commercial producers of papayas include the India (36 per cent), Brazil (18 per cent), Nigeria (7 per cent), Indonesia (7 per cent) and Mexico (6 per cent). One can commonly find papaya trees growing around homes in all tropical countries. However, the highest productivity had been noted in Indonesia (85.8 t/ha), followed by Brazil (54.5 t/ha). The largest papaya producing states in India is Andhra Pradesh, followed by Gujarat, Karnataka and West Bengal with production share of 27.1, 23.2, 10.5 and 7.7 per cent, respectively (Table 6.1). Because home-grown crops are unregistered, it has been impossible to obtain reliable estimates of total world papaya production, but annual production must be several million metric tons.

Table 6.1: State-wise Area, Production and Productivity of Papaya in India

State	Area (000'Ha)	Production (000'MT)	Productivity (MT/Ha)
Andhra Pradesh	14.2	1,138.4	80.0
Gujarat	17.8	974.0	54.7
Karnataka	6.1	440.2	71.9
West Bengal	11.1	324.2	29.1
Maharashtra	9.0	319.0	35.4
Chhatishgarh	10.6	247.1	23.3
Madhya Pradesh	2.0	227.0	115.5
Assam	7.4	134.4	18.2
Tamil Nadu	0.6	105.0	164.1
Kerala	16.2	100.8	6.2
Others	10.5	185.4	17.6
Total	105.6	4,195.5	39.7

Source: National Horticulture Board Database-2011.

COMPOSITION AND USES

Papaya is a very wholesome fruit, and ranks second only to mango as a source of the pre-cursor of vitamin A. However, yellow pigment in papaya is not carotene but 'caricaxanthin'.

It is also a rich source of vitamin C and riboflanin. On an average, 100 g of ripe papaya contains 2,500 I.U. of edible Vitamin A and 70 mg of ascorbic acid (Table 6.2). It improves digestion and cure chronic constipation, piles and enlarged liver and spleen. Papain is a valuable enzyme prepared from the latex of papaya, which has several industrial and pharmaceutical applications.

Table 6.2: Nutritional Composition of Ripe Papaya Fruit

Component	Per 100 g Edible Portion	Component	Per 100 g Edible Portion
Food energy	35.0 calories	Iron	0.7 mg
Moisture	90.7 g	Sodium	3.0 mg
Protein	1.5 g	Potassium	16.0 mg
Fat	0.1 g	Beta-carotene	1,160 µg
Carbohydrates	7.1 g	Vitamin B$_2$	0.1 mg
Ash	0.1 g	Riboflavin	40 mg
Calcium	11.0 mg	Niacin	0.1 mg
Phosphorus	3.0 mg	Vitamin C	70.0 mg

TAXONOMY AND BOTANICAL DESCRIPTION

Taxonomy

Papaya (*Carica papaya* L.) belongs to the family Caricaceae of dicotyledonous angiosperm. This family consists of four genera, *Carica* with about 21 species, *Jacartia* with about 8 species, *Jarilla* (Mocinna) with only one species and Cylocomorpha with 2 species. Only *Carica papaya* of genus *Carica* produces edible fruits of economic importance. The related species of papaya may be mentioned here since they are of interest to plant breeders.

C. candamarcensis: It is the mountain papaya, which is a small tree with cordate palmately 5-lobed leaves and small yellow 5-angled fruit about 3 to 4 inches in length.

C. erythrocarpa: It is similar to *Carica papaya* L., but the fruit has thin red flesh.

C. quercifolia: It reaches the height of 5 to 6 feets (2 metres maximum) and has oak like leaves and cluster of small ellipsoid fruits, 1 to 2 inches (2 to 5 cm) along with longitudinal strips, those change from white-to-yellow when ripe.

C. gracilis: It is a small, slender and ornamental species. It has compound leaves of 5 digitate, each leaflet having wavy indentations, the middle leaflet being 3 lobed.

C. papaya: This is a small tree, evergreen with a hollow soft-wood and generally unbranched. Trunk bears at the top a crown of large, long stalked, palm like leaves. Normally, the trunk is erect. Fruits edible, produced from leaf axils and are generally spherical-to-oblong in shape having central cavity where numerous seeds are attached to placenta. On ripening, the flesh colour generally turns yellow, deep-orange, pinkish or deep-red, depending upon cultivars.

Uptill now, the following species appears to be utilized in breeding programme for different objectives with special reference to disease resistance.

1. *C. monoica* Desf: Monoecious plant susceptible to virus.
2. *C. microcarpa* Jack.: Dioecious plant, susceptible to virus.
3. *C. candamarcensis* Hooker: Dioecious plant, resistant to virus.
4. *C. cauliflora* Jack.: Dioecious plant, resistant to virus.
5. *C. goudotiana* Solms., Luback: Dioecious plant, susceptible to virus.
6. *C. parviflora* Solms: Dioecious plant, susceptible to virus.
7. *C. pennata* Svensk: Dioecious plant, resistant to frost.
8. *C. pubescens* Lenne et Koch: Dioecious plant, resistant to distortion ring spot virus.
9. *C. stipulata* Badillo: Dioecious plants, resistant to distortion ring spot virus.
10. *C. horivitziana* Badillo: Dioecious plant, susceptible to virus.
11. *C. candicans* Gray: Dioecious plant, resistant to distortion ring spot virus.
12. *C. pentagona* Heilborn: Dioeious plant, resistant to frost.
13. *C. papaya* L.: Polygamous plant, susceptible to virus.

According to their crossability behaviour, some of these species may be arranged in 3 groups *viz.*,

1. *C. monoica* Desf, *C. microcarpa* Jacq, *C. candamarcensis* Hooker and *C. cauliflora* Jacq.
2. *C. papaya* L.
3. *C. goudotiana* Solms – Lquback.

All the species in group (i) are easily crossable with each other and produce viable seeds. Crosses between species in group (i) and (ii) do not form mature seed but immature embryo in most of the cases can be cultivated through the embryo culture. Crosses between group (ii) and (iii) have not been found successful.

Botany

The papaya is a short-lived, fast-growing, woody, large herb to 10 or 12 feet in height. It generally branches only when injured. All parts contain latex. The hollow green or deep-purple trunk is straight and cylindrical with prominent leaf scars. Its diameter may be from 2 or 3 inches to over a feet at the base. The leaves emerge directly from the upper part of the stem in a spiral manner on nearly horizontal petioles 1 to 3½ feet long. The blade, deeply divided into 5 to 9 main segments, varies from 1 to 2 feet in width, and has prominent yellowish ribs and veins. The life of a leaf is 4 to 6 months. The five-petalled flowers are fleshy, waxy and slightly fragrant. Some plants bear only short-stalked female flowers or bisexual (perfect) flowers also on short stalks, while others may bear only male flowers, clustered on panicles 5 or 6 feet long. Some plants may have both male and female flowers. Others at certain seasons produce short-stalked male flowers, at other times perfect flowers. This change of sex may occur temporarily during high temperatures in mid-summer. Male or bisexual plants may change completely to female plants after being beheaded. Certain varieties have a propensity for producing certain types of flowers. For example, the 'Solo' variety has flowers of both sexes 66 per cent of the time, so two out of three plants will produce fruit, even if planted singly. How pollination takes place in papayas is not known with certainty. Wind is probably the main agent, as the pollen is light and abundant, but thrips and moths may assist. Hand pollination is sometimes necessary to get a proper fruit set.

There are two types of papayas, Hawaiian and Mexican. The Hawaiian varieties are the papayas commonly found in supermarkets. These pear-shaped fruit generally weigh about 1 pound and have yellow skin when ripe. The flesh is bright orange or pinkish, depending on variety, with small black seeds clustered in the center. Hawaiian papayas are easier to harvest because the plants seldom grow taller than 8 feet. Mexican papayas are much larger than the Hawaiian types and may weigh up to 10 pounds and be more than 15 inches long. The flesh may be yellow, orange or pink. The flavour is less intense than that the Hawaiian papaya but still is delicious and extremely enjoyable. They are slightly easier to grow than Hawaiian papayas. A properly ripened papaya is juicy, sweetish and somewhat like a cantaloupe in flavour, although musky in some types. The fruit and leaves contain papain, which helps in digestion and is used to tenderize meat. The edible seeds have a spicy flavour somewhat reminiscent of black pepper.

Genetics and Cytogenetics

Sex Forms

Papaya is a polygamous plant and has many sex forms. It is, primarily, a trioecious with three basic sex forms: female, male, and hermaphrodite. Cymose inflorescences arise in axils of leaves. The type of inflorescence produced, depends on the sex of the tree. Varieties typically are either dioecious (with unisexual flowers and exclusively male and female plants) or gynodioecious (with bisexual and unisexual flowers and hermaphrodite and female plants). Male trees are characterized by long, pendulous many-flowered inflorescences bearing slender male flowers lacking a pistil, except for occasional pistil-bearing flowers at the distal terminus. Female trees have short inflorescences with few flowers bearing large functional pistils without stamens. Hermaphroditic trees have short inflorescences bearing bisexual flowers, which can be sexually variable.

In general, pistillate plants are stable while the staminate and hermaphroditic trees undergo frequent sex reversals, especially in the tropics. The reversion of hermaphroditic trees to pistillate trees during heat and drought stress is particularly common. Hofmeyr (1967) claimed that changes in photoperiod induce sex reversal in papaya. Chemical treatment of male papaya trees with morphactin, ethephon (2-chloroethane phosphoric acid) and TIBA (2,3,5-triiodo-benzoic acid) resulted in the conversion to female trees. Sex reversal was shown to be seasonal, and often accompanied by stamen carpellody and female sterility and consequently poor fruit quality and low yields.

Storey (1958) has discussed thoroughly the different sex forms and flower types in papaya. He classified papaya flowers in eight categories *i.e.* (i) staminate (ii) teratological staminate, (ii) reduced elongata (iv) elongata (v) carpelloid elongata (vi) pentandria (vii) carpelloid pentandria and (viii) pistillate. This classification of papaya flower has simplified numerous overlappings and removed confusions. Staminate flower is produced by male plant, whereas the teratological flower is produced by sex reversing male plants. Reduced elongata, elongata, carpelloid elongata, pentandria and carpelloid pentandria are produced noramally by hermaphrodite plants, and pistillate flower is produced by female plants only. Since sex expression of sex-reversing male and hermaphrodite plants depend on environmental conditions, the sex reversing male plants can produce all 8 types of flowers during the year. The hermaphrodite plants can produce 6 types except staminate and teratological staminate flowers. The female plant produces only pistillate type and in a very rare case, also produces bisexual flower. In totality, 15 classes of sex variation have been proposed by Storey (1958) for hermaphrodite plants on the basis of seasonal shift in female sterility or carpeloidy of stamens or in both. Likewise 15 comparable classes are found in male plants. Thus, there are 32 heritable sex forms in papaya.

Genetics of Sex

Hofmeyr (1938) and Storey (1941) suggested that sex in papaya is controlled by single gene with 3 alleles. The male and hermaphrodite are heterozygous for sex, and female was homozygous. Hofmeyr (1939) explained the sex determination in papaya by forwarding the genic balance hypothesis. Later Storey (1953) reported that sex in papaya was determined not by single gene but by a complex of genes which lie closely to differential segments occupying identical region on sex chromosomes. The sex determining segments behave in heredity as unit factors there is two independent sets of factors which modify sex expression in male and hermaphrodite under certain environmental conditions. One set is reposable for seasonal shift from female fertility to sterility and *vice-versa*. The other set cause stamens to become carpeloid usually with the fusion of pistil. Storey (1958) reported that it is largely the expression of those sets of factors either singly or in combination which account for the differentiation of fifteen sex forms in papaya. Similarly, within a given genotype, temperature is the recessive factor for the types of flower produced. Thus, the degree of sexuality of a given flower depend on the relative proportion of substances present in the initial bud primordial.

From the foregoing discussion, it appears that no clear picture emerges regarding the genetic basis of sex in papaya. The scheme for sex determination in papaya, as adopted by Storey an Jones (1941) and by Hofmeyr (1938) has been further modified by Ram *et al.* (1983) as below:

Gene

M_1^{RR} = dominant factor for homozygous sex reversing maleness.

M_1^{Rr} = dominant factor for homozygous sex reversing maleness.

M_1^{rr} = dominant factor for pure maleness.

M_2 = dominant factor for hermaphroditism.

m = recessive factor for femaleness.

Genetic Constitution

$M_1^{RR}m$ = Sex reversing homozygous male plant.

$M_1^{Rr}m$ = Sex reversing homozygous male plant.

$M_1^{rr}m$ = Pure male plant.

M_2m = Hermaphrodite plant.

mm = Female plant

Sex Inheritance

Obviously, sex in papaya cannot be identified in seed or in juvenile vegetative characters before flowering. But different sex ratio can be predicted in the progenies raised through controlled pollination.

The theoretical plant types shown in Tables 6.3 and 6.6 as non-viable failed to appear because of lethality in early stages of development of M_1M_1, M_1M_2, M_2M_2, $M_1^{Rr}M_1^{rr}$, $M_1^{RR}M_1^{Rr}$ $M_1^{RR}M_1^{Rr}$, $M_1^{RR}M_1^{RR}$ and $M_1^{RR}M_1^{rr}$. Comprehensive studies of sex inheritance in inter and intra varietal crosses of papaya give a satisfactory fit (X^2 test) to the expected sex ratio. But in most of the cases, male and hermaphrodite populations

Table 6.3: Sex Inheritance in different Sex Cross Combinations in Papaya

Cross and Self	♀ Plants (per cent)	♀ Plants (per cent)	♂ Plants (per cent)	Non-viable
mm x M_1m	50 mm	-	50 M_1m	-
mm x M_2m	50 mm	50 M_2m	-	-
M_2m x M_2m	25 mm	50 M_2m	-	25 M_2M_2
M_1m x M_1m	25 mm	-	50 M_1m	25 M_1M_1
M_2m x M_1m	25 mm	2 M_2m	25 M_1m	25 M_2M_1

Table 6.4: Sex Inheritance in different Sex Reversing Male Cross Combinations in Papaya

Cross and Self	♀ Plants (per cent)	♂ Plants (per cent)	♂ Plants (per cent)		Non-viable
			Sex Reversing	Pure Male	
M_1^{Rr}m x M_1^{rr}m	25 mm	-	25M_1^{Rr}m	25M_1^{rr}m	25$M_1^{Rr}M_1^{rr}$
M_1^{Rr}m x M_1^{Rr}m	25 mm	-	37.5$M_1^{Rr/Rr}$m	12.5M_1^{rr}m	25$M_1^{Rr}M_1^{Rr}$
M_1^{RR}m x M_1^{Rr}m	25 mm	-	50 M_1^{Rr}m	-	25$M_1^{RR}M_1^{Rr}$
M_1^{RR}m x M_1^{RR}m	25 mm	-	50$M_1^{r/Rr}$m	-	25$M_1^{RR}M_1^{RR}$
M_1^{RR}m x M_1^{rr}m	25 mm	-	50$M_1^{Rr/Rr}$m	-	25$M_1^{RR}M_1^{rr}$

are on the higher side. Studies conducted by several researchers have indicated that significant deviation from male and hermaphrodite plant from the expected ratio of female to male (1:1), hermaphrodite to hermaphrodite and hermaphrodite self (2:1), and hermaphrodite to male and sex reversing male to hermaphrodite (1:1:1) are also observed in the total progeny. Thus, some unknown factors may have been operative in reducing the number of 'm' gamet or the viability of mm zygote in early stages of development. In such studies, Hofmeyr (1938) and Storey (1941) assumed that M_1, M_2 and m represented differential sector on the sex determining chromosomes. However, the respective hypothesis differed in certain aspects. Storey's (1953) hypothesis represents respective differential segments as linked genes. But according to Hofmeyr (1938), M_1 and M_2 represent inactivated or inert region of the sex chromosome in which vital genes are missing. It can be explained by assuming that X chromosome carries a vitality gene (V) which is absent in Y chromosome (v). The combination (vv) is, therefore, lethal. The order of genes in the sex determining segments has been not yet investigated. However, Westergaard (1958) postulated the order of the genes in the sex determining segment on the basis of hypothesis of Hofmeyr (1938) explaining the lethality of M_1M_1, M_1M_2 and M_2M_2 genotypes. The above chromosomal structure in the sex determining segment has been modified for better understanding the different sex type. Further, it is postulated that male sex falls under two broad categories on the basis of multiple allelic gene along with presence of a gene SuF/suF as below:

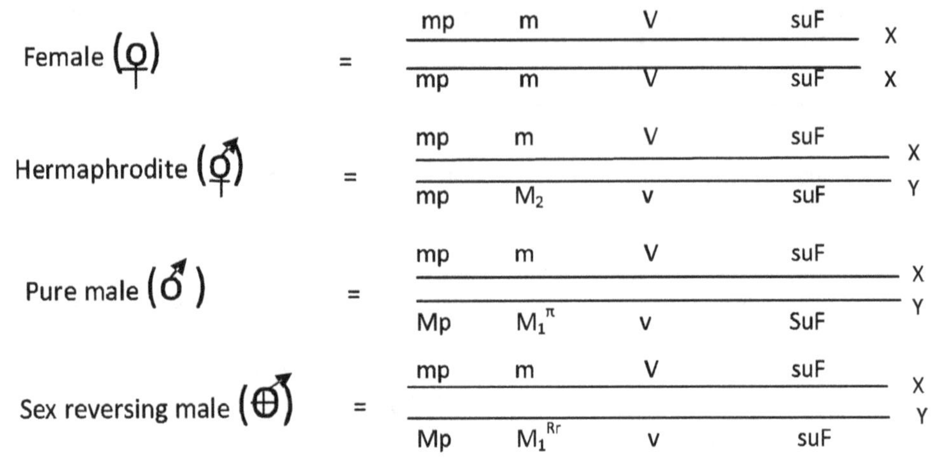

Female $\left(\underset{\text{+}}{\text{O}}\right)$ $=$ $\dfrac{mp \quad m \quad V \quad suF}{mp \quad m \quad V \quad suF}\begin{array}{l}X\\X\end{array}$

Hermaphrodite $\left(\underset{\text{+}}{\overset{\nearrow}{\text{O}}}\right)$ $=$ $\dfrac{mp \quad m \quad V \quad suF}{mp \quad M_2 \quad v \quad suF}\begin{array}{l}X\\Y\end{array}$

Pure male $\left(\overset{\nearrow}{\text{O}}\right)$ $=$ $\dfrac{mp \quad m \quad V \quad suF}{Mp \quad M_1^{\pi} \quad v \quad SuF}\begin{array}{l}X\\Y\end{array}$

Sex reversing male $\left(\overset{\nearrow}{\oplus}\right)$ $=$ $\dfrac{mp \quad m \quad V \quad suF}{Mp \quad M_1^{Rr} \quad v \quad suF}\begin{array}{l}X\\Y\end{array}$

Figure 6.1: Location of Genes for different Sexes in Papaya

It has been postulated that presence of gene Mp or mp is the principal distinguishing factor between male and hermaphrodite, respectively. The presence of lethal gene (v) in both males and hermaphrodites brings heterozygosity on the sex form. The dominant gene SuF is absent in sex reversing male, as a result, this plant also produces fruits.

Sex Identification

There is a great difficulty to determine the sex of papaya plants at before flowering and all attempts for identification of sex in papaya at an early stage were not successful. However, the karyological analyses indicate that there is a satellite chromosome in the male plant. This satellite chromosome determines the male sex in papaya. Further, it is reported that the leaves of male plants are richer in total carbohydrate, phosphorus and chlorophyll 'a' and 'b' than those of the female plants, which are rich in nitrogen and potash. Further female flowers are found to contain significantly more asparagin, arginine and histidine and less alanine and asparatic acid than the male flowers.

Cytogenetics

Chromosome Number

Several studies have been conducted to find out chromosome number in papaya. In 1921, it was established that chromosome (2n) of *C. chrysopetala, C. pentandria, C. candamarcensis* and *C. papaya* is 18. The same number was found for *C. guercifolia* by Storey (1941). Later, it has been observed that all the sex types of *C. papaya* has n=9. Hence, it is found that all the species have the same chromosome number, 2n=18.

Sex Chromosome

A number of cytological studies have been conducted to determine whether, as in other plants with unisexual flower, papaya has a heteromorphic pair of chromosome. However, most of the researcher could not find any heteromorphism in the somatic chromosomes of different sex types.

Karyotype Analysis

Very little attention has been paid to the karyotype morphological studies in *Carica*. No distinct morphological difference in the chromosome of different varieties has been found as all the chromosomes had median or sub median constriction.

Meiosis

Studies conducted by several researchers have reported that meiosis has been found to be normal in all the species, however, occurrence of multipolar and tripolar spindles in *Carica papaya* has been observed.

Sex Reversal

Sex reversal in papaya might be due to the interaction of certain genetic and environmental factors. Two types of male plant have been reported in papaya, the pure male and the sex reversing male. The sex reversing male and hermaphrodite readily change sex as a result of seasonal change in temperature, whereas pure male and the pistillate do not. It has been suggested that under warmer conditions, there do a trend towards masculinity. However, later it was established that higher temperature favoured femaleness and the lower temperature favoured masculinity. Singh *et al.* (1963) also pointed that the genetic make up of sex reversing male and pure male plants is entirely different. The recent studies on the genetics of pure male and sex reversing male reveales that the genetic make up of sex reversing male is entirely different than pure male. Hence, it may be a case of multiple allelism at M_1 locus itself, as there are two different forms of this locus.

Table 6.5: Sex Ratio of Crosses between different Sex Types in Papaya

Crosses	Sex Ratio			
	Female (mm)	Male (Mm)	Hermaphrodite (M^hm)	Non-viable ($M^{(h)}M^{(h)}$)
Male (*Mm*) selfed	1	2	0	1
Female (*mm*) × male (*Mm*)	1	1	0	0
Hermaphradite (M^hm) selfed	1	0	2	1
Female (*mm*) × hermaphradite (M^hm)	1	0	1	0
Hermaphradite (M^hm) × male (*Mm*)	1	1	1	1

The sex forms are not only morphologically distinct, but they are inherited in unexpected ratios that are due to a lethal factor associated with male dominant alleles (Table 6.5). Hermaphrodite papaya trees are primarily self-pollinated. However, seeds from selfed hermaphrodite trees always segregate into hermaphrodites and females in the ratio 2:1. Seeds from female trees segregate into hermaphrodite and female in the ratio of 1:1, if they are fertilized by pollen from a hermaphrodite tree, and a ratio of 1:1 male to female when fertilized by pollen from a male tree. Male trees are never produced when hermaphrodites are selfed or when hermaphrodites are used as a pollen source to fertilize female trees. However, male trees occur in a ratio of 2 male:1 female, when the occasional male fruit is selfed, or 1 male:1 hermaphrodite: 1 female when male pollen fertilizes the pistil of hermaphrodite trees. These unexpected sex form ratios have been the subject of extensive studies.

Sex Determination

Several hypotheses had been proposed to explain the phenomenon of sex determination and inheritance in papaya. Some of them are being discussed as below;

1. Sex Determination by a Single Gene with Three Alleles

Based on the segregation ratios from crosses among three sex types (Table 6.3), Hofmeyr and Storey independently proposed that sex determination in papaya is controlled by a single gene with three alleles, named as $M1$, $M2$, and m by Hofmeyr and M, M^h, and m by Storey. In the present chapter, Storey's designation will be followed for its convenience to separate the hermaphrodite allele M^h from male allele M. Male individuals (Mm) and hermaphrodite individuals (M^hm) are heterozygous, whereas female individuals (mm) are homozygous recessive. The dominant combinations of MM, M^hM^h, and MM^h are lethal, resulting in a 2:1 segregation of hermaphrodite to female from self-pollinated hermaphrodite seeds and a 1:1 segregation of male to female or hermaphrodite to female from cross-pollinated female seeds.

2. Genic Balance between Sex Chromosomes and Autosomes

Later, Hofmeyr (1967) suggested that the chromosomes bearing the M, M^h, and m alleles are 'sex chromosomes.' In this hypothesis, it is assumed that female sex determining factors predominate the 'sex chromosomes', while the male sex determining factors are in the autosome. It was further assumed that M and M^h represent an inactivated region of the sex chromosomes, where vital genes are eliminated, but that the inactivated region represented by M is longer than that represented by M^h. The different sex types were the results of genic balance between the sex chromosomes and autosome. Because vital genes are missing in the inactivated regions of M and M^h, any combinations of MM, MM^h, M^hM^h would be lethal, while Mm and M^hm would be viable because an m sex chromosome is present in each genotype.

SOIL AND CLIMATIC REQUIREMENTS

Soil

Papayas are adapted to practically any well-drained soil. The plant is shallow rooted and will not tolerate excessive wetness or standing water. Raised beds can partly overcome drainage problems. Although, papayas thrive best in full sun, some concession can be made to protection from wind or cold weather. Papayas can be grown successfully in shade, but the fruit is rarely sweet. Close to the South or South-East side of the house is the warmest location in most residential sites. Some wind protection provided by other plantings or structures is helpful. It is a tropical fruit and it has to be grown in areas free from frost. Well-drained medium black to red loamy soils are suitable. It can be grown on comparatively poor soils with heavy manuring and watering. It does not thrive well in calcareous and stony soils, which contain little organic matter. Papaya does not tolerate water-logging at all.

Climate

Papayas have exacting climate requirements for vigorous growth and fruit production. Papaya thrives best under warm, humid conditions, free from frost occurrence throughout the year. It is generally intolerant of strong winds and cold weather. Brief exposure to 32° F is damaging and prolonged cold without overhead sprinkling will kill the plants. Cold, wet soil is almost always lethal. Cool temperatures will also alter fruit flavour. Temperatures just below freezing can kill small plants to the ground, larger plants that are not killed outright will normally produce suckers to regenerate the plant and bear fruit within a year. Dry climate during flowering often causes sterility, while dry climate during fruit maturity adds to the sweetness of the fruit. Plant can be grown from sea level to elevation of 1,000 m above sea level but cannot withstand frost.

IMPORTANT CULTIVARS

Honey Dew or Madhu Bindu

Popular variety in western as well as northern parts of India. Fruits are medium-sized, oval in shape; these are very sweet in taste having pleasant flavour. The bearing is heavy. Keeping quality of fruit is medium and the fruit bears less number of seeds.

Pusa Majesty

It is a gynodioecious variety with 100 per cent productive plants with good fruit yield. Fruits are medium in size and round in shape. Fruit flesh is solid in texture and yellowish in colour having good keeping quality. Fruit are of excellent taste and good flavour. It is the highest papain yielder.

Washington

It is popular in western India. Productivity is high. The fruits are nearly round-to-ovate, medium to large sized, over 20 cm long. When ripe, the skin attains bright yellow colour. The flesh colour is yellow. It is sweet having pleasing flavour.

Pusa Delicious

It is a gynodioecious variety with high productivity. Fruits yield is good and the fruit size is very large. The taste of pulp is very sweet.

Pusa Giant

It is a vigorous variety. The fruits size is large. The fruits are suitable for vegetable making and canning industry.

Pusa Dwarf

It is a dwarf-statured delicious variety with good yield. The fruits are medium and the shape is oval. The flesh is melting, sweet and red-orange in colour and preferred by the consumers.

Pusa Nanha

It is an extremely dwarf variety suitable for kitchen garden, pots and roof-top cultivation. This is an ideal variety for high-density planting.

Coorg Honey Dew

It is a chance seedling of Honey Dew. The fruits are oblong. It is dwarf and heavy bearing variety. The flesh is thick with good flavor. The flesh is orange-yellow coloured and moderately soft and juicy.

Solo Sunrise

It is a famous variety from Hawaii Islands. The variety is named 'solo' as one man can easily consume one fruit. The fruits are small and pear shaped. Fruits are of excellent quality and free from unpleasant odour. The flesh is pink in colour.

Pink Flesh Sweet

Fruits are medium sized with pink flesh. It is recommended for fresh consumption.

CO-1

This variety has been bred by the continuous sibmating and selection from 'Ranchi' over a period of eight years. The plant is dwarf in habit producing the first fruit within 60-70 cm from the ground level. Fruit size is large, spherical, smooth skin, nipple slightly raised at the apex, flesh orange-yellow, soft, firm and quite juicy with good keeping quality. The fruits are practically free from objectionable papain odour.

CO-2

Plants are dioecious, vigorous and of medium height. Fruits medium to large (1 to 1.5 kg), producing 4 to 6 grams of papain per fruit (averaging 250 to 300 kg of dried papain per hectare), oblong, light green when ripe, shallow furrows numbering 5, skin smooth and glossy. Flesh orange, soft to firm, moderately juicy and has good keeping quality. It is one of the best varieties for papain extraction.

CO-3

This variety is a hybrid derivative between CO-2 x Sun Rise Solo. It is a gynodioecious variety. The fruits are smaller but larger than Solo weighing 450-600 g. Fruits are pyriform in shape with attractive pink colour flesh. The tree yields 90-120 fruits. Fruits are suitable only for table purpose and not for papain extraction.

CO-4

This is a dioecious variety developed as a hybrid derivative between CO-1 and Washington. Fruits are medium in size with round shape similar to CO-1. The plant characters are similar to Washington with purple coloured petiole and stem. Fruit weight ranges from 1.3-1.5 kg. The tree produces 80-90 fruits. Flesh is yellow in colour. It is suitable only for table purpose and not for papain extraction.

CO-5

This is a dioecious variety and it is a selection from Washington. This variety is grown for its high latex production. It produces 50-60g of wet latex/fruit. This variety yields 75-80 fruits per tree. The flesh is yellow in colour. Though the latex yield is higher than CO2, enzyme activity in the latex is lower than CO-2.

CO-6

This is a dioecious variety and is a selection from Pusa Giant. The fruits are larger in size weighing 1.5-2.0 kg. Fruits are round to oval in shape with yellow flesh. Tree yields 80-100 fruits. It is suitable both for table purpose and latex extraction.

CO-7

It is a gynodioecious variety. It has been obtained from a multiple crosses involving, Pusa Delicious, CO-3 and Coorg Honey Dew as parents. This variety is similar to CO-3 but the fruit size is bigger than CO-3. It yields 65-70 fruits per tree. Each fruit is weighing about 1.0 kg. This variety is exclusively suitable for table purpose and not for papain extraction.

Pant Papaya-1

It is a vigorous dwarf selection bearing first flower at a height of 45-60 cm with a total height of 125-135 cm in the first year.

Pant Papaya-2

Plants are vigorous of medium height (150-220 cm) in the first year bearing, first flower at a height of 90-100 cm from the ground. It produces medium to large size fruits (1-2 kg each) of good quality with high yield potential of 30-35 fruits per plant. It is tolerant to frost as well as water-logging.

Pant Papaya-3

The plants are of medium height (225-250 cm) in the first year, bearing first flower at a height of 115-130 cm from the ground on strong stem. Fruits small to medium size (0.5-0.9 kg each) of excellent quality. It bears large number of fruits per plant (45-60) although of small size. Plants are tolerant to frost and water-logging. Due to small sized fruits, it can be grown for distant and other export markets.

Surya

IIHR, Bangalore has developed a this hybrid papaya variety. It produces medium-sized fruits weighing on an average 600-800 grams each and the fruits are said to be sweet with good shelf life. Surya hybrid is said to yield 48-58 tonnes per acre. The fruits may be harvested 9-10 months of transplanting. Surya is considered ideal for distant markets.

HPSC 3

It is a hybrid between Tripura Local and Honey Dew. It has been developed at ICAR Research Complex, Tripura. It gives high fruit yield (197.7 t/ha). Papain yield is 5g/fruit. It shows resistance to papaya mosaic pote virus.

Kamiya

A selection from Waimanalo Solo type, released from the University of Hawaii. Small to medium-sized fruit. Distinct, blocky shape, very short neck. Deep yellow-orange skin and flesh, firm, juicy, very sweet. Dwarf and high-yielding plant.

Mexican Red

A rose-fleshed papaya that is lighter in flavour than Mexican Yellow. Medium to very large fruit. Generally not as sweet as Hawaiian types.

Mexican Yellow

A very sweet and flavorful, yellow-fleshed papaya. Medium-to-large fruit, can grow up to 10 pounds. Generally not as sweet as Hawaiian types.

Solo

Fruit round and shallowly furrowed in female plants, pear-shaped in bisexual plants. Weight of individual fruit is 1.1 to 2.2 pounds. Skin smooth, flesh firm, reddish-orange, very sweet, of excellent quality. Produces no male plants, only bisexual and female in a 2 to 1 ratio. Introduced into Hawaii from Barbados in 1911. Named 'Solo' in 1919.

Sunrise (Sunrise Solo)

Pear-shaped fruit with a slight neck. Averages 22 to 26 ounces depending on location. Skin smooth, flesh firm, reddish-orange, sweet, sugar content high. Quality similar to Solo. Seed cavity not as deeply indented as in other Solo strains, making seed removal easier. Plant precocious, maturing fruit about 9 months after transplanting. Fruit are at a height of about 3 feet.

Sunset (Sunset Solo)

Solo type. Small to medium-sized, pear-shaped fruit. Orange-red skin and flesh. Very sweet. Dwarf, high yielding plant. Originated at the University of Hawaii.

Vista Solo

Medium to large fruit depending on climate, 5 inches wide, up to 18 inches long. Skin yellow, flesh orange to yellow-orange. Hardy, compact Solo type producing high quality fruit. Needs fairly hot weather to develop sweetness. Self-fertile. Developed by Ralph Corwin at Vista, California, USA .

Waimanalo (Waimanalo Solo, X-77)

Fruit round with a short neck, average weight 16 to 09 ounces. Skin smooth, and glossy, cavity star-shaped. Flesh thick, firm, orange-yellow in color, flavor and quality high, keeps well. Recommended for fresh market and processing. Fruits of female plants rough in appearance. Average height to the first flower is 32 inches.

Known-You No.1

This hybrid is tolerant to papaya ring spot virus. Plants are thick, sturdy, early and heavy yielding. Fruit is large, weighing about 1.6-3 kg, with yellow flesh, sweet and good taste.

Red Lady

Early, vigorous, productive, and tolerant to papaya ring spot virus. Plants begin to bear fruit at 60-80 cm height and have over 30 fruits per plant in each fruit-setting season. Fruits are short oblong on female plants and rather long-shaped on bisexual plants, weighing about 1.5-2 kg. Flesh is thick, red, with 13 per cent sugar content, and aromatic. Good shipper.

Tainung No. 1

Plants are vigorous, prolific, and easy-to-grow. Fruit weighs about 1.1kg with red flesh and good aroma.

Tainung No. 2

Large fruit with pointed blossom end, weighs about 1.1 kg. Flesh is orange-red, tender with good taste and quality.

Rainbow

Rainbow is the premier example of a genetically engineered horticultural crop that made it to market. It is resistant to Papaya ring spot virus (PRSV). SunUp is another PRSV-resistant variety, which was developed through biotechnological interventions.

The papaya varieties cultivated in different states of India are given below:

State	Varieties being Grown
Andhra Pradesh	Honey Dew, Coorg Honey Dew, Washington, Solo, CO-1, CO-2, CO-3, Sun Rise Solo, Taiwan
Jharkhand	Ranchi selection, Honey Dew, Pusa Delicious and Pusa Nanha
Karnataka and Kerala	Coorg Honey Dew, Coorg Green, Pusa Delicious and Pusa Nanha
West Bengal	Ranchi selection, Honey Dew, Washington, Coorg Green
Odisha	Pusa Delicious, Pusa Nanha, Ranchi selection, Honey Dew, Washington, Coorg Green
Tamil Nadu	CO2, CO5, CO7, Pusa Delicious, Pusa Nanha

PLANT PROPAGATION

Most papayas are grown from seed because of the impracticality of vegetative propagation methods in nursery production. Seeds are extracted from fully ripe fruit, washed to remove gelatinous material and planted several seeds per pot. Germination is accomplished in approximately two weeks under full sunlight. The plants can be set out as soon as they are large enough (about 1 foot tall) to survive with minimal care. The pots of plants should be spaced 8 to 10 feet apart. Papaya seedlings begin flowering in five-to-six months, at which time they can be thinned to a single female or bisexual plant at each site. In the absence of bisexual plants, one male plant is needed for every eight-to-ten females.

Seed can be obtained from papaya fruit purchased at the local supermarket. If the fruit is from Hawaii, the chances are good that the resulting seedlings will be mostly bisexual. For nursery production of papaya seedlings with a high percentage of female or bisexual flowers, controlled pollination between desirable parents is essential. By seed and tissue culture; it is necessary to obtain pure seeds from reliable sources.

Raising Nursery

Prepare raised seed-beds of 2 m in length, 1 m in breadth and 10 cm height. Apply 10-15 kg. FYM and 1/2 kg of 15:15:15 fertilizer mixture per bed. Drench the beds with ceresan wet (2 g in 1 lit. of water) solution. Sow freshly extracted seeds at a depth of 2-3 cm. in rows with a spacing of 5 cm in the rows and about 15 cm. between rows. Sowing of seeds during March-April is desirable to facilitate availability of plants during June-July. About 250 g of seeds is enough for raising seedlings for 1 ha. Provide water regularly during summer. In about 3 weeks time, seeds germinate. In 2 months, seedlings with a height of 15-20 cm are ready for transplanting Papaya seedlings can be raised in polybag, which is convenient during transport and easy establishment.

PLANTING AND ORCHARD ESTABLISHMENT

Plough the land. Dig pits of 40 × 40 × 40 cm. cube at about 1.8 m apart either way and fill them with topsoil and compost. This would accomadate 3,000 plants/ha. Place the seedling in the centre and provide support. A closer spacing of 1.2 × 1.2 m. for cv. Pusha Nanha is adopted for high-density planting, accommodating 6,400 plants/ha.

Due to sex variations, about 40-60 per cent of the plant may turn to be male in the cases of dioecious varieties. Therefore, in such case, 2-3 seedlings per hole at 30 cm apart in the pit should be planted, so that when they reach the flowering phase, the unproductive male trees can be removed to keep the population ratio of one male tree for every 15-20 female trees. In the case of bisexual varieties, such contingency may not arise. The ideal seasons are June to October and January to March as the other months are either too hot or rainy.

INTERCULTURAL OPERATIONS

Manures and Fertilizers

Application of 25 tonnes of FYM per ha and 250 g N, 250 g P and 500 g K/plant/ year is recommended. Apply the entire quantity of N, P_2O_5 and K_2O in split application once in 2 months commencing from the 2nd month of planting. Fertilizer at the rate of one-quarter pound of ammonium sulfate (21-0-0) per plant should be applied monthly after planting, increasing to one half pound six months after establishment.

Weed Control

Weed and grass control within 3 to 4 feet of the papaya plant is essential for optimum growth and fruiting. Cultivation for weed control should be quite shallow, as the papaya's roots are concentrated near the soil surface. The use of organic mulches is highly recommended.

Irrigation

Papaya responds well to copious irrigation in well-drained soils. Regular irrigation helps fruit development and induces the tree to bear larger sized fruits. Water stagnation should be avoided. In most parts of India, papaya plants/orchards are irrigated once in 8 to 10 days. Irrigation should be applied to thoroughly wet the soil periodically as needed through the year. A fluctuating irrigation regime may retard growth and cause poor fruit set. The ring system of irrigation is better than bed or basin system because the ring system prevents 'run off' of water.

Removal of Male Plants

About 10 per cent of the male plants are kept in the orchards for good pollination where dioecious varieties are cultivated. As soon as the plants flower, the extra male plants should be uprooted.

Intercropping

When papaya is grown as a pure crop, vegetables can be profitably grown as intercrops for about 6 months from planting of papaya seedlings. Intercropping also protects the plant from direct contact with irrigation water, thus preventing the incidence of collar rot.

PLANT PROTECTION

A. Major Diseases and their Management

Fungal Diseases

Stem or Foot Rot

This disease is also called as 'collar rot' and is considered as the most serious fungal disease of papaya in every part of world, wherever papaya is cultivated. In India, the disease appears during the rainy season and is prevalent throughout the country. The disease may destroy entire plantations within a season under favourable climatic conditions and the fungi make the soil sick and unfit for replanting. The severity of disease depends upon the intensity of rainfall coupled with high temperature. This disease was earlier attributed to _Pythium butleri_ but later the name was changed to _P. aphanidermatum_. In India, several other species of _Pythium_ like _P. vexans_, _P. ultimum_, _P. complectens_, etc. are known to incite the disease; however, _P. aphanidermatum_ is considered the main pathogen responsible for foot rot and 'damping off' of seedlings. The pathogen is soil inhabitant and is capable of growing and surviving on the plant residue left in the soil where it produces abundant oospores in the decaying organic matter.

Symptoms

In case of nursery plants, 'damping off' phase is predominant, whereas, in adult plants, foot rot, collar rot or root rot symptoms are produced. The foot rot symptoms appear as water-soaked patches on the stem at ground level, which enlarge and girdle the base of the stem. The affected tissues turn brown then black and rot. The terminal leaves turn yellow, wilt and drop. Fruits if formed, also shrivel and drop off. The entire plant topples over and dies. The internal tissues look like a honeycomb. The affected roots deteriorate and lose connection from the soil. Foot rot phase of the disease is more common in 2 - 3 years old trees, and younger plants if get infected early, are killed quickly.

Management Practices

Papaya growers should adopt the under-mentioned practices to avoid the incidence of foot rot in papaya orchards:

1. Disease free and healthy papaya seedlings should be used for planting.
2. Ensure good drainage in the orchard.
3. Avoid direct contact of water with stem of plants. Injury to roots and basal or collar region of the stem should be avoided at the time of planting.
4. The badly damaged plants should be carefully uprooted and destroyed by burning to prevent further spread of the disease.
5. The infected areas on the stem, at an early stage of the disease, should be removed with a sharp knife and fungicidal paste should be applied.
6. Drench and spray the plants with Bordeaux mixture or copper oxychloride prior to commencement of rainy season. Soil application of other fungicides like captan, mancozeb and difolatan considerably reduces the disease incidence.

Powdery Mildew

This is also a devastating disease of papaya. Many fungi, such as, *Oidium caricae, O. indicum, O. caricola, Sphaerotheca fuligenia* and *Leveillula taurica* are responsible for causing this disease. The temperature in the range of 16-23°C, relative humidity of 65-86 per cent and sunshine duration of about 6.2 hours are favourable for the disease development.

Symptoms

The symptoms varies slightly with the type of pathogens involved, however, most common symptoms are superficially growth of the fungus on the under surfaces of the leaves as patches of whitish powdery material. On the upper side of leaves, there is development of yellow or pale green blotches, usually near the veins, which coalesce and cover the entire leaf. The infection results in distinct, symmetrical and creeping colonies on infected leaves, which spread readily on one or both the surfaces, producing numerous floury conidia. Badly infected leaves curl, become turgid, dry up, hang down and ultimately fall off. In the advanced stages of infection, severe die-back symptoms develop.

Management Practices

Adopt the following management practices for controlling the powdery in papaya:

1. Disease can be effectively controlled by spraying wettable sulphur (0.3 per cent). Bayleton, bavistin, benomyl and methyl thiophanate (each at 0.1 per cent) sprayed at monthly intervals are more effective than other fungicides.

2. Papaya species namely, *Carica monoica, C. goudotiana* and *C. cauliflora* have been found resistant to *Acrosporium caricae (O. caricae)*.

Anthracnose

Anthracnose can be highly destructive disease of papaya if not controlled in time. In India, the disease is prevalent throughout the country and has been reported to occur in Uttar Pradesh, Madhya Pradesh, Bihar and Kerala. Many species of *Colletotrichum* and *Gloeosporium* are known to be associated with anthracnose of papaya. However, *Colletotrichum gloeosporioides* Pewz. (*C. papayae*) perfect stage *Glomerella cingulata* has been identified as the main pathogen of papaya anthracnose. The fungus enters through the pores in the fruits while it is still green and develops further in the flesh during the ripening period.

Symptoms

The disease appears on leaves, petiole, flowers and fruits. On leaves, initially well-defined minute specks appear, which increase in size. The central tissue of the spot becomes thin, papery, white and finally fall-off, giving a characteristic shot-hole appearance. In advanced stages, leaves turn yellow, dry up and give blighted appearance. Petioles also get infected and produce pustule like spots. Infection of flowers also takes place, which later fall on the ground. Necrotic spots are produced on the stem.

On fruits, small dark areas appear near ripening. As the fruit ripen, these spots also get enlarged and turn light-orange or pink with darker margins. Gradually, the lesions coalesce and sparse mycelial growth often appears on the margins of such spots. In case of severe infestation, fruits get completely rotten and drop down. An encrustation of salmon-pink spores often arranged in a concentric pattern may develop on some of the older spots under humid conditions. The fungus also invades fruit tissues underneath producing rot and turning them soft and darker in colour. Finally, the whole fruit turns dirty dark-brown and rots. More number of infections on young fruit results in its mummification but sometimes, a white fungal growth can also be seen on the whole surface.

Management Practices

Following control measures should be adopted for the control of anthracnose in the papaya orchards:

1. Removal and destruction of infected leaves, petioles and fruits from the orchard.
2. Spray bavistin (0.1 per cent) at 45 days interval and daconil (0.2 per cent) at forthrightly interval.
3. The post-harvest phase of the disease can be controlled by submerging fruits in hot water at 46-49°C for 20 minutes.
4. The combined application of 2 per cent sodium bicarbonate or ammonium carbonate (3 per cent) in wax formulation and *Candida oleophila* (2x 10 [8] cells) results in a significant reduction of anthracnose incidence and severity in naturally infected fruits stored at 13.5 °C and 95 per cent RH.

Phytophthora Blight

Blight, caused by *Phytophthora nicotiana* var. *parasitica* is also an important disease of papaya. It infests both plants and fruits of papaya, and may cause great losses to the growers.

Symptoms

Fungus invades stem and often results in complete girdling of the stem. As a result, the plant top gets wilted and eventually dies. Sometimes, stem is not completely girdled, but it is weakened and plant is easily broken off by the wind. The fungus can infest fruits of any age and as the fruit grows, the fungus also grows. The fruit get shriveled, turn dark-brown and fall down to the ground. Mummified fruits ultimately rot and serve as a source of further infection.

Management Practices

Following measures are absolutely necessary for controlling blight in papaya orchards:

1. Remove and destroy the diseased fruits and plants.
2. Spray dithane M-45 or Foltaf (0.2 per cent) or ridomil (0.1 per cent) at fortnightly interval.

Viral Diseases

Papaya Mosaic

The occurrence of papaya mosaic has been reported from the USA, Hawaii, West Indies, Cuba, Brazil and India. In India, it was first reported from Maharashtra in 1948. Later, reports of its infestations have come from other states as well. The disease is spread by many species of aphids, like *Myzus persicae, Aphis gossypii, Aphis malvae* and *A. medicaginis*. The virus infects the plants at every age, but about a year old plants are more susceptible.

Symptoms

The symptoms usually depend on the age of plant, time of inoculation and strain of virus. Initially, there is development of fruit chlorotic spots, vein clearing, puckering and profuse mottling of younger leaves, followed by degeneration and marked reduction in growth. The younger leaves show intense yellowing, chlorosis with dark-green patches and large blister-like islands of deep-green areas on the leaf blade. The leaves are much reduced in size, distorted and modify into tendril like (shoe string) structures. The number of lobes of a leaf is increased, which become thin and distorted. In some leaves, only half of the lamina undergoes these modifications, while the other half remains normal. In young plants, growth is arrested and these fail to bear and die. The older leaves, showing marginal curling, turn brown, gradually bend and shed prematurely, making the plants denuded with only a small cluster of reduced and much distorted leaves at their tops. The dark-green spots and elongated streaks become prominent on petiole and stem of the diseased plants. However, the necrotic spots on the lamina, water-soaked areas or oil spots and elongated streaks on the stem and petiole are either absent or rarely seen. During summer months, the symptoms are masked, but are visible during winter. The fruits produced on diseased plants usually develop circular water-soaked lesions with solid spots in the center. In severe cases, the fruit size is severely reduced with deformed shape. Though, there is no reduction in the flow of latex, but its quality gets deteriorated.

Management Practices

Following remedial measures are considered necessary for the control of papaya mosaic:

1. Rogue out the diseased plants and destroy them.
2. Spray systemic insecticides to kill the vectors.
3. Spray groundnut oil (1 per cent) at weekly interval.
4. Cover the young seedlings with polyethylene bags.
5. Destroy the alternate host plants from the vicinity of papaya orchards.

Papaya Ring Spot

Papaya ring spot is a viral disease of economic importance. It is transmitted mechanically. In addition, the aphids like, *Aphis gossypii, A. citricola* and *Myzus persicae* are the most efficient vectors for its transmission.

Symptoms

The symptoms of ring spot are characterized by vein clearing, puckering or bulging of the leaf tissue between the secondary veins and vein lesions on the upper surface of the terminal leaves. The margins and distal parts of young leaves may roll downwards and inwards. The virus induces mosaic mottling, dark-green blisters, and necrosis of chlorotic areas, leaf distortion resulting in shoestring symptoms followed by stunting of the plants. On stem of young plants, mosaic or mottle symptoms also show dark-green spots and oily or water-soaked streaks. The fruits produced on diseased plants are smaller, deeply lobed and lopsided and have circular and concentric rings. Diseased fruits have lower sugar content and poor latex flow.

Management Practices

Growers should adopt the following measures for the prevention of ring spot of papaya:

1. Rogue out the diseased plants and destroy them.
2. Uproot the alternate host plants of Chenopodiaceae and Solanaceae family.
3. Control aphid vectors in time with suitable insecticides.
4. Spray the extract of plants like, *Argemone mexicana* and *Carum coprocum*.

Leaf Curl

Leaf curl of papaya is not a serious disease in many countries, but in India, it causes serious loss in some areas. It is transmitted by white flies (*Bemicia tabaci*) and grafting.

Symptoms

The disease is characterized by severe curling, crinkling and distortion of leaves accompanied by vein clearing and reduction in leaf size. The leaf margins are rolled downward and inwards to form inverted cup, followed by thickening of veins. The affected leaves become leathery and brittle and petioles get twisted in a *zigzag* manner. The inter-veinal areas are raised on the upper surface due to hypertrophy, which gives rugosity to the leaves. The affected plants fail to flower or bear fruit. In advanced stages, defoliation takes place and the plant growth is arrested.

Management Practices

Growers should adopt the following measures for the prevention of papaya leaf curl effectively:

1. Uproot and destroy the infected plants.
2 Keep the whiteflies (vector) under control by spraying suitable insecticides.
3. Remove alternate host plants like, tomato, tobacco, petunia, datura etc.
4. Spray the plants with suitable insecticides (malathion or metasystox @ 0.05 per cent) to avoid the incidence of whiteflies.

Distortion Ring Spot

In some parts of India, distortion ring spot infects papaya plants and cause severe losses. It is transmitted in nature by *Aphis nerii, Aphis gossypii and Myzus persicae* and through seeds.

Symptoms

The disease is characterized by severe mosaic mottling of the leaves, followed by great reduction in the size of axillary leaves showing variable degrees of chlorosis and malformation of lamina. The leaves are often reduced to strings (veins) or bear enation of various shapes and sizes on either sides and even on veins. Symptoms on stem, petiole and fruit consist of dark-green (oily or water-soaked spots) streaks and rings with or without central solid spots. Fruits usually exhibit circular coalescent, concentric rings and various degrees of apocarpy or a double fruit formation, *i.e.*, fruit within the fruit. Diseased plants may bear a few small fruits, which have tendency to fall prematurely. Quantity and quality of the papain (latex) and sugar contents are greatly or considerably reduced.

Management Practices

Growers should adopt the following measures for the prevention of distortion ring spot of papaya:

1. Uproot and destroy the infected plants.
2. Keep the aphids (vector) under control by spraying suitable insecticides.
3. Remove alternate host plants like, tomato, tobacco, petunia, datura etc.
4. Spray the plants with suitable insecticides (malathion or metasystox 0.05 per cent) to avoid the incidence of aphids.

Leaf Crinkle

It mainly occurs in some states of India like Maharashtra and Tamil Nadu. It is transmitted by grafting but not by the insects.

Symptoms

Leaves show crinkled and inverted appearance accompanied by thick veins. Stems show *zigzag* nature. Sometimes, crumpling of the leaves is observed on the side shoots.

Management Practices

Growers should adopt the following measures for the prevention/control of leaf crinkle of papaya effectively:

1. Uproot and destroy the infected plants.
2. Remove alternate host plants like, tomato, tobacco, petunia, datura etc.
3. Grafting equipments should be properly sterilized before grafting.

B. Major Insect-Pests and their Management

Fruit Fly (*Bactrocera cucurbitae*)

Most of the fruit flies affecting papaya are placed in genus *Bactrocera*. Melon fruit fly, *Bactrocera cucurbitae* has been reported to cause extensive damage to papaya fruits in India. The adult female flies lay eggs on fruits. Damage to papaya fruits is caused primarily by larval feeding. The maggots on hatching bore in to the fruits and feed inside, making them unfit for human consumption. Usually, the fallen infested fruits serve as the major reservoir for breeding of fruit flies. The following measures are recommended for the effective management of fruit fly in papaya:

1. Collect and destroy the fallen fruits.
2. Harvest fruits at $3/4^{th}$ maturity.

3. Monitor the flies with the use of methyl eugenol baited traps.

4. Wrap the fruits with semi-permeable shrink-wrap films.

5. Bait sprays of malathion combined with protein hydrolysate is the most commonly used treatment. With bait sprays, male and female flies are attracted to a protein source from which ammonia emanates.

6. Double dip hot water (immersion) treatment is a viable quarantine treatment for controlling fruit flies.

Thrips (*Thrips parvispinus*)

In certain localities, thrips can be very harmful to papaya. Thrips suck sap from the new leaves, which results in mottling or streaking of the leaves. On infection site, the saprophytic fungi, *Cladosporium oxysporum* makes its entry, resulting in the development of bunchy and malformed top. Thrips can easily be controlled by weekly spray of malathion (0.5 per cent) alternating with the spray of rogor (0.1 per cent).

Red Spider Mites (*Tetranychus telarius*)

Red spider mite may cause considerable damage to papaya in certain localities. It has a worldwide distribution. The pest has a very wide host range, including many temperate, tropical and subtropical fruits. Mtes feed on epidermal cells, causing leaf curling, chlorosis, and plant stunting. Both nymphs and adults remain protected under webs and suck the cell sap. Mite feeding results in the formation of yellow spots on dorsal surface of leaves. Mites can also cause premature yellowing and leaf drop as well as direct damage to small fruits. Mites feeding on developing fruits can cause fruit downgrading. In severe infestation, leaves become distorted. The bearing capacity of the affected plants is reduced drastically. In addition to direct feeding damage, the presence of mites is associated with infection of fungus (*Oidi* spp.). The combined effect of mites and fungal damage causes severe chlorosis. Mites can be controlled by the following practices:

1. Remove the infested leaves and burn them.

2. Spray malathion (0.2 per cent) or rogor (0.05 per cent) or sulphur or karathane (0.2 per cent).

3. Sulphur or karathane are highly effective giving complete control of mite population.

White Fly (*Bemisia tabaci*)

B. tabaci is found widely around the world including most parts of the tropics and subtropics. White fly feeds on a wide range of host plants and suck sap from the ventral surface of leaves. The affected leaves become yellowish, curl downwards, wrinkle and finally shed. This curling is due to tobacco leaf curl virus, which is transmitted by *B. tabaci*. Due to this virus, there is severe curling, crinkling and distortion of leaves. The leaf margins are rolled downward and inward giving an appearance of infested crops. The petiole after infestation gets twisted. This disease is transmitted only by grafting and white flies (*B. tabaci*). Following measures can be adopted for its control:

1. Remove weeds from papaya plantation.
2. Rogue out infested plants and destroy them.
3. Nurseries should be raised under protected conditions in a place free from whiteflies.
4. The nursery plants must be sprayed regularly with monocrotophos (0.05 per cent) to avoid any insect infestation.
5. Under field conditions, spray dimethoate (0.03 per cent) or malathion (0.05 per cent or monocrotophos (0.04 per cent).

C. Physiological Disorders and their Managment

Skin Abrasions

Skin abrasions result in blotchy coloration such as green islands (areas of skin that remain green and sunken when the fruit is fully-ripe) and accelerate water loss. Abrasion and puncture injuries are more important than impact injury for papayas. Fruit bagging is the best way to avoid abrasions.

Chilling Injury

Chilling injury occur it papayas are stored below 10°C. Symptoms include pitting, blotchy coloration, uneven ripening, skin scald, hard core (hard areas in the flesh around the vascular bundles), water-soaking of tissues and increased susceptibility to decay. Increased alternaria rot was observed in mature-green papayas kept for 4 days at 2°C, 6 days at 5°C, 10 days at 7.5°C, or 14 days at 10°C. Susceptibility to chilling injury varies among cultivars and is greater in mature-green than ripe papayas (10 vs. 17 days at 2°C; 20 vs. 26 days at 7.5°C) and store papayas at 13-14°C to avoid chilling injury.

Heat Injury

Exposure of papayas to temperatures above 30°C (86°F) for longer than 10 days or to temperature-time combinations beyond those needed for decay and/or insect control result in heat injury (uneven ripening, blotchy ripening, poor color, abnormal softening, surface pitting, accelerated decay). Quick cooling of fruits to 13°C (55°F) after heat treatments minimizes heat injury.

HARVESTING AND YIELD

Maturity Indices

Change of skin color from dark-green to light-green with some yellow at the blossom end (colour break) are important indicators of papaya maturity. Papayas are usually harvested at colour break to ¼ yellow for export or at ½ to ¾ yellow for local markets. Flesh colour changes from green-to-yellow or red (depending on cultivar) as the papayas ripen. A minimum soluble solids of 11.5 per cent is required by the Hawaiian grade standards.

Harvesting

The change of colour from green-to-yellow and the consistency of the latex from milky to watery indicates that the fruit is ready for harvest. The fruit, should be harvested individually by hand picking taking care to avoid all possible injuries.

Yield

Papaya fruits are ready for harvest in about 9-10 months after planting. Fruits are borne throughout the year. Yield varies from 75 to 100 tonnes per hectare. The economic life of papaya plant is only 2½ to 3 years. A tree with good management produces 25 to 40 fruits weighing 40 to 60 kg in the first 15 to 18 months.

POSTHARVEST MANAGEMNT

After harvsting, papayas should be sorted out, pre-cooled, graded into different grades on the basis of size and colour and then stored or transported for marketing.

Storage

Papaya can be stored at 10°C and 15°C for 16 days but 12°C has been recommended as an optimum storage temperature for 2 weeks. Storage below 10°C has been known to cause chilling injuries. Papaya ripens satisfactorily between 20°C and 25°C while temperature above 32.2 °C cause delay colouring and ripening, copious latex oozing and fruit surface bronzing.

Dipping of papayas in hot water (55°C for 2-3 minutes) helps to protect them from postharvest diseases during storage.

PAPAIN: USES AND EXTRACTION

The immature papaya fruit contains a milky latex containing papain. It has several uses in the industry *viz.*, food processing, tanning and textile industry. In India, CO-2 and CO-5 varieties of papaya are recommended for papain production. Papaya fruits, which are about 90-100 days old (fully grown but not mature) are selected for tapping. In the morning hours before 10 a.m, four longitudinal incisions are given on the four sides of the selected fruits from the stalk-end to the top of the fruit. The depth of the incision should be 3 mm. On incising, latex starts flowing, which is collected in suitable containers (Areca nut spates, aluminum trays or glass vessels). Care should be taken not to use any other containers for papain collection since it will react with papain rendering it unfit for any use. The latex that solidifies in the cuts should also be scrapped carefully and added to the liquid latex. This process of making four incisions in the untapped fruit surface at 3-4 days interval is repeated thrice or four times over a period of 12 to 16 days. The latex thus collected every time should be dried in the sun or in driers at temperature ranging between 50 °C to 55 °C. Potassium metabisulphite (0.05 per cent) is added to the liquid latex in small quantities before it is dried since this helps to extend the storage life of papain. The drying is continued until it comes off in flakes having a porous structure. The dried papain is powdered, sieved in a 10 mesh sieve and stored in polythene bags or any other suitable container. This can be stored for six months at 9 °C. After papain extraction, fruits can also be used for consumption. The yield of papain under good management system in the first, second and third year could be approximately, 250 kg, 150-200 kg and 75-100 kg per hectare, respectively. Approximately, 25g of latex is obtained per fruit. 5 kg of latex on drying yield 1kg of papain.

Factors Affecting Papain Production

Fruit Size

Fruit size plays an important role in tapping of papain. The yield of papain has been found to be high in oblong fruits. It increases with an increases in the size of fruit, and a significant and positive correlation has been established between girth of fruit and papain yield.

Fruit Maturity

The fruit maturity is another important factor influencing the yield of papain. It has been recorded that unripe but fully-grown fruits yield maximum papain. Thus, it is advisable that fully-grown fruits should only be tapped. Studies have revealed that 21-23 month old fruits yield the highest quantity of papain.

Season

It has been reported that the flow of latex in papaya is low if the temperature is below 10^0C. Studies at the Central Food Technological Research Institute, Mysore, showed that the yield of papain was the highest in the month of October. Although the yield of papain may fluctuate among different places, with the lowest yield in September.

Cultivars

Marked variations has been observed among the different cultivars for the yield of papain. In Sri Lanka, an improved strain called 'Botanist's Selection' has been recommended for the production of papain. Among Indian cultivars, CO_2, and CO_5 have been reported to be good papain yielder than other cultivars.

Use of Ethrel

Spraying papaya plants with ethrel @ 200 ppm or ethephon has been reported to significantly increase the flow of latex in Coorg Honey Dew.

CROP IMPROVEMENT

Objectives of Papaya Breeding

☆ Breeding early maturity, dwarf and low fruiting phenotypes.

☆ To produce high yielding gynodioceious lines with uniform fruit shape and size.

☆ Good fruit quality with better quality fruits, medium in size (max. 1.0-1.5 kg), firm and juicy pulp.

☆ Easily separable placenta, small fruit cavity, less seeded, good taste (low papain), desirable flavour, uniform pulp colour. Deep-orange to deep-yellow pulp is preferred.

☆ Good keeping quality (15-20 days after harvest).

☆ Variety should be resistant to biotic (collar rot, leaf curl, leaf mosaic, ring spot virus, damping off) and abiotic (frost and water logging) stresses.

☆ Production of variety for papain extraction and for processing (puree and tooti-frooti).

Problems in Papaya Breeding

1. Highly cross-pollinated crop and hence, extreme care should be taken to avoid contamination by foreign pollen while going for planned hybridization.

2. Lack of information on several QTLs (Quantitative Trait Loci), which do not give information regarding their linkage and inheritance pattern.

3. Existance of many intermediate sex forms and unpredictable sex expressions.

4. Lack of reliable morphological or genetic markers for differentiating sex at seedling stage.

Inheritance Pattern

Some of the characters for which the inheritance patteren F_1 and F_2 generation are studied in detail are listed hereunder:

☆ Albino (*a*) plants are recessive to normal green plants.

☆ Dwarf (*d*) plants are recessive to tall plants

☆ Dimunitive (*dp*) plants are recessive to normal tall plants and are characterized by short slender trunk with small petiole, flowers and fruits.

☆ Crippled (*Cp*) leaf is recessive to normal flat leaf.

☆ Rugose (*rg*) leaf is recessive to normal flat or smooth leaf.

☆ Waxy (*w*) leaf is recessive to normal flat bladed leaf.

☆ Yellow (*y*) flower is dominant over to normal white colour.

☆ Red pulp colour (*r*) is recessive to yellow pulp colour.

☆ Purple pigmentation (*p*) of stem and leaf petiole is dominant over green colour.

☆ Grey seed coat (*B*) is dominant to black seed coat.

☆ Fruit odour and flavour are multiallelic characters.

Breeding Methods

Back Cross Breeding

In papaya improvement programmes, backcrossing has very often been employed. Papaya being a highly cross-pollinated crop renders a huge variations in shape, size, quality, taste, colour and flavour of the fruit. With backcrossing, increased homozygosity for recurrent parent alleles will occur, leading to fair uniformity in desirable traits. Although backcrossing does have some specific advantages as a breeding technique, a serious weakness in the form of inbreeding depression, is involved in its use.

Recurrent Selection

The basic technique in recurrent selection is the identification of individual with superior genotypes, and their subsequent inter-mating to produce a population. Some type of progeny test, depending on the selection system could be necessary to assess the genotype value of the parents. This breeding technique can be used for selecting superior types in dioecious as well as gynodioecious lines in papaya. Coorg Honey Dew was the first gynodioceious variety developed through selection at IIHR, Bangalore. Likewise, Pusa Delicious, Pusa Majesty, Pusa Giant, Punjab Sweet, CO-5 and CO-6 are also derived through selection.

Hybrid Breeding

Despite the development of some papaya hybrids through inter-varietal or intergeneic crosses, there still exists ample scope for development of superior hybrids for better yield and quality. This is more relevant particularly to production of F_1 hybrid seeds. Two hybrids IIHR-39 (Sunrise Solo x Pink Flesh Sweet) and IIHR-54 (Waimanalo x Pink Flesh Sweet) developed at Indian Institute of Horticultural Research, Bangalore have been found highly promising on account of superior fruit quality. They bear medium sized fruits, sweet in taste with good shelf life. Both are gynodioecious. IIHR-39 has now been released as 'Surya'. Another promising hybrid 'HPSC-3' has been developed at ICAR Research Complex, Tripura by crossing Tripura Local x Honey Dew. It possesses high yield potential and resistance to mosaic virus. Characteristics of some of the papaya varieties developed in India are presented in Table 6.6.

Table 6.6: Characteristics of Important Papaya Varieties Developed in India

Variety	Characteristic Features
CO-1	Selection from progenies of var. Ranchi, plant dwarf, fruit round or oval.
CO-2	A pure-line selection from a local type, plant medium tall, fruit ovate and large in size, good for papain production.
CO-3	A hybrid between CO-2 x Sunrise Solo, a tall vigorous plant. Fruit medium size, sweet with good keeping quality.
CO-4	A hybrid between C-1 x Washington; plant medium tall, fruit large, flesh thick and yellow, good keeping quality.
CO-5	A selection from Washington variety, good for papain production, yields 1500-1600 kg papain/ha.
CO-6	A selection from Pusa Majesty, dioecious, dwarf in stature, fruit size large, good for papain production.

Contd...

Table 6.6—Contd...

Variety	Characteristic Features
CO-7	A hybrid between Coorg Honey Dew x CP85; high yielding, uniformity in fruit shape, gynodioecious with good edible character, least summer skip and stamen carpellody, yield 9 fruits/tree each weighing 1.15 kg.
Pusa Delicious	A selection from progenies of variety "Ranchi", gynodioecious excellent fruit quality, deep orange flesh; 13 per cent TSS.
Pusa Majesty	A selection from Ranchi, gynodioecious, gives better papain yield, than CO-2 and C0-6 (500 g/plant); resistant to nematodes.
Pusa Giant	A selection from Ranchi, dioecious, large size fruit, sturdy and tolerant to strong winds.
Pusa Dwarf	A selection from Ranchi, dioecious, very dwarf in stature, fruits medium in size and oval in shape.
Pusa Nanha	A mutant of a local type, very dwarf plant, more suitable fo high density planting, dioecious, good fruit quality.
Coorg Honey Dew	A selection from Honey Dew, gynodioecious, excellent fruit quality under south Indian agro-climate.
Pink Flesh Sweet	Fruits with pink flesh, TSS 12 to 14° Brix, fruit medium in size.
Surya	A gynodioecious variety developed by crossing Sunrise Solo x Pink Flesh Sweet at IIHR, Bangalore, fruits-medium in size weighing on an average 600-800 g each, pulp - red with thickness of about 3.0-3.5 cm, TSS 13.5-15° Brix, sweet in taste, fruit cavity small, keeping quality good, yield 48-58 tonnes/acre.
Pant 1	Selected at Pantnagar, 1-1.5 kg oblong fruits, good quality.
Punjab Sweet	Selection made at PAU, Ludhiana, Dioecious, fruits large, oblong, weighing more than 1.0 kg each, yield 50 kg/plant, frost tolerant.
HPSC 3	A hybrid between Tripura Local x Honey Dew, gives high fruit yield of 197.7 t/ha. Papain yield 5g/fruit, shows resistance to papaya mosaic virus.

Mutation Breeding

A dwarf mutant line was evolved by treating papaya seeds with 15 kR gamma rays. Initially, three dwarf plants were isolated from M2 population. Repeated sibmating among the dwarf plants helped in establishing a homozygous dwarf line which was later named as 'Pusa Nanha'. It is dwarf (106 cm) in height having a thinner trunk girth (25 cm) and shorter leaf length (86 cm) as compared to 213 cm tall parent plants having thicker trunk (36 cm) and larger (193 cm) leaves. The fruiting in this strain starts at lower height (30 cm). Thus, rendering it more suitable for high density planting and pot cultivation.

7

Pineapple

Pineapple is a tropical fruit plant known for its juicy and fragrant fruit. It probably received its name because the fruit looks like a large pinecone. Many people enjoy drinking the juice of the pineapple and eating the fruit as a dessert or in salads. Thailand grows more pineapples than any other country, producing about one-fourth of the world's pineapples. Pineapples are the only bromeliad fruit in widespread cultivation. It is one of the most commercially important plants, which carry out CAM photosynthesis. The largest producers in the world for pineapple are Philipines, Brazil and Costa Rica. India ranks sixth in pineapple production in world with 1.41 million tones from 0.89 million ha area. The major pineapple producing states in the country are West Bengal, Assam, Karnataka and Tripura with production share of 21.5, 15.6, 13.1 and 10.8 per cent, respectively.

COMPOSITION AND USES

Composition

Ripe pineapple fruit is a good source of carbohydrates, β-carotene, vitamin B, and minerals like phosphorus, iron and potassium (Table 7.1)

Pineapple contains a proteolytic enzyme 'bromelin', which breaks down protein. Pineapple juice can thus be used as a marinade and tenderizer for meat. The enzymes in raw pineapples can interfere with the preparation of some foods, such as jelly or other gelatin-based desserts. The bromelin breaks down in cooking or the canning process, thus canned pineapple can generally be used with gelatin. These enzymes can be hazardous to someone suffering from certain protein deficiencies or disorders, such as Ehlers-Danlos syndrome. Raw pineapples also should not be consumed by those with hemophilia or by those with kidney or liver disease, as it may reduce the time taken to coagulate a consumer's blood. Consumers of pineapple have claimed that pineapple has benefits for some intestinal disorders and others believe that it serves as a pain reliever; others claim that it helps to induce childbirth when a baby is overdue.

Table 7.1: Nutritional Composition of Ripe Pineapple Fruit

Component	Per 100 g Edible Portion	Component	Per 100 g Edible Portion
Food energy	45.0 calories	Iron	6.0 mg
Moisture	87.8 g	Sodium	1.4 mg
Protein	0.5 g	Potassium	31.0 mg
Fat	0.1 g	Beta-carotene	40.0 mg
Carbohydrates	10.6 g	Vitamin B_1	270.0 µg
Fibre	0.6 g	Vitamin B_2	0.1 mg
Calcium	0.4 mg	Niacin	0.1 mg
Phosphorus	24.0 mg	Vitamin C	0.1 mg

Uses

Pineapples are mostly consumed fresh. However, it can be processed into value added products such as: Confectionery, jelly, jam, juice, snack food, canned fruits and wine, vinegar. Pineapple plants also have several other uses. Various parts of the plant are used to make cattle feed, meat tenderizers and medicines. In the Philippines, people weave the fibers of the plant into a cloth called '*ping*'.

ORIGIN, HISTORY AND DISTRIBUTION

Many scientists believe that pineapple originated in Brazil. Christopher Columbus and his crew, who explored the West Indies in 1493, were probably the first Europeans to taste the fruit. Europeans later found pineapples throughout the South and Central America and the West Indies. They took the fruit to Europe and planted it in hothouses. It became a favourite fruit of royalty and the wealthy. Commercial production of pineapples began during the mid-1800's in Australia, the Azores, and South Africa. Today, the world's chief pineapple producers include Brazil, China, Cote d'Ivoire, Indonesia, Malaysia, Mexico, Philippines, South Africa, Thailand, and the United States. Plantations in Hawaii produce almost all the pineapples that are grown in the United States. The ten leading countries together with the cultivar/cultivars grown in each country are listed in Table 7.2.

In India, it is grown profitably in several states such as Wst Bengal, Assam, Karnataka, Tripura, Meghalaya and Arunachal Pradesh (Table 7.3). Among states, Karnataka has the highest productivity (62.0 MT/ha), followed by West Bengal (30.6 MT/ha), whereas Arunachal Pradesh has the lowest productivity (3.2 MT/ha.) (Table 7.0)

Table 7.2: Leading Cultivars of Pineapple and Countries where Grown

Country	Cultivars	Country	Cultivars
Hawaii	Cayenne and Hilo	Kenya	Cayenne
Philippines	Cayenne	Mexico	Cayenne
Malaysia	Singapore Spanish	Cuba	Cayenne and Red Spanish
Australia	Cayenne and Queen	Taiwan	Cayenne
Puerto Rico	Cayenne and Red Spanish	Brazil	Red Spanish and Abacaxi

Table 7.3: State-wise Area, Production and Productivity Scenario of Pineapple in India

State	Area (000'Ha)	Production (000'MT)	Productivity (MT/Ha)
West Bengal	9.9	303.7	30.6
Assam	14.0	220.7	15.8
Karnataka	3.0	186.1	62.0
Tripura	6.8	153.3	22.6
Bihar	4.9	129.4	26.5
Manipur	12.2	104.4	8.6
Meghalaya	9.7	86.0	8.9
Kerala	10.2	85.5	8.4
Nagaland	3.7	57.5	15.5
Arunachal Pradesh	10.9	34.4	3.2
Others	3.5	54.5	15.6
Total	88.7	1415.4	15.9

Source: National Horticulture Board Database-2011.

TAXONOMY AND BOTANICAL DESCRIPTION

Taxonomy

Pineapple belongs to genus *Ananas*, order Bromeliales, and family Bromeliaceae. The Bromeliaceae has adapted to a very wide range of habitats. They are monocots but are set apart from other monocots by several unique characters. Earlier, *Ananas* was considered as a monotypic genus including many cultivars but with only one species. Later, two genera *viz. Ananas* and *Pseudomonas* were established. In *Pseudomonas*, syncarp at maturity bears a minute inconspicuous coma of bracts and produces elongated stolons and no slips, whereas in *Ananas*, syncarp bears conspicuous coma of foliaceous bracts and plant produce slips but never produce stolon. *Pseudomonas* has no crown whereas *Ananas* bears a terminal crown. *Ananas* is diploid (2n= 2x = 50) and *Pseudomonas* is tetraploid (2n = 2x = 100).

Pineapple is the most important economic plant in the Bromeliaceae. The first botanical description of cultivated pineapple was by Charles Plumier at the end of the 17th century when he created the genus *Bromelia* for the plants called *karatas*, in honour of the Swedish physician Olaf Bromel and also described *Ananas* as *Ananas aculeatus fructu ovato, carne albida*. Linnaeus in 1753 in his Species Plantarum designated the pineapple as *Bromelia ananas* and *Bromelia comosa*. However, Miller (1754, 1768) maintained the name *Ananas*, with all six cultivated varieties. The genus *Ananas* and the closely related genus *Pseudomonas* are distinguished from each other by fruit.

From 1934 onwards, pineapple taxonomy was dominated by the views of L. B. Smith and F. Camargo, who divided the genus *Ananas* and renamed and multiplied species. Ultimately, this resulted in 2 genera and nine species recognised in 1979. This classification has been criticised on the basis of practicality and inconsistency with available data on reproductive behaviour and morphological, biochemical and molecular diversity and therefore, a much simpler and consistent classification has been prepared, taking the above information into consideration. The present classification is as follows;

☆ *Ananas comosus* var. *ananassoides* (formerly two species: *A. ananassoides* and *A. nanus*)

☆ *Ananas comosus* var. *bracteatus* (formerly two species: *A. bracteatus* and *A. fritzmuelleri*)

☆ *Ananas comosus* var. *comosus* (formerly *A. comosus*)

☆ *Ananas comosus* var. *erectifolius* (formerly *A. lucidus* (formerly *A. erectifolius*)

☆ *Ananas comosus* var. *parguazensis* (formerly *A. parguazensis*)

☆ *Ananas macrodontes* (formerly *Pseudananas sagenarius*)

☆ *Ananas monstrous* has been invalidated by Leal (1990) because the crownless fruit characteristic is not stable.

Botany

Pineapple plant grows from 2 to 3 feet (60 to 90 centimeters) tall, and the fruit weighs from 4 to 8 pounds (2 to 4 kilograms). The ripe fruit has a greenish-orange, yellowish-green, or dark-green *shell* (skin). At the top of the fruit is a group of small leaves called the *crown*. The flesh of the fruit, the part eaten by people, is firm and pale yellow, though it may be white. The most widely grown kind of pineapple, Smooth Cayenne, is seedless, but some varieties have small brown seeds beneath the shell.

A pineapple plant has blue-green, sword-shaped leaves that grow around a thick stem. The edges of the leaves of most varieties of pineapples have sharp spines. But the leaves of the Smooth Cayenne have spines only at the tips and bases. The pineapple plant has underground roots and small roots that grow above ground.

When the plant is 14 to 16 months old, an *inflorescence* (flower stalk with tiny flowers attached) appears in the center. The inflorescence resembles a small pink-red cone. After the inflorescence has grown about 2 inches (5 centimeters) high, blue-violet flowers begin to open. Each flower blooms for only one day. All the flowers open within 20 to 30 days. Each flower develops into a fruitlet. The fleshy parts of the fruitlets unite with the stalk to which they were attached. This combination of fruitlets is called a *multiple fruit*. The multiple fruit and the stalk forms the yellow center of the pineapple. The pineapple's shell develops from thick, hard, leaf-like structures called *floral bracts*. There are 100-200 individual fruits arranged spirally around the thick axis and the whole forms a broad, almost cylindrical multiple fruit. The average fruit is around 20 cm long and 14 cm broad in the middle; but fruit size varies appreciably among the cultivars. The fruit tapers towards the top where it is mounted by a rosette of short, stiff, spirally arranged leaves called the crown (Figure 7.1).

SOIL AND CLIMATIC REQUIREMENTS

Soil

Soil requirement specifications are minimal for pineapple cultivation. However, the plant is particularly sensitive to soil being waterlogged. Medium-to-heavy loams, rich in humus and having slightly acidic reaction are more suitable soils for pineapple. Plant prefers soil pH 5.0-6.0. Soils with a higher pH are unsuitable owing to the development of lime induced iron chlorosis. Too much water can harm them, but irrigation is necessary in some dry regions. The plastic strips keep the chemicals escaping from the soil. The plastic also conserves moisture, keeps the soil warm, and discourages the growth of weeds.

Climate

Pineapple plants need a warm climate. Extremes of climates such as occurrence of frost and intense solar radiations associated with very low humidity are not favourable for pineapple cultivation. The optimum temperature range for successful pineapple cultivation is between 15.6 °C and 32.2 °C. High temperature over 35 °C is unfavourable for the development of fruits, especially if the relative humidity is low. In general, pineapple needs a sunny climate, though there are no exact figures on hours of sunshine or of solar radiations required for flowering and higher productivity.

IMPORTANT VARIETIES

Pineapple cultivars were classified by Hume and Miller (1904), Samuels (1969) and by Leal and Soule (1977). Each effort has attempted to clarify the classification problem in light of the current information. According to Knight (1980), all cultivated pineapples may be placed according to their characteristics in 5 major groups. A key to the identification of pineapple varieties has been proposed on the basis of (1) presence of thorns on the leaf margins, (2) leaf colour, (3) fruit weight, (4) fruit shape, (5) external and internal colour of the fruit, (6) locule depth, and (7) fruitlet orientation. It allows 23 varieties from throughout the world to be distinguished. However, in general, pineapple varieties have been grouped into the following 5 gropus:

1. Spanish Group

Shape of fruit is globose, with large deep set eyes; fruit weighs from 0.9 to 1.8 kg,

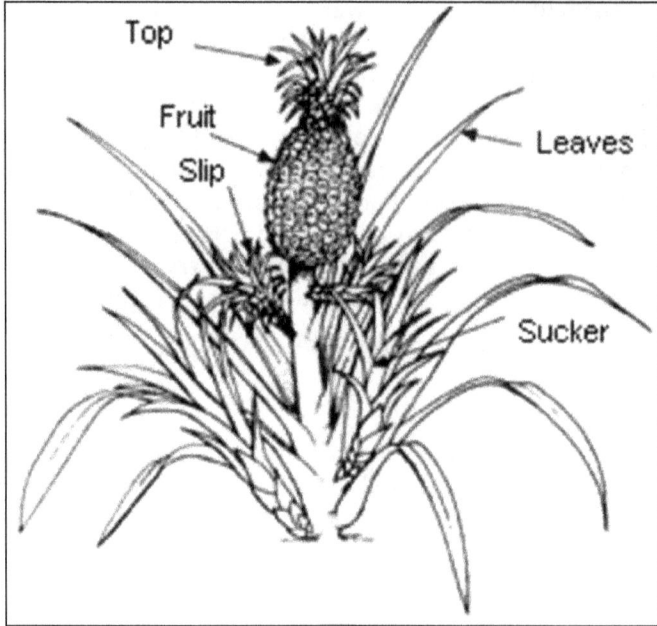

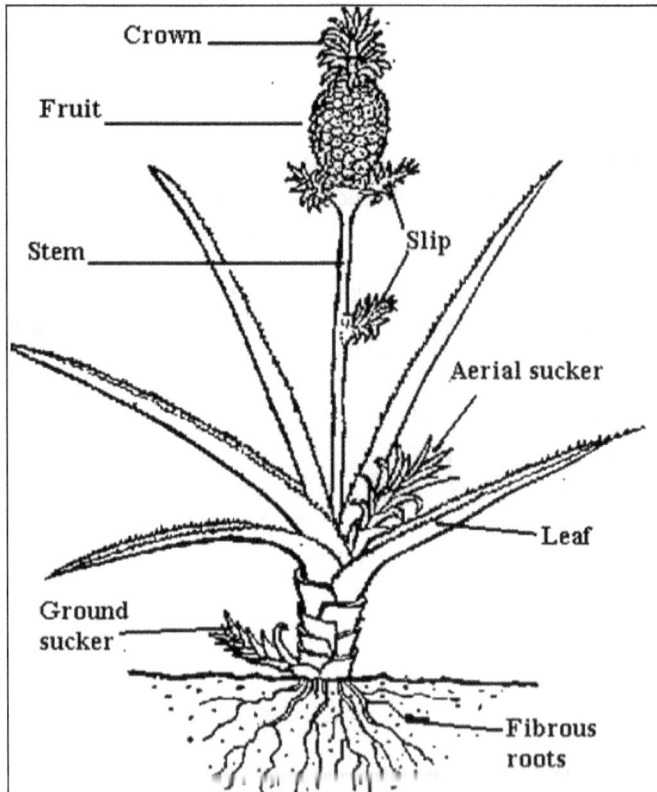

Figure 7.1: Parts of a Pineapple Plant

rind is deep reddish-orange coloured, flesh pale-yellow to white with a spicy-acid taste and fibrous texture; leaves are spiny. One advantage of this group is its resistance to mealy bug wilt, a disease apparently caused by a virus and spread by mealy bugs. It is, however, susceptible to gummosis, an exudation from cracks in the fruit, evidently caused by feeding of a larval butterfly, *Batrachedra* spp., on the fruit shell at blossom time. Members of this group are grown primarily for export and local consumption in the fresh condition. Red Spanish, Singapore Spanish, Green Selangor, Castilla, Cabezona and P.R. 1-67 are important cultivars of this group.

2. Queen Group

Fruit shape is conical with deep eyes, weighs 0.5 to 1.1 kg. Rind is yellow and flesh is deep-yellow in colour, sweet, less acid than Cayenne and low in fibre. Leaves are spiny. This group is more disease resistant than Cayenne group. Grown primarily for fresh consumption. Queen, MacGregor, Z Queen, Natal, Ripley and Alexandria are some important cultivars of this group.

3. Abacaxi Group

Fruit is conical in shape and weighs 1.4 kg; rind is yellow and flesh is pale-yellow or white, sweet, tender and juicy. Leaves are spiny. This group is disease resistant. The fruits do not process well nor survive export well, but are grown in large quantities in Brazil for domestic consumption in fresh form. The important cultivars of this group are Perola, Abakka, Sugar Loaf, Papelon, Amarella, Venezolana and Ananas Vermelho.

4. Cayenne Group

Fruits of this group are cylindrical in shape with a slight upward taper and flat eyes; thus better adapted to canning than other groups. Fruit weight is 2.3 kg on the average, rind is a dark-orange in colour and the flesh is pale-yellow to yellow, sweet, mildly acid, with low fibre and a tender juicy texture. Leaves are smooth with a few spines near the tip ('Spiny tip'). This group is very susceptible to mealy bug wilt. In addition to canning, the fruit is good for local fresh consumption and only fair for export. Cayenne (Smooth Cayenne, Cayenne Lisse), Boron Rothschild, Smooth Guatemalan, Typhone, St.Michael and Esmeralda cultivars belong to this group.

5. Maipure Group

Fruits of this group are cylindrical-ovoid to cylindrical, weighs from 0.8 to 2.9 kg and have a yellow-to-dark orange to red rind with flesh that may be white or deep-yellow in colour. It is sweeter in flavour than Cayenne, fibrous, but tender and very juicy. The leaves are completely smooth with 'piping" *i.e.* margins folded over. Members of the group are best adapted to local, fresh consumption. Maipure, Bumanguesa, Piamba de Marquita, Rondon, Perolera, Legrija and Monte Liro are cultivars of this group.

The above groupings by Knight (1980) do not include names of the cultivars grown in India. For example, popular cultivars like Giant Kew, Kew, Mauritius, Jhaldhup and Bakhat do not find mention in any of the 5 groups listed above. According to Sen (1985), Jhaldhup and Bakhat, grown commercially in Assam, fall in

Queen group. Jhaldhup has sweetness well blended with acidity and has a characteristic alcoholic flavour, while Bakhat is markedly sour. Giant Kew is grown extensively in the country and ideally suited for canning industry. It resembles with Cayenne or Smooth Cayenne in plant as well as fruit characters. It may be considered as a member of the Cayenne group. Similarly, Mauritius can also be included in the same group. In this variety, there are two types-yellow and red skinned. Yellow types are oblong, sloping slightly towards apex and the other one is rounded at the base and sloping abruptly towards the crown. The fruits of yellow type are green in colour when mature and gradually turn deep-yellow when ripe. The crown is large and spiny and the flesh is light-yellow, fibrous and medium sweet. The fruits of red type are red when ripe, flesh is reddish-yellow, fibrous but sweeter than the yellow type. The crown is medium in size and also spiny, leaves are closely serrated. Other members of the Cayenne group are Baronne' de Rothchild, which has spiny leaves and Hilo, which does not produce slips. Thus, though Smooth Cayenne is regarded as one cultivar, it is highly variable.

The description of some of the important varieties grown in India is as follows;

Kew

This variety belongs to Cayenne group and is considered best for canning in India. It is widely grown in most of the pineapple growing states. The fruits are of big size (1.5 kg to 2.5kg), oblong in shape and tapering slightly towards the crown. Eyes are broad and shallow, making the fruits more suitable for canning. The fruit is yellow when fully ripe and the flesh is light-yellow, almost fibreless, very juicy and of agreeable taste. The leaves often have a short sector of small margin of spines just behind the tip, and irregularly on the base near attachment to the stem.

Queen

The plant, fruit and shoots of Queen are smaller than Kew. This variety is mostly grown in Tripura and partly in Assam and Meghalaya. The fruit is rich yellow in colour and weighs from 0.9 to 1.3 kg. The flesh is deep-golden, less juicy than Kew, crisp textured with a pleasant aroma and flavour. Eyes are small and deep, requiring a thicker cut when removing the skin. The leaves are brownish-red, shorter and very spiny.

Jaldhup and Lakhat

These are the two indigenous types grown in Assam, both being named after places of their maximum production. Both fall in the Queen group, with fruits still smaller than Queen. Lakhat is markedly sour in taste, whereas Jaldhup has its sweetness well-blended with acidity. The fruits of Jaldhup again have a characteristic alcoholic flavour of their own and can be easily distinguished from the other fruits of Queen group on this character alone.

Mauritius

A mid-season variety grown in some parts of Kerala. Fruit is medium in size (1.5 to 2.5 kg). Two types with deep yellow and red coloured fruits are found. Yellow variety is oblong, fibrous and medium sweet compared to red type. This variety is especially grown for table purpose. Leaves are yellowish-green, spiny throughout the margin. Crown in both types is also spiny.

Charlote Rothchild

In Kerala and Goa, this variety is partly under cultivation, which is similar to Kew in fruit characteristics and taste.

Simhachalam

In Vishakhapatanam district of Andhra Pradesh, a local variety named Simhachalam is largely grown.

PLANT PROPAGATION

Propagation of pineapple like most other fruit crops is exclusively done by vegetative means. In case of hybrids, progenies evolved through seeds are also vegetatively propagated. Propagation can be done from any four parts of a pineapple plant: (1) *shoots*, (2) *slips*, (3) *crowns* and (4) *suckers*. Shoots grow from the main stem. Slips grow from the flower stalk just below the fruit. Crowns are the groups of leaves at the top of the pineapple. Suckers arise from the roots below ground. Among the types and sizes of propagules tried, slips and suckers weighing around 350 g and 450 g were found best in terms of yield and quality for Kew pineapple under Coorg conditions. In Jorhat (Assam), suckers weighing 501-750 g and slips weighing 301-400 g were found ideal planting material, while suckers weighing 501-1000 g were best for planting in Thrissur (Kerala).

PLANTING AND ORCHARD ESTABLISHMENT

Planting

Because of the low stature of pineapple plant and the need for frequent field access, land preparation is normally more exacting than for other tropical fruits. Soil is usually worked to a depth of 25-30 cm, chopping and ploughing in the field the old pineapple stand. Heavy soils may be ploughed to a depth of 40-50 cm and drains constructed to achieve adequate drainage. Growing legume crops prior to planting has been found beneficial in reducing nematode population in the field.

Planting System

The planting system depends upon the topography of land and rainfall. There are four planting system in vogue, *viz.* flat-bed planting (2-row or 3-row), furrow planting, trench planting and contour planting (Figure 7.2). Workers insert the shoots, slips, crowns, or suckers through the plastic strips by hand. They punch holes in the plastic with a planting tool. After planting, pineapple plants require careful cultivation.

Plant Density

Plant spacing or density of pineapple depends on the growth of the plant and system of planting. Adoption of low planting densities has been the major constraint in India, contributing towards high cost of production per tone of pineapple produced. One of the ways to reduce cost of production is to increase yield per unit area by following high density planting.

In high rainfall areas, the density may be around 40,000 to 44,000 plants per ha. In low rainfall areas with cool weather (*e.g.* Karnataka), even higher density of 63,000 to 64,000/ha is recommended under high density planting.

Pineapple is planted in double hedge system for convenient intercultural operations. For a density of 44,000 plants per ha, the spacing should be 90cm × 30cm × 60cm *i.e.* 90cm between two rows of adjacent beds, 30cm between plants in a row and 60cm between two single rows (Figure 7.2). High-density planting is the latest advanced technology applied in pineapple cultivation. The other spacing follows are 120cm × 60cm × 40cm (41,600 plants/ha.), 75cm × 30cm × 30cm (63,000 plants/ha), 60cm × 30cm × 45cm (64,000 plants/ha).

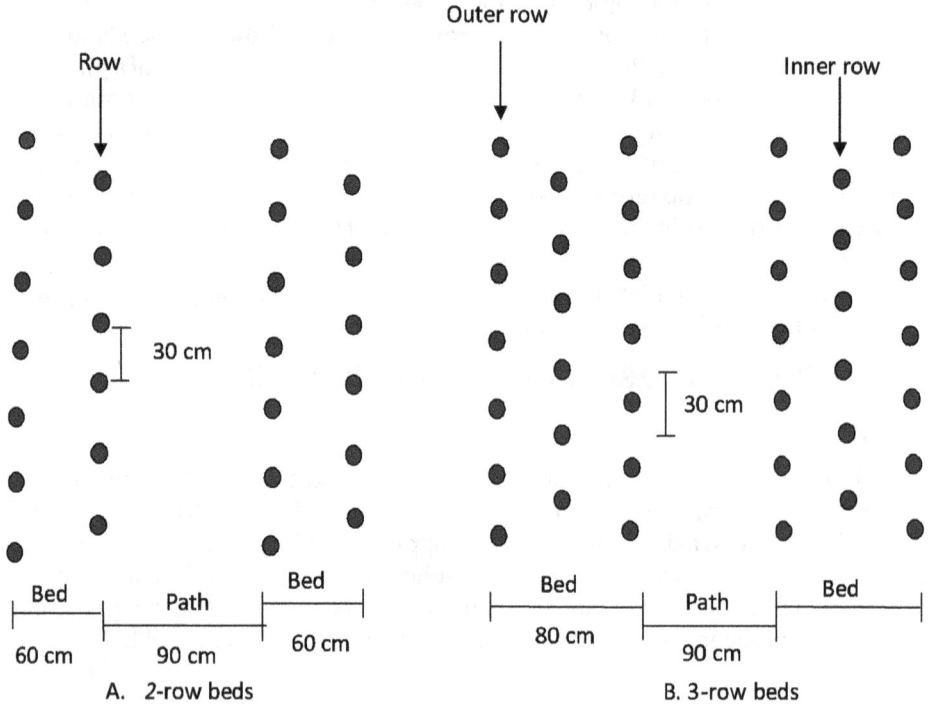

Figure 7.2: Planting Methods Adopted in Pineapple Planting.

Treatment of Planting Material

Before planting, the suckers and slips should be sun-cured and dry leaf scales at the base should be removed. Planting material should be dipped in ceresan solution (4g in 1 lit. of water) or 0.2 per cent dithane M-45 solution before planting to protect the plants against bud rot.

Planting Time and Method

The best time of planting pineapple is early rains or early winter. With irrigation, it may be planted at any time of the year. Suckers should be planted at 10 – 15cm depth in 15-20cm deep hole. About 500g FYM or cow dung is to be added to the soil of each hole. Planting may be done in single or double row following triangular or rectangular system. Pineapple may be planted using black polythene film as soil cover. This ensures total weed control and heavy crop. But fertilizers may be applied by spraying or drip method. Initial investment will be high but drip method is cost effective.

INTERCULTURAL OPERATIONS

Application of Manures and Fertilizers

The correct use of fertilizers is considered important because it will produce fruits of good size and quality. It is quite difficult to decide fertilizer doses for pineapple because these fertilizer requirements depend on several factors, such as, age of plant, variety, planting distance, climatic conditions, soil type, type of fertilizer to be used, nutrients available in the soil etc. For this, soil and foliar tests are mandatory. Hence, tissue analysis is the best guide for monitoring the actual need of plant.

For tissue analysis, D-leaf is preferred. This is the youngest and physiological mature whorl of leaves, which usually represent the tallest/longest leaf on the plant from the ground. D-leaf is used for analysis because, it is easily identified, it gives accurate and sensitive reflection of plant nutrients, its moisture content are relatively constant, as a result, nutrient content can readily be reflected on dry weight basis. However, in general the following doses of fertilizers are given to pineapple (Table 7.4.

Table 7.4: Recommended Fertilizer Doses for Pineapple

Time	Type of Fertilizer	Fertilizer Rate/ha
Before planting	Lime	2.5 tonnes
At time of planting	CIRP	200 kg
3 months	NPK 12:12:17:2	200 kg
6 months	NPK 12:6:22:3	500 kg
9 months	NPK 12:6:22:3	500 kg

Weed Control

The major factor, which contributs to the high cost of production of pineapple is the manual weeding, which accounts for 40 per cent of the total cost of production. Six months after planting, weeding has to be done every 3 months. This can be done manually or by the use of weed killer, paraquat. Pre-emergence application of bromacil (4 kg/ha) and diuron (2 kg/ha) can effectively reduce both dicot and monocot weed population.

Irrigation

Pineapple is grown mostly as a rainfed crop in heavy rainfall areas. Optimum range of rain needed for pineapple is 1,000 to 1,500 mm. However, some of the pineapple growing areas come under high rain fall zone, even to the extent of 2,000 mm. Although, pineapple is grown in India in rainfed areas, where sufficient rainfall is received, but it can also be grown successfully with a few irrigations during summer in the semi-arid tropics.

After Care of the Ratoon Crop

Desuckering immediately after fruit harvest is important. Keeping one or two suckers on the mother plant near the ground level, all others should be removed. Slips should also be removed. After desuckering, plants should be fertilised and earthed up.

FLOWERING AND CROP REGULATION

Flowering

Ripeness-to-flower stage is attained in 11-12 months after planting and formation of at least 40 leaves. Normally, 70 to 80 per cent of plants flower in a year. Sometimes flower does not appear on time. Even after 15-18 months of growth under optimal nutritional and environmental conditions, as low as only 50 to 60 per cent plants may come to flowering. In such cases, application of flower inducing chemicals is helpful. Yield may be increased by applying flower inducing chemicals. A pineapple plant produces only one fruit during its life time.

Crop Regulation

Pineapple plants dont' flower at one time. Inspite of best possible management practicies, about 60-70 per cent plants flower at one time. Hence, regulation of flowering becomes imperative to harvest all the pineapples at one time. Hence, studies have been conducted to use different strategies for crop regulation in pineapples, which indicated that unsaturated gases such as ethylene and acetylene, in the form of smoke, cause early flowering in pineapple. This can be achieved by applying acetylene saturated water or calcium carbide in the heart of the plant. About 50ml ethrel solution (10 ppm) in combination with 2 per cent urea and 0.04 per cent sodium carbonate can induce satisfactory flowering in pineapple. NAA@25 ppm also induces flowering.

Inflorescence development is initiated naturally by shortened day lengths and cool night temperatures. Usually, flower initiation begins in November and continues throughout the winter. Under natural conditions, flowering is highly irregular and some plants may fail to produce fruit. Thus, in commercial practice, growth regulators such as ethylene, and NAA are used to force plants to flower. Once inflorescence development has been initiated, further development occurs without interruption. Chemical forcing of flowering has several advantages:

1. All fruits become ready for picking at one time, thus eliminating the need for several harvests.
2. Mechanical harvesting is possible because of the simultaneous maturing of fruits.
3. The size of the fruit is controlled by adjusting the flowering time. Younger plants are made to flower when fruits of smaller size are desired and flowering is delayed when fruits of bigger size are required.

4. It helps in phasing of pineapple cultivation in such a way that fruits become available throughout the year for fresh consumption or for processing industries.

5. With crop regulation, higher yields per unit area can be obtained because with hormonal treatment, every plant can be brought into flowering and fruiting.

However, the chemical treatment should never be given too far ahead of the normal time of fruiting. A period of upto 6 months before the setting of the natural fruit appears to be satisfactory for the treatment. It has been reported that forcing treatment done at 10 month stage had an adverse effect on fruit yield while the treatment at 16 months after planting gives the largest fruit and the highest yield.

To delay harvest by a few days (10 – 15 days), 300 ppm planofix may be sprayed on the fruit just 60 to 70 days ahead of harvest. To ripen the fruit earlier by about 10 – 15 days, 500 ppm of ethrel may be sprayed on the fruit about one month before normal harvest.

Staggering of harvesting almost throughout the year is possible by the following means:

1. Using different planting materials (slips, suckers or crown).
2. Planting suckers and slips at regular intervals from July-December, and
3. Applying flower inducing chemicals at desired time as stated above.

Crop Cycle

One main crop followed by two ratoons is the usual crop cycle in pineapple. However, after the 4th year, the plot needs to be uprooted and replanted.

HIGH-DENSITY PLANTING

High-density planting is the latest advanced technology applied in pineapple cultivation. High-density planting, besides increasing the yield, is associated with other advantages like less weed infestation, protection to fruits from sunburn, increased production of suckers and slips per unit area and non-lodging of plants. Close planting also saves the cost of providing shade to fruits, as it provides natural shade through upright orientation of the apical leaves and eventually results in uniformly coloured lustrous fruits, which are cylindrical with low taper ratio, giving more canning recovery. Another advantage of high-density planting is the overlapping of basal leaves forming a sort of natural covering over the soil, preventing evaporation losses and thereby resulting in moisture conservation. Under dense planting, a microclimate with high humidity is created around the plant, which is congenial for growth and fruiting.

A few decades ago, the planting density followed for pineapple in India were only fifteen to twenty thousand plants per hectare, with a productivity of less than twenty thousand tones fruit per hectare, resulting in high cost of production. Under such low-density planting, 20 to 25 per cent fruits become unmarketable due to sunburn in the tropical region. Therefore, extensive studies were undertaken at the Indian Institute of Horticultural Research, Bangalore, during the 1970's and 80's to develop high-density planting of pineapple for increasing the productivity, fruit quality and reducing the cost of production.

These studies proved the benefits of increased planting density in the range of 53,000 to 63,500 plants per hectare. Two-row trench system of planting has been found to be the best for high density planting in the plains, whether the crop is grown with or without irrigation. The field is laid out into 22.5 to 30.0 cm deep trenches alternating with mounds. In each trench, two shallow furrows about 10 to 15 cm deep and 15 cm inside from the edge of the trench are opened and suckers or slips are planted in these furrows, so arranged that two plants will not be exactly opposite to each other.

Spacing of pineapple plants under high-density depend on the growth of the plant. Soil moisture and fertility influence plant growth and indirectly determine spacing required per plant and eventually planting density. In overall analysis, plant-to-plant spacing of 22.5 to 25.0 cm and row-to-row spacing of 45 to 60 cm are ideal. Where pineapple plants grow luxuriantly with long leaves, a wider spacing of 90 cm between the trenches is required, but in places where growth of the leaf is moderate, a trench-to-trench spacing of 75 cm is adequate. A plant density of 63,000 plants/ha (22.5 cm × 60 cm × 75 cm) has been found ideal in sem iarid mild tropical areas. In hot and humid tropical areas, plant density of 53,000 plants/ha (25 cm × 60 cm × 90 cm) performs well. However, decrease in fruit weight was quite evident when plant-to-plant spacing was reduced to 20 cm and row-to-row to 40 cm, irrespective of the spacing between the beds.

Initial establishment of planting material in the field is important to avoid gaps and the subsequent poor growth of replants under the competitive situation of high-planting density. Suckers weighing 500-600g and slips weighing 350-400g are the best for initial field establishment under high-density planting. July-August is found to be the best planting season for establishment and growth of plants. In order to obtain a good yield from high density planting, application of fertilizers and treatment with flower-inducing chemicals are considered very important. Nitrogen at 12g under irrigated and 16g under rainfed conditions along with potash at 12g per plant is recommended for the plant crop. For irrigated and rainfed ratoon crops, 10g nitrogen and 12g potash are recommended per plant. Ammonium sulphate is the best source of N.

The yields recorded under high-density planting are in the range of 85 to105 tones/ha, which are 55 to 85 tones more than the conventional planting densities, without adversely affecting the fruit size, quality and canning recovery. Two successive ratoon crops harvested at twelve monthly intervals amounted to 50.7 and 53.8 per cent of the plant crop yield at high-density planting under good management. Highest cost: benefit ratio of 1: 4.2 was observed in planting densities of 53,000 and 59,200 plants/ha. The following is the spacing required for different plant populations per hectare:

Plant Population/ha	Distance (cm)		
	Plant to Plant with in Row	Row to Row	Trench to Trench
43,500	30	60	90
53,300	25	60	90
63,700	22.5	60	75

CROP PROTECTION

A. Major Diseases and their Management

Fusariosis

Fusariosis, produced by *Fusarium subglutinans*, may cause losses upto 80 per cent of marketable pineapple fruit. The fungus infects approximately 40 per cent of propagules and kills about 20 per cent of plants prior to harvest.

Management Practices

Losses are reduced mainly by use of an integrated disease management programme including cultural, chemical and genetic control methods. Fungicide formulations such as captan, febuconazole and iminoctadine significantly reduce fusariosis severity on the fruits, suggesting integrated management of fusariosis is possible with new fungicides. The most important control measures are to use disease-free planting material and to protect the inflorescence and hence the following methods can be adopted (i) pathogen free plant material can be obtained by sectioning the stump plants after harvesting, (ii) to force plants to flowering in months with low rainfall, (iii) to protect the inflorescences of forced plants with benomyl (0.05 per cent) and (iv) to protect inflorescence with paper bags.

Base Rot

Base rot of pineapple is a serious disease prevalent in areas having humid and tropical to sub-tropical climatic conditions. In India, the disease is quite common in Assam. The pathogen causes three types of symptoms *i.e.* base rot, leaf spot and fruit rot. Leaf spot is of lesser economic importance. Base rot is the most serious phase of the disease and affects about 10 per cent of the plants in the field. Fruit rot phase is observed both in the field as well as during transit and storage. In the field, it is not very serious but causes tremendous losses during storage.

Causal Organism

The disease is caused by a fungus, *Ceratocystis paradoxa* (Dade). Suckers collected from the infested areas carry the pathogen with them to new areas.

Symptoms

The infected leaves exhibit spots with dark margins, which later turn to olive-brown or white colour. As the disease progresses, the tissues dry and leaves are distorted. Infection on suckers do not allow them to develop normally after planting and such plants do not respond to normal functioning of roots. The leaves on such suckers/plants remain underdeveloped, show yellowing and start withering. A closer examination of the plant exhibiting such type of symptoms show rotted black area on the base of the stem. Eventually, the plant/sucker starts dying due to extensive rotting of the base. This is the base rot phase, which is most serious as it does not allow the proper stand of the crop. On fruits, water-soaked yellowish lesions are developed, which finally turn dark and the fruits finally rot.

Management Practices

The following practices keep the incidence of base rot under control:

1. The diseased plants must be uprooted and destroyed to check the further spread of the disease.
2. For fresh plantings, suckers should always be collected from the disease-free fields.
3. Before planting, the disinfection of nursery soil with formaldehyde (1 per cent) and dipping of suckers in Bordeaux mixture (2:2:50) or borax provide protection against the soil borne inoculum.
4. The rot can be reduced by exposing the harvested fruits to the sun for 2 hours and dipping the cut ends of stalks in 10 per cent solution of benzoic acid dissolved in alcohol.
5. The yeast isolate mixture reduces spore germination, germ tube length and dry weight of *C. paradoxa*. Combining the isolated *Pichia* or yeast mixture with a half rate of fungicide results in complete control of base rot comparable to control achieved with a commercial recommended fungicide.

Heart or Stem Rot

It is a serious disease of pineapple in regions with high rainfall and low temperature, where 'Kew' cultivar is grown as mixed crop with mandarin. Poor physical condition of soil and poor drainage aggravate the problem.

Causal Organism

It is caused by fungus, *Phytophthora nicotiana* var. *parasitica*.

Symptoms

The initial infection may appear at the apical end of the stem, at the base of leaf, at the basal end of the stem or at the roots. This results in the loss of turgidity and a slight twisting of central leaves. In later stages, there may be withering and discoloration of the affected portion to yellowish-pink and brown. Inner whorl of leaves can be readily detached from the stem by a slight pull. The stem may show brownish discoloration of the margins between healthy and diseased tissue. Basal portion of leaves show signs of rotting and emits foul odour. Plants propagated through slips are more susceptible to stem rot than those propagated by suckers.

Management Practices

Following preventing measures are useful for the control of heart rot of pineapple:

1. Ensure proper drainage in the field.
2. Don't grow 'Kew' cultivar.
3. Use suckers as planting material.
4. Before planting, dip the propagating material with Bordeaux mixture (1 per cent) or copper oxychloride (0.25 per cent).
5. Planting may be done on ridges.
6. Spray Bordeaux mixture (1 per cent) during raining season.

Anthracnose

In certain areas, anthracnose cause heavy loss to pineapple. It is caused by *Colletotrichum ananas*, which mainly infects pineapple leaves.

Symptoms

Its infection results in the appearance of irregular blotches with minute fruiting bodies that originate mostly from the apex of the leaves. Sometimes, marginal spotting is also seen, which results in the withering and dropping of leaves and ultimately the growth of the infected culms and development of fruits is seriously affected. The disease is favoured by high humidity and takes about fortnight to develop on young leaves. During June-July, black specks are observed on leaves, which lead to withering.

Management Practices

1. Collect and destroy the affected leaves and fruits to avoid further spread of the disease.
2. Spray bavistin (0.1 per cent) or topsin-M (0.1 per cent) before the expected attack of the disease.

Leaf Blotch

It is a serious disease of pineapple in some parts of South India. It is caused by *Pestalotia microspora*.

Symptoms

On leaves, pale round or irregular small necrotic lesions appear, mostly originating from the tip or central part. Later, these lesions coalesce; turn dark-brown, having rough surface and irregular margins. The affected leaves look as if blotched. It assumes serious look in humid weather.

Management Practices

Following measures are considered necessary for the control of leaf blotch:

1. Destroy the source of inoculum *i.e.*, the fallen leaves.
2. New plantings should be done with disease-free planting material.
3. Spray dithane Z-78 (0.2 per cent) or blitox (0.4 per cent).

B. Major Insect-Pests and their Management

Mealy Bug (*Dysmicoccus brevipes*)

Pineapple mealy bug is regarded as a major pest of pineapple because of its association with pineapple mealy bug wilt (PMW). It has more than 50 hosts such as sugarcane, perennial grasses, banana and other crops. Mealy bugs cause wilting of the plants directly by feeding, injecting toxin and by introducing virus into the plants. The symptoms first appear on roots, which cease to grow and finally rot and collapse. The most prominent symptom is the wilting of leaves at their tips. Infested leaves change in color from green to red or pink and the margins reflect inwards. Severely infested plants become stunted. Fruits may fail to develop, or remain small, fibrous and sour. In mature plants, heavy infestations occur on slip bases, lower fruit surfaces around peduncles, and within fruit floral cavities. Among different cultivars, Cayenne is the most susceptible and Red Spanish and Queen being resistant to mealy bugs. Adopt the following measures to control this pest effectively:

1. Use healthy suckers for planting.
2. Before planting, remove all basal brownish leaves.
3. Remove grasses and other monocot weeds which serve as additional hosts for the mealy bug.
4. Dip the planting material in 0.02–0.04 per cent solution of methyl parathion.
5. Apply thimet 10 G @ 17.5 kg/ha at an interval of 100 – 125 days or apply disulfoton @ 20 kg/ha, at 45 days after harvest and after planting.
6. Destroy colonies of ants to check the spread of mealy bugs.
7. If possible, use natural parasite, *Anagrus coccioduorus* in the plantations.

Slug Caterpillar (*Latoia lepida*)

This is a polyphagous pest. The young caterpillars feed gregariously by scrapping the under surface of the leaves. Later, they feed voraciously on leaves leaving behind the midribs. The continuous feeding leads to loss of sap from plant tissues and ultimately, the vitality of plants is reduced. The affected leaves dry up. It can be controlled by the following practices:

1. Hand picking and mechanical destruction of gregarious caterpillars is reported at an early stage of infestation.
2. Dusting with methyl parathion (5 per cent) is effective and should be done when the caterpillars are still feeding gregariously.

Fruit-Eating Beetle (*Carpophilus dimidiatus*)

These beetles feed inside ripen fruits. Damaged fruits rot due to infestation of pathogens. To control these beetles, do not allow the fruits to ripe in the field. Birds and rats damaged fruits should be plucked and dumped in the pits.

C. Physiological Disorders and their Management

Chilling Injury

Exposure of pineapples to a temperature below 7°C (45°F) results in chilling injury. Ripe fruits are less susceptible than unripe or partially-ripe fruits. Symptoms include dull green color when ripened (failure to ripen properly), water- soaked flesh, darkening of the core tissue, increased susceptibility to decay, and wilting and discoloration of crown leaves. Do not store pineapples below 10°C to avoid chilling injury in them.

Black Heart

Blackheart of pineapple is also known as endogenous brown spot (EBS) or internal browning. It is an important physiological disorder of pineapple. The initial symptom of this disorder is the development of one or more brown, translucent spots at fruitlet bases, close to the core. In severe cases, the entire fruits can be affected by black discoloration. Low temperature or exogenous GA_3 can induce this disorder. Treatment with 150 or 300 ppm GA_3 could induce blackheart in summer fruits when stored at ambient temperature for 1 or 2 weeks after treatment. Heating fruits at high temperature (40°C) for 24 h reduces blackheart on cold-stored fresh pineapple. Waxing of fruits also reduces blackheart incidence and severity to a greater extent.

Multiple Crowns

Ordinarily, fruit bears a single crown but in some cases, fruit bears more than one. Consequently, the top of the fruit is flat and broad and fruit becomes unfit for canning. Such fruits taste insipid and are corky. It is supposed to be a heritable character, found mostly in Cayenne group to which the variety, Kew belongs.

Fruit and Crown Fasciation

Fasciated fruits are deformed to such an extent that they are totally useless. In certain cases, proliferation is so extreme that fruit is highly flattened and twisted with innumerable crowns. Fruits and crowns fasciation is associated with high vigour of plants, which take longer time to flower. High fertility of soil, warm weather and calcium or zinc deficiency may favour fasciation.

Collar of Slips

The collar of slips is typified by the presence of a large number of slips arising from stem close to the base of the fruit, or even directly from the fruits itself. The excessive slip growth is at the expense of the fruit, resulting in small, tapered fruits, often with knobs at the base. High nitrogen fertilization and high rainfall along with relatively low temperature are supposed to be congenial for such an abnormality.

Dry Fruit and Bottle Neck

The dry fruit and bottle neck fruit types are very similar and may be derived from the same parent. In dry fruit type, fruit is small, flowers are absent and fruitlets do not develop. In bottle neck, lower fruitlets develop normally and upper ones do not develop and give the same appearance as dry fruits. Suckers are freely produced from both the types.

Sun-Scald

This results when plant leans or falls over to one side, thus exposing one side of the fruit to direct sunlight. The cells of the exposed surface get damaged. Later, shell surface assumes a brownish to black colour and cracks may appear between fruitlets. Affected fruits soon rot and become infested with pests. They must be cut as soon as noticed and safely disposed of where they will not contaminate other fruits. In high-density planting, intensity of sun-scald is very much minimised. Under favourable climates where leaf growth is luxuriant, leaves can be tied around the fruits to protect them from sunscald.The other method is to cover sun-exposed portion of the fruit with dry straw or grass or with anyother locally available materials.

Besides, there are disorders such as Triad rot and Y-center rot, etiology of which is not known yet. Another disorder 'woody fruit' has also been encountered. The etiology is not known but it is widely credited to genetical attributes

HARVESTING AND YIELD

Maturity

About 15-18 months after planting, the pineapples are ready to be picked. A pineapple plant bears one fruit for the first harvest and may bear two fruits for the second or third harvest. Most planters replant fields after every two or three harvests. In most countries, pineapples are harvested by hand. The pineapple pickers grab the fruit by the crown and twist it from the stalk. They put the pineapples in baskets strapped to their backs or in canvas bags carried over the shoulder. Hawaiian pineapple growers use a *harvester conveyor* to simplify the fruit-picking. This machine consists of a long *boom* (metal arm) with a conveyor belt built into it. A truck moves the boom through the pineapple field, with the boom extending over many rows of plants. Pineapple pickers walk behind the boom. They pick the pineapples and drop them onto the conveyor belt, which carries them to the truck.

There are several maturity indices for harvesting of pineapple at right stage of maturity. However, the fruits are picked when one or two eyes turn yellow. Harvesting for local markets should be done at full maturity and for distant markets, at 75-80 per cent maturity stage. At maturity, lowermost eyelets show orange-yellow colour and eyes get flattened in the centre and bulge on side. There is a sharp decline in shell and flesh pH and respiration rate.The fruits are picked with 10cm of the stem and the parent plant is then cut 7-10 cm from ground level and the sucker is allowed to develop for the ratoon crop.

Yield

Pineapple flowers 10-12 months after planting and attains harvesting stage 15-18 months after planting, depending upon the variety, time of planting, type and size of planting material used and prevailing temperature during the fruit development.

The average yield for the first year is between 40-65 tonnes per hectare, depending on cultivars.

POSTHARVEST MANAGEMENT

Packaging and Transport

In several pineapple producing countries, pineapples are normally transported by road and rarely by air. The stalk of fruit should be trimmed to 10-30 mm and cut surface should be treated with a suitable fungicide. Then, fruits are grated on the basis of size, shape and colour and packed in fiberboard or wood containers for distant transport.

Ripening

Pineapple is a non-climacteric fruit. Hence, ripening of pineapple could be considered at the terminal period of maturation, during which, fruit attains most desirable quality. However, skin colour does not sometimes develops properly. For uniform ripening, ethrel treatment has been found effective to enhance uniform ripening in pineapple. Pre-harvest treatment of mature green pineapples with ethrel @ 500-1000 ppm helps to develop attractive yellow colour in pineapples.

Storage

Pineapple could be stored for about 20 days at 10-13°C. Controlled atmosphere (CA) storage (5 per cent O_2 and 10 per cent CO_2) helps to keep pineapples safe for 4 weeks. The ideal storage condition are pineapple are as under:

Table 7.5: Ideal Storage Conditions for Pineapples

Stage of Fruit	Temperature (°C)	Relative Humidity (Per cent)	Duration (Weeks)
Mature green	10-13	90	3-4
Turning	7-10	90	3-4
Ripe	7	90	2-4

Adequate ventilation is required to ensure arrival of fruits in perfect condition at the retail outlet if the fruits are to be transported for a shorter distance. Care should be taken to prevent bruising during harvesting and packing, and finally fruits have to be adequately protected against fungal infection.

Canning

At the cannery, the pineapples are washed and sorted by size. A machine called a *Ginaca* removes the shells, punches out the cores, and cuts off the ends of the pineapples. Then, the fruit is cut into slices or pieces, put in cans, syrup is added, and the cans are sealed. The unsweetened juice from the pineapple cores is also canned. Finally, the cans are heated to kill any micro-organisms that might cause spoilage.

8

Grape

The cultivation of grape is done from ancient times, which is believed to have originated near Caspian Sea. Fossil studies show that viticulture in Egypt is 6,000 years old. Grape had commercial importance in the Mediterranean and Middle East region even before the beginning of Christian era. However, the ancient Sanskrit literature, Arthashastra and Sushruta, reveal that grape was in cultivation in India even during the first century. Indigenous varieties known as 'Rangspay', 'Shonltu White' and 'Shonltu Red' are grown in Kinnaur, Himachal Pradesh even today. Though, grape was well known to people even in ancient India, but it was not commercially cultivated until the 14th century. Grape by nature is a deciduous fruit crop, growing extensively in temperate regions. Even five decades ago, grape cultivation was considered to be a non-viable proposition in the tropics. However, because of the extensive research efforts, the world's highest productivity has been recorded from the tropical region of the country and more than 90 per cent of the area under grapes in India falls in this region. It is mainly grown in Italy, France and Spain for vine making; in Italy, Turkey, Bulgaria, USA, Greece and Portugal for table delicacies; and in Turkey, Greece, Australia and the USA for raisin preparation. However in India, it is mainly grown for table use.

COMPOSITION AND USES

Composition

Grape is one of the most delicious, refreshing and nourishing fruits. Its fruits are a rich source of carbohydrates and vitamin A and B. Dried berries (*Munnakka* or raisin) are excellent source of sugars and minerals such as Ca, Fe (Table 8.1).

Table 8.1: Composition of Ripe Grape Berries

Component	Per 100 g Edible Portion	Component	Per 100 g Edible Portion
Carbohydrates	15-25 per cent	Thiamine	35-58 µg
Minerals	0.2-0.6 per cent	Riboflavin	20-25 µg
Iron	0.0003-0.0017 g	Pyridoxine	84-135 µg
Calcium	0.004-0.025 per cent	Pantothenic acid	70-78 µg
Potassium	0.15 to 0.25 per cent	Nicotinic acid	170-330 µg
Vitamin –A	100-8000 IU	Folic acid	4.2-10.2 µg
Vitamin –B	369-636 µg	Vitamin C	1-12.5 µg

Uses

The grape is utilized in many ways. Famous Indian medicine scholars, Sasruta and Charaka discussed importance of grape in their medical treatises entitled ' *Sasruta Samhita'* and *'Charaka Samhita'*, respectively. Approximately 71 per cent of world grape production is used for wine, 27 per cent as fresh fruit, and 2 per cent as dried fruit. However, in India, 90 per cent of the grape is used for table purpose, even though wine making has made strides. The rest of the grape is used mostly for raisin production. France, Italy and Spain are the leading produces of wine.

In India, some wild grapes grown in the North West Indian state (*e.g.,* Himachal Pradesh) were used to prepare a local wine, called as "*Angoori*". In India, grapes are mainly consumed as fresh fruit. Grapes are also used for the preparation of syrup, jam and vinegar. Tartaric acid, considered as a stimulant to kidneys, is extracted from grapes. Grape seeds contain 6–20 per cent oil, used for edible purposes, soaps, and as a linseed substitute. A number of byproducts of winery are also used, *e.g.,* pomace as cattle feed, to purify tannins and tart rates for making cream. The fruits are a rich source of sugars, particularly glucose and fructose, minerals, vitamins, tartaric and mallic acids and tannins. Grape provides 60-70- Kcal/100 g edible fruit parts. Grape berries also contain compounds which impart specific colour, odour and flavour. The pigments are mostly found in skin but in some cases these are both in pulp and skin (teinturier type) (*e.g.,* Pusa Navrang). The colour of grape berries is mainly due to four or five anthocyanins *viz.*, cyanidin (red colour), delphinidin (blue colour), petunidin (purple colour) and malvidin (blue colour). The yellow pigment in white and red grapes is quercetin – a flavone and its glycoside, the quercitrin – a flavonol. Some varieties develop distinct aroma during ripening. Muscat Hamburg and Bhokri cultivars have Muscat flavour whereas Concord has foxy flavour. The foxy flavour and aroma in grape is due to the presence of methyl anthranilate.

ORIGIN, HISTORY AND DISTRIBUTION

The American grapes consisting of *Vitis labrusca* and other species of *Euvitis* and *Muscadinia* are considered to have originated in the North American region. This region has the largest concentration of native *Vitis* species (about 70 per cent of all the species in the world). This region is commonly called as the vineland. The original home of European grape (*Vitis vinifera* Linn.) is Caucasus region between black and Caspian sea.

Grape cultivation first started in Asia minor. From there, it spread to West and East. Wine grapes soon spread throughout the central and southern Europe and raisin and table grapes to North Africa. Afterwards, grape cultivation moved to Turkey, Iran, Pakistan and India. The grape (*V. vinifera*) traveled into North American region in the 17th century. In USA, vine grapes were first introduced in the eastern states during 1616 and in California in 1697. The table and raisin cultivars of Europe were, however, introduced there by nurserymen and growers after 1850. The American grapes appear to have moved sporadically to Europe and Asia after the discovery of sea route to America. *Vitis vinifera* was introduced into Philippines from California in 1958.

Cultivated grapes are believed to have been introduced into the North India by the Persian invaders in 1300 AD, from where they were introduced into the South (Daulatabad in Aurangabad district of Maharashtra) during the historic event of changing the capital from Delhi to Daulatabad by King Mohammed-bin-Tughlak. Grape was also introduced in the South into Salem and Madurai districts of Tamil Nadu by the Christian missionaries around 1832 AD, and into Hyderabad province by HEH, the Nizam of Hyderabad in the early part of the 20th century. From Delhi, Daulatabad, Madurai, Salem and Hyderabad, grape cultivation spread to different parts of the country. The real boost to the Indian viticulture was initiated after the discovery of Anab-e-Shahi variety by Mr. Shankar Pilley in 1930, from the bunglow of Nawab Baquer Ali Khan. The commercial viticulture picked up slowly in other states like Karnataka with variety Bangalore Blue, in Tamil Nadu with Bhokri and in Maharashtra with Cheema Sahebi.

Grapes occupy a predominant position in terms of world fruit production. The total world production of grapes is estimated to be about 67.3 million tonnes, next only to citrus, bananas and apples. The major grape producing countries are China, Italy, U.S.A, Spain, France, Turkey, Chile, Argentina, India and Iran.

In India, grapes are mostly grown in the states of Maharashtra, Karnataka, Andhra Pradesh and Tamil Nadu. The major grape producing belts are Nasik, Sangli, Solapur and Pune in Maharashtra; Bijapur, Bangalore and Kolar in Karnataka; Theni and Cumbum valley in Tamil Nadu and Hyderabad and Rangareddy in Andhra Pradesh (Table 8.2).

TAXONOMY AND BOTANICAL DESCRIPTION

Grape has been classified under the genus *Vitis* of the family Vitaceae. The other important genera of Viticeae are *Ampelocissus, Cissus, Tetrastigma, Leea, Cayratia* and *Parthenocissus*.

Table 8.2 : Statewise Area, Production and Productivity Scenario of Grape in India

State	Area (000'Ha)	Production (000'MT)	Productivity (MT/Ha)
Maharastra	86.0	774.0	9.0
Karnataka	18.1	330.3	18.3
Tamil Nadu	2.7	53.0	19.3
Andhra Pradesh	1.3	27.6	21.0
Mizoram	1.6	20.4	12.9
Others	1.7	29.5	17.3
Total	111.4	1234.9	11.1

Source: National Horticulture Board Database-2011.

Vitis is a genus of deciduous, rarely evergreen, shrubby climbers. It has two subgenera (i) *Euvitis* and (ii) *Muscadinia*. *Muscadinia* has only three species namely, *Vitis rotundifolia*, *V. munsoniana* and *V. popeneii* and all other species have been grouped under *Envitis* sub-genera. The distinguishing characters of the sub-genera are as follows:

Characteristics	Euvitis	Muscadinia
Somatic chromosome No. (2n)	38	40
Tendril	Forked	Unified
Bark	Loose	Tight
Lenticels	Absent	Present
Seed	Beaked and pyriform	Beak absent
Diaphragm at the node of shoot	Present	Absent

Vitis has 60 species and some of them have popular names, such as, fox grape (*V. labrusca*), frost grape (*V. vulpina*), river bank grape (*V. riparia*), cat grape (*V. palmata*) and the bird grape (*V. munsoniana*). The classification of American grape is very confusing but the commercial cultivars of American grape are the direct derivatives of *V. rotundifolia* or *V. labrusca*. European grape (*V. vinifera*) is known for excellent quality but lacks resistance to insect-pests, diseases and hardiness to cold. This species is the main contributor of table grapes in the world.

Vitis species are considered as secondary polyploids, involving 3 basic sets in the combination of (6+7)+6 = 19. The Muscadinia species, on the other hand, are (6+7)+7 = 20. Cytogenetically, they behave as completely diploid species because of formation of regular bivalent pairing. The inter-specific hybrids within *Euvitis* genus are very fertile. However, the crosses between *Musca dinia* female and *Euvitis* male, do not set fruit. The reciprocal crosses set fruit but F_1 is sterile due to abnormal pairing and distribution of chromosomes.

The inflorescence of grape is called cyme, a much branched cluster each branch ending in a terminal flower. Botanically, fruit is a berry. The edible part of the berry is pericarp and placenta.

SOIL AND CLIMATE REQUIREMENTS

Soil

High soil fertility is not as important to grape as soil structure, conducive to the development of roots. The grape has a strong root system and can be grown on a wide range of soils but the best soil is sandy-loam that is well-drained and fairly fertile, having good amount of organic matter. Heavy clay, sand or slit are unsuitable for grape. However, sandy-clay, sandy-laterite, gravely-loam and red soils are suitable for grape cultivation. Grape is relatively tolerant to soil salinity and alkalinity. The upper salt tolerance limit of grape is 0.3 per cent. Excessive lime is harmful for grapes. In general, grape grows well when soil conductivity is 2.5 mmhos/cm ECe and exchangeable sodium is 45. However, the injury symptoms appear in some varieties even at lower concentrations. Among different varieties, Anab-e-Shahi and Perlette are considered susceptible and Thompson Seedless and Beauty Seedless comparatively resistant to salinity and alkalinity.

The details of soil types in major grape growing regions of India are being presented in the following table:

Major Grape Growing Regions	Soil Types
Uttar Pradesh and Haryana	Sandy loams, sandy clay loams
Andhra Pradesh	Red sandy soils
North interior Karnataka and Maharashtra	Shallow-medium deep black soils
South interior Karnataka and Tamil Nadu	Red loams

Climate

Grape is among the fruit crops, which can be grown in agro-climatic zones ranging from temperate, tropical to sub-tropical. It is most successfully grown at elevations ranging from 200-250m above m.s.l. While, most of the grape cultivation in the world is under temperate climate, in India, it is mainly confined to tropical region. Grapevines are deciduous, which require a long, dry and moderately hot season during cane maturity and ripening of berries, followed by cool winter. Area with annual rainfall not exceeding 900mm well-distributed throughout the year is ideal. Quantity, and seasonality of rainfall are important factors for successful viticulture. Rains during growing season are useful, but continuous rains, make it difficult to control diseases. Rains at the time of berry ripening are harmful. Even a single shower of rain during berry ripening can destroy the whole crop.

Similarly, vines do not grow well in humid summer due to the attack of several diseases particularly, during 30-110 days after forward pruning. At a high humidity, the vegetative growth of vines is vigorous, which affects the fruit size and quality. Grape can tolerate humidity better in cool regions than in warmer regions. Bright sunny days help in accumulation of sugars in berries. Quantity and quality of light play a vital role in synthesis and accumulation of sugars, acidity and pigments. Light intensity of 2,400-ft. candle is essential for optimum growth. However, low light intensities during the active growth stage (45-75 days after pruning) and fruit bud formation adversely affects the crop.

Vital vinifera can tolerate a temperature range of –12 °C to 41.6 °C. It is successfully grown in areas where the temperature range is from 15-40 °C. High temperatures above 40 °C during the fruit growth and development reduce fruit set and consequently the berry size. Low temperatures below 15°C followed by forward pruning impair the bud break leading to crop failure. The fruitfulness of buds is influenced by light.

Most of the American varieties are tropical. Vines require chilling hours to go to dormancy in winter, for about 60 days. However, heat units of 1,600-3,500 degree days (depending on early and late varieties) are necessary for flowering, fruiting and subsequent ripening of berries. For renewal of growth, a daily mean temperature of 10 °C is required, which attains peak at 30°C. In North India, the short growing season from flowering to ripening of berries does not allow the berries to attain high quality. Wine grapes require comparatively cool weather with less variation in temperature during the ripening as in Bangalore (India). Frost during spring and hails during fruiting season cause considerable damage to sprouting buds and fruits, respectively.

IMPORTANT CULTIVARS

There are about 10,000 varieties of grapes in the world, but only a few varieties have commercial status.

Commercial Classification of Varieties

Grape varieties have been classified into five major groups, *viz.* table, raisin, wine, juice and canning.

Table Grapes

The grapes, which are, utilized either as fresh for decoration purpose, are called as table grapes. These must be attractive in appearance, eating, good keeping and storage qualities and should be resistant to injury during handling. Large berries of uniform size, with firm pulp, tough skin, preferably seedless, stronger attachment with rachis and with Muscat flavour are best suited for table purpose. Important table varieties of grapes are Thompson Seedless, Pusa Seedless, Perlette, Beauty Seedless, Bhokri, Cardinal, Black Muscat, Tokay, Delaware.

Raisin Grapes

The grapes that produce acceptable dried product are included in this group. Seedless grapes with high TSS, possessing soft texture, pleasing flavour, large or very small size after drying and a little tendency to become sticky during storage, are classified as good raisin grapes. In general, large and non-sticky raisins are preferred in the world trade market. Raisins should not have more than 17 per cent moisture. In seeded raisins (*Munakka*), the seeds should be few, tasteless and soft. In India, yellowish raisins with greenish tinge are preferred. A few cultivars can meet all the criteria for raisin making. Some notable varieties for raisin making are Black Corinth, Thompson Seedless, Muscat of Alexandria, Sundekhani, Pusa Seedless and Kismish Beli.

Juice Grapes

Grape produce acceptable beverage, when it is preserved by pasteurization, germ proof filtration or other means. The juice must retain the original grape flavour. The cultivars having strong capacity to retain their muscat flavour during processing, are preferred for juice making. Early Muscat, Black Champa, Concord, Bangalore Blue, White Riesling, Arka Hans, Arka Shyam and Pusa Navrang are suitable for juice making.

Wine Grapes

Grape is mainly used for wine preparation in most parts of the world. Grapes of high acidity and low sugar are preferred for dry or table wines, while sweet or dessert wine is prepared from grapes with high sugar content and low acidity. White Riesling, Pinot Noir, Cabernet Sauvignon, Black Cheaper, Rubired, Madeleine Angevine, Cheema Sahebi and Pusa Navrang produce wine of good quality.

Canning Grapes

Seedless grapes are used for canning with other fruits, in fruit salad and fruit cocktail. Seedless white grapes with large berries are preferred for canning or bottling. Usually, a sugar syrup with 20-24 °Brix is used as a canning medium. Thompson Seedless, Pusa Seedless and Perlette can be used for canning purpose.

The chief characteristics of some important cultivars are described below:

A. Seeded Cultivars

Anab-e-Shahi

This variety has revolutionized the viticulture industry in South India. It has proved the most productive variety of grape. The vines are vigorous and productive. It has attractive large and loose clusters with large uniform and attractive greenish-yellow berries with good shipping quality. Berries are comparatively less sweet (TSS 11.0-16.0 per cent). This variety is highly sensitive to salinity and is susceptible to diseases like mildew, anthracnose and fruit rot.

Bangalore Blue

It is a commercial dessert cultivar of Karnataka. It is reported to be *V. vinifera* x *V. labrusca* hybrid and is resistant to insect-pests and diseases (mildews and anthracnose). The vines are vigorous, bearing small and compact bunches. The berries are medium (2.5-3.0 g), spherical, dark blackish, strongly adherent with pedicle and with foxy flavour. Apart from being a table grape, it is extensively used for juice and wine making.

Bhokri (*Panchdraksha*)

It is a famous cultivar of Maharashtra and bears well in North India. The vines are vigorous, long lived with high yielding capacity. It is a prolific bearer on bower system of training and does well on trellis and other systems of training. The bunches are large, compact, having small, well filled and compact berries of grass-green colour. The berries have strong muscat flavour. Ripening is late and uniform. Berries crack with rain and shipping quality is very poor. This variety has now been replaced by Thompson Seedless in Maharashtra and by Anab-e-Shahi in Tamil Nadu.

Cardinal

It is a cross between Tokay and Ribler, made by E. Snyder is USA. The vines are medium in vigour, bearing large but loose clusters of red and spherical berries. The colour of berries changes to dark red at full ripening. It suffers badly from fruit cracking if rains occur at ripening time.

Cheema Sahebi (Selection 7)

It is a selection made by Dr. G.S. Cheema from seedlings of a functionally female (male sterile) Pandhari Sahebi in Pune (India).The vines are vigorous and highly productive. The clusters are large, berries medium greenish yellow, loosely adherent, resulting in berry shattering during transit. The keeping quality is very poor. It is a late ripening cultivar.

Gold

It is an early to mid-season cultivar, bearing large, oval berries of attractive golden colour and mild muscat flavour. The bunches are loose and medium in size. Panicle drying is a serious problem of this cultivar.

Gulabi

It is also known as Karachi, Paneer, Black Prince and Muscat. It resembles Muscat Hamburg of Australia. It is the most important cultivar being grown around Coimbatore conditions. The vines are medium in vigour and yield. The bunches are small and loose bearing deep purple, small and spherical berries of muscat flavour. This variety can give 5 crops in two years. It withstand rains better than most *Vinifera* varieties. It has high tolerance to powdery mildew and anthracnose. The ripening is early and uneven.

Pearl of Casaba

It is an earliest ripening cultivar in North India as the ripening starts by 3rd week of May. The vines are poor in vigour, bearing small bunches of well filled light green spherical berries. Due to low yield, small bunches and high seed content, it is not suited as a commercial cultivar, but it is highly valuable parent for hybridization programme in North India, for developing early ripening and seedless varieties.

Pinot Noir

It is a famous cultivar of France and is used for making Burgundy wine. The vines are medium in vigour, bearing small and compact cluster of small, oval, black berries with large seeds.

Kali Sahebi

This variety is grown in the states of Maharashtra and Andhra Pradesh on a small scale. Berries are large, oval cylindrical, reddish-purple and seeded. The TSS is

22 per cent. Variety is susceptible to rust and downy mildew. Average yield is 12-18 t/ha. The variety is suitable for table purpose.

Madhu Angoor

It is a clonal selection probably, from cv. Carolina Black Rose. Vines are tolerant to downy mildew, bunches medium large, bold dark red seeded berries with mild muscat flavour and suitable for table purpose.

Arka Kanchan

It is a cross between Anab-e-Shahi and Queen of Vineyard, evolved at IIHR, Bangalore. Vines are very vigorous, bearing well filled large clusters (500-700 g). The berries are large (4-5 g), golden-yellow, seeded with muscat flavour and TSS range of 19-22 per cent. It has good yield potential and can produce fruits on head system also.

Arka Shyam

It has been evolved at IIHR, Bangalore by crossing Bangalore Blue and Black Champa. The vines are vigorous, cluster medium (250-350 g), well filled and compact, berry medium large (3.8 g), shining black, spherical, seeded with mild foxy flavour, and total soluble solids ranging from 22-25 per cent. Yield potential is high. It is fairly resistant to diseases.

Arka Hans

It is a cross between Bangalore Blue and Anab-e-Shahi, evolved at IIHR, Bangalore. The vines are vigorous bearing well filled medium clusters (330 g). The berries are seeded, medium large (3-3.5 g), yellowish-green, spherical with pleasant foxy flavour and T SS of 18-21 per cent. It responds well to head system of training.

Arka Soma

It is a cross between Anab-e-Shahi and Queen of Vineyards. Berries are greenish-yellow, round-to-ovoid. Pulp is meaty and has muscat flavour with 20-21 per cent TSS. The variety is tolerant to anthracnose, downy mildew and powdery mildew. Average yield is 40t/ha. It is good for preparing white desert wine.

Arka Chitra

It is a cross between Angur Kalan and Anab-e-Shahi. Berries are golden-yellow with pink blush, slightly elongated having 20-21 per cent TSS. Variety is tolerant to powdery mildew. Average yield is 38 t/ha. The berries are very attractive and thus suitable for table purpose.

Arka Majestic

It is a cross between Angur Kalan and Black Champa. Berries are deep tan coloured, uniform, round, seeded with 18-20 per cent TSS. Variety is tolerant to anthracnose. Average yield is 38 t/ha. Variety has a good export potential.

Arka Trishna

It is a cross between Bangalore Blue and Convent Large Black. Berries are deep tan in colour, round to ovoid with 22-23 per cent TSS. Variety is resistant to anthracnose and tolerant to downy mildew. Average yield is 26 t/ha. It is a good variety for wine making.

Pusa Navrang

It is a cross between Madeleine Angevine and Rubi Red, released by IARI, New Delhi. The vines are semi-vigorous bearing medium bunches (270 g) of well filled berries of attractive black colour and medium TSS (16-18 per cent). This hybrid is a teinturier type (both pulp and skin are coloured) and thus highly valued for making deep red coloured juice and wine. It is an early ripening variety and is resistant to anthracnose disease.

B. Seedless Cultivars

Beauty Seedless

It is a cross between Queen of Vineyards and Black Kishmish, evolved at California and recommended for commercial cultivation in North India. The vine is medium in vigour and is heavy yielder on bower, trellis and head systems of training. The bunch is medium (260 g), compact with bluish-black, spherical seedless berries, having TSS of 17-19 per cent. It is a basal bearer and responds well to spur pruning. Heavy pre-harvest fruit drop and uneven ripening of berries are the major problems in expanding its area under cultivation.

Pusa Seedless

It is a clonal selection of Thompson Seedless, released by IARI, New Delhi. It resembles Thompson Seedless in many characters except that berries are elongated. It can be commercially grown in most parts of North India. Vines are vigorous and heavy yielder on bower system of training. However, after 8-10 years of bearing, the vines show a slight decline in yield. Bunch is medium (260 g), well-filled with attractive light-golden coloured seedless berries, having high TSS (22-24 per cent). It responds well to gibberellic acid treatment. Apart from table purpose, a good quality raisin can also be prepared from it.

Perlette

It is a cross between Scolokertek Keralynoje and Sultanina Marble 26, evolved at California by Dr. H.P. Olmo. This is a highly suited cultivar for commercial cultivation in North India. The vine is of medium vigour and heavy yielder on bower and trellis systems of training. Bunch is medium in size (250 g), compact and attractive. The berries are of medium size (2 g), spherical with attractive greenish-yellow to golden-yellow in colour. The most striking feature of this variety is that the berries are translucent at maturity. Uneven ripening and presence of shot berries (small undeveloped berries) are two major drawbacks of this variety. However, good quality grapes can be harvested by berry thinning and GA application.

Thompson Seedless

It originated in Asia Minor and was first grown by William Thompson. It is also known as Oval Kishmish (Mediterranean region) and Sultanina (Australia and South Africa). The vines are vigorous, bearing large bunches (300 g) of attractive golden coloured and very sweet (TSS, 22-33 per cent) berries. Eating and keeping quality is excellent. Poor bud-burst and apical dominance are drawbacks in this variety. It performs well on bower system and responds to long cane pruning. It is a well known cultivar for table and raisin purpose. More than half of the total raisins of world are prepared from this variety. It has wider adaptability and is commercially grown in Maharashtra and Tamil Nadu.

Delight

This is a sister seedling of Perlette, evolved at California, by Dr. H.P. Olmo. The vines are of medium vigour, producing medium-sized compact bunches and bearing green, small and round berries. This cultivar is known for very early ripening habit and characteristic muscat like flavour. It has very good eating and shipping quality.

Himrod

It is a cross between Ontario and Sultanina. Vines are vigorous and heavy yielder. Bunches are attractive, bearing well-filled golden-green berries. The skin is thick and tough and slightly difficult to chew. It is a prolific bearer, excellent in quality and is resistant to many insect-pests and diseases.

Kishmish Charni

Vines are of medium vigour. The bunch is medium-to-large, conical, compact and well filled with brick red, spherical and slightly elongated berries. It is a very sweet cultivar with high keeping quality. It responds well to long cane pruning under bower system of training.

Sharad Seedless

It is a variety local to Russia called as Kishmish Charni. The berries are seedless, black, crisp and very sweet. The TSS is upto 24 °Brix. It responds well to GA and has a good shelf life. It is grown mainly for table purpose.

Arkavati

It is a cross between Black Champa and Thompson Seedless, released by IIHR, Bangalore. The vines are vigorous, clusters medium (300-400 g), well filled with spherical, yellowish-green, seedless and very sweet berries (TSS 22-25 per cent). It has good yield potential as all the buds are fruitful and apical dominance is less pronounced as compared to Thompson Seedless.

Arka Neelmani

It is a cross between Black Champa and Thompson Seedless. Berries are black, seedless with crispy pulp having 20-22 per cent TSS. The variety is tolerant to anthracnose. Average yield is 28 t/ha. It is suitable for wine making and table purpose.

Arka Shweta

It is a cross between Anab-e-Shahi and Thompson Seedless. Berries are yellow, ovoid, seedless with 18-19 per cent TSS. Average yield is 31 t/ha. This variety is suitable for table purpose and has a good export potential.

Arka Krishna

It is a cross between Black Champa and Thompson Seedless. Berries are dark coloured, seedless, round-to-ovoid with 20-21 per cent TSS. Average yield is 33 t/ha. The variety is suitable for juice making.

Pusa Urvashi

It is a cross between Hur and Beauty Seedless and has been released by IARI, New Delhi. It is an early ripening, basal bearer hybrid, productive both on head and bower system of training. The vines are of medium vigour, bearing loose and medium-sized bunches. The berries are oval in shape, greenish-yellow and seedless, having high TSS (22 per cent).

Some selections have been made by farmers. Tas-e-Ganesh, a selection from Thompson Seedless is popular in Maharashtra. It responds better than Thomson Seedless to GA for berry elongation. Dilkush, Manik Chaman and Sonaka have been selected from Anab-e-Shahi, in Hyderabad. In H.P., some local seedless types (Bargron, Shungron) having good berry quality are used for making a local wine called 'Angoori'.

PLANT PROPAGATION AND ROOTSTOCK

Propagation

Grape is propagated by seeds for raising hybrid seedling or rootstocks. Seed propagation is encountered with many problems like poor germination, long period for germination and long juvenile period etc. Asexual propagation ensures genetic purity. Among different asexual methods, propagation by hardwood cuttings is most commonly followed. However, grape is also propagated by budding, layering, grafting and by micropropagation.

a. Sexual Propagation

Freshly harvested seeds of grape have poor germination. Presence of abscissic acid and phenolic compounds has been reported to induce seed dormancy in grape. Therefore, seed should be stratified or treated with chemicals to break dormancy and to obtain high and uniform germination. Exposure of seeds to low temperature (5-6 °C) for 3-4 months is effective in breaking seed dormancy and to facilitate germination. Treatment with GA_3 (100 ppm) also promotes germination. Soaking of seeds in water for 24 hrs dilutes the concentration of phenolic compounds and other inhibitory substances and promotes seed germination.

b. Vegetative Propagation

The following methods of vegetative propagation are in vogue for producing new plants of a desired variety.

Cuttings

Hardwood stem cuttings is the most commonly used method of propagation of grape. Cuttings are taken from mature canes of healthy, vigorous and disease-free vines. Depending on the length of internodes in a cultivar, the mature bud-wood from the annually pruned shoots should be cut into pieces of 25-30 cm length so that each cutting has at least 4 buds. The cuttings should be of pencil thickness. The cut at the base of cutting should be prependicular to the length of cutting, just below the bud, while the upper cut should be slanting and about 1.5 cm above the apical bud. It facilitates orientation of cuttings during planting and avoids accumulation of water thereby infection by diseases. These cuttings can either be planted in the field immediately after preparation or stored for sometime or buried in moist sand or saw dust in a cool place. In general, cuttings taken from basal parts root better than those taken from terminal portion. Cuttings root even without treatment of rooting hormones (IBA), but if treated, root initiation is improved. The best time for preparation of cutting is North India is at the time of annual pruning in mid-January. Propagation by softwood cuttings can be done in mist chambers.

Layering

In some cases, propagation by layering is used. Mound layering is employed to replace missing plants in a row. Difficult-to-root grape varieties or rootstocks like Dog Ridge and Salt Creek are propagated by layering.

Budding and Grafting

Under certain specific conditions such as (i) to impart protection from soil borne diseases and (ii) adverse soil conditions (*e.g.* salinity), commercial varieties are budded or grafted on desired rootstocks 'T' and 'Chip' budding are successful methods for grape propagation. April and June are considered the best months for Chip and 'T' budding in grape. Different grafting methods (whip, cleft, side, notch, green) are also used for propagation of grape with varying degree of success on different rootstocks. In general, cleft grafting, 'T' and 'Chip' budding are mainly practiced for top working of old and unproductive vines. Green grafting is mainly practiced in USA during May when the atmosphere remains comparatively dry and mild. Green grafting is basically a method of budding where a bud (2.5-4.0 cm long) with white pith is removed from current season's shoot, provided with a sloppy cut, inserted in a rootstock of similar age and allowed to grow.

Micropropagation

Micropropagation can be used to multiply newly released variety/hybrid and for large scale multiplication of disease-free plants. Protocols of multiplication through tissue culture have been standardized at IARI, New Delhi and IIHR, Bangalore. Some private organizations like TERI, New Delhi, are supplying tissue-cultured grape plants. In India, shoot-tip and nodal-cuttings have been recommended to be the best explants and MS as culture medium with slight modifications. With repetitive double node cutting method, standardized at IARI, New Delhi, thousands of plants can be prepared in 3 months cycle.

Rootstock

Selection of proper rootstock is important for propagation of grape by budding or grafting. Rootstock affects tree vigour, planting density, fruit yield and quality besides ensuring the desired effects with respect to resistance to the adverse soil conditions. For the control of *Phylloxera* in North American grapes, grafting on tolerant rootstocks like Riparia Gloire, St. George, 1202, A x R1 or 99-R is done. Most of the *V. vinifera* cultivars are susceptible to nematodes. By grafting or budding the commercial varieties on Dog Ridge, Salt Creek, 1613, 1616 rootstocks, this problem can be avoided. Dog Ridge and Salt Creek rootstocks have also been found to be useful to impact resistance against salinity in soil. However, compatible combinations need to be developed. For example, Dog Ridge is compatible with Anab-e-Shahi and Thompson Seedless under Bangalore conditions but not in North India. Thus, the selection of the rootstock should be done by studying the compatibility of the rootstock and scion.

PLANTING AND ORCHARD ESTABLISHMENT

Orchard Layout

The grape berries are perishable and cannot be stored for long time. Therefore, areas located near the vicinity of a market are best suited for estabilishing vineyards. The locality should be well connected with roads, railway station, storage facility and preservation facility. The soil and climatic conditions of the location should be suitable for grape growing. Before layout, the field should be ploughed thoroughly, manured and leveled. Layout has to be done according to the spacing to be practiced considering the variety, climatic conditions and system of training. Normally, a spacing of 2 m × 2 m is recommended for head system, 3 × 3 m for trellis and bower system, for low and medium varieties. Vigorous varieties are trained on bower system and hence a spacing of 3.6 × 3.6 m is usually followed. Square system is followed in most cases but under certain specific system of training, rectangular system of planting is followed. After deciding the spacing, planting spots should be marked, dug out to a dimension of 90 × 90 × 90 cm. The pits are allowed to remain open for 3 weeks and then re-filled with 1 : 1 mixture of top soil and FYM. One kg of super phosphate, 500 g of sulphate of potash and 30 g chloropyriphos may be mixed and added in each pit and irrigated immediately. The poles should be fixed at a specific distance, depending on the training system.

Planting

Usually, one-year-old rooted cuttings are planted. Before planting, a small hole is made in which roots of cuttings are put and soil is pressed firmly and gently. After planting, a light irrigation should be given. The planting time differs from area-to-area. The planting in North India is done during January-February. In South India, it can be done in March-April and September-October.

INTERCULTURAL PRACTICES

Training

Training of grape plant is done to give proper shape and desired growth to the plant for getting good quantity and quality of fruiting. It includes the removal of undesired growth, providing support, bending, typing and pinching of the shoots. After planting, the plants start growing and if left untrained, these grow as bushy climbers. The different training systems are named as bower, head, kniffin, trellis, telephone, etc. The choice of training system depends on factors such as vigour of vine, variety, bearing zone, apical dominance, ease in cultural operations, climatic conditions and affordable cost. However, a training system, which facilitates different cultural operations (pruning, irrigation and plant protection, etc.), good leaf exposure, well spread out fruiting area on entire vine, vitality of the vine for a long period, and the least amount of wounds on the vine is considered as the ideal ones. The most commonly followed training systems in grape are detailed below:

(a) Head System

This system of training is best suited for the varieties producing fruitful shoots from basal buds, *e.g.*, Beauty Seedless, Perlette, Delight, Cardinal and Gold. It is the cheapest and easiest method of training grape. In this method, the vine is allowed to grow on single stem up to a height of 1-1.2 m. Although, initial props are provided using bamboo or wooden poles, which are subsequently removed. The vine is headed back at the height of 1-1.2 m induce side shoots. After keeping 4 laterals in four directions, the rest of the shoots are thinned out. These laterals are pruned to 2 basal buds at the first dormant pruning so that these produce secondary arms during the coming season. Generally, two arms about 20-30 cm long are kept on each lateral. After second pruning, normally 1-2 fruiting spurs are kept on each secondary arm. After 3-4 years, the vine looks like a dwarf bush and requires no support. It is a popular method of training grape in wine producing countries. It is also employed in kitchen gardens.

(b) Bower System

This system is also called as Arbour or Pergola system of training. It is best suited for vigorous cultivars like Thompson Seedless, Anab-e-Shahi, Cheema Sahebi and Bhokri. In this system, the vines are spread over a criss-cross network of wires, usually 2.1 to 2.4 m above ground, supported by pillars (concrete, stone or iron) and angle arms of iron. Vines are spaced 3 × 3 m or more and are allowed to grow straight, without any branch, upto the height of 2.4 m. When the vine reaches the wire, it is 'pinched off' to induce initiation of side shoots. Two vigorous shoots, opposite in direction are selected as primary shoots. On each primary arm, 3 laterals are kept on each side as secondary arms. These arms are allowed to produce 8-10 tertiaries, which become the fruiting canes. This system also provides a very high cost benefit ratio (1:2.09) (Table 8.3). In this system, however, cultural operations like spraying, pruning, etc. become difficult. In spite of these problems, this training system is being adopted on a large scale in almost all the grape growing regions of India.

Table 8.3: Cost Benefit Ratio of different Training Systems in Grapevine

Training Systems	Cost : Benefit Ratio
Bower	1 : 2.09
Telephone	1 : 1.71
Kniffin	1 : 1.42
Head	1 : 0.06

(c) Kniffin System

This system of training was developed by William Kniffin in 1850 in New York. In this system, two wires are stretched horizontally at a height of 0.9 m and 0.6 m, supported by iron angle poles, fixed at a distance of 4.8 m. The vines are spaced 2.4 m, between the poles. The distance between rows is kept 3 m. Initially, vine grows on single stem and one arm is allowed to develop horizontally along each wire on either side. Thus, each vine has four arms. The system is best suited for moderately vigorous cultivars. It is comparatively less expensive training system than bower system. The varieties like Beauty Seedless, Perlette, Banque Abyad, Pusa Seedless and Delight may be trained on this system. The chief disadvantage of this system is that the lower arms become unproductive and after some years, produces wood at extreme ends.

(d) Telephone System

This system is also known as Overhead Trellis or 'T' trellis. It is a little improvement over Kniffin and Bower systems since arms remain productive, allows proper ventilation and light penetration and cultural operations are easy to perform. It is less expensive than Bower but more than Kniffin. This system is suitable for cultivars with more apical dominance. In this method, 3 parallel wires are strung 30-45 cm apart through the cross-iron arms welded on the top of the angle iron poles at 180-200 cm above ground level. The arms of the vine are trained along the wires in the direction of the row.

(e) 'Y' Trellis

This system has good provision for light interception and favourable fruit bud formation. When the trellis is fully covered with foliage, both foliage and bunches are protected from the sun burn. 'Y' trellis consists of a vertical post, 120-135 cm above the ground and two inclined arms measuring 90-120 cm, placed at an angle of 90-110. The main stem is pinched at 120-135 cm above the ground level and a single pair of primary arm is developed on the wire. The secondaries and the canes are allowed to trail on wires fixed 10-15 cm apart on the inclined surface of 'Y'.

(f) Flat Roof Gable System

An interconnected 'Y' trellis forming a flat roof gable has been developed, combining the advantages of both bower and the extended Y systems and eliminating their disadvantages. It is mostly suited for vigorously growing vines where the shoots are exposed to sunlight for fruit bud formation while bunches hang below the canopy and thus are protected from direct sunlight. In this system, the length of the stem of

'Y' and both its arms is 1.2m. The canopy wires are spaced at 30-35 cm apart from each arm. Thick wires connect two arms of the 'Y' on either side of the rows. Two more wires are stretched interconnecting the thick wires forming a narrow bower between the two rows.

Pruning

Judicious removal of any plant part for increased productivity, facilitation of various cultural operation, regulation of crop and maintenance of vitality of vine is referred to as pruning. It is the most important and crucial operation in grape and should be done with great care keeping in view the growth pattern of the varieties under different climatic conditions.

Time of Pruning

In North India, the vine is pruned during dormant season, from late-December to end of January. In Maharashtra, North Karnataka and Andhra Pradesh, the vines are forced to undergo rest for about a month immediately after harvest. This helps in storing the food material in the mature parts of the vine. The canes are cut back in April by keeping 1-2 buds, which develops into canes in 4-5 months. The dried canes are also removed. Here, it is called 'back pruning' or 'growth' pruning. In the month of September-October, these canes are pruned for fruiting. This pruning is called 'forward pruning' or winter pruning. Vines, which have attained the age of one year, can be subjected to this pruning. The level of forward pruning depends upon the region, variety and vine vigour. Normally, the vines start yielding in about 5 months from forward pruning.

In Tamil Nadu, pruning is done during November-December for summer crop harvested during March- April. While pruning in May-June results in second crop during August-September. In the south interior Karnataka, the forward pruning is done during October-November for summer crop harvested during February-March and during April-May for the second crop harvested during July-August. It is important to retain the desirable number of fruiting buds on a vine after pruning for optimum yield and better quality fruiting. Retention of more canes on vine (light pruning) results in a heavy crop, while retention of less canes (severe pruning) results in a light crop. All canes in a vine cannot be equally fruitful. Canes that are away from the trunk are more fruitful than the ones nearer to the trunk. Hence, the former are pruned lightly than the later.

Shoot Pinching

Shoot pinching is a part of pruning, which is mainly done to promote fruitfulness and regulate the current season growth. Shoot pinching is done when the main shoot attains 7-8 leaf stage. During pinching, the tip of the mature shoot is 'pinched off' by retaining only five nodes. As a result, the terminal bud along with 1-2 laterals resumes growth. These laterals are called as sub-canes. Buds up to third node from the base on the sub-cane are invariably fruitful resulting in 2-3 clusters/cane.

Level of Pruning

The pruning mainly consists of thinning out and heading back. Heading back refers to cutting back of shoots or canes of certain height or bud number depending on the bearing habit of a variety (Table 8.4).

Thinning out refers to cutting back of shoot/canes to retain one or at the most two basal buds. All the diseased, criss-cross, overcrowding and damaged canes are also thinned out (Table 8.5). After pruning, a single spray of 0.2 per cent blitox should be done to avoid fungal attack on the cut portion of the vines.

Table 8.4: The Level of Pruning in Grape with Number of Buds to be Retained/Cane

Variety	No. of Buds/Cane
Beauty Seedless	2-3
Perlettte, Delight, Pusa Navrang, Pusa Urvashi	3-4
Cardinal, Gold	4-5
Pusa Seedless, Thompson Seedless	8-10

Table 8.5: Training System of Grape and Number of Canes to be Retained per Vine

Training System	Planting Distance (m)	No. of Canes/Vine
Bower or Pandal	3 × 3	40-50
Telephone	3 × 2	20-30
Kniffin	3 × 2	15-20
Head	2 × 2	8-10

Irrigation

Grape has a low water requirement. However, for optimum growth, yield and profitable production, judicious irrigation is required. The requirement of water depends on soil type, climate, time of ripening, stage of vine growth, etc. During initial years of planting, vine may be irrigated frequently. Full grown vines usually do not require any irrigation during the dormant season but during active growth, frequent irrigation is required. Irrigation must be stopped after veraison (colour development) stage as it may lead to fruit cracking and affect berry quality adversely. While stagnation of irrigation may lead to the mortality of vine, excess water from frequent irrigations results in excessive and rapid vegetative growth at the cost of fruiting. Therefore, judicious irrigation should be given when the moisture is depleted at the soil depth of 5 cm. However, vines have to be irrigated heavily, after pruning, to force them into active growth. In North India, the grape is irrigated at 7-10 days interval during growing season until beginning of sugar formation in berries and thereafter, irrigation frequency is curtailed to allow proper ripening of grapes. Generally, 12-15 irrigations are given to grape in South India. Though, economy is necessary in the quantum of irrigation water, the quality of water should be good

Water containing salts causes injury to the vine. In major grape growing areas of Maharashtra, Andhra Pradesh, Punjab, Haryana and Gujarat, where irrigation water is saline, low volume irrigation systems have been installed.

Manuring and Fertilization

Grape does not require a high fertility soil. It, however, removes an appreciable quantity of nutrients from the soil. Therefore, to maintain the soil fertility and to get consistent yields, it is necessary to supplement the soil with nutrients by application of manures and fertilizers.

When a nutrient falls short supply, its deficiency symptoms appear and causes direct or indirect losses. Deficiency of nitrogen (N) is most frequent. Its deficiency results in chlorosis and stunted growth of vine, adversely affecting vine productivity. In P deficiency, the margins of older leaves turn brownish with central portion remaining green. In acute cases, whole leaf may dry up. P is necessary for fruit bud differentiation and the element in combination with K, induces fertility of buds. K is responsible for apical dominance and cane thickness. K also improves retention of berries and increases berry size. Mg is required for chlorophyll synthesis in leaves and its deficiency results in interveinal chlorosis. Boron and zinc regulate starch metabolism and their deficiencies may result in hen and chicken disorder and little leaf, respectively.

The nutrient requirement of vine depends on different factors like variety, age of vine, agro-climatic condition, management practices, etc. The nutrient status of the vine also varies with the seasonal fluctuations. Though, the visual deficiency symptoms of the nutrients may be used to predict nutrient requirement, it may become difficult to ascertain which of the nutrient is really deficient. Therefore, the nutrient requirements of the vine should be ascertained by tissue analysis. In grape, leaf petiole is considered as the most reliable and ideal tissue for nutrient analysis.

As one fertilizer recommendation may not hold good for different conditions, the following doses may be used with slight modifications, depending on variety, zone or other conditions.

(a) Vines under the age of 3-5 years, should be given 40-50 kg well rotten FYM, and fertilizer combination of 500 g N + 300 g P_2O_5 + 700 g K_2O.

(b) Vines above 5 years of age, should be given 50-70 kg well rotten FYM and fertilizer combination of 500 g N + 700 g P_2O_5 + 1,000 g K_2O per year.

The N and K fertilizers should be applied in two split doses. The first dose of 60 per cent N, full P and 50 per cent K and full FYM should be applied soon after pruning. The rest of N and K should be applied at fruit set during the first year. The fertilizers should be applied 30 cm away from the trunk in circular ring. In the subsequent years, these should be applied in 15 cm deep furrows. The fertilizers should be mixed well in soil and the field should be irrigated afterwards. The nutrients can also be supplied to grapevine through foliar spraying. Foliar feeding is highly effective and efficient method of fertilizer use under adverse soil conditions.

Micronutrients like boron, zinc, Mn, Fe, etc., are generally supplied through foliar sprays. Foliar application of boron (0.2 per cent) at pre-bloom and during full bloom period is highly effective to obtain good yield of quality fruits. Similarly, application of magnesium sulphate (0.3 per cent), zinc sulphate (0.2 per cent) and iron sulphate (0.05 per cent) improves yield and quality of berries.

Weed Control

Weeds compete for nutrients and water with the vine and sometimes act as secondary host for insect-pests and diseases. Frequent weeding allows feeder roots to avail the nutrients and moisture without competition. In India, weeding in the vineyards is generally done mechanically. Bullock-drawn or tractor-drawn implements can be used for inter-cultivation and weed control, if sufficient space is available between the vines. In the vineyards, where close spacing is adopted, manual weeding or digging the plots with garden forks and lifting the weeds once in three months is a common practice. Diuron @ 3 kg/ha and simazine @ 2 kg a.i./ha can control all weeds except *Cyprus rotundus*. Atrazine @ 2-3 kg/ha is another effective herbicide in the vineyard. Glyphosate @ 10 ml/litre mixed with 5g of ammonium sulphate and detergent, as a post-emergence spray, is effective in controlling weeds for a period of 4-6 months. Mulching also controls weed population and conserves moisture in a vineyard. Mulching with *sarkanda* (*Saccharum munja*) and black polyethylene have been found to be good mulch materials for vineyards.

GROWTH, FLOWERING AND FRUIT SET

Growth

In sub-tropical and temperate climates, the vines shed their leaves and enter dormancy in winter. However, in tropical climate, it remains green and growth continues almost all through the year. As soon as the temperature rises, the buds begin to swell and burst. The vegetative buds give rise to a shoot, which bears only leaves, while reproductive and mixed buds give rise to shoot, which bears one or more cluster of flowers. The temperature during summer months governs bud burst, flowering and ripening as well as quality of berries. The temperature between 28-32 °C is the most ideal for the development of all parts of a vine. Bud burst takes place in February-March in North India and in April-May in the Himalayan region. In South India, bud burst for vegetative growth takes place in April-May and for fruiting in October-November. Some chemicals have been reported to hasten bud burst in grape. For example, application of thiourea (2 per cent) on dormant vines after pruning enhances bud burst by 2 weeks in Thompson Seedless and Pusa Seedless, which hastens ripening by a week or so. Similarly, application of dormex/hydrogen cynamide (1.5 per cent a.i.) on pruned vines enhances bud burst by 21, 18, 16 days in Pusa Seedless, Perlette and Beauty Seedless, respectively. Root pruning is also practiced in some areas to hasten bud burst.

Flowering

In North India, flowering takes place in March-April and in South India, it takes place during November-December. Flowering usually takes place at an average temperature of 20 °C. It gets almost retarted below 15.5 °C and above 35 °C. High humidity, rains, cloudy day etc., delay flowering.

Most of *Vinifera* grapes have perfect or hermophodite flowers. The female flowers have very small or reflexed stamens (Hur, Banqui Abyad, Khatta Kurghan, Angoor Kalan), which produce little or no pollen and are sterile. Male or staminate flowers have undeveloped pistils and small ovary, which can not be fertilized. Some grape species are dioecious (*V. rotundifolia*).

Pollination, Fruit Set and Berry Growth

Self pollination is a rule in *Vinifera* grapes. However, under certain circumstances, cross pollination as in reflexed stamen cultivars, is necessary and desirable. After successful pollination and fertilization, fruit set takes place. Grape varieties differ in the extent of fruit setting. Some cultivars set fruit without fertilization (parthenocarpy) on stimulus from pollen, *e.g.*, Black Corinth. In cultivars like Thompson Seedless, Pusa Seedless, Beauty Seedless and Perlette, fertilization occurs but embryo aborts after sometime. This phenomenon is known as 'stenospermocarpy'. In some cultivars, abortion is delayed and the seeds remain hard and empty.

There is a natural heavy berry drop (50-60 per cent) after few days of fruit set. It is considered desirable to keep the cluster loose. Too heavy drop causes heavy fruits losses. In cultivars like Perlette, the clusters are loose and attractive, allowing proper development of berries, but in others, these are small and compact. Such clusters develop many small and undeveloped 'shot berries'. This condition is called as 'millerandage' and may be aggravated by bad weather at the time of pollination or by heavy crop load.

The grape berry passes through four developmental stages. There are green, ripening stage, ripe and over-ripe stage. The green stage extends from berry set to start of colour change *i.e.* veraison. It is characterized by a rapid increase in berry size, the sugars remain low and acidity is high and the berries are hard. The ripening stage extends from the beginning of ripening until the berries are fully ripe. In this stage, berries start developing colour and start softening. The sugar content starts increasing and acidity decreasing. Usually, the berries at apical portion ripen at the last. At over-ripe stage, the berries lose the peak quality, which become prone to fungal and insect attack and may shatter.

CROP REGULATION AND QUALITY IMPROVEMENT

A good quality table grape should have medium sized clusters of uniformly large and seedless berries with characteristic colour, flavour and texture. Grape production is beset with a number of problems, owing to adverse agro-climatic conditions, nutrient imbalances, hormonal imbalance and related to management and plant protection measures. Some techniques can be used to regulate crop for getting good yield and quality of grape. These are described below:

Pruning and Thinning

While pruning, the number of fruiting canes should be properly distributed over the entire vine. Due to heavy bearing, the berries do not develop properly and their quality is also impaired. This can be regulated by berry or cluster thinning. In general, 60-70 clusters are considered optimum in Bower system with 3 × 3 m spacing. Similarly, 40-50 clusters in Telephone and Kniffin and 20-30 clusters on head system is a optimum load. Berry thinning helps in development of proper size, colour, ripening and quality in berries

Girdling

It consists of removal of a complete ring (0.5 cm) of bark from the shoot, trunk or cane. The stage of girdling depends upon the cultivar and the grower's interest. For example, to improve berry set and yield, girdling is done one week before flowering, for increasing berry size, it is done at berry set or just after set, and for advancing ripening, uniform colour and quality development, it is done at veraison stage. Girdling wounds heal within a month. This technique is very effective if integrated with pruning, thining or sprays of growth regulators.

Use of Growth Regulators

Growth regulators are used in grape for achieving one or other desirable effects. GA_3 has been found to be highly beneficial in loosening the bunches, increasing berry size and yield and for improving fruit quality in seedless varieties like Thompson Seedless, Beauty Seedless, Pusa Seedless, Perlette, Delight and Kishmish Charni. Sprays of 45 ppm GA_3 at full bloom in Pusa Seedless, 45 ppm at half bloom in Beauty Seedless, 30 ppm at half bloom in Perlette and 40 ppm in Thompson Seedless at full bloom stage have been recommended. GA_3 can be applied either by spraying at the recommended bloom stage or by dipping the bunches for half a minute soon after fruit set. Application of 4-CPA is beneficial for increasing berry set and reducing berry drop. SADH (2,500 ppm) if applied a week before full bloom, increases berry set and berry number. Ethrel is mainly used for uniform ripening and colour development. Application of ethrel (250 ppm), 5 weeks after anthesis gives desired effects. Dipping of bunches in ethephon (500 ppm) helps in early ripening, uniform colour development and reducing acidity in berries.

PROBLEMS IN VITICULTURE AND THEIR MANAGEMENT

Grape growing faces several problems in India. For example, monsoon showers during ripening spoil the crop is North India, diseases like powdery and downy mildew cause heavy losses in South India. In addition, some disorders described hereunder also cause serious losses.

Blossom End Rot

A black shrunken spot appears at the blossom end of the berry, which later spreads with water-soaked lesions around the berry. The defective berries may fall prematurely. Affected berries fetch poor price in the market. Calcium deficiency is the cause of this disorder. Spray of calcium nitrate (1.0 per cent) helps in checking this disorder.

Interveinal Chlorosis

This disorder is frequently observed in 'Anab-e-Shahi' grape around Hyderabad.

This is more pronounced at the fruit maturity stage. In this disorder, the area between, the veins becomes yellowish. In severe condition, the leaf tips may dry completely. It is believed that this disorder is developed due to the deficiency of magnesium probably induced by excessive fertilization with potash. To check this disorder, spraying with 0.5 per cent solution of neutralized magnesium sulphate during the fruit development stage is recommended. Two-three sprays at an intervals of 20-30 days are effective.

Flower Bud and Berry Drop

The drop of flower buds and developing berries may be serious in vigorous cultivars like Thompson Seedless, Pusa Seedless, Bhokri, Gold and Kishmish Beli. High nitrogen supply, water deficit, heavy crop load, improper fertilization of berries, uneven ripening and auxin deficiency at a particular stage of berry development are the major reasons attributed to the berry drop. The following measures are recommended for controlling bud, flower and berry drop:

1. Judicious supply of nitrogen.
2. Girdling of shoots/canes, 10 days before full bloom.
3. Dipping of bunches in NAA (10 ppm), 10 days before ripening.
4. Avoid both deficit of water and excessive irrigation during flowering.

Barrenness of Vine

Some commercial cultivars like Anab-e-Shahi in South India and Pusa Seedless in North India become erratic in bearing after good yields during first few years. This may either be due to failure of flower bud differentiation or due to the death of floral primordia but the real cause of this malady is still unknown. However, the menace can be reduced by discouraging excessive shoot growth by balanced nutrition limiting the excessive supply of N and checking sprouting of buds during and after the rains.

Cluster-Apex Wilt

In Thompson Seedless grape, green fruit appears soft to touch at the apex. At maturity, the apical portion of the cluster contains wilted or shriveled berries, which are sour and undeveloped. This is more evident on a weak shoot having poor leaf area. Over-bearing and insufficient transpiration may also be associated with the malady. Cluster pinching or berry thinning is recommended to reduce excessive crop load on the vines. Ensuring adequate irrigation during the berry development and protection of bunches from direct sunlight also help in reducing the incidence of cluster-apex wilting.

Pink Berry

Sometimes, 'off-pink' colour develops in Thompson Seedless berries, which turns black within a day after harvesting. Causes and control of this problem are still unknown. However, the incidence of pink berries is low in the early-season crop and increases with the rise in temperature late in the season. Indiscriminate use of ethrel for berry colouration can also cause this disorder.

Cracking of Rachis

It is characterized by a swelling, followed by longitudinal cracking of rachis, as a result of which, supply of nutrients and water to the berries is hindered. It may result in berry drying before full ripening and thereby adversely affecting the yield and quality. The cause and control of this malady are not still yet known.

Premature Defoliation

The problem is alarming in South India. The probable cause of this malady is said to be eco-physiological related to soil salinity, frequent use of pesticides or soil moisture deficit. This problem is more pronounced during summer rains when a period of cloudy and rainy weather promoting rapid top growth is followed by a hot sunny day.

Water Berries

Water berry is associated with fruit ripening and most often begins to develop shortly after berry softening. The affected berries become watery, soft, and flabby when ripe. They are almost normal in size but their flesh is not firm. They shrivel and dry by the time of harvest. Such berries mostly confine to the tip of the main rachis or its branches. This disorder occurs due to excessive cropping and inadequate nourishment available to all the berries in a cluster. Excessive irrigation and nitrogenous fertilizers should be avoided during berry development to reduce the water-berry formation.

Chicken and Hen Disorder

Boron deficiency results in chicken and hen disorder in grape. In this disorder, apical portion of the cluster has more number of undeveloped or shot berries. In severe deficiency, the flower clusters may dry up. Boron deficiency, incorrect stages of GA application and girdling are the known reasons for shot-berry formation. For correcting this disorder, 20-30 kg of borax per hectares should be applied or foliar application of boric acid (0.3-0.5 per cent) may also be used. Similarly, application of GA at proper stage should be ensured. Prebloom GA sprays of 10 ppm and 15 ppm are given, respectively on the 11th and 14th day after bud break for cluster elongation. Similarly, Rachies of the clusters are trimmed to retain only 8-12 clusters, depending on the number of leaves available per cluster.

Further, the clusters can also be dipped in GA solution of 40-50 ppm concentration once at 3-4 mm size of the berries and again at 7-8 mm size to improve berry size. When berry diameter is to be increased to more than 16 mm, clusters are dipped in a mixture of 10 ppm BA + 25 ppm GA or 2 ppm CPPU + 25 ppm GA or 1 ppm brassinosteroid + 25 ppm GA instead of GA alone at these two stages of berry growth.

Poor Cane Maturity

This is a very common problem in peninsular India. Shoot pinching after April pruning, closer planting, excessive use of N and irrigation and heavy crop load in previous season, are some factors associated with this problem. Judicious shoot pinching to check excessive vegetative growth; shoot thinning 30 days after summer pruning to prevent mutual shading of the shoots and promotion of light interception are some of the suggested remedial measures. Avoiding excess irrigation and nitrogenous fertilizers during 40-70 days after back pruning helps to overcome cane immaturity.

PLANT PROTECTION

A. Major Diseases and their Management

a. Fungal Diseases

Powdery Mildew

Powdery mildew is an endemic disease wherever the grapes are grown in the world. The disease has been reported from the American continent, Europe, Africa, Australia and Asia. In India, the disease is most common in Maharashtra, Gujrat, Andhra Pradesh and Tamil Nadu, the major grape growing areas. The disease causes extensive damage in whole of Europe and Western USA, sometimes destroying the crop completely. The disease not only reduces the yield and lowers the fruit quality but wine from infected fruits often develops off-flavour .

Symptoms

The fungus attacks all the green parts at all stages of vine growth. The fungus produces white-to-greyish powdery patches on the affected plant parts including fruits but young leaves are most susceptible and develop small whitish patches on upper as well as lower surface. These patches grow in size and coalesce to cover large areas on the leaf lamina. Malformation and discolouration of the affected leaves are also common symptoms, resulting in distortion. Similarly, floury patches are produced on the stem, tendril, flowers and young fruit branches. Diseased vines appear wilted and the stem parts turn brown. The infected blossom and berries turn dark in colour, irregular in shape and brittle. In advance stage of infection, berries may develop cracks and do not develop and ripe.

Causal Organism

Powdery mildew of grapevine is caused by *Uncinula necator*. Mycelium of the fungus is superficial, usually semi-persistent, very thin and effused or forming patches. Humidity plays a significant role in grapevine powdery mildew. The disease severity increases with the increasing humidity to an optimum near 85 per cent RH.

Management Practices

The disease is considered as one of the most damaging diseases that can result in complete loss of the crop and it poses a major threat to the grape production. The following measures are necessary for its control:

1. The use of training system, which allows proper air circulation through the canopy and prevents excess shading helps in reducing the disease.

2. Orchard sanitation is also important in reducing the disease pressure during the growing season.

3. Spray schedules of 7 to 10 days are usually required for effective control by sulphur. Dinocap has 10-14 days schedule while dimethylation inhibiting fungicides are commonly used at 14 to 21 days schedule.

4. Naturally occurring *Ampelomyces* hyperparasites play a role in the natural control of grapevine powdery mildew and might be reducing the over wintering inoculum of *U. necator*. Use such cultures, if possible.

Downy Mildew

The downy mildew of grapes is a historical disease, which led to the discovery of Bordeaux mixture by Prof. P.M.A. Millardet. Had Bordeaux mixture not been discovered at this stage, the wine industry in whole of Europe would have been in danger. The serious intensity of disease on the leaves leads to early defoliation and exposure of the canes to sunburn, which results in the reduced life of the plant. The overall losses due to the disease are thus tremendous as the quality and quantity of the fruit is reduced. Presently, the disease is known to be serious in many grape growing areas of United States, Europe, Asia and South America and has probably been recorded in more than 90 countries.

Symptoms

Symptoms of the disease appear on all the aerial parts of the vine. Leaves are most susceptible to infection during active growth. Symptoms vary according to the leaf age and prevailing weather conditions. On young leaves, circular yellow translucent oily lesions often surrounded by an oily brown halo are the first symptoms. This halo fades as the oil spots expand. Later, these spots may coalesce to cover entire leaf surface. Under favourable conditions, masses of white downy growth consisting of sporangiophores and sporangia develop on the underside of oily spots. During hot, dry weather or after sporulation, these spots turn brown and dry out. The growth on lower surface of the leaves becomes dirty-grey. On older leaves, small angular, yellow to reddish-brown spots appear, which are limited by veinlets, resulting in a mosaic pattern. Severely affected leaves fall prematurely. Young shoots and tendrils turn brown and are stunted and distorted, resulting in necrosis. Older shoots are rarely infected. Nodes are more susceptible than internodes. Flower clusters are highly susceptible. Infected inflorescence turn oily yellow-to-brown and may become whitish-brown during periods of high humidity at night. This brown colour develops due to production of sporangia. Infection on young peduncles results in the death of entire cluster.

Causal Organism

Grape downy mildew is caused by an obligate fungus, *Plasmopara viticola*. The best temperatures for sporulation are between 18-28°C with maximum sporulation at 23 to 28°C.

Management Practices

The management of downy mildew of grapes requires combination of different practices. All the methods including regulatory measures, cultural practices and spraying chemicals have to be combined.

1. **Regulatory measures:** Restriction on the movement of planting material from place-to-place *i.e.* on regional, national or international level should be imposed, since the pathogen spreads through dormant cuttings.

2. **Cultural practices:** All infected plant material and pruned parts must be removed and burnt before bud sprouting, so as to reduce primary inoculum. Even during growing season, plant debris must be avoided in and around the field. Careful attention should be paid to spacing of vines, row direction and placement of windbreaks, which will ensure maximum air circulation and minimum leaf wetness. Removal of leaves around berry clusters, trellies designs and pruning systems be practiced to encourage air movement within the plant canopy during the early development of vines. Careful disbudding and training of vines should be practiced to maximize distance between soil and foliage. The oily paste extracts of *Inula viscosa* leaves made with organic solvents are effective in controlling downy mildew.

3. **Chemical control:** Chemicals such as Bordeaux mixture, copper oxychloride, copper hydroxide, chlorothalonil are effective for the management of disease.

Anthracnose

The disease is known in Europe since ancient time, however, in India, the disease was first recorded in 1903 near Poona and is now widely prevalent in Rajasthan, Uttar Pradesh, Punjab, Haryana, Andhra Pradesh, Karnataka and Tamil Nadu. Under North Indian conditions, the disease appears only during rainy season.

Causal Organism

The anthrachnose of grapes is caused by *Gloeosporium ampelophagum*. Its perfect stage is *Elsinoe ampelina*.

Symptoms

The pathogen infects all the green parts of the vine including fruits. Small, irregular, dark-brown spots appear on the leaves, which turn grey in the centre. Sometimes, the necrotic tissues eventually drop out of the lesion, leaving a 'shot-hole'. The young leaves are more susceptible to infection than the older ones. On tendrils and shoots, small isolated, light-brown spots develop, which enlarge and become depressed in the centre. The ashy-grey centre of the lesions is surrounded by a dark-purple margin. During the humid weather, spore masses of the fungus appear as small, pink spots within the lesion. The affected shoots show restricted growth.

The lesions on the canes are elongated, sunken, dark-brown with dark-purple raised margins. On the berries, the spots are initially light-brown but later turn grey in the centre and are surrounded by reddish or purple margins. The well known 'birds eye' spots are produced on berries. The flesh of the affected berries may crack, exposing the seeds. Cankers produced on canes and leaves are the main source of primary infection and these can withstand the extreme summer and winter conditions.

Management Practices

1. Training of vines should be such that water splashes should not reach the foliage, canes and branches during rainy season. Ground level canes and branches should be removed.

2. Spray vineyards at the time of leaf emergence with fungicides like thiophanate methyl (0.1 per cent), baycor (0.1 per cent), benlate (0.1 per cent), bavistin (0.1 per cent), topsin-M (0.1 per cent) or Bordeaux mixture (0.8 per cent). At least four sprays of fungicides should be given during rainy season at fortnightly intervals.

3. Variety 'Delight' is tolerant whereas Bharat Early and Hussaini Chamber and Naibel are highly resistant.

Bunch Rot

Bunch rot, also known as grey mould or Botrytis rot is prevalent throughout the world wherever grapes are grown.

Causal Organism

The disease is primarily caused by fungus *Botrytis cinerea*, although the association of other fungi like *Alternaria alternata* and *Aspergillus* sp. has also been reported. Under favourable conditions, the fungus attacks all the fleshy tissues having high sugar, however, maximum damage is noticed in berries at harvest time as well as during transport and storage.

Symptoms

The disease symptoms appear on all plant parts *i.e.* leaves, shoots, flowers and berries. Infected flowers normally do not develop any apparent symptoms but necrosis of stamens, style and stigma can be observed microscopically. After the dehiscence of infected stamens, the solitary ovary can often be seen covered with tufts of sporulating mycelia. Both young and relatively older leaves are infected by the pathogen where irregular, necrotic spots are produced. The spots normally occur in the centre of leaves, however, under certain conditions, marginal necrosis is also evident.

The most prominent symptoms of the disease are found on the berries. In coloured grapes, the first signs are development of small, circular, faintly cleared spots. Infected berries become dark coloured and show typical greyish, hairy mycelium all over their surface.

Management Practices

1. Diseased vines, leaves and fruit grape mummies must be picked up and destroyed.

2. Fungicides like dicarboximide, procemidone, vinclozolin and iprodione are effective in disease control. Some of promising new botrycides in grapes are triazole, folicur and ergosterol biosynthesis inhibitors (EBI).Two (before berry touch and 3 weeks before harvest) or three treatments (before berry touch, beginning of veraison and 3 weeks before harvest) of pyrimethanil

(748 ml a.i./ha) and mepanipyrim (500 g a.i./ha) are highly effective in reducing the disease

b. Bacterial Diseases

Bacterial Canker

The disease was first observed on Anab-e-Shahi variety during 1969, but its seriousness was noticed only in 1984, when it appeared in epidemic form on almost grape-growing areas.

Causal Organism

Initially *Pseudomonas viticola* was identified as its causal organism, but later, it was established that *Xanthomonas campestris* pv. *viticola* is the real casual organism of grapevine bacterial canker. Some weeds, *Amaranthus* sp., *Glycine* sp., *Senna obtusifolia* are the alternative hosts of bacterium.

Symptoms

The bacterium infects all the aerial parts of the plant. Initially, small water-soaked spots appear on the lower side of the leaves, which are surrounded by yellow halo. These spots enlarge in size and become dark-brown, angular and cankerous. Sometimes, these spots coalesce and form larger patches. Leaves also show vein infection. Infected leaves after drying usually remain attached to the stem and give a blighted appearance. In advance stages of infection, stunting, cracking and abnormal growth of canes is also noticed. Lesions on berries are brown-to-black, cankerous. Severely affected berries become small and shriveled. It may cause pre-mature drop of berries. Further, the affected berries are not fit for marketing.

Management Practices

1. Destroy the affected vines.
2. Pruning should be done in late-October.
3. Alternate hosts like, neem, *Phyllanthus* spp. etc., in the vicinity of vineyards should not have any infection.
4. Use disease-free cuttings for propagation.
5. Avoid excessive use of irrigation.
6. Give prophylactic sprays of copper oxychloride or Bordeaux mixture before monsoon.
7. Usually antibiotics have poor efficiency, but streptomycin sulphate (500 ppm) may be sprayed at 5-7 days interval in areas receiving low rainfall.
8. Excessive use of nitrogenous fertilizers should be avoided.

B. Major Insect-Pests and their Management

A brief description of major insect-pests of grape, nature of damage and measures to control these pests has been given hereunder:

Flea Beetle (*Seclodonta strigicollis*)

The flea beetle is becoming a more regular and serious pest on grapevines. This beetle attacks grapes throughout the world, causing considerable losses to grapevines. It is a specific pest of grapevine and is widely distributed in India. In South India, Maharashtra, Gujarat and Rajasthan, it causes more losses.

The beetle is especially serious when it attacks buds following pruning. Adult beetles scrap the sprouting buds or eat them completely. The damaged buds or sprouts soon dry out. Many growers fail to detect it as the adults are nocturnal in habit and hide during day under loose bark or clods. But when the populations is fairly high, adults can be detected in the orchards easily. The beetles also feed on tender shoots, leaves and tendrils, which later may wither and drop down. Characteristic long hole in the leaf lamina is another indication of the presence of the beetle in the field. The grubs, which feed on roots of weeds live in the soil but do not cause any major damage to grape roots.

Management Practices

1. Remove the loose bark of the vine and kill the egg masses manually.
2. Put bundles of dry shreds of banana leaves on the pruned grape plants. Collect the beetles next morning, which are congregating beneath these shreds and kill them.
3. Apply aldicarb (1.0 kg a.i./hectare) or phosalone dust (2.0 kg a.i./hectare) to reduce the population.
4. Spraying of fenvalerate (0.02 per cent) or cypermethrin (0.02 per cent) or carbaryl (0.15 per cent) is equally effective.

Grapevine Thrips (*Rhipiphorothrips cruentatus*)

It is a major pest of grapevine in India and is highly polyphagous. Damage is caused both by nymphs and adults by rasping the lower surface of the leaf with their stylets and sucking the oozing cell sap. As a result of feeding or scraping of leaf surface, the tips of leaves become silvery. In case of severe infestation, the leaves may dry up and drop off the vine. As the leaves dry, thrips move to newer leaves. Thrips also suck sap from the blossoms and developing berries, which results in poor fruit set and reduction in yield. The affected berries develop a corky layer and become brown, which have poor quality, thereby fetching low price in the markets. Maximum damage is observed during April-to-June and August-to-November.

Management Practices

1. Remove infested leaves from the vines.
2. Expose the hibernating pupae of thrips to sun heat and natural enemies by digging the soil near vine roots.
3. Spray either phosphamidon (0.05 per cent) or carbaryl (0.1 per cent) or malathion (0.05 per cent) or enosulfan (0.07 per cent) or fenvalerate (0.02 per cent) during March and April. First spray should be given before

flowering and repeat it after fruit set. If required, spray should be repeated reduring August-November also.

4. Rear and release endoparasite, *Thripoctenus maculatus*, which is known to keep thrips under control.

Grapevine Girdler (*Sthenias grisator*)

The girdle beetle has been reported as a serious pest from South India and Sri Lanka. The beetles ring or girdle the young green branches with their strong mandibles. This is a unique device that invariably precedes egg laying and allows the girdled branch to dry up so as to enable the grubs to bore and tunnel the dry wood. Due to girdling, vines dry up above the point of girdle. Girdling is performed on branches of 1.25-2.5 cm in thickness that are 15 cm to 3 meters above the ground.

Management Practices

1. Collect and kill the beetles mechanically.
2. Cut the affected braches below the girdling point and burn them.
3. Spraying with monocrotophos (0.036 per cent) or carbaryl (0.1 per cent) is quite effective in mitigating this pest.

Leaf Roller (*Sylepta lunalis*)

It is reported as a serious pest of grapevines in Tamil Nadu, but now emerging as a pest of major concern in northern region as well. It is a specific pest of grapevine. The pest activity is more from August-to-October. The caterpillars roll the leaves and also feed on flowers. The larvae during their first three instars feed on epidermis of leaves and skeletonise them. The forth and fifth instar larvae roll up the leaves from margin inwards along midrib with one caterpillar in each roll. Larval feeding weakens the vine, hence production and quality of berries is affected.

Management Practices

1. Remove the rolled leaves, containing larvae or pupae inside and destroy them.
2. Spraying endosulphan (0.05 per cent) after appearance of initial symptoms of attack, gives relief from the pest. If required, repeat the spray after 20 days. A waiting period of two weeks may be observed.

Stem Boring Beetle (*Sinoxylon anale*)

Stem boring beetle is an important pest of grapevines in Punjab. Both adults and grubs cause damage. The adult beetles bore inside the stems and feed within, making longitudinal tunnels and a number of exit holes. Chewed-up wood particles are thrown out of these holes. The grubs also feed inside tunnels causing further damage. As a result of attack, the plant portion above the point of attack gradually dries and dies away. The pest is active only in dormant vines and prefers late-sprouting varieties like Anab-e-Shahi.

Management Practices

1. To prevent its infestation, clean cultivation including removal of loose bark, coupled with pruning and destruction of infested parts is very useful.
2. During the dormant period of vines, give a cover spray with any of the recommended insecticides.

Mealy Bugs

Among various pests of grapevine, mealy bugs are considered as the most serious in India, which cause serious losses in all major grape-growing regions.

Both nymphs and adults (females) cause damage by sucking cell sap from the tender shoots, leaves and developing berries. The affected shoots become crinkled and stunted and berries do not develop normally and are shrivelled, which may drop down. Sometimes, there is development of sooty mold on the affected leaves and berries. Such bunches loose market value, are poor in quality and become unfit for human consumption. It is estimated that mealy bugs alone can cause cent percent yield losses in some cases.

Management Practices

1. Debarking and rubbing of vine stems with stiff cloth after October pruning and pasting with a mixture of copper oxychloride and carbaryl is very effective.
2. Apply sticky substance like, 'Tacktrap' or 'Bird tanglefort' either to stem or branches. These substances reduce the climbing of the nymphs.
3. Spray of dichlorvos or monocrotophos (0.05 per cent) after October is quite effective to control the mealy bugs.
4. Dipping of bunches for two minutes either in phosalone (2 ml), monocrotophos (1.25 ml) or dichlorvos (3 ml) mixed with fish oil and rosin soap is equally effective to control mealy bug attack in bunches.
5. Predator, *Cryptolaemus montrouzieri* is quite effective in suppressing the population of mealy bugs in field. About 10,000-12,000 beetles are required per hectare. However, their numbers can be reduced, if these beetles are released only in spots having adequate population of mealy bugs.

Scale Insects

Different types of scale insects have been reported to cause damage to vineyards. These are polyphagous pests having a wide range of host plants. The losses to grapevine by a single species may be of minor importance, but usually more than one species are found damaging the vines and their collective damage invariably is substantial.

Scale insects feed on leaves, petioles, main veins and tender shoots of grapevines. As a result of sucking of sap, vine may show decline of one arm and bleaching of leaves in the early years. Later, more arms of the vine may be killed and gradually the whole vine may succumb. Borers may then infest the declining vines.

Management Practices

It is really very difficult to control the scales because these pests usually remain hidden in the crevices or under loose bark, trunk or arms of the vine. However, the following measures may be useful to reduce their infestation.

1. Remove the loose bark and kill the insects mechanically.
2. Do not use infested cuttings for propagation.
3. Scrap the encrustation made by the scales and spray trithion (0.05 per cent) or carbophenothion (0.05 per cent).

Wasps and Honeybees

Vespa orientalis, Polistes hebraeus and honeybees feed on sweet juice and spoil the ripening berries. These wasps are also a great nuisance to the pickers and other workers. In home vineyard, the bunches may be covered with olive-green muslin cloth bags. The nests of wasps should be destroyed either by burning or by spraying endosulfan @ 2ml/litre of water.

Nematodes

About 29 species of nematodes have been reported to be associated with grape. However, *Meloidegyne incognita* and *Rotylenchulus reniformis* cause the heavy losses. Aldicarb and carbofuran (6 kg a.i./ha) control the nematodes in the vineyard. Now, nematode resistant rootstocks like Dog Ridge, Salt Creek and 1613 are available and can be effectively used.

Birds

Birds cause extensive damage to ripening grapes. Crows, sparrow and mynah are the most troublesome. Bird scaring by drum beating or tying reflecting ribbons and bagging bunches with olive-green muslin cloth bags save the grapes from bird damage. The large vineyards may be covered with a nylon net.

HARVESTING AND YIELD

Harvesting

Grape is a non-climacteric fruit and does not ripen after harvest. Therefore, fully ripe fruits are harvested. The maturity standards at harvest depends upon the desired use of berries. Some standards such as physical appearance, heat unit accumulation after anthesis and TSS : Acid ratio in the fruits have been used. The criteria used to judge ripening differ according to the use of the grapes. For making raisins, harvesting at a late stage is preferred, to get more sugar in grapes for increasing the weight of the dried product. For all other purposes, ripening is judged on the basis of sugar: acid ratio for getting proper blend required for table purpose or wine making. The correct blend of sugar acid ratio should be between 25-30.

For harvesting grapes for export market, the following parameters are considered-Berry size- should be more than 16 mm in diameter. The other desired attributes are listed as below;

1. TSS- should be more than 17° Brix.
2. Bunch weight- should be between 300-750 g.
3. Bunch colour- milky green.
4. The selected bunch should not be compact.
5. All the berries should be of uniform colour and size in a bunch.
6. Less than 2 per cent sunburnt or sulphur bleached, bruised or crushed berries.
7. Pedicel should be fresh and green.

However, the farmers harvest the berries on the basis of glossiness and softness of berries. Harvesting of bunches is done by detaching them from the vine along with canes. Preferably, grapes should be harvested either in the morning or evening hours. However, for export purpose, harvesting has to be done, preferably, in early hours of the day. It should be stopped when the berry temperature rises above 20 °C. It is advisable to close harvest by 10 a.m., otherwise the berry temperature cannot be brought down to 4 °C within the stipulated time of six hours by precooling. The selected bunches are harvested by giving a cut above the knot present on the stalk of the bunch. Any type of mechanical injury to berries should be avoided while harvesting and handling. The injured portion of skin serves as the entry point for many fungi causing decay. After harvesting, the bunches should be kept in shade. Immature, rotten, cracked, diseased or deformed berries are gently removed. Grading is done considering size, colour and variety.

Yield

A properly managed grape orchard starts bearing after 3rd year of planting. Depending on variety, climatic conditions, cultural and management practices, grapevine may survive for 20-25 years, or even more. Generally, varieties in temperate climate have longer life than in tropical and subtropical climate.

The highest productivity of grapes (21.5 t/ha) is the world has been reported in India. Yields as high as 79.04 and 91.31 t/ha have been reported in Anab-e-shahi and Cheema Sahebi, respectively. Well-maintained vineyard of Perlette, Thompson Seedless may yield about 25-30 and 15-20 t/ha, respectively. However, the productivity not only varies among regions, cultivars, but also depends upon care and management, training and pruning and agro-climatic condition. In general, the yield in tropical climate is more as some varieties bear crops twice a year. Heavy cropping may be harmful for sustained growth and vitality of vine.

POSTHARVEST MANAGEMENT

Packaging

After harvesting, grapes are sorted out and graded into different grades. After grading and packing, grapes are sent to market. Improper packaging causes bruising in fruits, which fetches poor price in the market. In general, table grapes should be packed in such a way that these require minimum handling. Different types of containers are used for packing grapes but the most commonly used are wooden or CFB boxes accommodating 2-4 kg of grapes. These boxes are lined with newsprint paper. Fine paper shred or fine hay is spread at the bottom and top of the box for cushioning. The open flaps of the box are secured firmly by an adhesive tape. Table grapes for overseas market are packed in 5 ply corrugated boxes of size 500x300 mm accommodating 5 kg of grapes. The graded bunches are weighed into 5 kg lots in plastic trays. One or two bunches weighing between 350-650 g are placed in small and thin polythene pouches. Before placing the pouches in the carton, bubble sheet is spread with its rough surface facing towards the base of the box. A white and soft polythene liner is spread over the top of the bubble sheet. These pouches are arranged in a single layer in a slanting fashion in the carton. In Maharashtra, grape is exported to Gulf countries after fumigation and pre-cooling treatments, using special grape guard (a paper made of sodium metabisulphite) for packaging. This technique extends the marketing season from the present 3 months to 6-7 months.

Storage

For extending the marketing season and to avoid gluts, grapes should to be stored. Grapes can be stored for 40-50 days at 0 °C and 80-90 per cent relative humidity and for 4 weeks with potassium metabisulphite. However, pre-cooling of grapes at 2.2-4.4 °C for 6-24h is essential before keeping in cold storage. The pre-cooling treatment should be given as soon as possible and at least within 6 hours after the harvest. During storage, fruits may develop postharvest disorders, *viz.* weakening of pedicel attachment resulting in berry shattering and decay caused by *Aspergillus niger, Penicillium* and *Rhizopus* etc. Fumigation of berries with SO_2 before storage, helps to maintain fresh stem colour and reduce decay losses. Treatment of berries with captan (0.2 per cent) or benomyl, controls decay disorders in storage. Similarly, pre-harvest application of calcium nitrate (1 per cent) reduces weight loss and decay rot in stored grapes.

Marketing

In India, there is no organized marketing system for grapes. Usually, growers sell their produce to contractors, who in turn depend on the wishes of traders for their returns. The highly perishable nature of the crop allows traders to exploit the situation in their favour. Some Societies or Associations like Maharashtra Grape Growers Association, Mahagrape, Drakshakul Pvt. Ltd. etc. are now active in Maharashtra. These co-operatives have helped in boosting Viticulture in Maharashtra. They help the farmers by giving necessary inputs, provide storage facilities and help in marketing of the produce within and outside the country.

CROP IMPROVEMENT

Ideotype Concept in Grape

An ideal grape variety is visualized as a good size, high quality fruits borne on vigorous, productive, self-fertile vine, which are satisfactorily resistant to almost all biotic and abiotic stresses. It should be locally adapted and needed purpose or intended use. It should also be eco-friendly and involve low expenditure cost.

Therefore, while developing any variety of grapes, these points should be kept in mind by breeder(s). The important breeding methods for development of an ideal grape variety are discussed hereunder.

Introduction and Selection

During post-independence era, many *Vitis* species, cultivars and rootstocks have been introduced to India from different countries such as USA, the erstwhile USSR, Australia, Yugoslavia, Italy, France, Bulgaria, Canada, Greece, Japan, Romania and other countries. Some of the important introductions are *e.g.* Thompson Seedless, Perlette, Cardinal, Beauty Seedless, Pearl of Casaba, Black Muscat, Delight, Himrod, Hur etc. These have been found suitable for yield and fruit quality characters under our conditions were released for general cultivation. The most popular variety of Andhra Pradesh, Anab-e-Shahi, is also an introduction from Middle East made by Abdul Baquer Khan way back in the year 1890. Several selections have been made from open pollinated seedlings of these introductions. The examples are cv. Cheema Sahebi and Selection 94. Likewise, Pusa Seedless, a popular variety of North India, is also a clonal selection from cv. Thompson Seedless. The natural bud sports have been identified, which have revolutionized the grape industry. Some of the striking examples are Tas-e-Ganesh, Manik Chaman and Sonaka from Thompson Seedless, Rao Sahibi from Cheema Sahibi and Sharad Seedless from Kishmish Charni.

Hybrid Breeding

Crucial assesment of hybridization work carried out at IARI, New Delhi as early as 1968 showed that Banqui-Abyad, Hur and Angoor Kalan were the best female parents for the intended objectives of earliness, seedlessness and good fruit quality. The hybrid breeding programme was initiated with the objectives of development of early maturing variety with high yield, better quality, seedlessness and preferably resistance to diseases. When Beauty Seedless, Perlette and Pusa Seedless were used as male parents, they too imparted these characters in the offsprings. It was suggested that for inducing seedlessness in the progeny, the varieties like Banqui-Abyad, Katta Kurghan and Hur should be selected as female parent on account of their high seed index (*i.e.* high ratio of berry weight to seed weight). When both the parents were seeded, the percentage of seedless progeny was very low and hence seedless varieties are often used as male parent to incorporate seedlessness. In India, seedless cultivars like Thomson Seedless and Beauty Seedless have been widely used as male parent in hybridization programmes. The important grape hybrids developed in India are presented in Table 8.6.

Mutation Breeding

Unlike natural bud mutation, little has been contributed by induced mutation in development of superior grape variety. Cv. Anab-e-Shahi Seedless was developed by using EMS @0.1 per cent on sprouted buds.

Inheritance Pattern

✩ Yield is a polygenic trait.

Table 8.6: Important Characteristics of Grape Hybrids Developed in India

Hybrid	Chief Characteristics
Pusa Navrang [Madeleine Angevine x Rubi Red]	Extremely early, uniform in ripening, basal bearer, heavy yielder, teinturier (colour in skin and pulp), resistant to mildew, good for juice and coloured wine.
Pusa Urvashi [Hur x Beauty Seedless]	Early and uniform in ripening, basal bearer, more productive with loose bunches, berries uniform, oval, greenish-yellow and seedless with 20-22 per cent TSS, good as dessert grape and for raisins.
Arkavati [Black Champa x Thompson Seedless]	Prolific bearer with large cluster, berries yellowish-green, seedless, thin skin, TSS (23°B), acidity 0.7 per cent, good as table grape and for raisins.
Arka Kanchan [Anab-e-Shahi x Queen of Vineyards]	Prolific bearer with medium to large clusters, golden yellow berries, seeded and muscat flavour, TSS 200B acidity 0.7 per cent, good for table and wine purpose.
Arka Shyam [Bangalore Blue x Black Champa]	Moderate to heavy yielder, Medium clusters, big sized round berries, TSS 24°B, acidity 0.6 per cent, suitable for table and wine purpose.
Arka Hans [Anab-e-Shahi x Bangalore Blue]	Prolific bearer, bunches medium in size, yellowish-green berries, seeded, TSS 21°Brix, acidity 0.5 per cent, suitable for wine making.
Arka Sweta [Anab-e-Shahi x Thompson Seedless]	Prolific bearer, a hybrid for table use and raisin making, avg. bunch weight 260 g, Greenish yellow coloured; obovid, uniform, seedless berries, avg. berry wt. 4.08 g, TSS 18-19° Brix, acidity 0.5-0.6 per cent .
Arka Majestic [Angur Kalan x Black Champa]	Vigorous plants, prolific bearer, high yielding, berries deep-red, obovoid, bold and seeded, suitable for table use, avg. bunch wt. 370 g, 2-3 small seeds/berry, avg. berry wt. 7.7 g, TSS 18-20° Brix, acidity 0.4-0.6 per cent, pedicel attachment very good, ideal for export, all buds are fruitful.
Arka Neelamani [Black Champa x Thompson Seedless]	Vigorous plants, well filled to slightly compact bunches, weighing on an average 360 g, black coloured and seedless, avg. berry wt. 3.2 g, TSS 20-22° Brix, acidity 0.6-0.7 per cent, all buds on a cane are fruitful; does not require specific pruning.
Arka Chitra [Angur Kalan x Anab-e-Shahi]	Prolific, high yielding, berries golden-yellow with pink blush, seeded but attractive, suitable for table purpose, average bunch weight 260 g, avg. berry wt. 4 g, TSS 18-19° Brix.
Arka Soma [Anab-e-Shahi x Queen of Vineyards]	Heavy yielder, white berries, seeded, meaty pulp with muscat flavour, good for wine making.
Arka Trishna [Bangalore Blue x Convent Large Black]	Prolific bearer, berries deep-tan coloured, seeded, very sweet pulp, male sterile, good for wine making.
Arka Krishna [Black Champa x Thompson Seedless]	Prolific bearer, berries-black, seedless and sweet, more juicy and suitable for beverage industry.

- ☆ Muscat flavour is controlled by five complementary dominant genes, while methyl anthranilate segregation pattern were likely due to three dominant complementary genes.
- ☆ The large berry size is dominant over small one.
- ☆ Berry shape is a polygenically inherited trait.
- ☆ Black berry colour is dominant over red and white, while red is dominant over white.
- ☆ The inheritance of sex in grape shows that presence of hermaphrodite flower is heterozygous and dominant and female is recessive. Functionally females (reflexed stamens) normal male development is dominant over suppression of female organ development. Accordingly, genotypes for various sex types in grape are assigned as Sh Sh, Sh Sf for hermaphrodite, Sf Sf for female and SoSp/SoSp for reflexed stamen.

Problems in Grape Breeding

- ☆ Long juvenile pre-bearing phase.
- ☆ Narrow varietal base and heterozygous nature of crop.
- ☆ Small flowers, short flowering duration and stigmatic receptivity, tedious emasculation and excessive berry drop.
- ☆ Lack of genetical information on inheritance pattern of horticulturally important traits.
- ☆ Lack of reliable and simple detection methods for mutants.
- ☆ Seedlessness as a result of meiotic irregularities renders recovery of seeds, through conventional means, impossible in hybrid development programme.
- ☆ Polyploidy as encountered in some grape varieties further complicate the breeding of grape.

9

Litchi

The litchi (*Litchi chinensis* Sonn.) is one of the most important subtropical evergreen fruit crops, belonging to the soapberry family, Sapindaceae. Litchi fruit is famous for its excellent quality, pleasant flavour and attractive red colour. It is one of the most environmentally sensitive subtropical tree fruit crops, which is adapted to areas of the world, characterized by warm subtropics and elevated tropics having cool dry winters and warm-wet summers. The fruit consists of a single seed, coveted by an agreeable sweet-acid tasting, crisp, white, juicy, translucent aril or pulp, which is high in sugars and vitamin C.

It is liked very much in India, China, Japan, Australia, South Africa, Thailand, Myanmar, USA, West Indies and New Zealand. Because of its richness in nutrients and refreshing taste, it has got a wide industrial importance. Therefore, the cultivation of litchi, either at large scale or small scale, is highly profitable. In India, litchi is mainly grown in the states of Bihar, West Bengal, Assam and Jharkhand, which contributes 45.6, 17.1, 8.2 and 7.2 per cent, respectively (Table 9.1). The specific litchi growing belts in India are Muzaffarnagar, Saharanpur (U.P.), Dehradun (U.K.), Darbhanga and Muzaffarpur (Bihar), Gurdaspur and Hoshiarpur (Punjab), Hoogly (West Bengal), Kangra valley (Himachal Pradesh), Nilgiri hills of South India and Araku hills (Andhra Pradesh).

COMPOSITION AND USES

Composition

Litchi is a major source of sugars and acids. The contents vary depending upon variety and climate. The main constituents of litchi fruit are carbohydrate, organic acids and vitamins. Litchi fruits are considerably rich in sugars. Sugar content in Indian varieties vary from 6.74 to 18.0 per cent and acidity from 0.20 to 0.64 per cent. However, in general carbohydrate content varies from 13.3 to 16.4 g/100 pulp (Table 9.2). The malic acid is predominant (80 per cent) organic acid in it. Litchi fruit also contains small proportion of citric, succinic, glutaric, malonic and lactic acids. It is an excellent source of vitamin C. Similarly, litchi contains small proportion of minerals like calcium, phosphorus and iron. However, the content of protein, fat and pectin are very low.

Table 9.1: State-wise Area, Production and Productivity of Litchi in India

State	Area (000'Ha)	Production (000'MT)	Productivity (MT/Ha)
Bihar	31.1	227.0	7.3
West Bengal	8.6	85.1	9.9
Assam	5.2	40.5	7.8
Jharkhand	4.3	35.9	8.4
Punjab	1.6	23.9	14.7
Chhatisgarh	3.9	21.4	5.5
Odisha	4.4	19.2	4.4
Uttarakhand	9.3	18.7	2.0
Tripura	2.9	16.6	5.6
Others	6.3	9.1	1.4
Total	77.6	497.3	6.4

Source: National Horticulture Board Database-2011.

Table 9.2: Nutrient Composition of Ripe Litchi Fruit

Component	Per 100 g Edible Portion	Component	Per 100 g Edible Portion
Calories	63-64	Phosphorus	30-42 mg
Moisture	81.9-84.83 per cent	Iron	0.4 mg
Protein	0.68-1.0 g	Sodium	3 mg
Fat	0.3-0.58 g	Potassium	170 mg
Carbohydrates	13.31-16.4 g	Thiamine	28 mcg
Fiber	0.23-0.4 g	Nicotinic Acid	0.4 mg
Ash	0.37-0.5 g	Riboflavin	0.05 mg
Calcium	8-10 mg	Ascorbic Acid	24-60 mg

Uses

In India, litchi is mainly liked as a table fruit due to the translucent, flavoured aril or edible flesh, while in China and Japan, it is prized in fresh, dried or canned state. Dried litchi is known as 'litchi nut', which tastes like raisins. In USA, frozen litchi is preferred. Jelly can also be prepared from litchi fruits. Litchi squash is extremely popular during the summer months. Various other products, such as, pickles, preserves and wine are also made in China. Chinese also use the leaves, flowers, bark and roots for various purposes. The leaves are used for making poultices, the seeds as anodyne for the skin and flowers, bark and roots for making decoctions for throat gargle. In China, large quantity of honey is harvested from hives near litchi trees. Honey from bee colonies in litchi groves is light amber, of the highest quality, with a rich, delicious flavour like that of the juice, which leaks when the fruit is peeled, and the honey does not granulate. Further, if ingested in moderate amounts, the litchi is said to relieve cough and to have a beneficial effect on gastralgia, tumours and enlargements of the glands. In India, the seeds are powdered and, because of their astringency, administered in intestinal troubles, and they have the reputation there, as in China, of relieving neuralgic pains. Decoctions of the root, bark and flowers are gargled to alleviate ailments of the throat. The importance of litchi fruit in India is indeed, due to the fact that, it usually comes to market in May and early-June when very few fresh fruits are available.

ORIGIN, HISTORY AND DISTRIBUTION

The litchi is a native to low elevations of the provinces of Kwangtung and Fukien in southern China, where it flourishes especially along rivers and near the seacoast. It has been under cultivation for the last 3,000 years in China. It has a long and illustrious history having been praised and pictured in Chinese literature from the earliest known record in 1059 A.D. Cultivation spread over the years through neighboring areas of south-eastern Asia and offshore islands. From China, litchi reached eastern India via Burma by the end of 17th Century and thereafter by the end of 18th Century, it was introduced to Bengal and West India. From Bengal, it spread to other parts of India. It was introduced in Australia, South Africa and Hawaii by the end of 19th Century. At the same time, it was also introduced in countries like Thailand, Mauritius, and USA. In Florida, it was introduced from Saharanpur (UP) in 1883 with subsequent introduction into California in 1897. In India, litchi is mainly grown in the states of Bihar, West Bengal, Assam and Jharkhand (Table 9.1).

TAXONOMY AND BOTANICAL DESCRIPTION

Taxonomy

Litchi belongs to family Sapindaceae and sub-family Nepheleae. Its popular and standard binomial nomenclature is *Litchi chinensis* (Gaertn.) Sonn. The genus *Litchi* has two species *i.e. Litchi philippinensis* and *Litchi chinensis*, out of which, the former is wild plant grown in Philippines and mainly used as a rootstock. The basic chromosome number of litchi is 7 and somatic number is 28. The Sapindaceae or soapberry family to which litchi belongs, is a large family consisting of about 125 genera and more than 1,000 species. However, only a few members of this family are of much horticultural importance. These are longan (*Euphorbia longana*), rambutan (*Nephelium lappaceum*), pulasan (*Nephelium mutabile*) and spanish lime.

Botany

Litchi plants grow to a height of 6-10 m or more with rounded top. The leaves are compound, consisting of three pair of leaflets, glassy dark-green above and grayish-green underneath. The leaves are arranged alternatively, petiolate with an average length of 12 cm. The inflorescence is compound raceme, developing both from terminal and axillary buds. The flowers are unisexual, bisexual or hermaphrodite. Usually, flowers of different sexes do not open simultaneously but open in different flushes. Pistillate flowers usually bloom after male flowers start dropping. Sometimes, even hermaphrodite flowers behave either as male or pseudo-hermaphrodite. These conditions are sometimes responsible for poor fruit set in litchi.

Botanically, mature litchi fruit is called as *'one seeded nut'* and its edible portion is *'aril'*. The shape and size of the fruit depend upon the variety. Fruits appear in a bunch of 5-20 or more. The pericarp of the fruit is papillate like strawberry and turns pinkish-red on ripening. The edible portion, aril, is fleshy, succulent, translucent, white, soft in texture, which separates easily from the seed and pericarp. Due to excellent sugar acid blend, litchi is considered as one of the most delicious fruits.

SOIL AND CLIMATIC REQUIREMENTS

Soil

The litchi can grow in a variety of soils including alluvial soils, loam, heavy clays, organic soils, calcareous soils (with 30 per cent lime) and rock piles. However, fairly deep, friable, well-drained soil with high organic matter, are preferred. A sandy loam or clayey loam soil with pH range between 5.5 and 7.5 and sufficient depth, is an ideal soil for litchi cultivation. The water table should not be less than 1.5 to 2 m. The water should not stagnate near the roots as it may lead to root decay. It is also believed that litchi require mycorrhizal association for successful growth. Hence, it is suggested that new plants should be grown in soil taken from the vicinity of old trees for the introduction of mycorrhiza in the new site.

Climate

The litchi is exacting in its climatic requirements. In general, litchi flourishes best in areas experiencing moist atmosphere, abundant of rainfall (125 cm or above) with freedom from frost. Poor production is experienced in many areas because winters are not cool or dry enough to check growth before flowering. Winter frost appears to be a harmful factor for successful litchi cultivation. Generally, young plants are more sensitive to frost and require protection from frost for 3-4 years in areas having danger of frost. After the age of 5-6 years, the plants become less sensitive to frost. Periodic cold in winter between −1 and 5 °C seem to be necessary for fruit bearing. For proper vegetative growth, a temperature of 28-30 °C is best. However, for profuse flowering, a temperature below 7.2 °C (200 hrs) in autumn and winter is considered ideal. In litchi growing tracts of India, the maximum temperature during flowering varies from 21 °C in February to 38 °C in June. Humidity is another important factor for the successful cultivation of litchi. The dry hot winds in summer during fruit development are harmful and cause fruit cracking and subsequent damage to the pulp. Expansion of litchi cultivation has been adversely affected by occurrence of hot winds in parts of North India. Generally, a wet spring and summer, a dry fall and winter are considered to be desirable conditions for fruiting in litchi. In China, litchi grows satisfactorily in areas receiving an annual rainfall of 1,500 mm with a relative humidity between 69 and 84 per cent.

Intensity of sunlight also appears to be a vital factor for litchi cultivation. During young age (up to 2 years), the plants should not be exposed to sunlight. High sunlight during summer also causes sun burning and skin cracking in litchi fruits.

IMPORTANT CULTIVARS

China possesses the largest named varieties of litchi. There are about 150 cultivars grown in China but only 15 are commercial. Chinese find it extremely difficult to maintain good variety under a particular climate and soil conditions. Therefore, every district has its own recognized cultivar. The most dominant varieties of litchi grown in China are Wai Chee, Souey Tung, Heak Yip, Tai So, Chen Zi (Brewstar) and Kwai May.

In India, about 50 varieties of litchi are grown in different states, but most of them are not clearly distinguished. The same variety may be known under different names in different places. In Bihar alone, about 35 varieties have been reported but Shahi, Rose Scented and China are popular in Muzaffarpur, and Kasba and Purbi are the choicest cultivar in Darbhanga area of the state. Other cultivars like Early Bedana, Late Bedana, Desi, Maclean, Longiya Kaisailiya are grown on a small scale. In West Bengal, the varieties like Bombai, Elaichi Late and China are grown commercially. Dehradun is the mostly grown cultivar in UP, while Saharanpur, Dehradun, Rose Scented, Muzaffarpur, Calcuttia and Seedless Late are grown in Punjab and Haryana. The chief characteristics of some of the important cultivars of litchi grown in India are given below:

Early Seedless

This variety is mainly grown in UP and Punjab. This is an earliest maturing variety. Plants of this variety are of medium vigour and attain a height of 5 m and spread of 6 m. It is a medium to poor yielding variety but bears regularly. The yield varies from 50-60 kg/tree. Fruits are small-to-medium size (15-16 g), heart to oval in shape with carmine red tubercles at maturity. Pulp is creamy white to white soft, juicy (69 per cent) with good TSS (19.8 per cent). Seeds are very small, shrunken, glabrous, and dirty chocolate in colour.

Late Seedless

This is a late season variety, ripening by 3rd week of June. Plants are vigorous and attain an average height of 5.5 m and spread of 7.0 m. The yield varies from 80-100 kg/tree. The fruits are medium to large in size (25 g), conical in shape, with dark-blackish brown tubercles at maturity. The fruit is not completely devoid of seed. The seeds are rather shriveled, shrunken and small in size (0.85 g) having shape like the canine of dog. The pulp is creamy-white, juicy (65.4 per cent) with high TSS (20 per cent).

Rose Scented

This is one of the most popular varieties grown in India. In addition to high fruit quality, its aril has delicate rose aroma, hence, called 'Rose Scented'. Plants are vigorous and attain a height of 7.6 m and spread of 8.2 m. It is a mid-season variety and starts ripening by first week of June. The fruits are medium to large in size (20.4 g), heart shaped with red tubercles at maturity. Pulp is greyish white, soft, moderately juicy (54.8 per cent), very sweet with TSS of 20 per cent. Seeds are small, shining and blackish chocolate in colour. It is susceptible to cracking and sunburn.

Muzaffarpur

This is an important variety grown in Bihar and its adjoining areas. This variety bears profusely every year with an average yield of 80-100 kg/plant. It ripens in 1st week of May in eastern parts of India and during mid-June in northern parts. The fruits are large sized with an average weight of 20 g, oval shaped, moderately juicy (59.2 per cent) with high TSS (17.2 per cent) and moderate flavour.

Bombai

This is the most important commercial cultivar of West Bengal. It is an early variety and ripens by second week of May. The plants are vigorous and attain an average height of 6-7 m and spread of 7-8 m. It bears fruits in large bunches. The yield varies from 80-90 kg/plant. The fruits are large in size (15-20 g), heart shaped with carmine red tubercles at maturity. In this variety, each developed fruit has another tiny under-developed fruit attached to the fruit stalk. Fruit pulp is greyish white, soft, juicy with pleasant flavour and TSS of 18 per cent. Seeds are big, shining and light chocolate in colour.

Calcuttia

This variety has proved to be the best of all the varieties grown in northern parts of India. This variety is considered as partly resistant to '*loo*' (hot winds) and can be cultivated successfully even in hotter areas having provision of windbreaks and plenty of irrigation water. This is a regular variety and bears profusely with a yield of 80-100 kg/plant. It is a dwarf and a late season variety and ripens by last week of June. The fruits are however, large in size (22 g) with lopsided shape and dark tubercles at maturity. The pulp is dirty creamy-white, soft, juicy (65.4 per cent) with high TSS (20 per cent). Seeds are glaucous, medium-sized (3.4 g) and concave or plano-convex in shape.

Dehradun

This variety is mainly grown in UP and Punjab. In Bihar, it is cultivated under the name of 'Dehra Rose'. It is also a late-season variety and ripens by 3rd week of June. The plants are medium in vigour, attaining a height of 5 m and spread of 7 m. A plant may yield 80-90 kg fruits. The fruits are medium in size (15 g), heart shaped and attain bright rose pink colour at maturity. The pulp is greyish white, soft, juicy (61.9 per cent) with good TSS (17 per cent). The seeds are smooth or shrunken, oblong shaped and dark chocolate in colour. This variety is most susceptible to sunburn and fruit cracking.

China

This is one of the excellent cultivars of litchi grown in India. The plants are semi-dwarf with small leaves. The fruits are large (25-27 g), globose in shape and attain marigold- orange colour at maturity. The pulp is creamy white, very sweet (20.2 per cent TSS) soft, juicy and with pleasant flavour. The seeds are relatively small (3.2 g), smooth and shining. This variety is relatively less susceptible to sunburn and fruit cracking.

Deshi

A popular cultivar in Bihar and West Bengal. The plants are of medium vigour and regular in bearing. The fruits are medium in size (17.2 g), oval shaped with deep-red tubercles at maturity. Pulp is plenty, sweet and juicy.

Shahi

A highly popular variety of Muzaffarpur (Bihar). This is a mid-season variety and matures by 3rd week of May. Plants are of medium vigour and regular in bearing. The fruits are large sized (22-24 g weight), oval shaped and tubercles turning crimson red with uranium green background at maturity. The fruits are highly suitable for canning.

Gulabi

An important cultivar of North India. It is a late season variety and fruit-ripens in 4th week of June. The plants are of medium vigour and attain a height of 6.0 m and spread of 7.0 m. It bears profusely and regularly with an average yield of 90-100 kg/plant. The fruits are medium in size (20 g weight), variable in shapes, with mandarin red tubercles at maturity. The pulp is greyish white, firm, with high TSS (18 per cent). The seeds are smooth, cylindrical, medium sized (3.27 g), shining and chocolate in colour.

Swarna Roopa

It is a clonal selection from cultivars planted at CHES, Ranchi. Fruits have attractive deep-pink colour, small seed and high TSS/acid ratio. Fruits are highly resistant to cracking. Fruits mature a week later than the late cultivar, China.

CHES-2

It is selection from Bombai cultivar made at CHES, Ranchi. It is a late maturing variety and bears fruit inside the canopy. Fruits are free from the problem of sunburn and cracking.

Saharanpur Selection

It is a seedling selection from a local cultivar and released from GBPUA&T, Pantnagar. It is a late maturing variety with fruit TSS around 19.8 per cent and very low percentage of fruit cracking (2 per cent only).

Sabour Madhu

It is a cross between Purbi and Bedana, made at Regional Fruit Research Station, Sabour (Bihar) in 1980. It bears higher number of fruits (24) per panicle and ripens 8 days later than another late maturing cultivar, Kasba. It has higher TSS and aril percentage than Purbi but fruit shape resembles Purbi.

Sabour Priya

It is a cross between Purbi and Bedana, made at Regional Fruit Research Station, Sabour (Bihar). Fruits have better quality than Purbi, and fruit shape has combination of both the parents but fruit weight is higher than the better parent (Purbi).

PLANT PROPAGATION

Seed Propagation

Propagation of litchi by seed is not common, as the plants raised through seeds take 8-10 years to come into bearing. Further, the plants raised through seeds are not true-to-type and often produce poor quality fruits. However, seedlings for rootstock purpose are raised through seeds. Moreover, the hybrid seedlings are also raised by seeds. Seeds of litchi lose their viability soon (3-5 days) after extraction and must be sown immediately after extraction. The seeds can be kept viable for about two months by packing them in moist sphagnum moss grass mixed with ground charcoal.

Vegetative Propagation

Although litchi can be propagated by various asexual means, the most common and easiest method adopted allover the world is 'air layering'. The other successful methods are grafting, stooling, stem cuttings and budding, which have been briefly described hereunder:

(a) Air Layering

Air layering is also known as 'marcottage' in China and *goottee* in India. In this method, one to 1½ years old healthy and vigorous branches are selected. A ring of about 2-2½ cm wide bark is removed below the bud from the selected branches. This cut is covered with mixture of mud or sphagnum moss and wrapped with polyethylene sheet. The roots start emerging from upper end of the cut within a month's time. However, the layers should be removed after two months when sufficient number of roots develop. The success in layering depends on factors like age and thickness of shoot, time of operation, rooting media, wrapping material, use of growth regulators etc. The best time for layering in India is July-August when plants are in their active growth phase and there is high humidity in the environment. Use of IBA (2,000-5,000 ppm) to induce rooting in litchi layers has been suggested. Usually, shoots of 1.5 to 2.5 cm thickness root better than thinner shoots. A mixture of sand and soil in 1: 1 ratio is better rooting medium. Mud plaster or sphagnum moss as a rooting medium for litchi layer has proven equally good.

(b) Pot Layering

In this method, a low branch of mature wood is cinctured and then the cut surface is buried in a pot or other container filled with the rooting medium. The pot is watered regularly. The roots develop in the cinctured portion of the branch in 3-4 months. Then, the branch is detached from the main plant. No repotting is required before transplanting in the field. Application of IBA (2,000-5,000 ppm) improves rooting and survival of the layers.

(c) Stooling

In stooling, the plant is headed back to leaving a stump during January-February. The new shoots (stools) will emerge from the stump within two months. A 2-2 ½ cm ring of bark is removed from the base of the newly emerged shoots. IBA (5,000 ppm) made in lanolin paste is applied to the upper end of the cuts and then a mound of soil is raised around the shoots to encourage rooting. Profuse rooting occur in the stools within two months. These stools are ready for transplanting in July-August. In stooling, one must be careful in not allowing the soil mound to dry. Therefore, irrigation to stool bed should be given at weekly interval from April-to-June. Many research workers consider stooling as a better propagation method for litchi than air layering.

(d) Stem Cuttings

Attempts have been made to propagate litchi by stem cuttings. Litchi can be propagated by softwood, semi-hard and hardwood cuttings but success varies widely. Usually, cuttings of 15-20 cm length and 10-15 mm diameter root better. Girdling of the branches, 2-4 weeks prior to cuttings often promotes rooting, primarily because of starch accumulation. Treatment of cuttings with growth regulators also promotes rooting process. Soaking the cuttings either in 200 ppm solution of NAA for 24 hrs or dipping in 500 ppm IBA, increases percentage and uniformity of rooting. It may, however, be noted that higher concentration of these growth regulators or treatment for longer time is harmful as it may affect the rooting process adversely. Further, the rooting initiation, the quality of roots and survival of the cuttings is favoured under high humidity, partial shade and warm root temperature. Similarly, type of medium (sand, peat, vermiculite) affects the success of rooting and ultimate survival of the cuttings.

(e) Grafting

Grafting in litchi is mainly practiced for top-working seedling or unproductive or old trees. The apical, side and approach grafting methods are practiced. In apical grafting, 10 cm long scion-wood (non-terminal) with at least 2 slightly swollen buds gives better results. The technique of splice or tongue grafting is highly successful for getting best results with apical grafting. It is, however, recommended that the scion wood should be cinctured 3 weeks before grafting operation. Apical grafting has been commercially used for litchi propagation in China, Israel and South Africa. In side grafting, the scion-wood is inserted onto one side of the stock. Stock (seedlings) of two year's age with semi-hard scions of 3-5 cm length having 3-4 buds are usually preferred in side-grafting of litchi. The best time for side-grafting is April and August.

(f) Approach Grafting/Inarching

This technique gives best results if plants are in active growth stage. It has been extensively used for propagation of litchi in China and to some extent in Florida. In India, inarching has given good results but being the cumbersome, tedious and laborious technique, orchardists prefer marcottage *i.e.* air layering.

(g) Budding

Budding has not been widely used for propagation of litchi, though some success has been obtained with chip and shield budding in India, Florida (USA), Philippines and South Africa.

PLANTING AND ORCHARD ESTABLISHMENT

The land should be thoroughly ploughed and then leveled. The required manure and fertilizers should be mixed in soil during land preparation. Green manuring with *dhaincha* (*Sesbania aculeata*) and sunhemp (*Crotolaria juncea*) is useful. It helps in improving the physical condition and fertility of soils. The pits of $1 \times 1 \times 1$ m size, in a square or hexagonal system, at 8-10 m spacing should be dug, about a month before planting time. These pits should be refilled with a mixture of farm yard manure, top soil and manures and fertilizers and then irrigated. Adding a basket of soil per pit from a litchi orchard containing mycorrhizal fungi helps in better establishment and growth of newly planted trees.

The best time for litchi plantation is beginning of monsoon season. Planting can also be done in spring, if irrigation facilities are available. Planting should, however, be avoided, if the field is either too dry or too wet. Immediately after planting, a light irrigation should be given.

Litchi orchard needs to be protected from strong winds, which cause complete flower or fruit drop. During summer, the hot winds ('*loo*') causes cracking and sunburn of fruits. Therefore, a suitable windbreak should be raised around the orchard at a right angle to the direction of wind. A row of tall growing trees, such as seedling mango, *jamun*, eucalyptus etc., may be raised at least one year before the establishment of the orchard.

INTERCULTURAL OPERATIONS

Irrigation

In areas having well distributed rainfall of 125 cm or more, litchi can grow without irrigation. Litchi is a water loving plant but it does not tolerate water stagnation. The young as well as the bearing plants require frequent watering. The critical period for irrigation is from end of January until the onset of monsoon as this is the time when vegetative growth and fruit development occur. The plants should not be irrigated during December-January as the floral initiation takes place during this time. During fruit development (March-May), irrigation is necessary at regular intervals to avoid severe fruit drop and cracking. Irrigation of young plants should be done by basin system and the fully grown plants can be irrigated by furrow or basin system, depending on the availability and source of water.

Manuring and Fertilization

Shortage of nutrients adversely affects satisfactory growth and quality yield in litchi.

The deficiency symptoms of some major nutrients are as follows:

Nitrogen	:	Yellowing of leaves, stunted growth, poor flowering and small fruit.
Phosphorus	:	Tip and marginal necrosis of old leaves, leaf curl, desiccation of leaves and fall.
Potassium	:	Yellowing of leaves, necrotic leaf tips and margins, leaf fall, poor fruit set and stunted growth of plant and fruits.
Magnesium	:	Small leaves, leaf necrosis, leaf drop and poor flowering.
Calcium	:	Death of growing points.
Zinc	:	Bronzing of leaves, stunted growth of leaves and fruits.
Iron	:	Yellowing of leaves and die-back of twigs.
Boron	:	Smaller fruits.
Copper	:	Die-back of twigs, smaller fruits, reduced pulp recovery.

The amount of manure and fertilizers to be applied in litchi orchard depends upon the nutrient status, physical condition of soil, climatic conditions, age of plant, spacing and variety etc. It is advisable to get the nutrient status in soil and leaf tested to decide fertilizer doses. On the basis of foliar nutrient status, the following fertilizer schedule has been recommended for litchi (Table 9.3).

Table 9.3: Fertilizer Schedule for Litchi Based on Leaf Nutrient Content

Age of Plant (Years)	Nutrients per Plant per Year (kg)			
	FYM	CAN	SSP	Muriate of Potash
1-3	10-20	0.30-1.00	0.20-0.60	0.05-0.15
4-6	25-40	1.00-2.00	0.75-1.25	0.20-0.30
7-10	40-50	2.00-3.00	1.50-2.00	0.30-0.50
Above 10 years	60	3.50	2.25	0.60

The fertilizers should be applied in 2-3 split doses *i.e.*, during flowering, fruit growth and vegetative flush emergence. However, the fertilizer application should be withheld during the period of vegetative dormancy *i.e.*, during autumn-to-winter. The fertilizers should be applied 30 cm away from the tree trunk, as litchi is sensitive to fertilizer burn. After mixing the fertilizers with soil, light irrigation should be given.

Training and Pruning

Litchi plants do not require much training and pruning. However, some training is beneficial to achieve good shape and strong tree framework. Pruning of young as well as mature plants is a standard practice in China, Taiwan and Australia. The main objectives of pruning are :

1. To improve plant structure against wind damage.
2. To open the canopy for better light interception.
3. To increase bearing surface area during early years.
4. To reduce incidence of insect-pests and diseases.
5. To improve the canopy shape, and
6. To provide sufficient area for routine cultural practices like spraying, fertilization and weeding etc.

Old, criss-cross diseased and dead branches should be removed. Heavy pruning should be avoided as it may cause profuse vegetative growth, resulting in poor fruiting. Heavy pruning is, however, recommended when plants are too old and produce fruits of small size and poor quality. This improves productivity and quality of fruits by promoting new vegetative growth.

Intercropping

The litchi is slow growing plant and it takes about 5-6 years to come into bearing. Therefore, intercropping in young orchards will not only add to income of farmer in the pre-bearing period, but will also protects the young litchi plants, enriches soil, improves physical conditions of the soil and keeps the weeds under control. Leguminous crops are preferred as intercrops in litchi orchards. Near big towns/ cities, vegetable crops, pulses, quick growing fruits like papaya and banana can also be grown as intercrops.

Weed Control

Usual practices of hand weeding or hoeing are laborious and expensive. So, application of herbicides is recommended for controlling weeds. Diuron and atrazine both @ 2 kg/acre are quite effective for controlling weeds both in young and bearing litchi orchards. The systemic herbicides like glyphosate, are highly effective for controlling persistent weeds and paraquat for *Cyprus rotundus* and *Oxalis corniculata*. Mulching also controls weed population to a greater extent. Among various mulch materials, black polyethylene is very effective.

FLOWERING, FRUIT SET AND CROP REGULATION

Flowering

The inflorescence of litchi is determinate and is composed of several multiple branched panicles on the current season's wood. The time of floral initiation usually occurs between November-to-February in northern hemisphere and between June-to-September in southern hemisphere. In India, the floral bud differentiation starts in December and ends by January. Subsequently, the flower emergence takes place in January and continues up to the end of February and fruiting takes place in April-May.

It is generally believed that litchi plants need a period of vegetative dormancy for flowering. This can be induced either by moisture and fertilizer stress, application of growth regulators, root pruning or by girdling. Higher temperature accelerates panicle growth and flowering, whereas, a temperature below freezing may kill the entire inflorescence. The deficiency of macro or micronutrients also influences the floral initiation and subsequent growth and development of the panicle.

Floral Biology

The number of flowers per panicle varies greatly with the place, cultivar, plant age, vigour of the plant and panicle. The flowering duration (anthesis to pollination) ranges from 20-45 days, but it mainly depends upon seasonal conditions. The flowers are self-sterile and insect-pollinated. The anthesis and dehiscence continues throughout the day and night, with peak opening in the early morning hours. The stigma becomes receptive soon after it starts dividing into lobes and remains receptive for about 3 days.

Fruit Set

The litchi plants flower profusely but only a small percentage of flowers develop into fruits. The fruit set ranges from 1-50 per cent, depending upon variety and climate. There is a definite relationship between fruit set and sex ratio. The variety Calcuttia has maximum proportion of female flowers (48.7 per cent) and fruit set (23.2 per cent). Temperature and relative humidity and several other factors play vital role in fruit set in litchi. Erratic bearing of litchi is mainly due to the occurrence of high temperature throughout the year. High soil moisture prior to floral initiation promotes vegetative growth and suppresses flowering. Acute deficiency of nutrient elements during flowering affects fruit set and subsequent fruit growth. Cincturing has been found to enhance cropping in young litchi plants and promotes flowering and fruiting in bearing plants. This practice may however, lead to alternate bearing, leaf scorching and die-back. Application of growth regulators like IAA (40 ppm) at anthesis and 2,4-D (20 ppm) before flowering are reported to have positive effects on fruit set of litchi.

Fruit Retention and Drop

Initially fruit set is heavy but all fruits do not reach maturity. The pre-mature fruit drop commences soon after the fruit set and continues till fruit maturity. Most of the drop occurs in the first 2-4 weeks. The extent of drop varies with the genotype, locality, year, environmental and cultural practices. Major reasons attributed to the drop are failure of fertilization, embryo abortion, internal nutritional and hormonal imbalance, moisture stress, high temperature, low humidity and strong winds.

Application of growth regulators like NAA (30 ppm), GA_3 (20-50 ppm), 2,4-D (10-20 ppm) before flower opening, helps in minimizing fruit drop to a greater extent. The incidence of drop can also be reduced by foliar application of micronutrients like boron and zinc at flower initiation.

Crop Regulation

Low and irregular bearing is a major problem in commercial litchi cultivation.

Poor yield often occurs even if there is profuse flowering. Sometimes, flower production can be a limiting factor especially when plants are in juvenile phase, in 'off' year and in case if the chilling hour requirement is not fulfilled and vegetative dormancy has not been induced prior to floral initiation.

Low female-to-male ratio, poor pollination due to low bee activity, pre-mature flower and fruit drop, also limit the fruit yield. The environmental factors also hinders the different processes invariably. The following operations have been suggested for crop regulation in litchi:

1. Select a suitable cultivar.
2. Withhold nitrogen application during flowering and early fruit growth.
3. Maintain optimum moisture status to reduce flower and fruit abscission.
4. Use of growth regulators like NAA and 2,4-D is helpful for improvement of fruit set and retention.
5. Cincturing and root pruning are also helpful.

CROP PROTECTION

A. Major Diseases and their Management

Besides a few fungal diseases that cause spoilage and decay of litchi fruits after harvest, there is no serious disease in litchi. However, the following diseases may cause problems in some areas:

Red Rust of Litchi

Red rust of litchi was first noticed during 1981 in Kangra valley (H.P.). It is caused by algal parasite, *Cephaleuros parasiticus*. Some authors have reported that *C. mycoides* is associated with it.

Symptoms

The disease first appears on young-unfolded leaves, wherein, small lesions of velvety growth are formed on their lower surfaces. On upper surfaces, chlorotic patches are developed, just opposite to velvety growth. In later stages, there is conspicuous development of velvety growth; as a result, a larger area of the leaf is covered. The affected leaves may show a number of depressions and curling, resulting in distortion at many places. Lastly, the velvety growth becomes leathery and brittle. The food manufacturing capacity of the leaves is decreased considerably, resulting in decline in the tree vigour and yield.

Management Practices

1. Remove and destroy the affected leaves.
2. Six sprays of lime sulphur (3 sprays in autumn and 3 during spring season) at fortnightly intervals give satisfactory control of the disease.

Fruit Rot

Rotting of the fruits is a serious problem associated with organisms like *Penicillium* spp., *Collectotrichum* spp., *Rhizophus* spp., *Cladosporium* spp. and *Fusarium* spp.

Dipping of fruits in hot benomyl (0.05 per cent) at 52 °C for 2 minutes and then packing the fruits in polyethylene has been recommended as control measure for fruit rot.

Postharvest Browning

Postharvest browning of fruit occurs mainly due to injury to pericarp during harvesting. Later, fungi like *Penicillium, Aspergillus, Rhizopus* and *Fusarium* etc. may infest the injured fruits. Packing the fruits in polyethylene reduces the browning problem to a greater extent.

B. Major Insect-Pests and their Management

About 40 species of insects attack litchi, but only three namely, eriophyid mite, bark eating caterpillars and fruit borer cause serious damage, as discussed hereunder.

Eriophyid Mite (*Aceria litchi*)

It is found in all litchi growing countries of the world. In India, its incidence is severe in North Bihar and Mysore. Both nymphs and adults cause damage by sucking sap from young leaves, buds, inflorescence and developing fruits. The mites also puncture and lacerate the tissues of the leaves. The affected young leaves turn yellow or greyish yellow with velvety growth, which later turns brown. Severe infestation may result in gall formation, curling, twisting, thickening and pitting of the affected leaves. The upper surface of the leaves gives characteristic greyish or dried appearance. The leaves ultimately dry and fall. In addition to leaves, the mites may cause malformation of inflorescence. The affected flowers or buds show an enormous increase in size. Bombai, China and Kasba varieties are more susceptible for the attack of mite.

Two to three sprays of dimethoate (0.05 per cent) or metasystox (0.025 per cent) or kelthane (0.12 per cent) at fortnightly interval, starting from bud-break, control the incidence of mites effectively.

Bark Eating Caterpillar (*Indarbela quadrinotata*)

It is another serious pest of litchi. The damage is caused by the caterpillars during night by eating the bark and boring holes in the trunk or main stem. It results in complete girdling of plant. The growth of the heavily infested plants is completely ceased. For controlling the bark eating caterpillar, clean the affected portion by removing the web and then plug the holes either with mud or cotton, soaked in chloroform, formalin or petrol. The mites can also be effectively controlled by spraying dichlorvos (0.03 per cent) or endosulfan (0.05 per cent).

Fruit or Nut Borer (*Conopomorpha cramerella*)

It is a serious pest of litchi in rainy season and causes serious damage in late ripening varieties. The larvae enter the fruit from a pin head size hole near the attachment of peduncle to fruit and causes damage by feeding inside the fruit. Infected fruits are rendered unfit for human consumption and did not fetch good price in the market. Nut/fruit borer can be effectively controlled by three fortnightly sprays offenvelarate (0.05 per cent) or quinalphos (0.5 per cent). The first spray should be given at the initiation of fruit set.

C. Physiological Disorders and their Management

Litchi cultivation and its potential yield is affected by several physiological disorders like fruit cracking and splitting, flower and fruit drop, sunburn, irregular bearing and black spot, pericarp browning etc. Therefore, it is necessary to know the causes and symptoms to manage these disorders for increased quality production. The incidence and severity of these physiological disorders vary with locality, season, cultivar and orchard management practices.

Fruit Cracking

It is the most important disorder in litchi, occurring in almost all the important litchi growing tracts of the country or even in the world (mainly throughout Asia and the Pacific). It is a major problem in litchi in India, which causes loss as high as 5-70 per cent . This may occur due to varietal characters, orchard soil management, inappropriate levels of water at maturity stage, light, mechanical injuries, temperature and micronutrient deficiency. All cracked fruits lose their value for fresh market and they are used for processing only (especially for fruit juice) if they are not affected by fungus. Cracked fruits are susceptible to storage disease, have shorter storage as well as shelf-life.

In India, most of the cultivars are susceptible to fruit cracking. It occurs when trees are subjected to drought soon after fruit set and if the drought is severe enough, fruit development is affected, particularly the development of the fruit skin, resultantly the cell division is reduced and the fruit skin becomes inelastic, and often splits when the aril grows rapidly before harvest. This can occur after irrigation or heavy rain, or just an increase in relative humidity.

Factors Influencing the Fruit Cracking

(1) Environmental Factors

1. *Effect of temperature:* Temperature plays a very important role in the fruit cracking. Arid and semi-arid zone where temperature is more and humidity or rainfall is very low, favours cracking. In general, there is a linear increase in cracking with temperature increase. Temperature may also affect other factors such as permeability of the cell walls and bio-chemical processes of the cells etc.
2. *Wind:* In north India, occurrence of loo/hot wind during summer months, a common phenomenon. When such hot wind passes from the litchi fruit surface, due to water loss, it becomes hard and inelastic and in case of sudden fluctuation in moisture level leads to cracking in litchi fruit.

(2) Fruit Characteristics

1. *Maturity:* Over maturity leads to cracking of fruit epidermis.
2. *Fruit size.* It is generally supposed that large fruits are more prone to the cracking than the smaller ones.

3. *Fruit firmness*: It is found that firm fleshed varieties are more susceptible to fruit cracking than soft fleshed ones. Cracking of fruits is caused by excess uptake of water resulting in bursting of skin.

(3) Lack of Orchard Management

1. *Moisture stress:* Moisture imbalance and heavy rainfall or irrigation after a prolonged dry spell, sudden and high fluctuation in the water supply to plants may cause cracking of the fruits.

2. *Nutrient:* The deficiency of boron and calcium is responsible for cracking in litchi.

3. *Incidence of insect-pests and diseases:* In litchi, due to sunburn, there is the appearance of small dark water-shaped spots, which is finally assuming the shape of raised spots. These areas on the fruit develop longitudinal cracks and starts oozing out from the splits.

4. *Bagging:* It is also a remedial operation to escape the sunlight from the plant surface because water lost by transpiration though stomata.

5. *Early picking*: Early picking of fruits is also a remedial measure to overcome cracking. This does not allow to over maturity or over ripening, which causes cracking of fruit, however it is not practicable in litchi.

Management Practices

1. Select appropriate site for litchi cultivation, which has the history of no rain near harvesting time of litchi.

2. Application of calcium @ 2 m/l liquid formulations and gibberellins @ 20 ppm, reduces the activity of cellulose and thereby reduced cracking.

3. Spray of 2,4-D and NAA at concentrations 20 ppm or 20 mg/litre reduces cracking.

4. Maintain constant moisture and appropriate humidity in the orchard at the time of fruit maturity. Irrigation at 30-40 per cent depletion of available soil moisture is quite helpful in reducing cracking of fruits. Installation of drip as well as micro sprinkler below the canopy area has been reported to be effective in reducing the fruit cracking.

5. Mulching of litchi orchard and daily irrigation in such border crop creates congenial micro-climate, which reduces fruit cracking.

6. Planting windbreak around the orchard at a right angle to the direction of prevailing wind. A row of tall growing trees, such as seedling mango and *jamun* are suitable windbreaks.

7. Boron sprays in the form of borax or Boric acid @ 2g/l at the initial stage of aril development with enough soil moisture in the root zone checks fruit cracking significantly.

8. Use cracking resistant cultivars such as Swarn Roopa. It is reported to be tolerant to this disorder. Litchi cultivars, which have relatively thin skin, few tubercles per unit area and rounded to flat in shape are less prone to cracking. Early cultivars are more susceptible than late cultivars.

Sunburn

Sunburn also known as lesion browning or pericarp necrosis, is a serious problem in litchi producing areas. Climatic factors and cultivars in particular growing areas are determinants for incidence and severity of sunburn disorder. The damage caused due to sun burning is sometimes very high. Apart from environmental factors, varietal, hormonal, nutritional and soil moisture factors are associated with this disorder. This disorder is physiologically related with PPO (Polyphenol Oxidaze) activities and it also varies with cultivars.

Sunburn is pronounced in poorly managed orchards having sandy or sandy loam soils or light soils receiving/exposed to high temperature (>40°C) and very less RH (<50 per cent). It is a type of direct thermal injury and in case of higher temperature, the tissue coming in contact/exposure gets sunburnt/sunscalded.

Sunburn problem is more in early ripening cultivars. Fruits on shaded branches suffer less damage than those exposed to sun. Lower translocation of calcium in the pericarp region also found to favour sunburn disorder.

Management Practices

1. Irrigate the orchards at regular intervals during the fruit growth and ripening stage.
2. Plant a row of windbreak trees around the orchard. It provides protection from desiccating hot winds, which lowers the incidence of sunburn.
3. If possible, use drip pr sprinkler system during of irrigation. It increases humidity, cools the orchard atmosphere, thus decreases the incidence of sunburn.
4. Increase the frequency of irrigation in light and sandy soils.
5. Use recommended doses of manures and fertilizers.
6. Planting of maize or sugarcane around the litchi orchard and daily irrigation in such border crops creates congenial microclimate, which reduces sunburn.

Black Spot

The occurrence of black spot is not wide spread in litchi but it has been found infesting and damaging the quality harvest of litchi. In Indian condition, this disorder has been observed in existing plantation mainly in vicinity of urban areas and there are no other reports in litchi about its occurrence. It is caused due to deleterious effect of smoke fumes, which contain sulphur dioxide, acetylene and carbon dioxide as found in case of mango.

The symptoms of black spot become apparent when the fruits attain certain size and remain in developmental phase *i.e.* from March end to mid-May. The first symptom appears as the development of a small etiolated area at the distal end of the fruit against the normal green colour of the fruit pericarp, which gradually spreads, turns nearly black and covers the distal end completely. The infested fruit shows discolouration and overall slow pace of development, resulting in very poor development of aril, seed, size and colour.

Management Practices

1. Apply recommended doses of manues and fertilizers.
2. Apply regular irrigations particularly during fruit growth and aril development stage.
3. Spray micronutrients like zinc ($ZnSO_4$ @ 0.2 per cent) a month prior to flower panicle initiation and boron (borax @ 0.2 per cent) during fruit growth.

Pericarp Browning

Pericarp browning is a serious problem in harvested litchi, which occurs during the first few days after harvest. This problem has been reported in several litchi producing areas and involves the appearance of necrotic spots, which may expand and affect a significant portion of the fruit. Fruits start to brown once they lose a few percent of the harvested pericarp fresh weight. Browning of the pericarp occurs at ambient temperatures of 20-30°C within 24 hours of harvest. Water loss (desiccation) of litchi results in brown spots on the bright-red shell (pericarp). Other factors also cause the fruit to brown, including mechanical stresses of various sorts (tugging the pedicel at harvest, sliding the fruit down a rough picking bag, dropping fruit from short heights); microbial and insect attack; and extremes of temperatures.

This occurs as a result of damage caused to the parenchymatous cells of the mesocarp. This damage speeds up a characteristic hypersensitive reaction, leading to necrosis of these cells, which then spreads to the epicarp and endocarp. Skin-browning is associated with an increase in polyphenol oxydase and perioxydase enzyme activity and with ascorbic acid oxidation. Losses may be considerable, although the aril is not affected; the fruit is not suitable for marketing. In extreme cases, the pericarp may crack. Through the use of fungicides and refrigeration, litchi fruit have a storage life of about 30 days. However, when they are removed from storage, their shelf life at ambient temperature is very short due to pericarp browning and fruit rotting.

Management Practices

1. Pack the fruit into moisture-proof (plastic) bags and punnets, which substantially reduce water loss and slows down the rate of browning.
2. Store the fruits in cool temperature. Low temperature slows down the evaporation as well as respiration and thus tissue senescence.
3. Fruits treated with polyamines, suspected anti-senescence agents, then wrapped and stored at 5°C, had lower membrane permeability and less browning than controls.
4. A controlled atmosphere of 3 to 5 per cent O_2 and 3 to 5 percent CO_2 has also been shown to slow water loss. Fruit stored under such an atmosphere for 30 days at 1°C lost only a quarter of the water in comparison to water lost by the controls.

5. Sulphur dioxide fumigation has been used extensively in South Africa and Israel. There have also been many experiments in China and Thailand. However, sometimes the fruit are tainted. There are also concerns about high sulphite residues in relation to sulphur-sensitive individuals. Sulphur dioxide fumigation effectively reduces pericarp browning, but some countries do not permit its use due to concerns over sulphur residues in fumigated fruit.

6. Israel has recently promoted a hot water brush/acid/prochloraz treatment, but acid treatments sometimes give an artificial and persistent red colour to the fruit that masks poor eating quality.

HARVESTING AND YIELD

Harvesting

Litchi is a non-climacteric fruit. Therefore, the fruits should be harvested at correct stage of maturity when these have typical taste and flavour of the variety. The various criteria recommended for judging fruit maturity are : (i) days after fruit set, (ii) development of colour on fruit, (iii) firmness of tubercles and smoothness of epicarp, and (iv) chemical changes (TSS and acidity) in fruit.

Number of days taken by the fruit to mature, varies with variety and environment, but in general 50-55 days after fruit set are considered optimum for fruit maturity. The development of colour on fruit is more dependable maturity index, though it varies from variety-to-variety. However, in most varieties, colour changes from greenish to pinkish on maturity. When fruits have attained bright pinkish red colour, they can be harvested. Shape of tubercles also indicates maturity of fruits. At full maturity, tubercles become somewhat flattened. TSS, acidity and specific gravity of fruit are also taken as maturity indices for proper harvesting of fruits.

Litchi fruits are harvested in bunches along with a portion of a branch and a few leaves. This helps to prolong postharvest life of fruits. If the fruits are harvested individually, they may rupture at stem end and fruits rot quickly. All the fruits do not ripe at one time, hence, the fruit panicles are spot picked several times. Harvesting should be done on a bright sunny day in the morning and evening hours. It should, however, be avoided in rainy days as it may lead to spoilage. The harvesting time in India is May and June.

Yield

Litchi plants start bearing after 5-6 years of age. Commercial bearing, however, starts from 15th year. Yield depends on variety, age of plant, environmental conditions and management practices. At the initial stage, 100-150 fruits/plant are obtained. On an average, a full bearing plant bears about 80-150 kg fruit.

POSTHARVEST MANAGEMENT

Pre-Cooling, Sorting and Grading

Fruits should be packed soon after harvesting otherwise their quality deteriorates when kept in sun for longer period. Pre-cooling of fruits in water before packing is a useful practice for maintaining fruit firmness and flavour. Similarly, before packing, damaged, sun burnt and cracked fruits should be sorted out and graded before packing. Fruits of one variety and one grade should be packed in one box.

Packaging

After harvesting the fruits at appropriate maturity, the fruits should be kept in cool, dry and well-ventilated rooms because the litchi fruits ripen under high atmospheric temperature. For sending the fruits to distant markets, fruits in the package should not be loose and air should circulate freely inside the container.

In India, litchi fruits are usually packed in shallow bamboo baskets or wooden crates. These containers are lined with litchi leaves or other soft packing material. Litchis can also be packed in light wooden boxes having small holes.

Storage

Litchi deteriorates rapidly after harvesting. At a temperature of 25-30 °C, it loses its bright, red colour within 24 hours of harvesting. Since, the harvesting season of litchi is very short (hardly a month), proper storage is important. Litchi fruits can be stored for 4 weeks at a temperature of 2-3 °C and relative humidity of 80-85 per cent. Fruits can be kept in perforated polyethylene bags and stored at a low temperature (5 °C) for 5 weeks.

Fruit decay is a major problem in litchi during storage. Fungicide treatment (benomyl @0.05 per cent) and hot water treatment at 52 °C for 2 minutes to the fruits before packing is beneficial for checking fruit decay during storage.

Value Addition

In India, litchi is consumed fresh. The fruit being highly perishable, its large quantity goes waste. Hence, products like squash, jelly, juice and dried or canned litchi should be made to utilize the surplus produce during peak harvesting season.

Canned litchi is of excellent quality and has great demand in the market. In this case, discoloration of canned litchi fruits is a major hurdle. The problem of discoloration can be avoided by dipping peeled pulp in 1 per cent alum solution for 30 minutes prior to canning. The varieties like Early Large Red, Early Seedless, Rose Scented, Shahi and Purbi are highly suitable for canning purpose.

Freezing the whole fruit is one of the best and easiest methods to preserve fresh flavour and quality of litchi fruits for a long time. Freshly harvested fruits when readily cooled and kept at –25 °C remain in excellent conditions for 12 months.

Litchi juice can be used to prepare beverages like squash, syrup, nectar and carbonated drinks. Litchi drying (dried litchi/litchi nut) in sun or charcoal is very popular in China and a large quantity of the produce is dried and exported all over the world. In India, dehydration of litchi at 54-60°C for about 16 hours with 50 per cent humidity in the beginning and 5-10 per cent during the end-product, has been recommended. The best quality dried product is obtained if the fruits are blanched for 2 minutes at 100 °C, sulphited with 1 per cent $NaHSO_3$ and dipped in 0.5 per cent solution of citric acid for 10 minutes, before drying.

Ideal jelly flavour is obtained with a delicate balance of sweetness, tartness and intensity. A good litchi jelly can be prepared with the juice of 16 per cent or more soluble solids and combined by weight with sugar in a 50:50 ratio. The pH of jelly should be adjusted to 3.2 with citric acid. Addition of 1 per cent commercial slow set pectin to litchi jelly is sufficient for good gel formation.

CROP IMPROVEMENT

A large number of litchi varieties are grown in the country. The improvement in litchi appears to be confined to selection of improved chance seedlings or genotypes. Saharanpur selection is a chance seedling, which is late in maturity. It has found to have low fruit cracking percentage. Through intensive survey conducted in the state of Bihar, 'Swarna Roopa' has been identified as a promising cultivar having attractive fruit with rose blush, high pulp content, small seeds and resistance to cracking. This variety is suitable for extended harvest. The breeding programme involving 9 cultivars *viz.*, Deshi, Early Bedana, Ojholi, Dehrarose, Purbi, Shahi, China, Kasba and Late Bedana at Agricultural College, Sabour resulted into the development of two hybrids namely 'Madhu'/'H-105' and 'H-73'. Madhu is a cross of Purbi and Bedana. It has higher TSS and aril percentage than Purbi. Fruit shape resembles with Purbi. Hybrid 'H-73' (Priya) is a product of cross between Purbi x Bedana. Its fruit weight is higher than the Purbi.

10

Coconut

The coconut palm is considered to be one of the five legendary divine trees as stated in Indian classics and is extolled as *Kalpavriksha* – the all giving tree. Coconut is one of the ten most useful trees in the world, providing food for millions of people, especially in the tropics. At any one time, a coconut palm has 12 different crops of nuts on it, from opening flower to ripe nut.

ORIGIN, HISTORY AND DISTRIBUTION

Coconut is presumed to have originated in the Old World, somewhere in the South-East Asia or Pacific Ocean Islands. At the time of the discovery of the New World, coconuts (as we know them today) were confined to limited areas on the Pacific coast of Central America, and absent from the Atlantic shores of the America and Africa. Coconuts drifted as far North as Norway, are still capable of germination. The wide distribution of coconut has no doubt been aided by man and marine currents as well. India occupies the premier position in the world with an annual production of 10.8 billion nuts. Area, production and productivity of coconut in India have increased during the past three decades (Table 10.1). During the year 2010-11, 10840 million nuts were produced in India from 1.89 million ha with an average productivity of 5,718 nuts/ha, which is the highest in the world. Kerala, Tamil Nadu, Karnataka and Andhra Pradesh are the four major coconut producing Indian states accounting for more than 90 per cent of total area and production (Table 10.2).

Kerala is the largest producer of coconut in India followed by Tamil Nadu. As per available statistics for 2010-11, the Kerala state accounts for 36.8 per cent of coconut production alone followed by Tamil Nadu (34.1 per cent).

Table 10.1: Area, Production and Productivity of Coconut in India

Particulars	Years				
	1969-70	1979-80	1989-90	1998-99	2010-11
Area ('000 ha)	1,033	1,075.8 (4.14)	1,472.2 (36.85)	1,908.1 (22.85)	1,896
Production (million nuts)	5,859	5,636.0 (−3.81)	9,358.8 (66.05)	14,924.8 (37.29)	10,840
Productivity (nuts/ha)	5,672	5,239 (−7.63)	6,357 (21.34)	7,821 (18.72)	5,718

Source: NHB Database-2011. Figures in brackets indicate percentage change to the previous column.

Table 10.2: State-wise Area, Production and Productivity of Coconut in India (2010-11)

Particulars	Andhra Pradesh	Karnataka	Kerala	Tamil Nadu
Area ('000 ha)	104.0	419.0	788.0	390.0
Production (million nuts)	667.0	1497.0	3992.0	3692.0

Source: NHB Database-2011

COMPOSITION AND USES

The wet meat or kernel of coconut is rich in fat, carbohydrates, protein, fibre and moderate in mineral content. Coconut water is rich in minerals and has a caloric value of 17.4 per 100g of water. It is also a rich source of fats, carbohydrates and minerals like phosphorus and calcium (Table 10.3).

Table 10.3: Nutritional Composition of Coconut

Component	Per 100 g Edible Portion	Component	Per 100 g Edible Portion
Moisture	36.3 g	Protein	4.5 g
Fat	41.6 g	Total carbohydrates	13.0 g
Fiber	3.6 g	Ash	1.0 g
Calcium	10 mg	Phosphorus	24 mg
Iron	1.7 mg		

Coconut forms an important component in the socio-economic and cultural life of every Indian household, particularly in South India. It is a versatile tree crop and no other tree crop, can match coconut palm in its usefulness. It provides food, fuel, timber, fibre and numerous products. In addition, it gives a refreshing drink of the tender nut. The tender coconut water is recommended in cases of gastroenteritis, diarrhea and vomiting against dehydration. It is also a urinary antiseptic and eliminates poisons through kidneys in case of mineral poisoning. The dried kernel of the matured nut yields edible oil, which mainly contains saturated fatty acids. Coconut is used for culinary purposes and for making products like copra, cream coconut milk powder, sweet coconut flex, coconut milk based consumer products, spray-dried coconut milk powder, coconut jam, *nata de coco* and coconut vinegar.

The trunk of the palm is used for house construction and for making furniture and handicrafts. The leaves are used for thatching sheds and as fuel. A clump of unopened flowers may be bound tightly together, bent over and its tip bruised. Soon it begins to 'weep' a steady dripping of sweet juice, which is rich in ascorbic acid. The cloudy brown liquid is easily boiled down to syrup, called coconut molasses, then crystalized into a dark sugar. Sometimes, it is mixed with grated coconut for candy. Left standing, it ferments quickly into a beer with alcohol content up to 8 per cent, called '*toddy*' in India and Sri Lanka. After a few weeks, it becomes vinegar. '*Arrack*' is the product after distilling fermented '*toddy*'. Nut has a husk, which is a mass of packed fibers called coir, which can be woven into strong twine or rope, and is used for padding mattresses, upholstery and life-preservers. Fiber is resistant to sea water and is used for cables and rigging on ships, for making mats, rugs, bags, brooms and brushes; also used for fires and mosquito smudges. Coconut roots provide a dye, a mouthwash, a medicine for dysentery, and frayed out make tooth burshes; scorched, used as coffee substitute. It is believed to be antiblenorrhagic, antibronchitis, febrifugal, and antigingivitic. Considering its multipurpose uses, coconut is regarded as tree of heaven or *kalpavriksha*.

TAXONOMY AND BOTANICAL DESCRIPTION

Taxonomy

Coconut is is one of the important members of mocotyledons. It is botanically, called as *Cocos nucifera* L. and belongs to family Arecaceae (Palmae). The basic chromosome number (n) is 16 and 2n = 32.

Botany

Coconut is a tall palm, attaining a height of 20-30 m or more, bearing crown of large pinnate leaves; trunk stout, 30-45 cm in diameter, straight or slightly curved, rising from a swollen base, surrounded by mass of roots; rarely branched, marked with rings of leaf scars; leaves 2-6 m long, pinnatisect, leaflets 0.6-1 m long, narrow, tapering; inflorescence in axil of each leaf as spathe enclosing a spadix 1.3-2 m long, stout, straw or orange coloured, simply branched; female flowers numerous, small, sweet-scented, horne towards top of panicle; fruit ovoid, 3-angled, 15-30 cm long, containing single seed; exocarp a thick fibrous, husk, enclosing a hard, bony endocarp or shell. Adhering inside wall of endocarp is testa with thick albuminous endosperm, the coconut meat; embryo below one of the three pores at end of fruit, cavity of endosperm filled in unripe fruit with watery fluid, the coconut water, and only partially filled, when ripe. Flowering and fruiting take place year round in tropics.

IMPORTANT VARIETIES

Coconut varieties can be classified into two groups – the tall and the dwarf – each comprising a few varieties and forms distinguished by differences in size, number, colour and shape of fruits and the bearing capacity of palm. The fruit colour varies from dark-green to deep-orange or brick-red. The size and shape of fruits show striking variations. Some are small, elongated and distinctly triangular in cross section, while others are big and nearly globular.

Tall Varieties

The palms which are commercially grown at present belong to this group. These are hardy, long-lived (80-90 years) and thrive under a variety of soil, climate and cultural conditions. They attain a height of 15-18 metres and begin to flower in about 8-10 years after planting. The nuts of tall cultivars are generally medium to big with good quality and quantity of copra and fairly high oil content (68-70 per cent).

Tall varieties differ in shape, size, colour and quantity of fruits. Some are esteemed for high yield of copra and others for the quality of tender nut water or for '*toddy*' obtained by tapping. There are also a few varieties with peculiar characters but of little economic importance. For example 'Thairu Thengai' yields nuts in which the cavity of nuts is filled with gelatinous and highly palatable curd. These nuts do not normally germinate, but a few normal fruits produced by the same tree germinate on planting. Variety 'Kaithathali', has soft and fleshy husk due to poor development of fibre and can be eaten raw. Variety 'Spikeless' or 'Spicata' has no branches or spikes in the inflorescence unlike the ordinary coconut palm. The mature nuts of this variety are smaller. The tall cultivars largely grown in India are West Coast Tall, East Coast Tall and Tiptur Tall.

The following tall cultivars of coconut have been found suitable in different states of India:

West Coast Tall	:	Kerala, Tamil Nadu, Karnataka, Madhya Pradesh, Lakshadweep, Odisha and Tripura.
Andaman Ordinary	:	Andamans, Kerala, Andhra Pradesh, Tamil Nadu, Pondicherry, Bihar, Assam, West Bengal and Tripura.
East Coast Tall	:	Tamil Nadu, Andhra Pradesh, Bihar, Madhya Pradesh, Odisha, Pondicherry, Andamans and West Bengal.
Tiptur Tall	:	Karnataka.
Benaulim Tall	:	Maharashtra and Goa.
Philippines Ordinary	:	Coastal Kerala and Andhra Pradesh.

The chief characteristics of some of the tall coconut varieties grown in the world are briefly described hereunder:

☆ **West Coast Tall**: This variety is mainly grown in the West coast of India. Under favourable conditions, it comes to bearing within 6 to 8 years after planting. The average yield of nuts range from 70 to 80 per year. Its copra content is about 176 g per nut, which produces about 68 per cent oil.

☆ **Lakshadweep Ordinary :** This variety is a native of Lakshadweep islands and nearly similar to West Coast Tall in several aspects. The average yield of nuts per palm per year is about 98. The copra content per nut is 176 gms with an oil content of 70 per cent by weight. Highly suitable for toddy tapping.

☆ **Lakshadweep Micro :** It is very popular variety in all the coconut producing countries because it produces large numbers of very small nuts, primarily because of production of large number of female flowers, high setting percentage, low shedding of buttons, and high oil content. The average copra content per nut is 72 g and average yield is about 180 nuts per year with 75 per cent oil content. The nuts are used for table purpose.

☆ **Andaman Ordinary:** The palms are huge with high level tolerance diseases. It produces about 94 nuts per year with average copra content of 160 g but produces quite low oil content (66 per cent).

☆ **Laguna:** This is a widely grown in Philippines. It produces reddish or deep orange nuts at the younger stage. It produces about 88 nuts/year with an average weight of 1.26 kg, with the mean copra content of 259 g and oil content of 66 per cent.

☆ **Kappadam:** It is main cultivar in Trissur district of Kerala (India). It has a long pre-bearing period (8 to 10 years). It produces large nuts in good number (about 90/palm/year), weighing 283 g and with an oil content of 76 per cent.

☆ **Macapuno:** This is a very peculiar type variety, grown in isolated patches of Philippines. One or two nuts in a bunch are filled with a soft and tender jelly like endosperm, which is considered as a delicacy in Philippines. The other nuts in the bunch are like ordinary coconuts.

☆ **Spicata:** This variety has a sporadic distribution in all the major coconut growing countries. This variety bears an unbranched inflorescence having one or two spikelets of only female flowers, yet fruit-setting percentage is very low. The meat content of a matured nut is about 430 g.

☆ **San Raman:** It is also a high yielding commercial variety, of Philippines. The nuts are bigger and contain copra weighing about 350 g with 68% oil. The average yield of nuts per year is 64.

Dwarf Varieties

Dwarf varieties have short stature and a shorter life span of 40-50 years. They start bearing 3-4 years after planting. The size of the nut and the quality of copra are inferior with lower oil content (66-68 per cent) compared to tall palms. These palms are more homozygous due to self pollination. These varieties have three fruit colours-green, yellow and orange. These are cultivated for ornamental purpose or for tender nut water. However, at present, these are are being utilized on large scale for the production of hybrids. The common varieties found in India are Chowghat Green Dwarf, Chowghat Orange Dwarf and Gangabondam. Some exotic dwarf varieties, Malayan Yellow Dwarf, Malayan Orange Dwarf and Malayan Green Dwarf have also been planted in seed gardens for the production of D × T hybrids.

Lakshadweep Ordinary (released as 'Chandrakalpa' in 1985 by CPCRI, Kasaragod) gives 33 per cent more number of nuts and 30 per cent higher copra yield than local tall varieties under rainfed conditions. Benawali Green Round – a popular cultivar of Goa state (released in 1985 under the name 'Pratap' by the Konkan Krishi Vidyapeeth, Dapoli) gives 62.3 per cent higher nut yield than West Coast Tall (Table 10.4-a).

Table 10.4a: Performance of Released Cultivars in Comparison with West Coast Tall

Cultivar	Nut Yield		Copra Yield			States for which Recommended
	No. of Nuts/ Palm/ Year	Per cent Increase Over WCT	Mean/ nut (g)	Mean/ Palm/ Year (kg)	Per cent Over WCT	
Lakshadweep Ordinary (CHANDRAKALPA)	100	25	176	17.6	25	All states
Benawali Green Round (PRATAP)	150	88	152	22.8	62	Goa, Maharashtra
Philippine Ordinary (KERACHANDRA)	110	30	198	21.8	55	Coastal Kerala, AP and Konkan Region
Chowghat Orange Dwarf (COD)	65	Released as tender nut variety				All coconut growing states
Local Tall (WCT)	80	-	176	14.1		Kerala, Tamil Nadu, Karnataka, MP, Lakshadweep, Odisha, Tripura

The chief characteristics of some of commercially grown dwarf coconut varieties in the world are described briefly hereunder:

☆ **Chowghat Dwarf Green:** The palms of this variety start bearing after 3 years of planting. It is completely a self-pollinated variety and produces about 66 nuts per year. This variety produces dark-green, oblong small nuts, weighing 92 g. The copra is of poor quality as it is hard and leathery. The oil content is 73.5 per cent by weight of copra and the tender green nuts yield about 170 to 225 ml of sweet water.

☆ **Chowghat Dwarf Orange:** It produces nuts of spherical, medium sized and orange colour. The yield of tender coconut water is about 340 to 510 ml of sweet water. The average yield of nuts is 91 per palm. The copra is hard and leathery.

☆ **Malayan Dwarf:** This is precautious in bearing and produces nuts of smaller size having 3 distinct colours, such as yellow, red and green. The average yield of nuts is 102 per palm. The variety is grown on a large scale in Kanyakumari district of Tamil Nadu and used for toddy tappings.

☆ **Rath Thembli:** This is a self-pollinated variety, commercially cultivated on a large scale in Sri Lanka and is locally known as 'king coconut'. It produces ovoid and narrow-shaped nuts, from which desiccated coconut is produced, which is high demand in confectionery and other food industries the world over.

☆ **Coco Nino:** This is very dwarf variety and grown commercially in Philippines. The nuts are green and small, producing about 100 nuts per palm per year. This variety is commonly used as ornamentals and toddy tapping as it produces high quality toddy in plenty.

☆ **N'uleka:** This is an early bearer, cross-pollinated variety, commercially grown in Fiji islands, which produces about 150 small-sized nuts per palm per year with high copra content.

☆ **Mangipod:** This is a self=pollinated distinctly dwarf variety, which bears about 50 to 60 nuts per bunch per year. Interestingly, the matured nuts contain very little or no water. As a result, they become lighter and hang around the trunk in brown coloured clusters.

Hybrid

Hybrid vigour in crosses between tall and dwarf coconut cultivars was first reported from India in the year 1937 by J.S. Patel. Kerasankara (WCT × COD) is a T × D hybrid while Chandrasankara (COD × WCT) is a D × T hybrid. The hybrids like Kerasankara, Chandrasankara, Chandralaksha, Lakshaganga, Ananthaganga, Keraganga yield 19-42 per cent higher than their parents. Lakshaganga and Chandralaksha show tolerance to drought. VHC-2 released by the Tamil Nadu Agricultural University is a T × D hybrid (East Coast Tall × Malayan Yellow Dwarf) and gives about 26 per cent more nut yield. Performance of released hybrids in comparison with West Coast Tall is given in Table 10.4(b).

Table 10.4(b): Performance of Released Coconut Hybrids Under Rainfed Conditions in Comparison with West Coast Tall

Hybrid	Nut Yield/ Palm/ year	Copra Yield Mean/ Nut (g)	Mean/ Palm (kg)	Oil Content (Per cent)	State for which Recom- mended	Agency Responsible for Release
Chandrasankara (COD × WCT)	116	215	24.9	68	Kerala	CPCRI
Kerasankara (WCT × COD)	108	187	20.2	68	Kerala, Coastal Maharashtra, AP	CPCRI
Chandralaksha (LCT × COD)	109	195	21.2	89	Kerala	CPCRI
Lakshaganga (LCT × GBD)	108	195	21.1	70	Kerala	KAU
Anandaganga (ADOT × GBD)	95	216	20.5	68	Kerala	KAU
Keraganga (WCT × GBD)	100	201	20.1	69	Kerala	KAU
Kerasree (WCT × MYD)	130	216	28.0	66	Kerala	KAU
VHC-1 (ECT × DG)	98	135	13.2	70	Tamil Nadu	TNAU
VHC-2	140	150	21.0	68	Andhra Pradesh	APAU
WCT	80	176	14.1	68	–	–

Recently, CPCRI, Kasargod has released 4 new varieties of coconut trees namely, 'Kalpa Prathibha', 'Kalpadenu', 'Kalpa Mithra' and 'Kalpa Raksha'. Of the four, 'Kalpa Prathibha' has been giving constant good yield. Another speciality about 'Kalpa Prathibha' variety is that it can withstand normal droughts and can give yield of over 15,874 coconuts per hectare. Due to its features, it is named by the scientists as ' a variety specially made for the coastal belt'. Rest of the varieties too have great features. 'Kalpa Mithra' can give yield of about 13,973 coconut per hectare, 'Kalpa Raksha', which has been developed from Malayan Green has the capacity to withstand the stem borer incidence.

SOIL AND CLIMATIC REQUIREMENTS

Soil

The coconut, though essentially belonging to the humid regions of the tropics, is a very adaptable palm, which can tolerate a wide range of soil conditions. However, the palm prefers a well-drained, light soil permitting free root development and aeration. It grows best on coastal alluvial and sandy soils. Low lying areas, subjected to flooding and which cannot be drained, should be avoided. Coconut can also be grown in laterite soils without hard pan up to a depth of 2 meters and red sandy-loam soils with a pH ranging from 5.2 to 8.0.

Climate

The coconut palm draws its characteristics from the fact that it is essentially a tropical plant and therefore, such as latitude, rainfall, temperature, humidity, sunshine, etc., should be given thorough consideration. Coconut palm thrives in the tropical zone with major coconut growing regions between 20°N and 20°S having an average temperature of 27°-32°C and a diurnal variation of about 7°C and a well distributed rainfall of 1,270-2,500 mm per annum. It can tolerate considerably higher precipitation if the soil is well-drained. In regions receiving rainfall below 1,000 mm per annum, it can be grown only with irrigation. In regions receiving rainfall over 2,500 mm per annum, it can be grown if there is adequate drainage. In general, the coconut palm likes a climate typified by warm and humid conditions. However, excess humidity is not considered good for the palm as it reduces transpiration, thereby reducing the uptake of nutrients. Besides, providing congenial conditions for the rapid spread of the fatal diseases of the palm, *viz.*, 'bud-rot' etc. The leaf disease of the coconut palm in Kerala is found to spread rapidly during rainy months when the atmospheric humidity is high. Coconut does not thrive in areas with continuous cloudy weather since it requires plenty of sun light and cannot tolerate shade even from 'vegetation'.

PLANT PROPAGATION

In coconut, propagation is usually done by seed. In seed propagation, selection of mother plant from which nuts have to be collected, has great bearing on the productive life of palm. Hence, utmost care must be exercised while selecting, sowing and during subsequent care and handling of seedling palm.

Selection of Mother Palms

Owing to the fact, coconut is propagated by seedlings raised from fully mature fruits, seeds should be selected from high-yielding stock with desirable traits. In order to identify promising mother palms, 10 per cent of the best palms in a garden are selected, which have consistently high nut production (above 80), high weight of husked nuts (about 650 g), high percentage of flower set, large number of spikelets with one or two female flowers on each, inflorescence with about 100 female flowers along with the other desirable morphological characters such as straight trunk and even growth, with closely spaced leaf-scars, short fronds, well oriented on the crown and short bunch stalks. Plants which exhibit drooping or erect crown should be avoided. Mother palm should be in the age group of 25-50 years. Palm should have 30-40 fully overed leaves and 12-15 bunches with a high setting of female flowers.

Selection of Seed Nuts

As the maximum development of kernel is attained with the nut is fully mature, fully mature nuts should only be used for seed purpose. There is no difference in performance of seedlings raised from 11 and 12 months old seednuts. On west coast of India, seednuts collected in March-to-June result in early germination and vigorous seedlings. However, this varies from region-to-region and the season of nut harvest assumes importance only in those areas where nursery raising depends on rainfall. Seed nuts should be medium-sized and nearly spherical in shape; long nuts usually have too much husk in relation to kernel. It has been observed that heavy seednuts with good copra content float vertically in water with stalk end up. Such seed nuts result in more vigorous seedlings. Thus, seed nuts with thin husk and higher husked nut weight are likely to give more vigorous seedlings.

Sowing of Seed Nuts

Seed nuts are stored for a month under shed after harvest prior to sowing. This results in early and high germination. However, often it is not possible to sow seed nuts after one month of storage and on such occasion, harvested seed nuts are kept in open under shade till the husk becomes dry. The seed nuts are arranged with stalk end up and covered with sand up to about 5-7 cm. This way, the seed nuts can be stored for over 3 months. The coconut nursery is generally raised under shade and sometimes even under the coconut plantation. Seeds planted in nursery facilitate selection of best to put in field, as only half will produce a high-yielding palm for copra. Also, watering and insect control is much easier to manage in nursery. Raised seed beds (raised about 22 cm) are preferred to avoid drainage due to stagnation of water during rainy season. The seed beds are drenched with chlordane (5 per cent) dust @ 120 kg/ha and seed nuts are sown at a spacing of 30 × 40 cm during May-June either vertically or horizontally at a depth of 20-25 cm. Irrigation is essential during the summer months. Nursery beds must be kept free from weeds and regular inspection should be made to prevent incidence of pests and diseases, particularly root grub.

In most countries, manuring of coconut nurseries is not recommended. But fertilizer application was found to improve seedling vigour in Ivory Coast. At CPCRI, Kasargod, there was no significant difference in vigour of seedlings by fertilizer applications.

Selection of Seedlings

Selection of seedlings is important to obtain quality planting material. Rigorous culling of seedlings is essential. All late germinators and very slow growers are discarded. Early germinated nuts having faster rate of leaf production result in early flowering and high nut production. Seed nuts germinated within 3 months after sowing should be selected for planting and those sprouting 5 months after sowing should be rejected. Seedlings with short stem, good girth at collar, producing large number of dark green leaves, splitting early are characteristics of quality seedlings, which subsequently develop into high yielding palms.

Transplanting of Seedlings

Seedlings of 9-12 months are generally transplanted. Pruning of roots of seedlings up to 12 months age does not cause any damage. In parts of Karnataka and Andhra Pradesh, 2-3 year old seedlings are also planted particularly in flood prone areas having poor drainage. But due to considerable root damage, seedlings show delayed establishment and growth retardation.

Land preparation for field planting is done considering topography, soil type and water table. In undulating and sloppy lands, under growth is cleared and soil conservation measures are adopted to prevent soil erosion. Adequate drainage must be provided in water logging areas to prevent direct contact of root of young coconut palms with water. In such areas, planting is commonly done on raised mounds or bunds.

Optimum spacing is adopted so that the canopies of the palms do not touch each other 8 to 20 years after planting. Best spacing depends upon soil, terrain, variety, intercropping and mixed cropping and the fact that the active roots of coconut are observed to utilize only 25 per cent of the available land. In square system of planting, generally adopted in India, a spacing of 7.5 m or 9 m is adopted for tall varieties to respectively accommodate 1 75 and 124 palms/ha. In triangular system at a spacing of 9 m, about 140 palms can be accommodated. Hedge system of planting is also adopted for establishment of seed-gardens by planting dwarf and tall palms in alternate rows to facilitate hybridization. In the single hedge system, planting is done at a distance of 9 m between rows and 5 m within rows.

The pits for planting are prepared during the summer months. Pits of 1 m × 1 m × 1 m are prepared in Karnataka and coastal Maharashtra. In India, 300-400 husks are burned in each pit, providing 4-5 kg ash per pit. This is mixed with topsoil. Two layers of coconut husks are put into bottom of pit before filling with the topsoil mixed ash. Furadan @ 5 g/pit is also added. The seedling is planted in the centre of the pit and firmly pressed. Care should be taken not to cover the collar of the seedling by soil. As plant develops, trunk may be earthed up, until soil is flush with general ground level. Usually 7-8 month old seedlings are used for transplanting. In some instances, plants up to 5 years old are used, as they are more resistant to termite damage. If older plants are used, care must be taken not to damage roots, as they are slow to recover. The most appropriate time of planting is during the beginning of monsoon (May-June) and during October-November in low lying areas.

After planting, the young seedlings are given some support. During the first year, the plants are provided shade, regular irrigation during summer months and constant weeding. At end of first year after transplanting, vacancies should be filled with plants of same age held in reserve in nursery. Also any slow-growers, or disease damaged plants should be replaced.

INTERCULTURAL OPERATIONS

Application of Manures and Fertilizers

For sustained productivity, the palms should be manured every year right from the first year after planting. The optimum dose of fertilizer recommended for an adult palm is 0.5 kg N, 0.32 kg P_2O_5 and 1.2 kg K_2O/year. The first dose of fertilizer is applied 3 months after planting and the dosage is increased every year and the full dose of the fertilizer is applied 4th year onwards (Table 10.5).

Table 10.5: Fertilizer Recommendations for Coconut Gardens (g/tree)

Age of Palm	May-June			September-October		
	N	P_2O_5	K_2O	N	P_2O_5	K_2O
First year	-	-	-	50	40	135
Second year	50	40	135	110	80	270
Third year	110	80	270	220	160	540
Fourth year onwards	170	120	400	330	200	800

The fertilizer should be applied under the optimum soil moisture condition. One third of the fertilizer is applied immediately after the onset of South-West monsoon and the remaining $2/3^{rd}$ dose should be applied towards the end of the monsoon. A shallow trench (20-25 cm deep) of 1.8 m radius is made around the base of the coconut after the receipt of rain towards the end of May and the fertilizer is broadcasted and covered with organic manure and soil. Farm yard manure @ 50 kg/palm is applied in addition to inorganic fertilizers. Rock phosphate has been found to be the cheapest and best phosphatic fertilizer particularly for acidic soil. Field trial conducted at CPCRI, Kasargod has indicated that phosphorus application can be skipped for few years if the available soil phosphorus is more than 20 ppm. Crops like *Calapogonium mucunoides* and *Mimosa invisa* during the monsoon season can generate up to 25 kg green manure for incorporation into the basin. Application of lime is not generally recommended. There is no evidence that salt is beneficial as is sometimes claimed. Coconut palm can withstand a degree of salinity, about 0.6 per cent, which is lethal to many other crops. Palms seem to need some magnesium, but are extremely sensitive to its excess.

The hybrids (particularly D × T) are found to be most efficient users of applied nutrients than the tall cultivars both under rainfed and irrigated conditions and give higher yield even at lower levels of fertilizer application.

Irrigation

Response of coconut palms to irrigation is location-specific, depending upon climate, soil, topography, ground water table, etc. Moisture stress causes leaf fall, lowers growth and leaf area, resulting in reduced light interception and thereby lower number of female flowers/inflorescence, fruit set and yield. The size of nut and copra content are also reduced due to moisture stress. Water transpired by a coconut palm is estimated to be 53 litres/day in coastal Kerala and 90 litres/day in Tamil Nadu (28 to 200 litres/day). Under the circumstances, where scarcity of water and increasing cost of labour and energy are deterrents in adopting traditional irrigation systems, drip irrigation seems to be the most suitable system of irrigation to coconut as it not only saves water but also enhances plant growth and yield, saves energy and labour and reduces weed growth and improves efficiency of fertilizers.

During summer months in coastal Kerala and Karnataka, application of 200 litres water once in four days in the basin of 1.8 m radius has been recommended. Drip irrigation is found to be the most economical. Application is done at 3-4 points keeping the drippers at a depth of about 30 cm below soil, 1 m away from the base to supply water at the root zone and prevent surface evaporation. For drip irrigation, about 30-40 litres of water per day is found to be sufficient under West Coast condition. Sprinkler system of irrigation has been recommended under inter cropping system giving about 75 per cent water of the pan evaporation. Mulching with coconut husk, coir dust, green and dry coconut leaves improves water retention and prevents soil erosion.

Inter-cultivation

Inter-cultivation is done by ploughing the land once in one or two years. Inter-cultivation increases the yield substantially. The inter-cultivation also helps in controlling weeds and conservation of moisture in the soil. However, excessive inter-cultivation causes rapid loss of soil organic matter and adversely affects the soil structure. For laterite, sandy and red sandy-loam soils, two ploughings or diggings in May-June and September-October can be adopted, followed by one raking in January. In areas where surface run off is more, mounds may be prepared in September-October, which can be leveled in November-December. Care should be taken to keep trees clear of weeds, especially climbers. In general, a circle 1-2 m in radius should be weeded out with mattock several times a year, the weeds left can serve as mulch. Cover-crops such as *Centrosema pubescens, Calopogonium mucunoides,* or *Pueraria phaseoloides,* are used and turned under before dry season. Catch-crops like cassava (*Manihot utilissima*), and green gram (*Vigna aureus*) and cowpea (*Vigna unguiculata*), bananas and pineapples, may also be used. Sometimes bush crops, in addition or instead of, ground covers are used as green manures, as *Tephrosia candida, Crotalaria striata, C. uraramoensis, C. anagyroides,* all fast growers. *Gliricidia sepium* and *Erythrina lithosperma* may be grown as hedges or live fences and their loppings later can be utilized as green manure. Response of West Coast Tall palms to different management practices at CPCRI, Kasargod is given in Table 10.6.

Table 10.6: Response of West Coast Tall Palms to different Management Practices at CPCRI, Kasargod

Sl.No.	Treatment	Nut Yield/Palm/Year
1.	Organic and inorganic manuring and tillage	110.4
2.	Inorganic fertilizers and tillage	96.7
3.	Inorganic fertilizers and forking of basins	90.5
4.	Tillage	53.1
5.	Weed control by herbicides	27.3
6.	Control (no manuring and no tillage)	9.7

COCONUT BASED CROPPING SYSTEMS

Coconut exhibits the unique feature of accommodating any crop combination in the inter-spaces. A well-spaced coconut garden renders sufficient inter-spaces to let a variety of crops, both seasonal and perennial, successfully flourish. When annuals or seasonal crops are grown in coconut holdings, it is designated as inter-cropping and when perennials are grown, it is called mix cropping. A combination of inter-crops and mixed crops raised together are referred to as a 'multi-storeyed' cropping system. During the pre-bearing period, especially up to three years after planting, the entire area could be made use of because of the negligible shade effect. Thereafter, depending on the age of the palms and canopy coverage, suitable crops or a combination of different crops could be selected for growing in the coconut gardens.

When planted at 7.5 × 7.5 m to 9.0 × 9.0 m spacing, coconut tree does not fully utilise the available soil area and sunlight. The remaining area could be profitably used by intercropping under the diffused sunlight between coconut plantation, a number of shade tolerant crops can be grown.

Tropical tuber crops like cassava, elephant foot yam and spices like turmeric and ginger are the most popular intercrops grown in coconut gardens. The productivity of tuber crops and rhizomatous spices raised as intercrops varies from 8 to 15 tonnes per hectare. Cereals, pulses, oilseeds, fruit crops and vegetables can also be grown as intercrops. Likewise, fodder crops such as hybrid napier and guinea grass can also be grown in coconut plantation.

A number of perennial crops like cocoa, clove, nutmeg, pepper are often grown as mixed crops in coconut plantation. Among these, cocoa is an ideal mixed crop. The productivity of coconut is also improved as a result of cocoa as a mixed crop. Although productivity of cocoa raised in double hedge system in a coconut garden planted at 7.5 × 7.5 m spacing is slightly higher than in single hedge system, but the latter system is more desirable.

Multi-storey cropping system is an intensive four crop combination of coconut, black pepper, cocoa and pineapple. The system requires irrigation during summer months and ensures efficient foraging of soil without undue competition among the component crops and intercepts solar radiation at different heights. Pineapple performs well in such a system in the initial five or six years but it has to be removed when cocoa plants develop full canopy.

High density multi-species cropping system grows a large number of crops to meet diverse needs of the farmer such as food, fuel, timber, fodder and is ideally suited for smaller land units to obtain maximum production per unit area of land, time and inputs with minimum or no deterioration of land. The component crops may vary based on local preference, for different soil area and to make maximum use of space and sunlight.

In the interspaces of coconut plantation, shade tolerant forage crops can be grown for integrating animal enterprises. Fodder grasses like hybrid napier, guinea grass yield about 50 to 60 tonnes of green fodder/ha under coconut shade, which is sufficient to maintain five milch animals. Additional income from the mixed farming can be generated by training pepper on to coconut, and raising intercrops like banana, cassava, ornamental plants (*Heliconia, Anthurium, Jasminum pubescence* and Marigold), vegetables (snake gourd, bottle gourd, amaranthus, coccinia, brinjal and bitter gourd) and medicinal plants (long pepper and *Patchouli*) along the borders and field bunds. Cowdung and urine from the system can be used for a biogas plant to obtain gas for cooking and domestic electricity. The cowdung slurry from the gas plant and urine can be recycled into the plantation.

COCONUT WILT: A THREAT TO COCONUT INDUSTRY

Coconut palm is a tropical fruit plant, which is grown commercially in a number of countries. It is considered as a very sacred fruit plant and hence used in many religious functions. It is an excellent source of nutrients and provides refreshing drink, edible oil and fiber for commercial use. In addition, its different plant parts can be used as fuel and for timber purposes, whereas, shell is used for many industrial purposes. Hence, it is conclusive enough to state that each and every part of coconut plant is useful to the mankind.

It is a versatile tree-crop and adapts well to divergent farming conditions and agro-climatic situations. It has survived for centuries in the tropics under neglected conditions, thus showing tolerance to diseases. However, under humid conditions and due to stagnation of water around the coconut roots etc., predisposes the palm to various infections. Palm pathologists have described severak diseases of coconut. Of which, coconut wilt is considered as the most important because the incidence and spread of this malady is very confusing. Butler (1908) was the first to refer this malady as 'root disease' in coconut, because of root rotting, and later referred it to as 'root (wilt) disease'.

Symptoms

Diseased palms show flaccidity (abnormal bending or ribbing of leaflets), foliar yellowing mostly in older palms, marginal necrosis, paling and conspicuous bending of middle and outer whorls of leaves. Abnormal shedding of female flowers and immature nuts and drying up to spathes from tip downwards is also observed. Intensity of symptoms is increased according to the age of the palms. For example, with the increase in age, necrosis and yellowing become prominent and flaccidity remains consistently high; leaf area and size of the crown is reduced. However, in general, a symptom of the foliage, especially flaccidity is the consistent visual symptom of root wilt in coconut. Some workers have also reported the sterility of high percentage of pollen, and meiotic irregularities in wilt-affected palms. Wilt-affected palms show abnormalities in roots, leaves, stomatal behaviour, and reproductive behaviour with some physiological changes as described briefly hereunder:

a. Underground Symptoms

In diseased palms, roots are mostly affected and root sap emits foul smell. They produce less number of active roots and their regeneration capacity is reduced. Similarly, meristematic activity is also reduced drastically and thus roots become stunted. Vascular tissues of roots disintegrate in high percentage (about 60 per cent). Phloem tissues show chromophylly, necrotic reduction in size; and xylem vessels also develop tylosis and become smaller. Permeability of root tissues is damaged, water absorptive capacity of the roots becomes about 60 per cent low and water uptake is hindered greatly. Some workers have reported that 60 per cent of apparently healthy roots show internal browning and about 92 per cent roots start rotting.

b. Foliar Symptoms

In affected palms, abnormal symptoms in leaves are spectacular. Mature leaves become thinner; wall thickness of sclerenchymatous tissue and cuticle in abaxial surface is reduced; xylem vessels are abnormally small and phloem tissues also become smaller, though they produce more number of cells. Number of epidermal cells per unit area increases and transverse division of these cells is accelerated. There may be curtailing in longitudinal division in the leaves, as a result, leaves show downward curling.

There is degeneration in chlorophyll content of leaves, thus they appear chlorotic and have reduced photosynthetic activity. The leaves of a diseased palm appear dehydrated and show fast degradation and aging processes. Frequency of stomata on abaxial side of leaves and transpiration rate in leaves is increased, which is usually very high in first and second leaves. Respiration rate of the leaves is also higher than normal leaves. Water contents in the leaf tissue are decreased, and thus water economy is imbalanced. Leaves of diseased palms had longer stomatal resistance and leaf water potential than apparently healthy palms. There is almost complete loss of turgor potential, which results in flaccidity in the leaves. The wilt-affected palms fail to close their stomata in response to soil and atmospheric drought. Thus, the stomatal regulation is adversely affected by disease.

Certain chemical changes have also been observed by the scientists in the leaves and roots of diseased palms. For example, chloroplastids in mesophyll cells are degenerated and there is a poor deposition of lignin in leaflets; abnormal deposition of protein and high sugar content in leaves, indicating derangement of sugar transmission. Total carbohydrate and starch, C/N ratio in leaf and roots and zinc are reduced; while tannin and phenol content in the middle outer leaves are increased. An altered nitrogen metabolism with an accumulation of amino acids and argenine in leaf area of diseased palms are observed. But, in root sap, the argenine and phenol contents are found to be low. Vascular sap from the diseased palms had higher pH, but reduced osmotic concentration and an imbalance in the water absorption mechanism. Abnormal shedding of female flowers and lack of ability to produce more female flowers are mostly observed in affected palms. High percentage of palms become sterile, pollen grains become irregular and they also show meiotic irregularities. Flowering in the affected palms is drastically delayed and/or reduced.

Disease Identification

Field identification of the diseased palms is mainly based on visual symptoms, considering flaccidity as the primary symptom and yellowing and necrosis as other associated symptoms. Since reliable identification of diseased palms becomes difficult under certain conditions, the need for a diagnostic test, which could detect the disease status even before the foliar symptoms are evident, was felt necessary. For this, a serological diagnostic test and a physiological test based on stomatal resistance have been standardized for detecting the disease. With these aids, the diseased condition could be diagnosed about 6 to 20 months before the expression of foliar symptoms.

Spread of Disease

Following factors have been found responsible for the spread of the disease in different localities:

1. It is reported that the disease is mostly spread in coastal tracts close to the riverbanks and canals where high water table exists or in water-logged conditions. It is suspected that the disease-causing organism gets along water and afterwards transmission takes place, perhaps through soil or insects.

2. This disease is spreading in all directions in all soil types; but remains mostly concentrated in sandy and sandy-loam soil areas than in laterite soils. The infected soil area is normally acidic with low content of exchangeable bases and poor base-exchange capacity. Soils with low pH, water logging, lack of micronutrients like, Zn, Mn, B etc. and deficient in Ca and Mg, may result easy susceptibility in the palms.

3. At high water table, there is a constant association of the species, *Rhizoctonia solani* in the roots of diseased palms. The disease becomes complicated when it is associated with leaf rot (a fungal infection) during rainy season and under improper drainage condition. Correlation between concentration of polyphenol oxidase and root (wilt) index, based on foliar symptoms is highly significant and positive.

Causative Factors

A lot of research work has been done about the possible causes of this malady. However, it is not conclusive enough to state that a particular factor is involved for its cause. The sporadic occurrence and spreading nature of the disease suggests the possibility of association of some pathogens. The possible association of different factors has been described below:

Fungi

As early as 1906, the root disease (wilt) was first reported to be fungal in nature, because certain fungi, *Botryodiplodia* and *Rhizoctonia* were isolated from the roots of the affected palms. But, when pathogenicity experiments were conducted, these fungi could cause rotting of roots in coconut, but failed to produce characteristic symptoms of the disease both is field and lab conditions. Based on various experiments, it was later concluded that possibility of involvement of fungi with root wilt of coconut is negligible.

Bacteria

Research work conducted on the possible involvement of bacteria with root wilt of coconut reveal that some scientists have isolated *Pseudomonas* spp., and *Enterobacter cloacae* bacteria from the roots of diseased palms and speculated their involvement in the causation of wilt. However, inoculation of the coconut seedlings with these bacteria failed to produce any symptom of the disease, and hence, they ruled out the possibility of involvement of bacteria in the cause of this disease.

Nematodes

Possible involvement of soil borne pathogens or a pathogen transmitted through soil borne vectors warranted nematological studies. Scientists isolated a number of nematodes from the infested roots. In general, they observed that infested roots yield maximum number of *Radopholus similis* (burrowing nematode) during October-November and minimum during March-July. On sandy-loam soil, its occurrence was also very high. On inoculation, this nematode produced small, reddish-brown elongated, cortical lesions resulting in severe rotting of roots. Reduction in growth and intensity of root lesions and rotting are directly proportional to the increase in nematode population. However, Weischer (1967) opined that low population density, their widespread occurrence and the general distribution pattern of disease could exclude the nematodes from being considered as a primary cause of wilt disease in coconut. However, considering the notoriety of *Radopholus similis* in citrus and banana, some scientists conducted pathogencity experiments and inoculated coconut seedlings with *R. similis*, but it failed to reproduce roots with symptoms, thus excluding its association with this disease.

Physiological Factors

Some earlier research workers attributed this malady to various malfunctions encountered in the diseased palms. They reported that root wilt is the result of lack of nutrients, excess or inadequate water or combination of both, which disturbs the physiology of palms and thus predisposing the disease. No doubt that scientists have observed low concentration of many nutrients, high respiration rate, mismanagement in water absorbing capacity in the diseased palms, but these malfunctions or disturbances in physiology of diseased palms are rather suggestive of a pathogen-mediated altered host metabolism than of a physiological disorder.

Soil Condition and Nutritional Aspects

The impact of soil conditions and associated nutritional factors on the disease incidence has been under investigation since 1939. Based on a survey and study of the disease in relation to the soil conditions, it was concluded that the disease-affected palms are generally poor in major nutrients (especially potash) and the soils are mostly acidic. Later, detailed studies were conducted in different coconut growing areas of Kerala and other states and it was concluded that though most of the wilt affected palms are in a state of unbalanced nutrition than healthy palms but application of appropriate doses of major and minor nutrients could not reduce its incidence. Hence, the scientists ruled out the possibility of the direct involvement of any major nutrients in the incidence of the disease.

Virus and Virus like Organisms

Based on the systemic nature and resemblance of symptoms of wilt with other known plant virus diseases, a theory on viral etiology for the disease was proposed. This theory gained further significance with positive transmission of the disease to coconut by sap inoculation and through the insect vector, *Stephanitis typica* under field conditions and under insect-proof conditions. Later, cowpea was reported its alternate host, because inoculated plants of cowpea produced almost similar symptoms of the disease. However, all studies conducted on this aspect in the later years led to conclude that virus or virus like organisms are not involved as a primary cause of root will of coconut.

Mycoplasma like Organisms (MLOs)

Although biological agents like fungi, bacteria, nematodes, virus as well as physiological and nutritional factors have been implicated, but the cause of this malady has not been specified earlier by anyone. The evidence now available through electron microscopic examination of tissue transmission and chemotherapy indicate that the disease is caused by mycoplasma-like-organisms (MLOs). The presence of MLOs is further confirmed with fluorescence microscopy, which helped in identifying MLOs presence in periwinkle (*Vinca rosea*), dodder (*Cassytha filiformis*) and diseased coconut seedling, used in transmission studies and thus brought out the presence of MLOs in the vegetative vector and test plant, periwinkle.

MLOs have also been observed in the salivary glands and brain tissues of the lacewing bug (*Stephanitis typica*) earlier, and later detected in the plant hopper (*Proustista moesta*), which are suspected as possible vectors. The disease could be transmitted from diseased to healthy coconut seedlings grown in insect-proof cage through lace bug, thereby predicting the vector role of the insect. A linear correlation of abundance of lace bugs and fresh incidence of root (wilt) disease has been observed, suspecting the presence of lace bug in endemic areas to transmit the disease. Further, chemotherapeutic treatments with tetracycline type antibiotics at different concentrations produced temporary remission of symptoms, thereby confirming the involvement of MLOs in the etiology of the disease.

Involvement of Toxins

Microorganisms like *Enterobacter cloacae* living around root zone, produce certain wilt-inducing toxic substances, indicating the possible involvement of toxins. Similarly, fusicoccin, a phytotoxin produced by the fungus, *Fusicoccin amygadali* has been reported, but its actual role in the causation of the disease awaits elucidation.

Management Practices

The scientists have developed a package of practices, which includes multi-disciplinary approach to keep the palms all time healthy. By adopting this set of practices, which consists of proper nutritional and water management, addition of organic matter, raising green manure crops in the basin, recycling of the organic matter through mixed-farming, inter-and mixed-cropping with compatible crops etc., keep the disease under check to a great extent, so as to maintain the productivity of the affected palms at a satisfactory level for a sufficient long period. These measures have been described briefly hereunder:

Cultural Practices

a. Water Management

Assured water supply (supplementary irrigation) during dry period is very important to avoid the incidence of coconut wilt. Installation of drip irrigation would be the most ideal method of irrigation to palms. Drip irrigation unit supplies water to the plants as per requirements.

b. Fertilizer Management

Application of recommended doses of organic (50 kg FYM/palm) and inorganic (1 kg urea, 2 kg superphosphate, 2 kg murate of potash/palm) fertilizers applied in two split doses along with summer irrigation from December-to-May can retain productivity of the disease affected palm for a longer period. The yield potential of palms of early and middle stages of disease symptoms can be delayed by regular fertilizer management and imposition of irrigation.

c. Use of Soil Amendments

Besides normal doses of fertilizers, addition of 1 kg magnesium sulphate and 1 kg slaked lime may prevent the development of foliar yellowing. Similarly, regular application of magnesium sulphate @ 3 kg/palm helps to retain more buttons on the palm, which may increase the nut yield by about 30-40 per cent in the early bearing period.

d. Intercropping

Indiscriminate use of intercropping should always be avoided. However, raising tuber crops, particularly elephant-foot-yam in the coconut interspaces have resulted in an increase in the main yield of palms with a decline in severity of the disease. However, other intercrops should be avoided.

e. Mixed Cropping

Mixed cropping of coconut with cocoa has been reported to be very useful. It not only avoids the incidence of wilt in the coconut plantations, but increasea yield substantially.

f. Adopt Mixed Farming

Some studies have indicated that mixed farming is very useful in coconut plantations. Mixed farming of growing grasses and legumes in the interspaces of aged palms, rearing milch animals with a recycling of organic wastes over a period of 5 years can increase nut production substantially in root (wilt) affected areas.

g. Eradication of Affected Palms

Eradication of affected palms from the sporadic areas of incidence showed negligible recurrence of the disease, suggesting thereby that removal of root (wilt) affected palms should be taken up on a continued basis from the isolated and mildly affected areas. Palms affected in the pre-bearing age do not flower and produce nuts. Soon palms might not bear at all and hence should be removed on priority. Similarly, those palms, which have advanced stage of the disease, need to be eradicated with replacement of healthy ones.

Chemical Control

In nematode infested areas, application of phorate or aldicarb @1 g a.i./planting pit and @10 kg a.i./ha in infested nursery is very useful. It is also suggested not to raise nematode susceptible crops like banana in or near the coconut gardens. Spraying with 1 per cent Bordeaux mixture regularly on the crown, particularly during pre- and post-monsoon period, may reduce the incidence of leaf rot, superimposed on the root (wilt).

Use of Resistant Cultivars

The hybrid, D x T with good fertilizer management has been found to be more productive (almost double the nut production) with a lower incidence of root (wilt) as compared to the existing seedling palms. The disease-free palms in 'hot-spot' areas should receive maximum attention along with selected germplasm materials. The possibility of selection of resistant lines from good mother palms and raising of seedlings by tissue culture techniques should be preferred.

PLANT PROTECTION

Coconut is susceptible to the attack of various diseases and insect-pests. The major pests to coconut in India are rhinoceros beetle, red palm weevil, leaf-eating caterpillar and rats and the major diseases are bud rot, root rot, thanjavur wilt/ganoderma, tatipaka, leaf rot, stem bleeding and crown chocking.

A. Major Diseases and their Management

Bud Rot

This is one of the most fatal diseases of coconut, affecting palms of all ages, but the young palms are more susceptible. It is serious in palms growing in marshy, and water-logged areas, and in environment having high humidity. It attacks palms during monsoon season when relative humidity is high and temperature is below 24°C. It is caused by *Phytophthora palmivora*.

Symptoms

The initial symptoms of the disease include the discoloration and yellowing of the youngest leaf, withering of the central spindle of the crown, rolling and death of growing point *i.e.*, terminal bud and the surrounding tissues. The tender leaf bases degenerate into a slimy mass of putrified material, emitting foul smell that attracts flies. When the base of the bud is badly affected, the palm finally succumbs. Bud rot is highly infectious and is spread by the wind during rainy season.

Management Practices

Following measures should be adopted to keep this diseases under control:

1. Remove the affected parts and destroy them.
2. Regular spraying with 1 per cent Bordeaux mixture just before and after monsoon to adjacent healthy palms is an effective preventive or prophylactic measure.
3. If the disease is detected in the early stage, apply Bordeaux paste on the crown, after thorough cleaning and removal of infected material.
4. The treated portion should be given a protective covering with polyethylene sheet to prevent 'washing off' of the paste.

Root rot or Ganoderma Wilt

This disease was first recorded in coastal areas of Thanjavur district of Tamil Nadu, India after cyclones in 1952 and1955, hence it is also called as Thanjavur wilt. However, now it is present in all the coconut growing areas of India. A root-infecting fungus, *Ganoderma lucidum*, causes it. This malady is largely seen in sandy and sandy-loam soils; waterlogged soils or under ill-drained conditions and in neglected orchards or orchards having improper sanitation.

Symptoms

The fungus makes entry to the trunk and the symptoms become visible on the crown. The characteristic symptoms of the malady are withering and drooping of the older leaves, which remain hanging around the trunk for several months. The new leaves become reduced in size and yellowish in colour, and the inflorescences become suppressed and the palms remain barren. The crown size is reduced and stem becomes tapering. Excessive shredding of nuts may also take place.

Management Practices

Following measures are helpful in reducing the incidence of root rot or Ganoderma wilt in coconut plantations:

1. Remove and destroy the diseased palms.
2. Bleeding patches should be chiseled completely, followed by hot coal tar application.

3. Isolation trenches (1cm x 30 cm) may be dug up around the diseased palms to prevent root contact between the palms.

4. Cultural practices should be followed judiciously. For example, repeated ploughing in the affected areas may be minimized, closer planting should be avoided, good drainage to be provided, adequate irrigations need to be given, mulching or green manuring should be practiced at the basin of the palms.

5. Application of 5 kg neem cake along with sufficient organic matter and 500 g phosphate + 1,200 g potassic fertilizer may be applied per palm per year, so as to keep proper nutrients in the soil.

6. Drenching with 40 l of Bordeaux mixture (1 per cent) and stem injection or root feeding of 2 g aureofungin solution + 1 g copper sulphate in 100 ml water, thrice a year, for only one year may reduce intensity of this malady.

7. Raising banana as intercrop wherever irrigation is possible.

Stem Bleeding

This is also a serious disease of coconut of all ages, but younger plants are more susceptible. It is caused by a fungus, *Thielaviopsis paradoxa*, though many other factors, like high water table, excessive acidity or alkalinity of soil are also associated with the causation of this malady.

Symptoms

The typical symptom of the disease is the exudation of a dark reddish-brown fluid through the cracks of the outer tissues or wounds of lower parts of the trunk. The tissues inside the trunk are decayed and they dry up soon. In advanced stage, the bark peeling leads to the formation of cavities on the stem. The crown shows reduced growth; as a result, the yield of the palm declines. In severe cases, palms may die.

Management Practices

Following measures have been recommended to keep this disease under control:

1. The rotting tissues should be cut away with a sharp chisel and the exposed surfaces should be painted with hot coal tar or 10 per cent Bordeaux paste.

2. Cavities should be filled up with cement for reinforcing.

3. Since the pathogen is a wood parasite, mechanical injury to the infected palms should be avoided.

4. The organic matter content of the soil should be increased with proper application of FYM or other amendments. Similarly, application of neem cake @ 5 kg/palm is beneficial.

5. Improvement in drainage and soil conservation in drought areas is useful and essential.

6. Apply bavistin (0.2 per cent) to the soil, to check the spread of the fungus.

Leaf Rot

Leaf rot is a serious disease of coconut palm, which is caused by *Bipolaris halodes*. However, some scientists believe that the fungi like, *Gliocladium roseum* and *Gloeosporium* sp. are also associated with leaf spot of coconut.

Symptoms

The visible symptoms of the disease are blackening and rotting of leaves and shriveling of the leaflets of the inner whorls of younger leaves. On drying up, the affected parts are 'blown off' easily by the wind. The affected leaves consequently attain 'fan-like' appearance. Dark-brown patches appear on the tender leaves, which slowly enlarge in size. The inflorescences fail to develop properly, causing the palms barren or there is considerable decline in yield.

Management Practices

Following measures have been considered absolutely necessary for the control of leaf rot of coconut:

1. Remove and burn the affected plants, specially the bottom-most leaves, which are severely affected.
2. Ensure proper drainage in the plantations.
3. Apply adequate quantity of manures and fertilizers.
4. Follow adequate cultural practices in the plantations.
5. Spray Bordeaux mixture (1 per cent) or dithane-M-45 (0.3 per cent) or copper oxychloride (0.5 per cent) in a sequential manner at quarterly interval, starting from January.

Leaf Blight or Gray Leaf Spot

Leaf blight or gray leaf spot disease of coconut is caused by *Pestalotia palmarum*, which is widespread in the major coconut growing countries of the world. The incidence of the disease is usually influenced by the nutrient status of the soil and the palm as well. Young palms are mostly susceptible to this disease if soils are deficient in potash and rich in nitrogen.

Symptoms

Leaf spot symptoms develop only on the mature leaves in the form of small yellowish-brown spots on the leaflets, which gradually become oval in shape with a grayish band. The center of the spots subsequently turns grayish white and the band darkens, surrounded by a yellow-green ring. In advanced stage, the affected portion of the leaflets shows a burnt or blighted appearance.

Management Practices

Following control measures have been suggested to be useful for the control of this disease:

1. Spray Bordeaux mixture (1 per cent) during the expected attack of this disease.

2. Ensure proper drainage in the plantations.

3. Remove and burn the affected leaves.

4. Apply recommended doses of manures and potassic fertilizers.

5. Spray with Bordeaux mixture (1 per cent) during pre-monsoon period.

Fruit rot or 'Mahali'

Fruit rot is also called as 'Mahali'. It is caused by the fungus, *Phytophthora palmivora* during monsoon season. It affects immature as well as mature nuts. Dropping of buttons becomes more virulent after the rains when the atmospheric humidity is high and the temperature is comparatively low.

Symptoms

Shedding of female flowers and immature nuts in large number are the major symptoms of the disease. Initially, a water-soaked area develops near the flower and fruit stalk during the monsoon period. The flowers and fruit will appear dark-green at first and in course of time, will turn brownish and appear as depression due to decay of underlying tissues. Later, such flowers and nuts fall down. The fruit rot sometimes extends into the husk and further into the kernel cavity.

Management Practices

1. Spray the young bunches with 1 per cent Bordeaux mixture, during the pre and post-monsoon periods.

2. Irrigation during summer months is particularly important for retention of buttons.

3. Regular manuring and proper cultural practices should be followed.

Tatipaka Disease

It is a serious disease of coconut in Andhra Pradesh. However, the diseases can be kept under control by adopting the recommended package of practices. Stem bleeding is characterized by dark gummy exudation from the trunk. This disease is caused by fungus, *Thilaviopsis paradoxa*. Control measures include chipping off the infected tissues, wound dressing and root feeding with 5 per cent calyxin.

Crown Chocking

Crown chocking is commonly observed in Assam and West Bengal. It is characterised by emergence of shorter leaves with fascinated and crinkled leaves. In many cases, it gives a choked appearance to the frond. Ultimately, the affected palm dies. Application of 50 g borax at half-yearly intervals (Feb-Mar and Sept-Oct) along with recommended fertilizers in the basins will control the disease when it is in the early stage.

B. Major Insect-Pests and their Management

Red Palm Weevil (*Rhynchophorus ferrugineus*)

Red palm weevil is one of the most destructive pests of coconut, oil palms and ornamental palms and is reported from all coconut growing regions of the country. Young coconut palms between the age group of 5-20 years are worst affected by this pest. The weevils are attracted to the rotting smell emmitting from the palms having cut injury, rhinoceros beetle damage, leaf rot and bud rot diseases.

Adult weevil scoops out small cavities on the injured portion of the palm near the crown or soft leaf axils and lays eggs. It breeds inside the palm and the grubs cause damage inside the stem or crown by feeding on soft tissues and often cause severe damage when large number of weevils bore into the soft growing parts. In case of severe infestation, the inside portion of the trunk is completely eaten and becomes full of rotting fibres and ultimately the plant dies. Since the pest remains inside, it is difficult to identify the infested palm in early stages of attack. However, the diagnostic symptoms of infestation are the presence of holes on the palm trunk with chewed up fibrous material extruding outside, wilting of inner middle whorl of leaves and splitting of leaf bases. In case of young palms, the top withers, while in older plants, the top portion of the trunk bends and ultimately breaks at the bend. Sometimes, the gnawing sound produced by the grubs feeding inside the trunk may be audible. Recently, the use of an electronic sensor has been suggested for detecting the presence of grubs inside.

Management Practices

It is very difficult to kill the pest as it remains hidden inside the palm trunk. However, the following measures have been suggested to be very important for the control of this pest:

1. Practice clean cultivation. Cut and remove damaged palms and decaying stumps in the garden. Such palms should be split open and the different stages of pest inside be destroyed.
2. Avoid injury to the trunk as the pest lays eggs in these wounds. Wounds should be treated with a mixture of carbaryl/endosulfan and soil.
3. While cutting leaves, retain at least 1 m of petiole.
4. Clear and plug the holes with coal tar and some insecticide.
5. Attract weevils to insect trap prepared by smearing the longitudinally split coconut stems with coconut toddy and a few grains of yeast. Then, destroy the trapped weevils daily.
6. The food baited pheromone traps (loaded with ferrugineol) can be used for attracting weevils, kill the collected ones using bucket trap. In such areas, young plants around the leaf axils should be treated with endosulfan (0.07 per cent) to prevent egg laying by the attracted weevils.
7. Injection of carbaryl (1 per cent) to the infested trunk through a hole by using a funnel is very useful to control weevil infestation. Then, this hole

should be plugged with mud. If need be, repeat the treatment after one week.

Black Headed Caterpillar (*Opsina arenosella*)

Black headed caterpillar is a known pest of coconut palms in India, Sri Lanka, Bangladesh and Myanmar. This pest assumes serious proportions periodically in coastal and back water areas and in the vicinity of water bodies in the internal parts of peninsular India. Leaf damage is caused by the caterpillars, which live in galleries on the under surface of leaflets. The larvae feed voraciously on green matter of the leaves and the infested leaflets slowly turn brown, showing characteristic scorched appearance. In badly infested palms, the functional leaf surface is considerably reduced and thereby reducing nut yield appreciably. There is reduction in photosynthetic activity, especially of lower fronds, which affects the nut development particularly in lower branches. Severity of the attack is seen during January-to-May. The older leaflets are preferred and infested trees can be recognized from a distance due to lower pale-brown damaged leaves. Heavily infested gardens present a burnt appearance.

Management Practices

1. Biological control is quite effective against this pest through the use of larval parasitoid, *Goniozus nephantidis*, prepupal parasitoid, *Elasmus nephantidis* and pupal parasitoid, *Brachymeria masatoi*. The release of these bioagents is made at fixed norms of 20, 49 and 32 per cent per 100 larvae, prepupae, and a pupae, respectively at fortnightly intervals, depending on the pest population in the field.
2. In case of severe attack, cut the affected leaves and destroy them by burning.
3. Spray malathion (0.05 per cent) or dichlorvos (0.02 per cent) on the lower surface of leaves.

Coconut Eriophid Mite (*Aceria guerreronis*)

In India, this mite was reported for the first time in 1998 from Ernakulam (Kerala) and in a very short spell, it has spread to be Tamil Nadu, Karnataka and Andhra Pradesh. This mite is widely distributed in South America, Brazil, Caribbean Islands, East Africa and Sri Lanka. The damage is found to the maximum during dry periods. The mite infests buttons during early stages of growth soon after pollination. Developing nuts harbour a large population of mites under the perianth. Appearance of white longitudinal patches just below the perianth and their development through triangular, patches are the early symptoms of pest infestation. As the nut grows, these patches turn brown and longitudinal fissures and wartings appear on the nut surface. There is drying and shedding of buttons and young nuts. The mite neither affect the female flowers prior to fertilization nor near the maturity of the fruit. The initiation of mite infestation occurs on nuts of 4-6 weeks age and symptoms are observed in young nuts at an age of 8-12 weeks.

Management Practices

Control of eriophid mite is very difficult because of its cryptic breeding behaviour beneath the tightly pressed bracts. However, the following practices are quite beneficial:

1. Appreciable control can be achieved by stem injection and root feeding techniques with monocrotophos, triazophos, methyl demeton and cyhexatin. Endosulfan (0.1 per cent), dicofol (0.1 per cent) and carbosulfan (0.05 per cent) are also effective against this mite. Owing to short life and high fecundity of mites, use of chemical alone for their management is less useful. Moreover, spraying should cover perianth area as there is very less probability of reaching the spray fluid at target area.

2. Some predatory mites like *Bedelle distincta* and *Amblyseius* spp., and a tarsonemid may be exploited for management of coconut mite.

3. Field application of mite specific strain of fungus, *Hirsutella thompsonii* is very effective.

4. Varieties with tight tepal remain almost free from mite infestation and must be encouraged in mite infested areas.

Coreid Bug (*Paradasynus rostratus*)

The coreid bug is a nut crinckler pest. Besides coconut, it also infests cashew, cocoa, guava, passion fruit and tamarind. The adult and nymphs suck sap from female flowers, buttons and tender nuts and thereby results in the production of malformed or crinkled nuts, having characteristics crevices on the lower side of perianth. Later, exudation of gum also takes place from the crevices. Severe infestation may result in the development of barren nuts.

Management Practices

1. Spray endosulfan (0.1 per cent) or carbaryl (0.05 per cent) on the fully opened inflorescence after the receptive stage of female flowers.

2. Placement of phorate 10G @5g/palm in two polythene sachets in the stalk region of the youngest two bunches also provides effective protection from this pest.

3. Management of the pest on collateral host plants is important to reduce infestation on coconut.

Coconut Weevil (*Diocalandra stigmaticollis*)

Coconut weevil attacks coconut palm, date palm and other palms, causing significant losses in coconut growing regions of world. Both grubs and adults are harmful, which attack all parts of the coconut palm, especially roots, leaves and fruit stalk. The larvae burrow into the soft tissues of the bark. In infested palms, the production of leaf and spathe is delayed. Button shedding occurs and setting is affected up to 75 per cent. The adult weevil infests the cut and wounds on the palm trunk and petiole. Tunnelling by the weevils within the leaf petiole that is close to the trunk, results in loosening of fibres and the petiole loses its grip and hangs from the trunk.

Management practices

1. Collect and destroy the affected palms.
2. Swabbing the trunk with quinalphos (0.05 per cent) after removing the hanging petioles gives effective control of this pest.
3. Spray carbaryl (0.01 per cent) or fenitrothion (0.05 per cent) to control this pest.

Rhinoceros Beetle (*Oryctes rhinoceros* (Linn.)

This is a ubiquitous pest of coconut palm causing considerable yield losses. *O. rhinoceros* attacks all types of palms (oil palm, palmy palm and wild date palm) besides attacking pineapple and banana. Isolated trees of irregular heights are more prone to attack. Peak adult emergence of beetles is observed during June-September after the pre-monsoon showers in the west coast of India. The beetles are normally very active in the early hours of the night. Adult beetles usually cause damage, which bore into the tender fronds and spathes, cutting and chewing the soft tissues of unopened leaves and inflorescences. The characteristic visible symptoms of damage are the occurrence of small limps of chewed up fibrous material at the entrance of the bored holes, and fan like appearance of cut off leaflets. When such fronds unfurl, holes in the axil of leaves are also seen. In severe cases of attack, the shape of the crown is disfigured. The affected leaves on emergence present a characteristic geometric cut on the body. As a result of severe damage, the yield of the palm is adversely affected.

Management Practices

1. Orchard sanitation checks the pest development. Burn all dead trunks, leaves and logs.
2. Mechanically hook out the beetles from the attacked palms.
3. As a prophylactic measure, fill up the topmost three leaf axils with sevidol 8G (25g) + fine sand (200g), thrice in April, September and December, or place 10 g naphthalene balls in the leaf axils and cover it with fine sand.
4. Spraying carbaryl (0.1 per cent) in the breeding sites of the beetles helps to kill the larvae.
5. Adult beetles can be attracted and trapped in castor cake suspension (1 kg castor cake and 5 litre water) placed in earthen pots.
6. Biological control can be achieved by using the baculovirus of *Oryctes*, which reduces longevity and fecundity of adult beetles and kills grubs in 15-20 days. Incorporating *Metarhizium anisopliae* (Metch.) in breeding sites of the pest is also helpful.

White Grub (*Leucopholis coneophora* Burm.)

The adults feed on the leaves, which show peculiar stepwise pattern on the cut edges. Grubs cause damage, which remains inside the soil and feeds on the roots of palm. Continuous feeding by the grubs on mature palms results in yellowing of leaves, premature nut fall, delayed flowering, retardation of growth and reduction in yield. The infested palms present a sickly appearance and there will be shedding of buttons. Peak grub population is present in the coconut basins during September.

Management Practices

Following measures have been suggested to control white grubs in coconut:

1. Deep ploughing during pre and post-monsoon periods exposes the grubs to predators.
2. Setting up of light traps to collect the beetles and their destruction during peak period of emergence.
3. Apply phorate 10G @ 100g/palm during May-June and September-October to reduce the population of grubs in soil.

C. Physiological Disorders and their Management

Leaf Scorch Decline

This disorder has been encountered in coconut plantations of Sri Lanka. The characteristic visible symptom of the Leaf Scorch Decline (LSD) disorder is leaf scorch that starts from the tips of the leaflets and advances towards the midrib of the frond. Scorch first appears on the physiologically less active mature fronds (mild stage of LSD), and advances progressively to the younger fronds of the middle whorl of the canopy (moderate LSD) and to still younger fronds (severe stage). Obstructions to the movement of water, by physical blockages in the vascular system, is a possible cause of the Leaf Scorch Decline (LSD) disorder of coconut palms. The other physiological disorders, the causes of which until now could not be ascertained, are bristle top, dry bud rot, finschafen disease, Malaysia wilt and frond rot.

HARVESTING AND YIELD

The tall varieties of coconut start flowering from 5th year and the dwarfs and hybrids flower from 3rd to 4th year. The average yield of coconut per tree per year is 44 nuts. With scientific management, the West Coast Tall yields on an average 80 nuts/palm/year in coastal Kerala and Karnataka and the hybrids yield from 100 to 140 nuts/palm/year

Coconut palm can produce 12 inflorescences/year, but generally number of inflorescence per palm are less than 12. In the West Coast of India, 6-12 harvests per year are usually taken. Coconut ripens in about 12-13 months after opening of the inflorescence. For obtaining maximum yield of copra and oil, only fully mature nuts should be harvested. By harvesting immature nuts, copra loss is to the extent of 6–33 per cent and oil loss is 5-33 per cent. Superior golden brown quality fibre with elastic and good tungsten strength is obtained from 10 month old nuts.

In scientifically managed garden, there is a possibility of monthly harvest but in many places in Kerala, coconut harvesting is done once in 45 days. Harvesting is done by experienced climbers by using a rope ring round the ankles. The climber examines the stages of maturity of nut by tapping and the mature bunches are cut and lowered down. In recent years, climbers are difficult to find and harvesting is done with the help of poles particularly in young plantations. A climber can harvest a maximum of 50 palms/day. With poles, 250 palms can be harvested. The harvested nuts are stored in heaps under shade for few days since the stored nuts are easy to husk and the moisture content of the meat decreases and thickness of the meat layer increases. However, storage is beneficial only when fully mature nuts are harvested.

POSTHARVEST MANAGEMENT AND PROCESSING

Of the total production of coconuts, about 5 per cent is consumed in the tender form for drinking purposes. The rest is utilised as mature nuts for household and religious purposes and for the production of edible copra, milling copra and desiccated coconut. Coconut oil production in the country is nearly 4.5 lakh tonnes. Of this, 40 per cent is consumed for edible purposes, 46 per cent for toiletry uses and 14 per cent for industrial uses.

Dry Processing

Coconut is converted into copra and coconut oil. Coconut husk is used for the manufacture of coir mat, cushions, etc. Two forms of copra is manufactured, edible copra and milling copra. Major portion goes for the preparation of milling copra. Edible copra has two forms; (1) ball copra and (2) edible quality cup copra. Ball copra is produced by storing fully mature unhusked nuts for 8 to 12 months on a raised platform usually made of bamboos. When the water dries out, the nut is dehusked and the shell is carefully broken to remove the copra in a ball form. For preparation of edible quality cup copra, fully mature nuts are stored for long periods and selected nuts are dehusked, cut into cups and dried in open sun until good quality copra is obtained. Cup copra is used for household edible preparation in northern states of India.

Preparation of milling copra is the most popular coconut processing activity particularly in southern states. In Kerala alone, 60 to 65 per cent of the coconut produced is converted into milling copra. The common method adopted for making milling copra is sun drying often combined with kiln drying during the monsoon period. A number of economically feasible models of copra dryers using sunlight, farm wastes as fuel and even electrical dryers have been adopted. Various capacity dryer models are being fabricated and marketed by Kerala Agro Industries Corporation.

About 55 per cent of the copra output in Kerala is used for oil milling. Rotary and Expeller mills are used for copra crushing. Rotary mills have a product recovery of 62.5 per cent oil and 35 per cent cake. In expeller mills, the product recovery is 64 per cent oil and 32 per cent cake. The coconut oil is used for household edible and toiletry purposes in crude form. Coconut oil is also refined in small quantity for use in food industries, perfumery and toiletry. Desiccated coconut is prepared is small scale units in Karnataka. Desiccated coconut is a partially defatted product and also yields superior quality coconut oil.

Coconut Byproducts

The processing of husk into coir and coir products is very popular in Kerala and coastal Tamil Nadu and Andhra Pradesh. White and brown fibres are produced from coconut hust. In Kerala, white fibre is made by natural retting of husk. In other states, only brown fibre is manufactured. Coir yam, door mats, matings, rugs, etc., are made from white fibre. Brown fibre is used for making brushes, brooms, mattresses, upholstery, cushions etc.

Wet Processing of Coconut

Coconut Cream

Coconut cream is the concentrated form of milk extracted from fresh matured coconuts. This is an instant product, which can either be used directly or diluted with water to make curries, sweets, desserts, puddings, etc. It can also be used in manufacturing bakery products and flavouring foodstuffs. Processed and packed coconut cream has shelf-life of 6 months. Once opened, it should be stored in refrigerator for subsequent use.

Desiccated Coconut Powder

Desiccated Coconut Powder (DCP) is obtained by drying ground or shredded coconut kernel after the removal of brown testa. It finds extensive use in confectioneries, puddings and many other food preparations as a substitute to raw grated coconut. A good quality DCP should have moisture 2-3%, fat 65 - 68%, solids not fat (SNF) = 30 – 32%. This is a very commercial product having demand in confectionery and other food items world over. Sri Lanka and Philippines are major DCP producing countries. In India the product is manufactured by small scale units scattered in Karnataka, Tamil Nadu, Kerala and Andhra Pradesh.

Coconut Milk Powder

It is prepared by spray drying the coconut milk along with homogenizer and emulsifier and without any preservative. It is very hygroscopic in nature.

Virgin Oil

For preparation of virgin oil, coconut milk is filtered and concentrated and then cream is separated by centrifugation. The cream is stirred vigorously to get the virgin coconut oil by a process called phase inversion. The oil thus obtained is very clear, nutritious and has longer shelf-life.

Medium/Low-Fat, Desiccated Coconut

After extraction of milk, the residual coconut cake can be dried and sold as medium/low fat, desiccated coconut, which may be used in bakery and preparation of low-calorie foods.

Nata-de-Coco

Nata-de-coco a cellulosic white to creamy yellow substance formed by *Acetobacter aceti* subspecies Xylinium, on the surface of sugar enriched coconut water / coconut milk / plant extract / fruit juices or other waste materials rich in sugar. It is popularly used as a dessert. It is also used as an ingredient in other food products, such as ice cream, fruit cocktails, etc. It is mainly produced in Philippines. In India, it is produced at laboratory scale but not commercially.

For the preparation of Nata-de-Coco, initially coconut water is strained and mixed with sugar and glacial acetic acid in stipulated proportions. Boil for ten minutes and cool. Add the culture solution and distribute the mixture in wide mouthed glass or plastic jars, cover the jar with a paper or a thin cloth to protect from dust. It is then kept aside undisturbed for two to three weeks. After this period, the white jelly like thick surface growth is harvested, washed thoroughly to remove all the acids and sliced into cubes. It is then immersed in flavoured sugar solution, again boiled and packed in glass jars or retortable pouches, sterilized and sealed.

Tender Coconut Water

The water of tender coconut, technically the liquid endosperm, is the most nutritious wholesome beverage that the nature has provided for the people of the tropics to fight the sultry heat. It has caloric value of 17.4 per 100gm."It is unctuous, sweet, increasing semen, promoting digestion and clearing the urinary path," says Ayurveda on tender coconut water (TWC). The medical properties of TWC are as under:

☆ Good for feeding infants suffering from intestinal disturbances.

☆ It is a good oral rehydration medium and keeps the body cool.

☆ Application on the body prevents prickly heat and summer boils and subsides the rashes caused by small pox, chicken pox, measles, etc.

☆ Cures malnourishment, checks urinary infections and very effective in the treatment of kidney and urethral stones.

☆ Excellent tonic for the old and sick, and can be injected intravenously in emergency case.

☆ Found as blood plasma substitute because it is sterile, does not produce heat, does not destroy red blood cells and is readily accepted by the body.

☆ Aids the quick absorption of the drugs and makes their peak concentration in the blood easier by its electrolytic effect.

The major chemical constituents of coconut water are sugars and minerals and minor ones are fat and nitrogenous substances as given below (Table 10.7):

Coconut Cheese

Coconut milk is allowed to stand for 8hr. until the cream is collected at the top. The cream is slowly scooped out and the skimmed milk heated with vinegar to coagulate the proteins. The coagulated protein is mixed with the cream and kneaded with salt.

Coconut Syrup

After homogenination of coconut milk, an equal quantity of sugar and 0.05% citric acid are added and steam cooked to a total soluble solids content of 65-68%. The boiled hot syrup is poured in to lacquered tin cans, sealed and then cooled under running water.

Table 10.7: Analysis of Mature and Tender Coconut Water

Content	Mature Coconut Water	Tender Coconut Water
Total solids (%)	5.4	6.5
Reducing sugars (%)	0.2	4.4
Minerals (%)	0.5	0.6
Acidity mg (%)	60.0	120.0
pH	5.2	4.5
Potassium (mg)	247.0	290.0
Sodium (mg)	48.0	42.0
Calcium (mg)	40.0	44.0
Magnesium (mg)	15.0	10.0
Phosphorous (mg)	6.3	9.2
Iron (mg)	79.0	106.0

Coconut Honey

To the coconut milk, 60% of brown sugar and 30% of glucose are added by weight and then boiled in steam-heated containers until a thick consistency is reached. The product is then hot filtered in lacquered tin containers or bottles and sealed. The final products is a golden coloured, thick paste with a nut flavour. This can also be used as an excellent base for soft drinks.

Intensified research efforts are required to diversify the products through innovative production technology. Coconut Development Board has sponsored research for the development of coconut powder, cream, vinegar, preservation of tender coconut water etc.

CROP IMPROVEMENT

Objectives of Coconut Breeding

Research on coconut improvement is mainly aimed at identifying superior cultivars and at developing hybrids, which can give higher nut and oil yield, besides possessing short stature, precocity and resistance to biotic and abiotic factors. In coconut, natural populations consist of 2 morphological forms, the talls and dwarfs. While talls are characterised to grow 25-30 m high, have a pre-bearing period of 6-10 years and are largely cross-pollinated and heterozygous in nature. Dwarfs are short in stature, early bearing (3-4 years) and predominantly self-pollinated due to total overlap of male and female phases. While making selections or developing hybrids, the characteristics of these 2 traits need to be looked into.

Breeding Methods

Selection

Based on the preliminary evaluation of indigenous and exotic cultivars at Kasaragod and subsequently in research centres in Andhra Pradesh, Tamil Nadu, Kerala, Maharashtra, cultivar Lakshdweep Ordinary (LO) was selected and released by CPCRI for commercial cultivation in these 4 states under the name Chandarkalpa. Subsequently, another cultivar, Bhanwali Green Round from Goa region was selected and released in 1987 by Konkan Krishi Vidyapeeth, Dapoli for cultivation in Konkan Coast under the name of Pratap. Similarly, the Agricultural Research Station, Ambajipeta, East Godavari, Andhra Pradesh has released another selection of Philippines Ordinary in the name of Double Century at AICRIP workshop held during 1995. A selection of Chowghat Orange Dwarf was also released during 1991 for tender nuts.

Mother palm selection is the most important factor for coconut improvement. In India, palms yielding 80 nuts and more with a copra out turn of 20 kg/palm/year consistently over a period are usually considered as mother palms. It is now recommended that selection process should also be exercised towards weight of copra in addition to yield of nuts. Studies on path coefficient and regression analysis have indicated that average number of female flowers, number of functional leaves in the crown, internodal distance at fixed mark, total leaf production up to 3 years after sowing and percentage nuts set and time taken for flowering are important components, showing are the largest direct influence on yield and thereby indicating their value in selection programme.

Hybridization

Tall x Dwarf Hybrids

The discovery of hybrid vigour in coconut between the West Coast Tall (female) x Chowghat Green Dwarf (male) crosses made at Nileshwar Coconut Research Station (now in KAU) is a significant landmark in the history of coconut improvement. The comparison of 25 years old Tall x Dwarf hybrids planted during 1935-36 at Nileshwar showed that they were early bearing, high yielding and attained steady bearing earlier than tall parents with higher number of functional leaves and rate of leaf production. Based on their studies with different Tall x Dwarf hybrids suggested proper choice of parents for efficient exploitation of hybrid vigour in coconut. Later, studies by using the 2 common dwarf parents, namely, Chowghat Orange Dwarf (COD) and Chowghat Green Dwarf (CGD), as male parents, indicated that COD is a better pollen parent than CGD for the production of promising hybrids.

For the first time, two coconut hybrids, viz., COD x WCT (Chandra Sankara) and LO x COD (Chandra Laksha) were released during 1985. Later, three more hybrids from Kerala viz., LO x GB (Laksha Ganga), WCT x COD (Kera Sankara), AO x GB (Andhra Ganga) and WCT X GB (Kera Ganga) and ECT x DG (VHC 1) and ECT x MYD (VHC 2) from Tamil Nadu and ECT x GB from Andhra Pradesh were released based as the higher yield performance than local WCT and ECT. The comparative performance of coconut hybrids developed in India are given in Table 10.8.

Dwarf x Tall Hybrids

The Dwarf x Tall hybrids produced by controlled pollination between COD as the female parent and WCT as the male parent were markedly superior both in nut production and copra outturn. A survey conducted in Kerala during 1978-79, also indicated that among Tall x Dwarf and Dwarf x Tall hybrids, the latter were superior. Although, there is a good response and high demand of these hybrid coconut seedlings, in view of their superior performance, one of the apprehensions among the coconut growers is the shorter economic life-span of these hybrids. However, studies carried out at different research stations have shown that Dwarf x Tall and Tall x Dwarf hybrids continued to give heavy yields even after 49 years.

Table 10.8: Comparative Performance of some Coconut Hybrids Developed in India

Hybrid	Nut Yield/ Palm/Year	Copra Yield (g) Mean/Palm/Year		Oil Content (per cent)	State for which Recommended
Chadra Sankara (COD x WCT)	116	215	24.9	68	Kerala, Karnataka and Tamil Nadu
Kera Sanskara (WCT x COD)	108	187	20.2	68	Kerala, coastal Maharashtra and coastal A.P.
Chandra Laksha (LO x COD)	109	195	21.2	89	Kerala and Karnataka
Laksh Ganga (LO x GB)	108	195	21.2	70	Kerala
Anand Ganga (AO x GB)	95	216	20.5	68	Kerala
Kera Ganga (WCT x GB)	100	201	20.1	69	Kerala
Kera Sree (WCT x MXD)	141	216	–	–	Kerala
VHC-1 (ECT x DG)	98	135	13.2	70	Tamil Nadu
VHC-2 (ECT x MYD)	107	152	16.3	69	Tamil Nadu
ECT x GB	140	150	21.0	68	Andhra Pradesh
WCT (Large Tall)	80	176	14.1	68	Kerala

Encouraged by the superior performance of Dwarf x Tall and Tall x Dwarf hybrids, intervarietal crosses involving promising exotic and indigenous varieties were made at the CPCRI, Kasaragod, the KAU and four other research centres. These are under evaluation.

Other aspects, which are receiving attention of the breeders, are breeding for drought and disease resistance/tolerance and nut water quality. Accordingly, all the identified drought tolerant varieties are currently being utilized in the breeding programme at CPCRI, Kasaragod, to evolve high yielding hybrids, possessing drought tolerance. Similarly, programme of breeding for root wilt tolerance is also receiving priority. For this purpose, survey was initiated in the disease endemic area in 1988 and so far 95 disease tolerant WCT and CGD palms was identified in 1998 and the evaluation of the hybrids for tolerance to root wilt disease is progressing from 1991 onwards. These are now being utilized in the hybridization programme in hot spots.

Presently, 204 mother palms free from diseases have been used in the breeding programme.

Heritability

Heritability value estimated for nut yield was found to vary from 0.47 to 0.63. One novel method is the identification of pre-potent palms among the talls, which help in improving the yield substantially.

11

Pomegranate

Pomegranate is a delicious table fruit, rich in vitamins, minerals, carbohydrates and proteins. Pomegranate aril juice provides about 16 per cent of an adult's daily vitamin C requirement per 100 ml serving, and is a good source of vitamin B_5 (pantothenic acid), potassium and antioxidant polyphenols. The fruit is used as a special diet for sick and aged persons.

The most abundant polyphenols in pomegranate juice are the hydrolyzable tannins called 'punicalagins', which have very high free-radical scavenging properties. The presence of many seeds and of tannin in the rind and membranes detracts from its attractiveness. The tree can withstand considerable drought than several fruit plants.

COMPOSITION AND USES

Composition

Pomegranate is considered as a highly nutritious fruit. Its edible portion is about 68 per cent. It is a very good source of carbohydrates, potassium, calcium, phosphorus and vitamin C (Table 11.1).

Apart from being used as table fruit and for juice making, several value added products are made from pomegranate. Grenadine syrup is thickened and sweetened pomegranate juice used in cocktail mixing. Wild pomegranate seeds are sometimes used as a spice known as *anardana* (which literally means pomegranate (*anar*) seeds (*dana*) in Persian, most notably in Indian and Pakistani cuisine but also as a replacement for pomegranate syrup in Middle Eastern cuisine. The seeds are separated from the flesh, dried for 10–15 days and used as an acidic agent for *chutney* and *curry* preparations.

Table 11.1: Composition of a Ripe Pomegranate Fruit

Component	Per 100 g Edible Portion	Component	Per 100 g Edible Portion
Moisture	78 per cent	Riboflavin	0.10 mg
Carbohydrates	14.5 per cent	Penthothemic acid	12 mg
Proteins	1.6 per cent	Potassium	236 mg
Fat	1.2 g	Calcium	10.0 mg
Dietary fibre	4.0 g	Phosphorus	70.0 mg
Ascorbic acid	16.0 mg	Iron	0.3 mg

ORIGIN, HISTORY AND DISTRIBUTION

Pomegranate is native of the Mediterranean region. It is commercially cultivated in Iran, Afghanistan, Russia, Israel, North and Latin American countries, Africa and India. It is widely cultivated throughout India and the drier parts of South-East Asia, Malaya, the East Indies and tropical Africa. The tree was introduced into California by Spanish settlers in 1769. In this country, it is grown for its fruits mainly in the drier parts of California and Arizona. In India, it is cultivated in Maharashtra, Karnataka, Gujarat, Rajasthan and Andhra Pradesh. Maharashtra being the largest producer of pomegranate in India, had produced 4.92 million tones fruit during 2010-11 (Table 11.2). The other important pomegranate producing states are Karnataka, Gujarat and Andhra Pradesh. The highest productivity had been recorded in Tamil Nadu (27.6 MT/ha) and lowest in Maharashtra (6.0 MT/ha).

Table 11.2: State-wise Area, Production and Productivity of Pomegranate in India

State	Area (000'Ha)	Production (000'MT)	Productivity (MT/Ha)
Maharashtra	82.0	492.0	6.0
Karnataka	13.6	142.6	10.5
Gujarat	5.8	60.3	10.4
Andhra Pradesh	2.8	27.8	10.0
Tamil Nadu	0.5	12.7	27.6
Rajasthan	0.8	5.5	6.6
Others	1.8	2.2	1.2
Total	107.3	743.1	6.9

Source: National Horticulture Board Database-2011.

TAXONOMY AND BOTANICAL DESCRIPTION

Taxonomy

Pomegranate (*Punica granatus* L.) belongs to family Punicaceae. *Punica* being perhaps the only known genus of the family. The genus *Punica* has 2 species, *Punica granatum* (cultivated) and *Punica protopunica* (wild). *Punica granatum* has 2n = 2x = 16 and 18 chromosomes. The varieties like Dholka, Ganesh, Khandhari, Muscat White have 2n = 16 whereas Double flower, Vellodu, Kashmiri have 2n = 18.

Botany

The pomegranate is a neat, rounded shrub or small tree that can grow to 20 or 30 ft., but more typically to 12 to 16 ft. in height. Dwarf varieties are also known. It is usually deciduous, but in certain areas, the leaves will persist on the tree. The trunk is covered by a red-brown bark, which later becomes grey. The branches are stiff, angular and often spiny. There is a strong tendency to sucker from the base. Pomegranates are also long-lived. There are specimens in Europe that are known to be of over 200 years of age. However, the vigour of a pomegranate declines after about 15 years. The pomegranate has glossy, leathery leaves that are narrow and lance-shaped.

The attractive scarlet, white or variegated flowers are over an inch across and have 5 to 8 crumpled petals and a red, fleshy, tubular calyx, which persists on the fruit. The flowers may be solitary or grouped in twos and threes at the ends of the branches. The pomegranate is self-pollinated as well as cross-pollinated by insects. Cross-pollination increases the fruit set. Wind pollination is in-significant.

The nearly round, 2.5 to 5 inch wide fruit is crowned at the base by the prominent calyx. The tough, leathery skin or rind is typically yellow overlaid with light or deep pink or rich red colour. The interior is separated by membranous walls and white, spongy, bitter tissue into compartments packed with sacs filled with sweetly acid, juicy, red, pink or whitish pulp or aril. In each sac, there is one angular, soft or hard seed. High temperatures are essential during the fruiting period to get the best flavour. The pomegranate may begin to bear in 1 year after planting, but 2½ to 3 years is more common. Under suitable conditions, the fruit matures in 5 to 7 months after bloom. Botanically, its fruit is called as 'Balausta', whose edible portion is fleshy aril.

SOIL AND CLIMATIC REQUIREMENTS

Soil

It can be grown on diverse types of soil. Although, pomegranate is not specific to soil requirement, but it is sensitive to fluctuations in soil moisture, particularly during the fruit bearing stage. Loam soils with medium texture. having good moisture holding capacity is preferred. It can thrive well on comparatively poor soils where other fruits fail to grow. Pomegranate can also be grown in medium and black soils. It is considered as salt-hardy fruit plant.

Climate

Pomegranate can grow up to an elevation of 1,800 m above mean sea level. Under tropical and sub-tropical climate, it behaves as an evergreen or partially deciduous plant. It thrives well in semi-arid and arid regions, having marginal agroclimate. Warm and cool nights help in the development of good colour and sweetness in the aril. Under humid conditions, the sweetness of fruit is adversely affected. The tree requires hot and dry climate during the period of fruit development and ripening. The optimum temperature for fruit development is 38 °C. The tree can not produce sweet fruits unless the temperature is high for a sufficient long period. High vapour pressure deficit owing to high temperature and low humidity lowers fruit quality. High humidity coupled with high temperature makes it susceptible to diseases. Aridity and frequent anomalies of the climate cause leaf shedding and fruit cracking.

IMPORTANT CULTIVARS

The popular cultivars of pomegranate are Ganesh and Muskat in Maharashtra, Bassein Seedless in Karnataka, Dholka in Gujarat, Kabul Red and Vellodu in Tamil Nadu, and Jalore Seedless and Jodhpur Red in Rajasthan. Some promising clonal selections have now been made from these varieties, *e.g.*, G-137 from Ganesh, P-23 and P-26 from Muskat in Maharashtra, Jyothi from Bassein Seedless in Karnataka and Yercaud-1 and CO-1 in Tamil Nadu. Some promising hybrids such as Mridula, Ruby, etc. have also been developed. For *anardana*, seedling selections having high acidity have been made. Important characters of some hybrids and cultivars are as follows:

Ganesh

It is an improved variety known as GBG No. 1. Ganesh is a seedling selection by Dr. G.S. Cheema at Pune. It is a selection from Alandi and considered to be the best variety. The fruit is medium in size. It has soft seeds. Ganesh is a high yielding variety and is a good cropper. The flesh is pinkish and has juice with agreeable taste.

Alandi

Fruit is medium in size, fleshy testa, blood red or deep pink, aril with sweet but slightly acidic juice. Seeds very hard. The variety is named after the name of village where it was grown extensively.

Dholka

Fruit large sized, rind grenish white, fleshy testa, aril is pinkish white or whitish with sweet juice. Seeds soft. Juice is acidic. It is medium cropper and an important variety of Gujarat.

Kandhari

It produces large fruits. The rind is deep red. The aril is dark red or deep pink. The juice is slightly acidic. The seeds are hard. The variety is successfully grown in Himachal Pradesh.

Jalore Seedless

This cultivar is commonly grown in arid region. The fruits are large, reddish green with sweet (15.2 °Brix TSS) and pink aril and soft seeds.

Bassein Seedless

It has spreading growth habit, bears red fruits with light pink aril, soft seeds and 16.2 °Brix TSS. A selection from this cultivar, Jyothi, has been made at IIHR, Bangalore which produces fruits having deep red aril colour.

Muskat

This variety is also largely grown in Karnataka. Fruits small-to-medium in size. Rind in somewhat thick. Fleshy testa, with moderately sweet juice. The seeds are rosy in colour and fruits are tasty.

Mridula

Mridula is a hybrid between Ganesh and Gulsha Rose Pink. It has dark red fruit and blood red aril, soft seeds and high TSS.

Ruby

This is a hybrid developed at IIHR, Bangalore as a result of complex crosses between F_2 of Ganesh × Kabul × F_1 × Yercaud and Ganesh × Gulsha Rose Pink. The red coloured fruits are of about 250 g having dark red aril and soft seeds.

Phule Arakta

The 'Arakta' variety of pomegranate presently under commercial cultivation in various regions of Maharashtra. Pre-released in the year 1989, it has now been released as 'Phule Arakta' for its cultivation by the Mahatma Phule Krishi Vidyapeeth, Rahuri. It is heavy yielder and possesses desirable fruit characters. The fruits are bigger in size, sweet with soft seeds and bold red arils. It also possesses glossy, attractive, dark red skin. It is less susceptible to fruit spots and thrips.

Amlidana

Amlidana is an F1 hybrid (Ganesh × Nana). It grows well under tropical climate. With quality fruit attributes, Amlidana is superior to sour variety Daru, whose trees come up naturally in temperate regions of North India. Its fruits provide more acidic (16.18 per cent) *anardana* and higher fruit yield/tree. In addition, short-statured trees are suitable for high-density planting, giving increased fruit yield/unit area. Hence, its cultivation is recommended.

Goma Khatta

The variety has been developed for *anardana* purpose from Central Horticulture Experiment Station, Godhara. Yield potential is 7.0 kg/plant and *anardana* yield is 1.18 kg/plant. Seed hardness is medium. Fruit having 46.7 per cent of juice and TSS is 14.5°Brix. Acidity is 7.3 per cent.

Bhagava

The 'Bhagawa' variety of pomegranate, which presently occupy a large area under commercial cultivation, is known by different names *viz.,* 'Shendari', 'Ashtagandha', 'Mastani', 'Jai Maharashtra', and 'Red Daina'. It is a clonal selection and has been recommended for cultivation by the Mahatma Phule Krishi Vidyapeeth, Rahuri. This variety matures in 180-190 days with average yield of 30.38 kg fruits/ tree. Bigger fruit size, sweet, bold and attractive arils, glossy, very attractive saffron coloured thick skin makes it suitable for distant markets. This variety is less susceptible to fruit spots and thrips as compared to other varieties of pomegranate.

PLANT PROPAGATION

Pomegranate plants raised from seed vary widely and are undesirable. Thus, they must be raised vegetatively. Among the vegetative methods of propagation, cuttings are universally used for raising pomegranate plants on commercial basis. Semi-hard and hardwood cuttings treated with 1,000 ppm IBA as basal dip are employed for multiplication of pomegranate. The best time of making the cuttings is December-January when the plants shed their leaves. The cuttings made during September-October can also root satisfactorily. The cuttings are planted in polythene tubes filled with a mixture of soil, FYM and sand in equal proportion. Cuttings taken from the mature, 6-12 mm thick branches emerging from the base of main stem gives better rooting. For propagating by air layering (Gootee), application of 10,000 ppm IBA with lanolene paste at the upper side of the ring and tying with moist moss grass and white polythene gives better rooting. Ground layering is another method used for multiplication of pomegranate plants.

PLANTING AND ORCHARD ESTABLISHMENT

Land is prepared thoroughly and levelled prior to pit digging. The layout is done following square or hexagonal system. Planting density is the most important yield contributing factor, which can be manipulated to attain the maximum production per unit area. The optimum spacing is important for the maximum utilization of land and good income over a long period. Pomegranate is planted at 5 × 5 m spacing. Pits of 60 × 60 × 60 cm size are dug about one month before planting and filled with top soil, pond silt and FYM mixture in 1: 1: 1 proportion and adding 50 g methyl parathion for protection agains termites. Rainy season is the best time of planting. In northern India, planting can also be done towards the end of winter when the plants are in dormant condition. In southern India, where plants remain evergreen, onset of monsoon is the best time for planting. Under arid and semi-arid regions of Rajasthan, planting is done during rainy season.

High-density plantation (5 × 2 m) with higher water and fertilizer inputs (22 irrigations and 625 : 250 : 250 g NPK/plant) under semi-arid agroclimate of Rahuri gave nearly 2.5 times fruit yield in comparison to planting at 5 × 5 m spacing.

INTERCULTURAL OPERATIONS

Water Management

Although pomegranate is a drought hardy fruit plant but to obtain good yield and fruit quality, assured irrigation is essential. The newly set plants require regular irrigation so that the roots become well established and the plants can start growth. Regular irrigation is essential from flowering to ripening of fruits, as irregular moisture condition results in dropping of flowers and small fruits. Water requirement of pomegranate largely depends upon the desired *bahar*. For *ambe bahar*, 13 irrigations (7 cm) are considered enough for good growth and yield. For *mrig bahar*, 9 irrigations (54 cm) are found to be sufficient. In *ambe bahar* crop of Ganesh, regular irrigations from March to July at 7-10 days interval increased the fruit yield. In arid region, due to scarce irrigation resources, *mrig bahar* crop is preferred to take advantage of the moisture available during monsoon. If long dry spell occurs, irrigations may be required even for *mrig bahar* crop.

Drip irrigation on 20 per cent wetted area basis at alternate days improves water use efficiency. Fruit cracking can be minimized by regular irrigations at 75 CPE.

In low rainfall areas, *in situ* water harvesting by providing 5 per cent catchment slope has been found to increase growth and fruit yield. Practices such as placement of subsurface barriers, weed control and mulching by available organic waste materials and polythene, etc. help to conserve soil moisture. Water losses from plant surface can be minimized by planting shelter belts and use of antitranspirants such as spray of 10 per cent kaolin (a radiation reflectant), 10^{-5} M phenyl mercuric acetate (PMA), 1.5 per cent power oil and 1 per cent liquid paraffin wax.

Nutrient Management

Doses of manures and fertilizers for application in pomegranate orchard depend on the fertility status of soil. In northern India, manures are applied during February, whereas in other areas, manuring may be done just before the start of monsoon in case of young plants. The one-year-old tree should be manured with about 10 kg of farmyard manure and 150 to 200 g of ammonium sulphate. The amount is increased by the same amount every year so that a five-year-old tree gets 50 kg of farmyard manure and one kg of ammonium sulphate. Upto non-bearing stage, fertilizers are applied in three split doses during January, June and September. After fruit bearing starts, time of fertilizer application should suit the *bahar* to be taken. Generally, nitrogenous fertilizers are applied in two split doses, one at the time of first irrigation after *bahar* treatment and the second after three weeks of the first application. Full doses of phosphorus, potassic fertilizers and FYM are applied at one time after *bahar* treatment.

Deficiencies of micronutrients such as iron, zinc and boron are commonly observed in pomegranate orchards. These can be overcome by two foliar sprays of zinc sulphate (0.06 per cent), ferrous sulphate (0.4 per cent) and borax (0.2 per cent) separately or in combination during flowering and fruit set periods.

Training and Pruning

Pomegranate is bushy in growth habit and thus produces considerable number of shoots from the base. Retaining all these shoots would create crowding, leading to infestation by shoot borers. Since single stem training of trees takes the fruiting area too high, of 3 to 4 well spaced stems are kept at the ground level. Pomegranate bears fruits on terminal and maxillary short spurs, arising from the mature shoots and thus does not require regular annual pruning. However, water sprouts, diseased and pest affected or dried branches should be removed. Pomegranate does not usually require pruning except for removal of suckers, dead and diseased branches and developing a sound framework of the tree. It is essential to remove the suckers as soon as they arise. The fruits are borne terminally on short spurs produced all along the slow growing mature wood. These bear fruits for 3 to 4 years. Therefore, only a limited pruning of bearing tree is required. Annual pruning in winter during dormant period should be confined to shortening of the previous season's growth to encourage fruiting.

For getting a good crop, a set of new shoots should be allowed to develop every year on all sides of the tree and gradual growth of new shoots should be encouraged by restricted cutting back of the bearing shoots.

Intercropping

Intercropping in pomegranate orchard is highly desirable because it takes about 6 – 7 years to come to commercial bearing. During these initial years, intercropping with leguminous crops like cowpea, clusterbean, gram and vegetables like chillies, cabbage, cauliflower, peas, tomato, carrot, radish, brinjal, etc. not only gives additional income but also improves the soil physical properties. The intercropping should be stopped when the pomegranate trees attain full growth and start commercial yield. However, intercrops can be continued for another 3 to 4 years after the plants had started bearing. It is best to grow a green manure crop during the monsoon and burry it in the soil, when it has completed its vegetative phase and started flowering. Regular weed control is done by interculture operations or by application of herbicides such as two sprays with dimon (larmax) @ 2 kg/ha.

FLOWERING AND CROP REGULATION

Pomegranate has three main flowering and fruiting seasons or *bahars*, *ambe bahar* (spring season flowering), *mrig bahar* (June-July flowering) and *hasth bahar* (September-October flowering). Pomegranate flowers continuously when watered regularly. The plants under such conditions may continue bearing flowers and bear small crop irregularly at different period of the year, which may not be desirable commercially. For commercial production, only one crop in a year is desirable. Therefore, by crop regulation, the tree is forced to rest and thereby produces profuse blossoms and fruits during the required *bahar*. Selection of the *bahar* depends mainly on the availability of irrigation water, risk of damage by diseases and pests and market factors. In dry areas of north-western India, with limited irrigation resources, *mrig bahar* is preferred to utilize the water available during the monsoon period. As the fruits develop during the rainy season and mature during winter, the colour and sweetness of the fruit is affected. In irrigated parts of Maharashtra and Gujarat, *ambe bahar* and *hasth bahar* is preferred, respectively. In *ambe bahar*, the fruit development takes place during dry months, they develop an attractive colour and quality, which are suitable for exports. Similarly, due to dry weather, the incidence of pests and diseases is also limited. The operation, thus, maximizes production from the available inputs and also avoids fruiting during the period when insect-pests and disease infestation are common. However, *ambe bahar* can be taken only in areas having assured irrigation facilities. For *bahar* tratment, operations like withholding irrigations, root exposure, root pruning and spray of chemicals (thiourea, NAA or potassium iodide) are practiced to induce leaf drop and cessation of growth during the period of the unwanted *bahar*. In general, the irrigation is withheld for two months prior to the *bahar*, followed by light earthing up in the basin. This facilitates the shedding of leaves. The trees are then medium pruned, 40-45 days after withholding irrigation. The recommended doses of fertilizers are applied immediately after pruning and irrigation is resumed. This leads to profuse flowering and fruiting. The fruits are ready for harvest in 4-5 months after flowering.

In order to increase proportion of good size grade fruits, number of fruits on a tree is regulated to retain 50-60 fruits on one bush by hand removal or by chemical floral thinning or by spray of 2,000 ppm ethephon or 500-3,000 ppm Alar.

PLANT PROTECTION

A. Major Diseases and their Management

Bacterial Blight or Oily Spot

Bacterial leaf spot is a serious bacterial disease of pomegranate in certain areas. Initially, *Xanthomonas punicae* was identified as its casual agent, but later, the name was changed to *Xanthomonas campestris* pv. *punicae* and finally to *Xanthomonas axonopodis* pv. *punicae*.

Symptoms

The disease appears as few to numerous small deep-red spots of 2-5 mm diameter with indefinite margins on leaf blade. The affected leaves may be distorted or malformed, which may shed in severe infestation. On fruits, there is development of raised, dark-brown lesions of indefinite margins. Later, these spots turn dark-red. Symptoms may also appear on branches and stem. Cracking of the affected plant parts has also been observed in severe infestation.

Management Practices

Adopt following measures to control bacterial blight of pomegranate:

1. Destroy the affected plants or their parts.
2. Avoid injury to the plants.
3. Apply Bordeaux paste to the cuts after pruning.
4. Spray of copper fungicides provides some control. Paushamycin, streptocycline and K- cycline are effective in controlling the disease.
5. Botanicals, such as, extracts of *tulsi*, *patchouli* and *miswak* moderately effective against the disease.

Leaf Spots

Pomegranate is attacked by a number of leaf spot diseases, which are caused by *Colletotrichum gloeosporiodes*, *Fusarium fusarioides* and *Phomopsis ancubicola*, of which, *Colletotrichum* spot is very serious in Uttar Pradesh, *Fusarium* spot in Maharashtra and *Phomopsis* spot in Rajasthan.

   a. *Colletotrichum* **leaf spot**: Minute, dull, violet-black spots appear on leaves, which are surrounded by yellow region. Later, these spots coalesce and become bigger, causing extensive leaf drop in severe infestation. It can be easily controlled by spraying tillex (0.2 per cent) as and when the first symptom of the disease is noticed.

   b. *Fusarium* **leaf spot:** In this disease, there is development of circukar to irregular minute brownish specks towards the leaf margins. These spots coalesce to form big dark-brown necrotic blotch. It can be controlled by spraying difolatan or miltox (0.2 per cent).

   c. *Phomopsis* **spot:** Dark-brown to black spots appear on leaves near to their margins. In severe infestation, these spots coalesce and form lesions. The affected leaves may drop off. It can be controlled by spraying benlate (0.2 per cent).

Fruit Rot

Fruit rot sometimes cause great losses to the grower, if suitable control measures are not adopted in time. It is caused by a fungus, *Glomerella cingulata*.

Symptoms

Initially, fruits start developing 'off' colour on their lower part. After a few days, the affected area becomes brown-to-black, which later enlarges and half or the whole fruit may get rotten.

Management Practices

Following measures are considered necessary to keep fruit rot of pomegranate under control:

   1 Collect and destroy the affected/fallen fruits.
   2. Give 2-3 sprays of bavistin (0.5 per cent) or dithane M-45 (0.025 per cent) at fortnightly interval.

B. Major Insect-Pests and their Management

About 50 insect-pest species attack pomegranate, but *anar* butterfly or fruit borer is the obnoxious enemy, which alone is responsible for the failure of crop in certain areas. Other insect-pests, like bark eating caterpillar, aphids, thrips, scale insects, fruit fly, leaf defoliators etc., are of minor importance. A brief description about the major pests of pomegranate and their management has been given below:

Anar Butterfly [*Deudorix* (*Virachola*) *isocrates*]

It is the most destructive pest of pomegranate, which is widely distributed in

India and is found in every corner of the world wherever pomegranate is grown. It may destroy 50-80 per cent of the fruit in certain locations. It is a polyphagous pest, having a wide range of host plants. Although, pomegranate is the most preferred host, yet fruits of *aonla, ber,* tamarind, mulberry, guava, *loquat,* sapota, litchi, citrus, peach, plum, apple are also attacked. *D. epijarbas,* a related species is reported to damage pomegranate in Himachal Pradesh and Uttrakhand. The mode of feeding of both the lycaenid borers is almost identical and there is resemblance in their biology, behaviour and life history.

The damage is caused by the caterpillars, which on hatching bore into the developing fruits. They feed on seeds and arils and fill the fruits with excreta. The conspicuous symptoms of the damage are offensive smell and excreta of caterpillars coming out of the entry holes of the fruits. The excreta is found stuck around the holes. Such holes predispose the fruits to bacteria, fungi and scavenging insects and birds, which lead to rotting. Often the anal segment of the larva can be seen plugging the borer hole and at times more than 3-4 larvae can be seen feeding inside a single fruit. Later, the fungi or bacteria may attack the infested fruits. The affected fruits ultimately fall down and are unfit for human consumption even if these are harvested before falling down to the ground.

Management Practices

Following management practices are important to control *anar* butterfly effectively in the pomegranate:

1. Collect and destroy the fallen infested fruits to prevent the build up of the pest.
2. If numbers of fruit trees are limited, bagging of the fruits before maturity is advised to reduce the fruit damage. Bagging is very effective but laborious and expansive, and often resulting in colonization of mealy bugs.
3. Spray phosalane (0.1 per cent) in December after pruning, followed by two sprays of monocrotophos (0.1 per cent) in February and April.
4. Spray metacid (0.01 per cent) or carbaryl (0.2 per cent) at fortnightly interval in February and April.
5. Release egg parasitoid, *Trichogramma chilonis* to reduce the population of caterpillars.
6. Avoid growing highly susceptible cultivars like Kagzi Anar, Achik, Jalore Seedless in butterfly endemic areas.

Bark Eating Caterpillar (*Indarbela tetraonis*)

Like many other fruit crops, bark eating caterpillar is a serious pest of pomegranate, causing serious losses in certain localities. Damage is caused by the caterpillars, which bore the bark and feed inside the old or neglected trees/orchards. Usually, one caterpillar is found/plant, which may bore 10-12 holes. Badly infested plants do not bear any crop.

Management Practices

Following measures have been proved useful for reducing the incidence of bark eating caterpillars in the pomegranate orchards:

1. Keep the orchards clean and avoid overcrowding of the trees.
2. Destroy the infested plants.
3. Clean the holes with a wire and insert cotton soaked in petrol/diesel or carbon bisulphide and seal the holes with mud.
4. Inject fenvalerate (0.01 per cent) or qunialphos (0.05 per cent) into the holes.

Mealy Bug (*Planococcus citri*)

It is a polyphagous pest. The activity period, nature of damage and life cycle is discussed in detail 'citrus'.

Management Practices

1. Prune and destroy the affected parts in initial stages to prevent further spread and build up of the pest.
2. Control ants in the orchard by destroying the ant colonies by digging/ploughing the soil around trees frequently.
3. Release of *Cryptolaemus montrouzieri* is effective in reducing the pest population.

Pomegranate Shot Hole Borer (*Xyleborus fornicates*)

This is a major pest on pomegranate in Karnataka. The adult beetles bore holes on the roots and later on lower parts of main trunk. The early stage of infestation in an orchard begins with a mild yellowing of lateral branches on one or more trees, usually in a contiguous patch. Within a week, the whole tree becomes yellow followed by drying of branches. Some infested trees show heavy bearing but with reduced size and premature ripening. The main trunk from a foot above the soil shows small pinholes, which may or may not be seen with powder coming out of it. However, if the infestation is due to shot hole borer, subterranean pinholes in the root region is a sure symptom. From the infested tree, adults migrate within a month to the nearest healthy trees for further infestation. The infested patches of trees, if kept untreated, become a powerful source of high population. The rate of spread of infestation at this time will be rapid, and a whole orchard can show symptoms in 3-6 months. From one orchard, the infestation can spread to neighbouring orchards.

Management Practices

Early diagnosis is a must, so regular visits to orchards by growers is suggested. Signs of lateral branch yellowing to quick drying of full tree, should be immediately brought to notice of specialists and treatments should be undertaken as follows:

1. Drench the soil around main trunk with a mixture of chlorpyriphos @2.5ml + tridomorph @1ml per/litre of mixture/tree. After three weeks, repeat with monocrotophos @1.5ml + carbendazim @1g per litre.

2. If incidence is severe, repeat the above drenching after a month.

3. If infestation is low, drench with azadirachtin (0.15 per cent) @ 3 ml/litre around main trunk @2-3 litres of mixture/tree with either of the above insecticides.

4. Avoid water-logging and keep soil raked and aerated.

5. Infested trees should be uprooted and burnt.

6. Pits of uprooted trees should be drenched with chlorpyriphos @2.5ml/litre.

7. Drench the soil with chlorpyriphos (0.05 per cent) around all uninfested trees prophylactically once in 6 months, followed by a spray on trees with quinalphos (0.06 per cent) followed by azadirachtin (1500 ppm) @3 ml/litre. Avoid leaving infested trees in the field after uprooting.

Aphid, White Fly and Thrips

Pomegranate aphid (*Aphis punicae*) and white fly (*Siphoninus phillyreae*) cause damage by sucking cell sap from leaves and tender twigs. Both nymphs and adults of aphid suck cell sap and the plants are devitalized. The affected plant parts get discoloured and disfigured. Sooty mould grows on honeydew secreted by these insects and hinders the photosynthetic activity of the plant. As a result, the flowers and developing fruits may drop down.

Various species of thrips like *Retithrips syriacus* and *Rhipiphorothrips cruentatus* feed on leaves and some species infest flowers. Nymphs and adults lacerate the leaves, flower, stalks, petals and sepals and rasp the sap that oozes out, resulting into leaf tip curl and flowers are shed.

Management Practices

1. Spray dimethoate or oxydemeton methyl @ 0.05 per cent at 10-15 days interval.

2. Conserve natural enemies like coccinelids and syrphids.

3. White fly has gained resistance to systematic insecticides and hence neem based biopesticide or lambda cyhalothim (0.0025 per cent) should be sprayed in the evening.

4. In case of aphid attack, one spray of neem seed kernel extract (2 per cent) at the initiation of new flush is highly beneficial.

5. In case of whitefly attack, apply little more nitrogen to infested plants and irrigate regularly.

C. Physiological Disorder and their Managment

Fruit Cracking

It is the most serious physiological disorder of pomegranate, which limits its cultivation. In young fruits, it could be due to boron deficiency but fully-grown fruits crack due to moisture imbalances as there are very sensitive to variations in soil moisture and humidity. Prolonged drought causes hardening of peel and if this is followed by heavy irrigation or down pour, then the pulp grows and the peel cracks. This problem can be overcome by (a) maintaining soil moisture and not allowing wide variations in soil moisture depletion, (b) cultivation of tolerant varieties, (c) early harvesting not allowing the fruits to crack, and (d) spray of calcium hydroxide on leaves and on fruit set.

Sunburn

Sunburn is a common problem of pomegranate in India and several other part of world. It occurs in cultivars or areas where temperatures are high and skies are clear. Heavy crops that cause branches to bend over mid-season can increase sunburn incidence as a sudden exposure to heat and sun promotes sunburn development. Initial symptoms are development of brownish-black patches on the sun exposed side of the fruit. With severe skin damage, injured areas can turn dark-brown on the tree. Affected fruits look ugly and are difficult to sell in the market, and if sold, farmers get very low price for their produce. It can be reduced by the following management practices:

1. Avoid sudden exposure of fruit to intense heat and solar radiation.
2. Adopt proper tree training and pruning to avoid excessive sunburn.
3. Avoid water stress in cropped orchard trees to reduce heat stress.
4. Careful sorting to remove affected fruit upon packing is the only solution once the injury has occurred.

Husk Scald

Scald is a brown superficial discoloration restricted to the husk. There are no internal changes of the arils or the white segments as occurs with chilling injury. At advanced stages, the scalded areas become moldy. The scald symptoms usually become evident after 8 weeks storage at 2°C(36°F). Symptoms will appear earlier at higher storage temperatures. Usually, husk scald development of pomegranate may be due to phenolic oxidation. Its incidence may be reduced by:

1. Late harvested fruit is less susceptible than earlier harvested fruit.
2. Scald development may be delayed up to 6 weeks at 2°C(36°F).

Chilling Injury

The incidence and severity of chilling injury depend upon storage temperature and duration. The minimum safe temperature is 5°C (41°F) for up to 8 weeks. Chilling injury can be a major cause of deterioration of pomegranates during marketing following exposure to temperatures below 5°C (41°F) during storage and transport for longer than 4 weeks.

External symptoms include brown discoloration (scald) of the skin, pitting, and increased susceptibility to decay. Internal symptoms include brown discoloration of the white segments separating the arils and pale color (loss of red color) of the arils. Its incidence can be avoided by not storing pomegranates below 5°C (41°F).

HARVESTING AND YIELD

Pomegranate bears male, female and hermaphrodite flowers on spurs and intermediate shoots. Only the bisexual flowers produce fruit. Fruits generally ripen 6-8 months after fruit set. Being non-climacteric, tree ripe fruits are harvested. Change in rind colour from light-green to yellowish-pink or red with waxy shining surface and a cracking sound of grains on pressing the fruit or make a metallic sound when tapped, indicate fruit maturity. The fruits must be picked before over maturity when they tend to crack open, particularly when rained on. Ripe fruits are individually picked. A full grown pomegranate bush normally produces 40-50 fruits. However, as high as 100 fruits per bush can be obtained under good management.

POSTHARVEST TECHNOLOGY

Sorting and Grading

After harvesting, the diseased, cracked, insect or disease infested fruits should be sorted out and the healthy ones be graded. Pomegranate fruits should be graded on the basis of their weight, size and external (rind(colour. The grades are:

- ☆ **Super-sized:** Fruits, which have good, attractive bright red colour, weight more than 750g each and without any spot on the peel, are graded as super-sized fruits.
- ☆ **King-sized:** Fruits, which are free from spots, having an attractive red colour and weight between 500-750g, are graded as king-sized fruits.
- ☆ **Queen-sized:** Fruits between 400 and 500g weight, having bright red colour and free from spots, are rated as queen sized fruits.
- ☆ **Prince-sized:** Those fruits, which are fully ripe fruits, weighing between 300 and 400g with red colour, are graded as prince-sized fruits.

In addition of above mentioned grades, pomegranates are also graded into two more grades such as 12-A and 12-B. The fruits weighing between 250 and 300g and having some spots on the peel, belong to 12-B grade, whereas fruits of 12-A grade are generally preferred in southern and northern India.

Packing and Storage

After proper sorting and grading, pomegranates are packed in suitable containers for transportation, storage or marketing. The size of packages depends on the grade of fruits. In general, corrugated fiber-board (CFB) boxes are used for packaging as these packages are light in weight, cause less or no bruising injury to fruits, are cheap and are easy to handle. In a single box, 4-5 queen-sized fruits, 12 prince-sized and some of 12-A and 12-B grades may be packed. The white-coloured boxes having 5 plies are generally used for export purpose, whereas red-coloured boxes having 3 plies are used for domestic markets. The red-coloured boxes are cheaper than white-coloured ones. The size of super-sized, queen-sized, prince-sized, 12-A and 12-B grades are 13" x 9" x 4", 15" x 11" x 4" and 14" x 10" x 4", respectively. The cut pieces of waste paper are generally used as cushioning material between and below the fruits. The graded fruits are placed on cushioning material, followed by an attractive red-coloured paper on the boxes.

Storage

Pomegranates can be stored for some time if appropriate storage conditions are available. It has been demonstrated that safe temperature for storage of pomegranate in cold storage is 5°C. At this temperature, pomegranates can be stored safely for 8-10 weeks. If stored at low temperature, fruits may suffer from chilling injury. For longer storage, fruits should be stored at 10°C and relative humidity of 90-95% to avoid chilling injury and weight loss.

Processing

Pomegranate is a juicy fruit, which can be processed into different beverages with the addition of sugar and preservatives. Sundried grains from cultivars having high acidity, known as '*anardana*' are used for garnishing curries and for culinary purpose.

CROP IMPROVEMENT

Objectives of Pomegranate Breeding

The major objectives of improvement in pomegranate are selecting/breeding cultivars with high yield potential, better fruit quality with respect to fruit colour, soft seed, red coloured arils and resistance to pest and diseases.

Breeding Methods

Selection

The research work on pomegranate improvement has been underway in India since 1905. The first cultivar, which was released for commercial cultivation in 1936, was a selection from an open pollinated seedlings population from cultivar Alandi named as GBG-1. This was renamed as Ganesh in 1970. The cultivar has pinkish and sweet aril and soft seeds unlike the deep pink and sour aril and hard seed in Alandi. However, this cultivar was adopted by farmers only in 1973 and is now the most popular cultivar not only in Maharashtra but also in other states of India.

At MPKV, 4 Muskat types superior in yield, quality and consumer acceptance have been identified. Of these types, P-23 and P-26 were released for commercial cultivation in 1986.

At TNAU, Coimbatore, a soft seeded selection CO-1 was made from a collection of 140 plants of 28 accessions.

At UAS, Bangalore, evaluation of seedling population raised from Bassein Seedless and Dholka varieties resulted in identification of seedling GKVK-1 with soft seeds and pink aril. This was later released as Jyoti.

Out of the four superior clones of Ganesh. The clone G-137 was superior to Ganesh in aril colour, size and TSS, and released in 1986, also reported a clone (ACC No 455) with medium-sized fruit, easily peelable rind, soft seeds and attractive deep purple aril. This clone was released for commercial cultivation in Tamil Nadu as Yercaud 1.

Hybridization

Hybridization of pomegranate has also been taken up by MPKV, Rahuri. Through hybridization of Russian cultivars Shirin Anar, Gulsha Rose Pink and Gulsha Red with Ganesh, 3 hybrids No 5, No 61 (Mridula) and No 242 have been produced, which have deep red coloured aril alongwith the traits of soft seededness and high TSS.

At IIHR, Bangalore, crosses were made between some deciduous cultivars having deep red aril colour and sweet and evergreen cultivars. This has resulted in Hybrid No 15-9-94, which has dark-red non-sticky and bold arils, soft seeds with high sweetness and low tannin. The physico-chemical characteristics of these selections are given in Table 11.3. Besides, for anardana, Amlidana an F1 hybrid (Ganesh x Nana) has been developed, which grows well under tropical climate. With quality fruit attributes, 'Amlidana' is superior to sour variety Daru, whose trees come up naturally in temperate regions of north India.

Table 11.3: Physico-chemical Characteristics of Pomegranate Selections and Hybrids

Selection/ Hybrid	Fruit Wt. (g)	Fruit Size (LxB)	Aril Colour	Seeds (Mellowness)	TSS (^{0}Brix)	Acidity (per cent)
Selections						
GKVK-1	219	7.9 x 7.5	Deep red	Soft	15.0	0.58
IIHR selection	195	6.6 x 7.6	Light pink	Soft	13.7	0.55
G-137	269	7.9 x 7.5	Light pink	Soft	15.5	0.36
P-23	385	7.8 x 7.5	Light pink	Medium	16.0	0.42
P-26	379	7.5 x 7.5	Light pink	Soft	15.5	0.41
Hybrids						
No. 5	414	9.2 x 8.9	Pink	Soft	16.3	0.36
Mridula	244	7.5 x7.7	Blood red	Soft	17.9	0.47
No.242	221	7.6 x 7.4	Red	Soft	19.1	0.55
Ruby	249	7.3 x 7.8	Dark red	Soft	17.2	0.64

Inheritance of Characters

During crosses made for pomegranate improvement at Bangalore, it was found that pollination by a soft seeded cultivar slightly decreased seed hardness in hard seeded cultivars, while pollination either by a hard or soft seeded cultivar, increased seed hardness in the soft seeded cultivars.

12

Sapota

Sapota, sapodilla or *Chiku* is another delicious tropical fruit of India, tropical America, South-East Mexico, Guatemala and other countries. Immature fruits are astringent, while ripe fruits are sweet and are used as desert fruit. Now, the crop has attained the status of major fruit industry of India. It is an important adjunct in ice cream and milk shake in its fresh form. Fruit can also be used for preparing liquor and alcohol because of its richness in sugar. The largest sapota producing state is Karnataka with 3.77 million tones production followed by Maharashtra with 3.22 million tones fruits (Table 12.1). The highest productivity has been observed in Tamil Nadu with an impressive 27 MT/ha sapota production, which is thrice of the national average productivity (8.9 t/ha) (Table 12.1).

ORIGIN, HISTORY AND DISTRIBUTION

Sapota is believed to have originated in Mexico and Central America and being a tropical crop, grow throughout the tropics. In India, it is cultivated in Maharashtra, Gujarat, Andhra Pradesh, Karnataka, Tamil Nadu, Kerala, Uttar Pradesh, West Bengal, Punjab and Haryana (Table 12.1). Among different states, highest acreage is in Maharasthra (70 thousand ha), followed by Karnataka (30.8 thousand ha) and Gujarat (28.8. thousand ha). However, the highest productivity is in Tamil Nadu (27 MT/ha) (Table 12.1).

COMPOSITION AND USES

Sapota is a rich source of carbohydrates; and minerals like calcium and potassium (Table 12.2). Sapodillas are nutritious and mostly eaten as fresh fruit. Sherbets, milk shakes and ice cream can be made from fresh pulp. In Chile, the latex obtained from the bark of the tree is used as the principal ingredient of chewing gum. Because of its beauty and tolerance to neglect, sapodilla trees are also used as an ornamental for landscaping in South Florida. It is considered as useful fruit for febrile attcks.

Table 12.1: State-wise Area, Production and Productivity of Sapota in India

State	Area (000'Ha)	Production (000'MT)	Productivity (MT/Ha)
Karnataka	30.8	377.8	12.3
Maharashtra	70.0	322.0	4.6
Gujarat	28.8	288.0	10.0
Tamil Nadu	9.0	242.3	27.0
Andhra Pradesh	12.2	122.1	10.0
West Bengal	4.0	43.6	10.9
Odisha	3.4	17.0	5.1
Others	1.9	11.3	6.0
Total	160.0	1424.1	8.9

Source: National Horticulture Board Database-2011.

Table 12.2: Nutritional Composition of Ripe Sapota Fruit

Component	Per 100 g Edible Portion	Component	Per 100 g Edible Portion
Moisture	73.7 per cent	Calcium	28.0 mg
Carbohydrate	21.4 g	Phosphorus	27.0 mg
Protein	0.2 g	Iron	2.0 mg
Fat	1.1 g	Ascorbic acid	6.0 mg

TAXONOMY AND BOTANICAL DESCRIPTION

Taxonomy

Sapota or sapodilla (*Manilkara achras* Mill.) belongs to family Sapotaceae, the 'naseberry family'. Sapotaceae includes trees or shrubs comprising about 70 genera and 800 species. The genus *Manilkara* comprises some 85 species of tropical and warm-temperate, fruit-bearing plants. The basic chromosome number $(x) = 7$, but it ranges between 9 and 13 in the Sapotaceae family.

Botany

Sapota is a medium-to-large tree. The long-lived trees grow slowly but after many years, can reach 60 to 100 ft in height. It is well adapted to subtropical and tropical climates. The tree has ornamental value and may be used for landscaping. Branches are horizontal or drooping. Milky latex exudes from all tree parts. This latex is known as chickle and in old days was used to make chewing gum. The foliage is evergreen. Leaves are 5-20 cm long, stiff, pointed and clustered at the ends of shoots. The leaves are light-green to pinkish when newly emerged, becoming dark-green at maturity. Flowers are off-white, bell-shaped, small, bisexual, borne at the leaf axils and measure about 9.5 mm in diameter. Based on nature of branches and colour of foliage, sapota trees (varieties) have been grouped in the following 4 types of growth habits:

Erect Growing Habit

Branches appear in whorls, foliage deep-green, broad and oval, fruits large in size with yellow peel, pulp butter like and sweet.

Drooping Growth Habit

Branches appear in whorls, foliage light-green, narrow and elliptical, fruits small with rough brownish peel, pulp of inferior quality.

Spreading Growth Habit-I

Irregular branching, foliage deep-green, broad and oval, fruits have smooth and yellow peel, pulp butter like and sweet.

Spreading Growth Habit-II

Irregular branching, foliage light-green, narrow, elliptical, fruits have rough peel, pulp of inferior quality.

The fruit is variously described as 'a small potato', 'a small tomato', 'a round kiwifruit', or 'a soft elongated tan egg'. Fruit development follows a sigmoidal pattern. The fruit is a large ellipsoid berry, 4-8 cm in diameter, very much resembles a smooth-skinned potato, containing seeds. The fruit has a scurfy, brown peel. Fruit may be round to oval-shaped or conical, 5-10 cm in diameter and weigh 50 to 200 g. The pulp is light-brown, brownish-yellow to reddish-brown, with a texture varying from gritty-to-smooth. The pulp has a sweet to very sweet (19-24 °Brix), pleasant flavour. Seed number varies from 0 to 12. Seeds are dark-brown to black, smooth, flattened, shiny, and 1.9 cm long. When fruit reaches maximum size, it may be picked and allowed to ripen off the tree. From experience, a grower can judge maturity of fruit of a particular variety or selection by its size and appearance.

SOIL AND CLIMATIC REQUIREMENTS

Soil

Sapota is well-adapted to a wide range of soils but grow best in well-drained, light soils. Trees are especially well-adapted to the rocky, highly calcareous soils of south Florida. It comes up well in alluvial soils of the riverbanks, sandy loams near coastal areas, red laterite soils of the heavy rainfall area and medium black soils.

Climate

Sapota is a tropical fruit crop. It prefers a warm and moist weather and grows in both dry and humid weather. Coastal climate is best suited. Sapodillas are adapted to tropical and warm sub-tropical climates. Young trees may be killed or injured at temperatures of 30 to –1 to 0 °C. Large trees can withstand temperatures as low as –3 °C for few hours with only minor damage. Optimum temperature for growth is between

15 to 35 °C, and at higher temperature above 43 °C during summer, the flower and fruitslets may drop. Areas with an annual rainfall of 125 to 250 cm are highly suitable.

IMPORTANT CULIVARS

Many varieties are popular among the growers, and chief characteristics of some of them are given in Table 12.3.

Table 12.3: Important Varieties of Sapota and their Chief Characteristics

Variety/Hybrid	Chief Characteristics
Cricket Ball (Calcutta Large)	Fruits are large (300- 350 g) and are round in shape, pulp is gritty and granular and moderately sweet (TSS 16-20 per cent), 2-12 seeded.
Baramasi	Fruits are medium sized, round in shape, medium sweet (TSS 20-23 per cent), 2-8 seeded, does not bear throughout the year.
Thagarampudi	Fruits are medium sized, round or oval shaped, soft flesh, melting, very sweet, (TSS 22-23 per cent), 1-2 seeded.
Badami	Fruits are small, soft, gritty, medium sweet (TSS 18-21 per cent), 1-4 seeded.
Dwarapudi	Fruits are large, round, soft flesh but gritty, mild flavour, very sweet (TSS 23-27 per cent), 1-6 seeded.
Oval	Fruits are small to medium sized, egg shaped, slightly gritty, medium taste (TSS 18-22 per cent), 2-6 seeded.
Kirtibharathi	Fruits are small to medium sized, oval or egg shaped, soft, gritty, medium taste (TSS 21-22 per cent), good keeping quality.
Pala	Fruits are small to medium sized, oval or egg shaped, melting, slightly gritty, good taste (TSS 18-21 per cent).
Guthi	Fruits are small, eliptic in shape, soft flesh, gritty, good in taste, (TSS 20-26 per cent).
Culcutta Round	Fruits are large, flesh is gritty, moderate quality.
CO-1	Hybrid between Cricket Ball x Oval, suitable for table purpose. Bearing after 4 years of planting, fruits large, oblong, sweet, flesh reddish brown, weight 125 g, TSS 18 per cent.
CO-2	Selection from 'Baramasi', suitable for table purpose. Tree medium in height, fruit oblong-round, medium sized, flesh soft, juicy, sweet, gritty, light brown in colour with pleasant aroma, TSS 23 per cent, average wt. of fruit is 125-150g.
CO-3	Hybrid between 'Cricket Ball' and 'Vavivalasa'. Trees are of intermediate stature. Bearing commences from the fourth year of planting. Fruits are dull brown, oblong, sweet, average annual yield of 157 kg fruits/tree
CO-4	Named 'PKM (Sa) 4', an open-pollinated clone of PKM-1, compact tree canopy, which is highly suited for high density planting. Bears fruits in clusters, spindle shaped fruits of 11 to 13 cm in length are suitable for the production of dry flakes, very sweet flesh, with 24-25 per cent. The fruits have just two or three seeds only. On an average a tree can yield about 100 kg in a year, and it is about 138 per cent more than that of PKM-1, according to the scientists, who developed this variety. The per hectare production worked out to 20.8 tonnes a year.
PKM-1	Selection from Guthi, suitable for table purose, profuse bearing, fruits medium sized, elliptical-oblong with thin skin, rich in sugars, TSS 23 per cent, average fruit wt. 80g.

Contd...

Table 12.3–*Contd...*

Variety/Hybrid	Chief Characteristics
PKM-2	Released in the year 1992. PKM.2 (H-2/4) is a hybrid between Guthi and Kirthibarthi varieties. High yielder – mean yield of 80 – 100 kg of fruits/tree during fifth year of planting. Fruits are bigger in size (95g) as compared to PKM.1 (84.0 g). Oblong to oval shaped fruits. Fruits are of good quality with higher TSS (26.3°brix), total sugar (14.25 per cent), reducing sugar (0.46 per cent) and sugar/acid ratio (593.8).
PKM-3	Released in the year 1994. Hybrid between Guthi x Cricket Ball. Adaptable to tropical plains of Tamil Nadu. Yield – 14t/ha. Fruits bear in clusters with oval shaped large fruits. Vertical growth habit of tree allows high density planting. The fruits mature earlier than other varieties in the season thus fetching higher return. Tolerant to leaf spot and leaf webber.
PKM-4	Released in the year 2003. Open pollinated clone of PKM-1 (MA-1) with compact tree canopy. Suitable for high density planting. High yielder (100.4 Kg/tree/year, 20.08 tonnes/ha) with spindle shaped fruits in clusters. Fruit length ranging from 11-13 cm and suitable for production of high grade dry flakes. Attractive pulp with honey brown colour. Crisp and sweet flesh with TSS of 24-25° Brix. Suitable for export market.
DHS-1	This variety was developed at Agriculture College, Dharwad. It is a hybrid of Kalipatti X Cricket Ball. Fruits are round in shape, which weigh around 210 g. TSS is 23.0 °brix.
DHS-2	It is also a hybrid of Kalipatti X Cricket Ball developed at Agriculture College, Dharwad. Fruits are oblong in shape, which weigh around 154 g. TSS is around 26.0 °brix.

Important varieties cultivated in different states of India are listed below (Table 12.4).

Table 12.4: State-wise Distribution of Sapota Varieties in India

State	Varieties being Grown
Andhra Pradesh	Cricket Ball, Kalipatti, Calcutta Round, Kirthibharathi, Dwarapudi, Pala, PKM-1, Jonnavalasa I and II, Bangalore, Vavi Valsa
Bihar	Baramasi
Gujarat	Kalipatti, Pilipatti, Cricket Ball, PKM-1
Karnataka	Cricket Ball, Kalipatti, Calcutta Round, DHS-1, DHS-2
Maharashtra	Kalipatti, Dhola Diwani, Cricket Ball, Murabba
Orissa	Cricket Ball, Kalipatti
Tamil Nadu	Pala, Cricket Ball, Guthi, CO-1, CO-2, PKM-1
Uttar Pradesh	Baramasi
West Bengal	Cricket Ball, Calcutta round, Baramasi, Baharu, Gandhevi Barada

PLANT PROPAGATION

Although seeds can be used for propagation and are used for selection of superior types, they should not be used for commercial plantings. Marcottage (air layering) has not been an effective propagation method. Side-veneer and cleft grafting on to seedling sapodilla rootstock are the most common grafting methods. Chip budding can also be used. Scions or budsticks are chosen from young terminal shoots. Cover the grafted scions completely with grafting tape. The best time to graft is late summer and early fall.

Inarching in sapota is the commercial method of propagation practiced over forty years. This method is tedious, cumbersome and time consuming. One of the advancements in propagation of sapota has been the commercial exploitation of softwood grafting. About one-year-old rootstocks of pala (*Manilkara hexandra*) growing outdoors are selected and all side branches are removed. The stock is then cut down to a height of 20 cm above soil level. Scions should be 8 to 10 cm long and of pencil thickness with bulging tips. The colour of the scion should be turning from green-to-brown. For precuring, leaf blades are clipped off seven-to-ten days before grafting. When the petioles dry and drop off, the scions can be detached from the tree and used for grafting. Longitudinal cut of 3-4 cm is made on the stock and the scion wedge is inserted to the softwood portion of rootstock and tied with a polythene strip of 200 gauge thickness. The grafts are kept in shade or mist chamber for 15 days and later transfer the successful grafts to open place. Treating the scion with IAA at 750 ppm increase the percentage of success. Higher percentage of (more than 95) success can be obtained from this method.

PLANTING AND ORCHARD ESTABLISHMENT

Land Preparation and Spacing

After preparation of soil, pits of 60 cm^3 are dug out. In light soils, a spacing of 10 m × 10 m and in heavy soils, a spacing of 13 m × 13 m is good. Although sapodillas grow slowly, trees eventually need a wide space between rows and in the row, as they will develop a large canopy. Spacing between rows should be 7.3-9.1 m and 4.6-9.1 m in row, from tree-to-tree. Groves with closely spaced trees will result in higher yields in the early life of the planting (up to 10 years). However, when tree crowding begins, pruning and/or removal of every other tree are recommended. A rectangle pattern with a North-to-South row orientation is recommended. Sapodilla trees in the home landscape should be planted 7.6 m or more away from the nearest tree and/or structure.

Planting

Plough the land. Dig pits of 1 m × 1 m × 1 m at a distance of 10 m × 10 m or 13 × 13 m apart depending on soil type. Fill the pits with top soil and compost. Plant the grafts in the middle of the pit, keeping the grafts joint above the ground level. The best time for planting is early monsoon. After planting, the soil around the roots is pressed firmly and then irrigate lightly.

INTERCULTURAL OPERATIONS

Manures and Fertilizers

Every year, FYM @ 50 kg/plant should be given every year. Manure the plants in the beginning of rainy season. The following doses of manures and fertilizers are recommended for profitable cultivation of sapota (Table 12.5). Apply manures and fertilizers, 60-90 cm away from the trunk.

Table 12.5: Recommended Doses of Manures and Fertilizers for Sapota in India

Sl.No.	Age of the Tree (yrs.)	Nitrogen (g/plant)	Phosphorus (g/plant)	Potash (g/plant)
1.	1-3	50	20	75
2.	4-6	100	40	150
3.	7-10	200	80	300
4.	11 Yrs. onwards	400	160	450

Irrigation

When plants are young, irrigation should be given throughout the year, depending on the soil conditions the primary arms and secondaries not allowing them to grow more than 60 cm at a time. As they grow, the shoots are tied with jute twine and all tendrils are removed.

Weeding

Weeds compete for water and nutrients with main crop. Weeds may be controlled by herbicide applications of registered materials and/or by mulching or hoeing. Weeds and lawn grasses should be removed within a 2 to 4 ft (0.6-1.2 m) radius around the tree trunk and under the canopy. A, 2 to 4 inches (5-10 cm) thick layer of mulch may be used to reduce soil drying and population of weeds. Keep mulch 8 to 12 inches (20-30 cm) away from the trunk.

Training and Pruning

Young Trees

The development of a strong limb framework is important to allow sapodilla trees to carry large crops of fruit without limb breakage. If the tree is leggy and lacks lower branches, remove part of the top to induce lateral budbreak on the lower trunk. In addition, shoot-tip removal (1 to 2 inches) once or twice between spring and summer, will force more branching and make the tree more compact. Remove any limbs that have a narrow crotch angle, as these may break under heavy fruit loads or due to strong winds.

Mature and Bearing Trees

As trees mature, most of the pruning is done to control tree height and width and to remove damaged or dead wood. A, 6 to 8 ft (1.8-2.4 m) wide space between rows will facilitate equipment traffic. When cutting the sides of trees (hedging), an angle of 10 to 15° from the vertical should be used so that trees will have a pyramid shape. This facilitates light penetration to the lower parts of the trees. The rule of thumb about the proper tree height is that tree height should be about twice that of the between-row middle space, *i.e.*, if the middle is 6 ft., tree height should be about 12 ft;

if it is 7 ft, then the height should be about 14 ft. Trees should be kept at a maximum of about 15 to 16 ft high. If the canopy becomes too dense, removing some inner branches will help in air circulation and light penetration. There are many plans for pruning. You may top and hedge every year or may top all rows every year but hedge alternate rows every year. Another pruning objective is the removal of dead, damaged or diseased branches. Low branches should not be cut, however, unless they touch the soil. Cultural practices *e.g.*, picking, spraying, and pruning are easier in small trees.

PLANT PROTECTION

A. Major Diseases and their Management

In general, sapota is considered as a hardy fruit crop, and no serious disease has been reported on it. However, the following diseases may cause some losses in sapota.

Leaf Spot

Leaf spot is considered as the most important disease of sapota. It assumes serious form in almost all the sapota growing areas of our country. It affects all varieties of sapota with variable incidence. It is caused by *Phaeophleospora indica*, with high incidence in dry regions.

Symptoms

The disease is characterized by the appearance of numerous small, circular pinkish-to-reddish brown spots with white centers on mature leaves. Usually, the fungus does not infest younger leaves. In severe infestation, the diseased leaves fall prematurely, giving barren look to the branches. As a result, fruit yield is reduced drastically.

Management Practices

Adopt the following measures for the control of leaf spot in sapota:

1. Grow resistant varieties like Cricket Ball, CO-1, CO-2 and Kalipatti.
2. Spray dithane Z-78 (0.2 per cent) or topsin-M (0.1 per cent) or bavistin (0.1 per cent) at monthly interval.

Flat Limb

'Flat limb' is commonly called as 'fasciation'. This malady is also a serious problem of sapota in all areas. It is caused by *Botryodiplodia theobromae*.

Symptoms

The major symptoms of the disease are that the branches get flattered and twisted. Their leaves become thin; small and yellow. The flowers and leaves also get clustered at the tip of such affected branches. Flowers in the affected twigs usually remain infertile and don't set fruits, and if fruit setting takes place, the fruits don't develop properly and remain undersized, are hard and fail to ripen properly. In severe infestation, the flowers, leaves and fruits of the affected twigs drop prematurely, thereby reducing the yield considerably.

Management Practices

1. Prune the affected branches and destroy them.
2. Give two sprays of phosphoric acid (500-1,000 ppm) after rainy season at fortnightly interval.

Sooty Mould

This is a fungal disease caused by *Capnodium* sp. in which there is development of blackish mass on the leaves and fruits, as a result, photosynthetic functions of the leaves get affected and fruits get disfigured. Spray starch or maida solution (1 kg boiled in 5 liters of water and diluted to 20 liters). When mild attack of disease appear, given one spray of plain water.

B. Major Insect-Pests and their Management

About 25 insect-pests have been reported to cause damage to sapota. Out of these, chiku moth is the most destructive, followed by chiku bud border and fruit fly. Other pests, like mealy bug, leaf webber, stem borer, whiteflies, leaf miner etc., are of minor importance.

Chiku Moth (*Nephopteryx eugraphella*)

It is a major pest of sapota and occurs widely in many parts of the world including India. The pest is active throughout the year, but maximum activity is found during June-July. Damage is caused by the caterpillars, which feed on leaves, flowers, buds and sometimes on tender fruits. They web the leaves together and feed on leaf grey matter, leaving behind a fine network of veins. Caterpillars also bore into flower buds and tender fruits, which wither away, and the larvae move on to the next bud or fruit, thus damaging many of them. Generally, first larval instar completes within the floral bud and then it comes out of bud and produces a parchment. Older larvae feed voraciously on tender leaves, buds and young fruits, causing considerable losses. The infestation of this pest can easily be detected by the presence of clusters of dried leaves hanging on webbed shoots and there is appearance of dark brown patches on the leaves and clusters of dead leaves. The bearing capacity of the affected plants is drastically reduced. Fruits are of smaller size with poor quality.

Management Practices

1. Remove and destroy all infested leaves, buds and fruits to reduce the infestation.
2. Installation of traps containing black tulsi (*Ocimum sanctum*) leaf extract + dichlorvos (1 trap/two trees) during April-September has been recommended to attract and kill male moths.
3. Spray monocrotophos (0.04 per cent) or endosulfan (0.03 per cent) or carbaryl (0.2 per cent) or phosphamidon (0.05 per cent) following economic threshold level of one larva/twig.

Chiku Bud Borer (*Anarsia achrasella*)

Chiku bud borer is also a major pest of vegetative and flower buds of sapota in almost all regions, wherever sapota is grown. The pest is active throughout the year with peak in the beginning of monsoon. The damage is caused by caterpillars, resulting in considerable losses. In severe infestation, insect has been found to be responsible for shedding of a large number of flowers. The larvae bore into flower buds and feed on ovary and petals, thereby adversely affecting the production. Sometimes they feed on leaves also.

Management Practices

1. Remove and destroy all infested leaves, buds and fruits to reduce the infestation.
2. Spray dimethoate (0.05 per cent) or dichlorvos (0.03 per cent) or endosulfan (0.07 per cent) or malathion (0.2 per cent) coinciding with flower bud setting on chiku trees. Generally two sprays are required for effective control.
3. Installation of traps containing black tulsi leaf extract + dichlorvos (one trap/two trees) during April-September is generally recommended to attract and kill male moths.

Stem Borer (*Indarbela tetraonis*)

It is a serious pest of many fruits including sapota and causes serious damage in the old or declining or neglected orchards. Damage is caused by the grubs, which bore into the bark of the trunk making circular galleries and feed on the living tissues of the inner bark. The presence of the insect can be detected from the chewed bark thrown out of a hole in the trunk. The borer can also be traced by cutting dead bark along the hollow tunnels with a knife. The affected plants are girdled, as a result, the supply of nutrients and water is hindered to the aerial parts of the plant. Affected plants may bear for a year or so and fall down afterwards.

Management Practices

1. Crop sanitation and planting of trees at proper distance prevents the occurrence of this pest.
2. Keep a regular watch from July onwards and as soon as attack is observed, thresh a stiff wire into the tunnel and kill the grubs.
3. Plug the hole with cotton soaked in petrol or kerosene or with some insecticide (*e.g.* malathion) and plaster it with mud.

FRUIT MATURITY, HARVESTING AND YIELD

Maturity

For the beginner, fruit maturity is difficult to judge. Immature fruit may not soften for many days, not develop optimum sweetness and flavour, and contain pockets of coagulated latex within the flesh. Fruit picked over-mature may soften (ripen) within

2 to 3 days and be difficult to pack, ship, and market. Fruit picked at optimum maturity usually ripen in 4 to 10 days.

Sapota is a climateric fruit and it improves in quality after harvesting but immature fruits should not be harvested. It takes about four months to mature its fruits after flowering. The fruits to be harvested must be fully mature and the maturity can be judged by several external symptoms as expressed below:

- ☆ The color of the skin turns from green-to-yellow when rubbed or scratched.
- ☆ Fruits at full maturity develop a dull orange or potato colour, with a yellowish tinge.
- ☆ The end tip needle easily drops; the skin gets less rusty, sandy and scurfy. Brownish scales diminishes as the fruits mature.
- ☆ Ripe fruits have brownish-red skin similar to the skin colour of potato.
- ☆ As the fruit matures, the oozing of latex decreases.
- ☆ The fruit stalk (panicle) separates easily from mature fruits when touched without oozing much latex.

Fruits can be harvested early at Index 2. The fully matured fruits are severedwith the stalk as such, individually by giving a twist and collected without bruising. The fruits thus harvested are spread in a thin layer on bamboo mats under shade for an hour or two. To avoid bruising of fruits, they are better collected in gunny bags and lowered to the ground carefully. The peak harvest periods are January-February and May-June in the West coast of Maharashtra, March-May and September-October in Andhra Pradesh and Karnataka.

Harvesting

Sapodilla trees may have harvestable fruit year round, though there is a main season for each cultivar. Fruit can be harvested by hand, using a pole with a basket, or using machines or platforms that place the picker close to the fruit. Removing the fruit with a hook is not advisable as many of them hit the ground before they can be caught in the air. Keeping tree height at 4.3-4.9 m facilitates harvesting and other operations. Fruit should be handled carefully from harvesting through packing and shipping.

Yield

In sapota, bearing starts from 4th year onwards and economical yields can be obtained from 7th year. Plants flower almost throughout the year. It takes 4-6 months from flowering to fruit maturity. March-to-May and September to October are the two distinct seasons of harvest. A full bearing tree may produce 2,500 to 3,000 fruit in a year (Table 12.6).

Storage

Sapodilla fruit take about 4 to 10 days from picking to ripening (soften). As the season for each cultivar advances, the ripening time decreases. Hence, fruits should be stored at 10-13°C and 85-90 per cent relative humidity for extending availability in the market.

Table 12.6: Yield Pattern of Sapota Trees Over the Years

Sl.No.	Age	Yields
1.	4-5th year	250 fruits per plant
2.	6-7th year	800 fruits per plant
3.	8-20th year	1,200-1,500 fruits per plant
4.	By 30th year	2,500-3,000 fruits per plant

CROP IMPROVEMENT

Objectives

The objectives of improvement in sapota are large fruit size, good dessert and keeping quality and dwarf trees.

Breeding Methods

Selection

Improvement work in sapota was taken up at TNAU, Coimbatore in 1954 and later shifted to Periakulam. This resulted in the release of superior clone *e.g.*, CO-2 from cultivar Baramasi. A clonal selection 'PKM 1' from Guthi was made at HRS, Periakulam. The tree is dwarf stature and fruit weighs approxmately 100 g.

Hybridization

Planned hybridization work has been taken up in sapota at a few centres *e.g.* Coimbatore and Periakulam and UAS, Dharwad. The important characters of 7 cultivars, 2 selections and 5 hybrids released are given in Table 12.7.

Table 12.7: Characteristics of improved cultivars of sapota developed in India

Location	Cultivar	Parents	Fruit Shape	Fruit Wt. (g)	TSS (per cent)	Tree Size
HYBRIDS						
TNAU,	CO-1	Cricket Ball x Oval	Oval	125	18.0	Large
Coimbatore	CO-3	Cricket Ball x Vavilavalasa	–	–	–	Intermediate stature
HRS,	PKM-2	Guthi x Kirtibharti	Oblong	81	26.2	Medium,
Periakulam	(Hybrid-2/4)					spreading
	PKM-3	Guthi x Cricket Ball	Oval	–	–	Erect
	Hybrid-7/1	Kalipatti x Cricket Ball	Oblong	80	26.1	Erect
UAS,	DSH-1	Kalipatti x Cricket Ball	Round	210	23.0	–
Dharwad	DSH-2	Kalipatti x Cricket Ball	Oblong	154	26.0	–
SELECTIONS						
TNAU,	CO-2	Baramasi	Round	40	23.0	Large
Coimbatore						
TNAU,	PKM 1	Guthi	Oblong	80	33.0	Dwarf
Periakulam						

Many hybrids involving Guthi, Cricket Ball, Kirtibharthi, Thagarpudi and CO-2 are under assessment.

Constraints in Breeding

Narrow genetic base and lack of understanding about pattern of inheritance of characters has limited crop improvement programme in sapota. Efforts to broaden the genetic base are essential.

13

Cashewnut

Cashew is the most versatile of all nuts cultivated worlover. The name, *Anacardium* refers to the shape of the fruit, which looks like an inverted heart (*ana* means 'upwards' and *–cardium* means 'heart'). In the Tupian languages *acajú* means 'nut that produces itself'. It was introduced in India during the later half of the Sixteenth Century for the purpose of afforestation and soil conservation. From its humble beginning as a crop intended to check soil erosion, cashew has emerged as a major foreign exchange earner next only to tea and coffee. It is an important crop earning foreign exchange and having considerable employment potential.

COMPOSITION AND USES

Composition

Cashewnuts are rich source of fats, proteins and carbohydrates. They also contain an ample amount of mineral like phosphorus and calcium and vitamins like thiamine and riboflavin (Table 13.1).

Uses

Many parts of the cashew plant are used. Fruits or seeds of the cashew are consumed whole, roasted, shelled and salted in Madeira wine, or mixed in chocolates. Shelling the roasted fruits yield the cashewnut of commerce. Seeds yield about 45 per cent of a pale yellow, bland, edible oil, resembling almond oil.The nuts are nutritious and contain fats, proteins, carbohydrates, minerals and vitamins. These are used in confectionery and desserts. Cashew proteins are complete with all essential and non-essential amino acids and can be considered equal to peanut and soybean. The shells contain high quality oil known as cashew nut shell liquid (CNSL), which has got wide industrial uses as waterproofing agent, as a preservative in the painting of boats, fishing nets and light wood work. It is also used in the manufacture of brake liners, paints, varnishes, lacquers, and insulators. The cashew 'apple,' the enlarged fully ripe fruit may be eaten raw, or preserved as jam or sweetmeat. The juice is made into a beverage (Brazil cajuado) or fermented into a wine. Further, in India, it is mainly used in making 'feni', an alcoholic drink. It is made from fermented apple juice. This drink is widely taken by the people in Goa. Cashew apple juice is good for stomach. Cashew apple is a fairy good source of ascorbic acid. The sap from the bark provides an indelible ink. The wood is used as firewood. The plant has a number of medicinal properties. Root infusion is an excellent purgative. The oil obtained from the shell is macerated in syrup and is used to cure cracks on the sole of the feet.

Table 13.1: Nutritional Composition of Cashewnuts

Component	Per 100 g Edible Portion (Kernel)	Component	Per 100 g Edible Portion (Kernel)
Moisture	5.2 g	Carbohydrates	29.3 g
Fat	45.7 g	Ash	2.6 g
Fiber	1.4 g	β-carotene equivalent	60 mg
Calcium	38 mg	Riboflavin	190 mg
Phosphorus	373 mg	Thiamine	630 mg
Iron	3.8 mg	Niacin	1.8 mg
Protein	17.2 g		

ORIGIN, HISTORY AND DISTRIBUTION

Cashew is widely cultivated throughout the tropics. Cashew (*Anacardium occidentale* L.), a native of eastern Brazil was introduced to India just as other commercial crops like rubber, coffee, tea etc., by Portuguese during the 16th Century. The first introduction of cashew in India was made in Goa from where it spread to other parts of the country. In the beginning, it was mainly considered as a crop for afforestation and soil binding to check erosions. It is also cultivated in the coastal regions of South Africa, Madagascar, Mozambique, the West Indies and in South East Asia from Sri Lanka to the Philippines.

In India, it is commercially cultivated in Maharashtra, Karnataka, Andhra Pradesh, Odisha, Goa and Kerala (Table 13.2). During 2010-11, country produced 674.6 thousand tonnes of cashewnuts from a total area of 953.2 thousand hectares with an average productivity of 700 kg/hectare.

TAXONOMY AND BOTANICAL DESCRIPTION

Cashewnut (*Anacardium occidentale* L.) belongs to family Anacardiace, to which mango and *Pistachia* belong. Anacardiaceae comprises of about 60 genera and 400 species. In India, about 50 species under 22 genera have been recognized. *Anacardium* is a small genus and has about 20 species, majority of them exist in Central and South

America.The important species of *Anacardium* are : A. *macrocarpa, A. phemilium, A. brasiliense, A. encardium, A. giganteum, A. humile, A. nanum, A. pyrifolium, A. negrense, A. tenuifolium, A. microcarpum* and A. *corymbosum. A. nanum* is referred as 'Dwarf', and, *A. giganteum* as 'Giant'. In earlier studies, it has been reported with polymorphic chromosome number 2n = 24, 30, 40, 42. However, recent cytological studies established a diploid and haploid chromosomes of 2n = 42 and n = 21, respectively in cashewnut.

Table 13.2: State-wise Area, Production and Productivity of Cashew in India

State	2005-06			2006-07			2007-08			2010-11		
	A	P	APY	A	P	APY	A	P	APY	A	P	APY
Kerala	80	67	900	80	72	900	84	78	900	78	71	91
Karnataka	100	45	700	102	52	700	103	56	710	119	57	47
Goa	55	27	690	55	29	690	55	31	700	56	24	42
Maharashtra	160	183	1300	164	197	1500	167	210	1500	181	208	115
Tamil Nadu	121	56	640	123	60	670	123	65	700	135	65	48
Andhra Pradesh	170	92	880	171	99	890	171	107	900	183	107	58
Orissa	120	78	860	125	84	860	131	90	860	149	91	61
West Bengal	10	10	950	10	10	1000	10	10	1000	11	11	10
Gujarat	4	4	900	4	4	900	4	4	1000	7.1	21.3	300
A&N Island	14	10	640	15	11	700	15	12	750	1.1	0.3	27
TOTAL	837	573	815	854	620	820	868	665	860	953.2	674.6	71

Source: NHB Database, 2011. A: Area; P: Producton; APY: Average Productivity Yield.

Botany

Sapota is spreading evergreen perennial tree to 12 m tall; leaves simple, alternate, obovate, glabrous, penninerved, to 20 cm long, 15 cm wide, apically rounded or notched, entire, short petiolate; flowers numerous in terminal panicles, 10–20 cm long, male or female, green and reddish, radially symmetrical nearly; sepals 5; petals 5; stamens 10; ovary one-locular, one-ovulate, style simple; fruit a reniform achene, about 3 cm long, 2.5 cm wide, attached to the distal end of an enlarged pedicel and hypocarp, called the cashew-apple; apple shiny, red or yellowish, pear-shaped, soft, juicy, 10–20 cm long, 4–8 cm broad; fruit reniform, edible, with two large white cotyledons and a small embryo, surrounded by a hard pericarp which is cellular and oily, oil is poisonous causing allergenic reactions in some humans.

SOIL AND CLIMATIC CONDITIONS

Soil

Cashew can be grown successfully in almost all the soil types, from sandy-to-lateritic soils. Cashew is insensitive to depth, stoniness and fertility of soil and availability of water. However, it cannot withstand water-logging and excessive salinity or alkalinity. In India, cashew is mainly grown in laterite, red sandy-loam and coastal sandy soil in the states of Kerala, Maharashtra, Goa, Karnataka, Tamil Nadu, Andhra Pradesh, Odisha and West Bengal.

Climate

Cashew is essentially a tropical crop and grows best in the warm, moist and typically tropical climate. The distribution of cashew is restricted to altitudes below 700 m where the temperature does not fall below 20 °C for prolonged periods. It can tolerate temperature of more than 36 °C for a shorter period but the most favourable temperature lies between 24 °C to 28 °C. High temperature (39-42 °C) during stage of fruit set and development causes fruit drop. Cashew is grown in areas with rainfall ranging from 600–4,500 mm per annum. Fruit setting in cashew will be good if rains are not abundant during flowering and nuts mature in a dry period. Cloudy weather during flowering enhances scorching of flowers due to tea mosquito infestation. The major limiting factor in cashew cultivation is that it can't tolerate frost and extreme cold for a long time. Cashew is a sun-loving tree and does not tolerate excessive shade. Cashew is regarded as 'essentially coastal tree' but that is not true as it also grows well at a considerable distance from the sea coast.

IMPORTANT VARIETIES

There are many high yielding cashew selections/hybrids suitable for different agro-climatic regions. Characteristic features of important varieties of cashew are presented in Table 13.3. However, several varieties have been recommended by the scientists for different states (Table 13.4).

PLANT PROPAGATION

Seed Propagation

Cashew plantations established prior to 1980 were raised through seedling progenies. For seed propagation, seeds from high yielding mother trees with the following criteria were used.

1. Compact and intensive branching tree with high percentage of flower bearing laterals should be selected.
2. 12-25 year old trees with yield of 10-15 kg nuts per year are the best.
3. Trees, which bear 4-5 fruits per panicle and have nuts of medium size (8 to 10 g) are selected.

For propagation through seeds, the seednuts should be soaked in water for 12-24 hours before sowing in order to get good germination. At the time of sowing, the soil should be moist and loosened in the polythene bag. Seeds are sown in the centre of the bag, stalk-end upwards, with a depth of not more than 2.5 cm and covered with little soil. Thereafter, polythene bags are watered. The seednuts usually germinate within 15-20 days after sowing. Seed bed may be mulched with paddy straw till germination takes place and partial shade may be provided during summer months. In order to control pests while seeds germinate, malathion (5 per cent dust), or spraying of chloropyriphos (0.05 per cent) should be applied.

Table 13.3: Cashew Varieties and their Characteristic Features

Name of Variety	Parentage	Institution	Year	Yield (kg/tree)	Nut Wt. (g)	Kernel Wt. (g)	Shelling (per cent)	Export Grade
Kanaka (H 1598)	BLA 139 x H3-13	KAU, Madakkathara	1993	12.80	6.80	2.08	30.58	W 280
Dhana (H 1608)	ALGD-1 x K30-1	KAU, Madakkathra	1993	10.66	8.20	2.44	29.80	W 210
Amrutha (H 1597)	BLA 139 x H3-13	KAU, Madakkathara	1998	18.35	7.18	2.24	31.58	W 210
Priyanka (H-1591)	BLA 139-1 x K-30-1	KAU, Madakkathara	1995	17.03	10.80	2.87	26.57	W 180
Madakkathara-2	Neduvellur Material	KAU, Madakkathara	1990	17.00	7.25	1.88	26.20	W 210
Vengurla-1	Ansur 1	KKV, Vengurla	1974	19.00	6.20	1.39	31.00	W 240
Vengurla-4	Midnapur Red x Vettur 56	KKV, Vengurla	1981	17.20	7.70	1.91	31.00	W 210
Vengurla-6	Vettur 56 x Ansur 1	KKV, Vengurla	1991	13.80	8.00	1.91	28.00	W 210
Vengurla-7	Vengurla 3 x M-10/4	KKV, Vengurla	1997	18.50	10.00	2.90	30.50	W 180
BPP-4	9/8 Epurupalam	ANGRAU, Baptala	1980	10.50	6.00	1.15	23.00	W 400
BPP-6	T No.56	ANGRAU, Baptala	1980	10.50	5.20	1.44	24.00	W 400
BPP)8 (H2/16)	T1xT39	ANGRAU, Baptala	1993	14.50	8.20	1.89	29.00	W 210
Vridhachalam-2 M 44/3)	T1668 of Katterpalli	TNAU,Vridhachalam	1985	7.40	5.10	1.45	28.30	W 320
Vridhachalam)3 (M 26/2)	Edayanchavadi	TNAU,Vridhachalam	1991	11.68	7.18	2.16	29.10	W 210
Ullal-1	8/46 Taliparmba	UAS, Ullal	1984	16.00	6.70	2.05	30.70	W 210
Ullal-3	5/37 Manjeri	UAS, Ullal	1993	14.70	7.00	2.10	30.00	W 210
Ullal-4	2/77 Tuni	UAS, Ullal	1994	9.50	7.20	2.15	31.00	W 210
Chintamani	8/46, Taliparamba	UAS, Chintamani	1993	7.20	6.90	2.10	31.00	W 210
UN-50	2/27 Nileshwar	UAS, Ullal	1995	10.50	9.00	2.24	32.80	W 180
NRCC-2	2/9 Dicheria	NRCC, Puttur	1989	9.00	9.20	2.15	28.60	W 210
Jhargram-1	T. No.16 of Baptala	BCKVV, Jhargram	1989	8.50	5.00	1.50	30.00	W 320
Bhubaneshwar –1	WBDC-5(V-36/3)	OUAT,Bhubaneshwar	1989	10.50	4.60	1.47	32.00	W 320
Goa-1	Balli-1	ICAR Res. Centre, Goa	1999	7.00	7.60	2.20	30.00	W 210

Vegetative Propagation

Since the yields were lower and variable from seed propagated progenies, efforts were made to establish cashew orchards through air layers. But inadequate root system without tap root resulted in poor establishment of orchards. In the East Coast, the plantations established through air layers were prone to damage during cyclones. Other methods of vegetative propagation like mound layering, stooling, side grafting, patch budding etc. gave varying degrees of success, but none could be adopted on a commercial scale.

Table 13.4: Cashew Varieties Recommended for different States

State	Varieties Recommended
Kerala	Madakathara –2, Amritha, Dhana (H-1608), Priyanka (H-1591), Kanaka (H – 1598) Selection-2, V4, V6, V7, VRI – 3,VRI – 2, Ullal-1, Ullal-3, Ullal-4
Maharashtra and Goa	Maharashtra V1, V4, V6, V7, VR1 – 2, VRI – 3, Ullal –1, Ullal – 3, Ullal – 4, BPP – 8, Goa-1
Tamil Nadu	VRI-1 (M 10/4), VRI-2 (M 44/3), VRI-3 (26/2)
Puducherry	VRI –2, VRI – 3
Karnataka	Ullal – 1,Ullal – 2, Ullal – 3, Ullal – 4, UN – 50, Selection-2, Chintamani – 1, VRI – 2, VRI – 3,V4
Andhra Pradesh	BPP-4 (EPM 9/8), BPP-6 (T No. 56), BPP-8 (hybrid)
Odisha	Bhubaneswar-1 (WBDC-v) VRI-2 (M 44/3), BPP-2
West Bengal	Jhargram-1, BLA 39/4
Assam	Ullal – 3, Ullal – 4, V1, V4, V7, VR1 – 2
Manipur	VRI-2, BLA 39-4
Nagaland	VR1 – 2, Ullal – 3, Ullal – 4, V1, V4, V7
Tripura	Ullal-3, Ullal – 4, V1, V4, V7, VRI – 2
Madhya Pradesh	VRI-2, V-1, V-4, V-6, BPP-4, BPP-6, BPP-8
Andaman and Nicobar Islands	VRI-2, V-1, V-4, Ullal-1, Sel. 1, Sel. 2, BLA 39-4

During early eighties, epicotyl grafting was used for quick multiplication, but the grafts had high mortality due to collar rot disease. Hence, softwood method of grafting was standardized by the scientists. Softwood grafting is practiced commercially after later half of eighties. In this method, 40-60 days old seedlings are used as rootstock, after retaining one or two pairs of leaves, and grafting the scion in the softwood portion of the rootstock. The leaves on the rootstocks are removed after the scion started sprouting. New grafting can be done even with green scion and stocks. Thus, softwood grafting can now be done round-the-year. However, during summer months (January-May) grafted plants should be protected by providing partial shade by erecting pandal of dry coconut fronds or nylon nets. Bordeaux mixture spray (1 per cent) may be given at 10 days interval during rainy season to control fungal infection of tender seedlings and grafted plants.

The old senile plantations of cashew give poor yield inspite of high input incurred. The productivity of such orchards can be improved by adoption of rejuvenation practices. The rejuvenation of unthrift cashew plantations through top-working involves beheading of trees, allowing juvenile shoots to start-out and taking up of *in-situ* grafting using procured scions of high-yielding varieties. Periods from November-to-March and February-to-June have been found to be ideal for beheading and *in-situ* grafting, respectively. It has been observed that the top-worked trees within a period of two years not only put forth a canopy of 3-4 m in diameter and 5-6 m in height (as that of 8-10 year old trees) but also give an yield of 3 to 5 kg nuts per tree in their first bearing itself.

PLANTING AND ORCHARD ESTABLISHMENT

The land is prepared with the onset of pre-monsoon showers by clearing bushes and wild growth, digging of pits, terracing the base of the trees etc. The square system of planting can be followed. The ideal time for planting is usually during monsoon season when the moisture in air surcharged (June-August) both in the West coast and East coast. If irrigation facilities are available, planting can be done throughout the year except winter months. Planting is done at 7.5 × 7.5 m or 8 × 8 m spacing to accommodate 156-175 plants per hectare. The pits for planting of 60 × 60 × 60 cm should be prepared with the start of monsoon and filled, with top soil and compost up to 45 cm level. Planting of grafts/seedlings is preferably done during July-August in the centre of the pit and covered with soil and pressed gently. Care must be taken to see that the graft joint remains above the ground level. Newly planted seedlings/grafts should be staked and the basin around the graft is mulched with green leaves to conserve moisture. Gap filling may be done as required.

In sloppy areas, terracing should be made to prevent soil erosion. To each tree, a flat basin of about 2 m radius is provided. Mulch the basins of plants with organic waste materials during early years. Mulching the tree basins will help in conservation of soil moisture, prevention of soil erosion, suppression of weed growth and regulation of the soil temperature. Under sloppy areas, soil and water conservation practices can be done by making trenches of 30 cm width and 60 cm depth. And convenient length may be taken in between rows along the contour. This will not only conserves soil and moisture but will also enable to improve the growth of cashew trees.

INTERCULTURAL OPERATIONS

Weeding

Maintenance operations like weeding, clearing land etc., are done regularly. Weeding is done twice a year before heavy rains to conserve available soil moisture during dry months. Weedicide can be applied 15-20 days after slashing, before the heavy rains. Agrodur-96 (2, 4-D) @ 40 ml per 10 litres of water or gramoxone @ 50 ml in 10 litres of water is sprayed. Weeding can also be repeated in the post-monsoon season.

Intercropping

Keeping in view, the long pre-bearing period and low economic returns in the early period of bearing and fluctuations in the yield and price from year-to-year, it is advisable to take up intercropping in cashew plantations. Intercrops should replace weeds and as such, should not compete for light, moisture and nutrients with cashew. Intercrops, which can be harvested very early in the dry season or at the end of rainy season, are very apt for cultivation in cashew gardens. Tall growing and nutrient exhaustive intercrops like certain varieties of sorghum and millets should not be encouraged between young cashew plants. Leguminous crops, such as, groundnuts and beans are very suitable for intercropping. Besides the annual crops, the arid zone fruit crops having less canopy especially anona, *phalsa*, etc., can also be taken up depending on the suitability.

Cover crops like *Calapogonium mucunoides* and *Controsema pubscens* may be sown in the beginning of rainy season. In Andhra Pradesh, legumes like horsegram, cowpea, groundnut are grown as intercrops. In Kerala and Karnataka, tapioca and pigeon pea are grown. Casuarina is a tree species planted in cashew plantations in Andhra Pradesh and Odisha. In Kerala and central Karnataka, pineapple is grown as an intercrop during the initial years of orchard establishment.

Manures and Fertilizers

Application of 10-15 kg farm yard manure or compost per plant is beneficial. The fertilizer recommendation for full bearing cashew tree is 500 g N (1 kg urea), 125 g P_2O_5 (800 g superphosphate), and 125 g K_2O (200 g muriate of potash) per tree applied twice a year, *i.e.*, during pre-monsoon (May-June) and post-monsoon (September-October) period.

Fertilizer should be applied in circular trenches around the plants. Trenches should be dug at 15 cm depth and at 50 cm, 75 cm and 1.0 m away from the seedlings during the first year, second year and third years of planting, respectively. Soon after the fertilizer application, the trenches are covered with soil. From fourth year onwards, fertilizers can be applied as broadcasting at 1.75 m width around the basins, after clearing the weed growth and the soil is forked. For leaf analysis, 4th leaf from top of matured branches is taken at the beginning of flowering.

Foliar application of nitrogen in the form of urea (2 per cent) alongwith endosulfan (0.05 per cent) two or three times at the emergence of flushes and at panicle initiation gives higher fruit set and can control major pests as well.

Irrigation

Cashew needs irrigation during summer months particularly at seedling stage. Protective irrigation especially in summer months and during January-March at fortnightly intervals @ 200 litres/plant improves fruit set, fruit retention, thereby increasing nut yield. It cannot withstand water stagnation and hence adequate drainage should be provided.

Training and Pruning

For strong framework and wanted canopy shape, training of cashew trees is necessary. However, no systematic training system is followed in India. Similary, regular pruning is not practised in cashew. Unwanted, weak, dried, diseased and criss-cross branches are removed. Single, clean stem up to 0.75 to 1 m height is allowed by removing side branches. Thereafter, three-to-five strong branches may be allowed to grow. Weaker and dried branches are removed. This should be done during June and July, when there is rain and care should be taken to smear 10 per cent Bordeaux paste to cut end. Pruning is carried out for the first two years after planting. First fruiting should be allowed in the third year after planting.

PLANT PROTECTION

A. Major Diseases and their Management

'Damping Off' of Seedlings

It is a serious disease of cashew plants in the nursery where drainage conditions are very poor. Different fungi causing 'damping off' in different areas are, *Pythium* spp., *Fusarium* spp., *Scterotium* spp., and *Phytophthora* spp. However, *Phytophthora palmivora* is the most important fungus associated with damping off of the seedlings of cashew.

Symptoms

The fungus attacks either the root or collar region or both of the tender seelings, causing 'damping off' and thereby causes serious loss. The affected seedlings turn pale and show water-soaked girdles of darkened tissue around the tender stems. The affected seedlings later coalesce and eventually rot. In severe cases, leaves also exhibit water-soaked lesions, involving sometimes the entire lamina.

Management Practices

Adopt the following measures for the control of damping off of the seedlings in cashewnut:

1. Avoid water stagnation in the nursery or polyethylene bags.
2. Avoid excessive shade in the nursery.
3. Treat the growing/propagating media with Bordeaux mixture (1 per cent) or ceresan (0.1 per cent) before transplanting the seedlings.

Inflorescence Blight

Inflorescence blight is a serious malady, causing 30-40 per cent yield losses in cashew. The fungi like, *Gloeosporium mangiferae* and *Phomopsis anacardii* in association with tea mosquito (*Helopeltis antonii*) cause inflorescence blight.

Symptoms

The disease is characterized by drying of floral branches. The initial symptoms appear as minute water-soaked lesions on the main rachis and/or secondary rachii.

Exudation of gum is seen at the lesion site, which turns pinkish-brown in a day or two and then enlarges in size within 3 days. The affected inflorescence thus dries up and present scorched appearance. The incidence of this malady becomes aggravated if cloudy weather prevails for a longer period.

Management Practices

This disease has strong relationship with the population of tea mosquito. Hence, disease can be kept under control if the population of tea mosquitoes is kept under control. For this, spray endosulfan (0.05 per cent) or cabaryl (0.01 per cent) or phosalone (0.01 per cent) at the time of the emergence of inflorescence.

Anthracnose

This is one of the most widespread and serious diseases of cashew in many parts of world, but not so serious in India, except in Tamilnadu. It is caused by *Colletotrichum gloeosporioides*, which affect leaves and fruits.

Symptoms

Initially, reddish-brown shiny, water-soaked symptoms appear on affected plant parts, followed by exudation of resinous material from them. The affected flowers turn black, wither and fall off. The lesions grow longitudinally, resulting in ultimate killing of shoots. The tender leaves when affected, become crumpled and covered with tiny necrotic patches at the tip. Nuts and apples become shriveled, decayed or dried out. The nuts show small necrotic spots on their epicarp; while apples turn black and mummified. The fungus enters the fruit through the floral stigma. The symptoms vary according to the organs attacked.

Management Practices

1. Remove and destroy the affected plant parts.
2. As windbreak, grow fast-growing species like, casuarina or eucalyptus around the boundary of the cashew plantations.
3. Keep the population of tea mosquito under control by spraying endosulfan (0.05 per cent) or carbaryl (0.02 per cent).
4. Spray Bordeaux mixture (1 per cent) just after the rainy season at the time of emergence of new flushers.
5. Spray dithane M-45 (0.2 per cent) and repeat the same spray atleast 3 times at 3 weeks interval.

Pink or Dieback Disease

In our country, dieback or pink disease of cashew is caused by *Corticum salmonicolor* or *Pellicularia salmonicolor*, which may cause heavy losses during monsoon season in certain localities.

Symptoms

The affected branches and shoots show white or pinkish growth of the fungus on the bark. In advanced stages, the bark splits and peels off. In due course, the affected shoots start drying up from tip downwards and hence the name 'dieback'. On the affected branches, the leaves turn yellow and fall off, giving barren appearance to the affected portion of the tree.

Management Practices

Growers should adopt the following measures to control dieback in cashewnut effectively:

1. Prune the affected branches/shoots and destroy them.
2. Apply Bordeaux paste to the cut portions of the shoots.
3. Give prophylactic spray of Bordeaux mixture (1 per cent) twice, first during May-June before the on set of monsoons and second during October.
4. In case of gummosis, application of vitavex (1 per cent) is useful.

B. Major Insect-Pests and their Management

Major insect-pests of cashewnut and their management have been described briefly hereunder:

Tea Mosquito Bug (*Helopeltis antonii*)

Tea mosquito bug is considered as the most serious pest of cashew in Kerala, Karnatka, Goa, Maharashtra and Tamil Nadu, where it causes more economic losses than any other pest. It is estimated that this pest alone is responsible for a damage of nearly 25-30 per cent of shoots, 30-40 per cent of inflorescence and 15-20 per cent of tender nuts. Thus, reduction in yield caused by this pest may be upto 30-40 per cent. Because, tea is known to be an alternate host of this species (from which it was first identified) and it also resembles mosquito in sitting position and hence this pest has been named '*tea mosquito*'. Both nymphs and adults suck sap from the leaves, young shoots, inflorescence, developing young nuts and apples. The peak period of infestation is during the flushing, flowering and fruiting season. The adults also inject phytotoxins to the tender shoots and inflorescences at the time of feeding. Affected shoots show long black lesions and may die in severe cases. Infested inflorescence usually turns black and die and the immature nuts may drop off. Feeding on tender leaves causes crinkling. Heavily infested trees show scorched appearance, leading to death of the shoots and killing of the growing tips, thereby causing extensive secondary branches, developing into a multi-stemmed tree. The trees are covered with dead leaves and look as if scorched. It may also attract fungal infection, which results in blossom blight. Immature nuts shrivel and dry up, whereas, the old nuts and apples develop a characteristic 'scabby' appearance.

Management Practices

1. The best approach to reduce infestation of tea mosquito bug is to go for three rounds of spraying during flushing, flowering and fruiting season either with endosulfan or monocrotophos (0.05 per cent) or carbaryl (0.1 per cent) or quinalphos (0.05 per cent).
2. The recommendation of three rounds of spraying is based mainly on the fact that its population density is very high during these periods.

Stem and Root Borer (*Plocaederus ferrugineus* L.)

It is a serious pest of cashewnut throughout East and West coasts of India. Besides cashewnut, it has been found feeding on drumsticks, sapota and citrus trees. The pest occurs throughout the year, however, severe infestation has been reported during March-May in Karnataka and during June-July in Andhra Pradesh. The trees of 10-15 year age having thick bark tissue at collar region are more prone to attack. The grubs cause the damage, which on hatching, bore into the fresh tissues of bark feed on sap-wood tissues and make tunnels in irregular directions. As a result of injury to cells, resinous material oozes out, which on exposure to air gets hardened. The affected trees show different degrees of foliar yellowing and during later stages; there is shedding of leaves, drying of twigs and gradual death of the tree. Its infection could be identified by the presence of small holes in the collar region and oozing out of the gum from damaged portions. Bore holes and tunnels are plugged with chewed fibre, frass and excreta. Due to boring of sapwood by a number of grubs in each tree, the bark all around collar region withers away. The tree is killed in 1-3 years, depending upon the grub load.

Management Practices

Trees can be saved if infestation is detected at an early stage of attack. Hence, orchardist should try the following measures for controlling the dreaded pest of cashew:

1. Adopt phytosanitary measures in the orchard. Remove and burn all the affected trees or its parts to avoid further spread of the pest.
2. Locate the loose bark and bore holes, spike out or chisel out grubs and kill them periodically. After extraction of grubs, swab the trunk with carbon disulphide (1 per cent). Paint the trunk with a mixture of coal tar + kerosene (1:2) twice a year during March and November.
3. The dead trees and those beyond recovery should be uprooted and burnt as they serve as breeding ground for the borers.
4. Insert cotton swab soaked in carbon disulphide in bore holes and plug with cow dung or mud.
5. Drench the soil around the tree trunk with carbaryl (0.2 per cent).
6. Inject 10 ml of dichlorvos (0.1 per cent) per hole and plaster them with mud.

Flower Thrips (*Rhynchothrips raoensis*)

Flower thrips attack inflorescence of cashew and cause maximum damage to nuts at mustard stage of their development (30-40 per cent), followed by pea stage (20-30 per cent). The rasping and feeding injury made by the thrips results in the development of scab of floral branches, apples and nuts. There may be formation of corky lesions on the affected parts and subsequently shedding of flowers, improper filling up of the kernels, malformation of nuts and even immature fruit drop of the developing nuts take place.

Management Practices

Spray of dimethoate or monocrotophos or phosalone (0.05 per cent) once at the time of flowering and again during fruit set control this pest effectively.

Apple and Nut Borer (*Thylocoptila paurosema*)

The apple and nut borer may cause 40-50 per cent crop loss during the years of severe infestation in certain areas. Apple and nut borer is an actively moving caterpillar, which bores into tender apples and nuts and causes severe damage. The young larvae of apple borer, attack the apples of all stages, spoil them and cause heavy crop losses. Entry holes are plugged with excreta. The larvae feed inside and go on tunnelling the apple. As a result, the apples become soft and subjected to secondary infection by fruitflies and drop off along with immature nuts. The young larvae move to the joints of nut and apple, scrap the epidermis and then bore into them. The borer affected nuts do not develop and they get shriveled and dried up.

Management Practices

1. Collect and destroy the affected apple and nuts.
2. Spray monocrotophos (0.05 per cent) or endosulfan (0.05 per cent) or phosphamidon (0.05 per cent) at third round of spraying at fruit setting.

Leaf Miner (*Acrocercops syngramma*)

Leaf miner a is a major pest of cashew, which is commonly observed in the post-monsoon period flushes attacking tender leaves of new flushes; as a result, nearly 25-30 per cent of freshly emerged leaves are damaged. When the tender leaves change their colour, the attack or leaf miner is very prominent. Nursery seedlings and young plantation are more prone to the infestation of this pest. In certain areas, it may damage 80 per cent of leaves. The caterpillar causes the damage, which after hatching, start mining the epidermal layer on the upper surface of the tender cashew leaves, leaving tortuous markings. Later on, the thin epidermal mined areas swell up, as a result, the affected areas form blistered patches of greyish white colour. Normally, 2 to 8 blisters and as many as eight caterpillars are observed on a single leaf. When the infested tender leaves mature, big holes are manifested in the damaged areas. The results of injury are the permanent damage to the young leaves, which are shriveled, dried and shredded and prematurely drop off from the trees. In case of severe infestation, complete defoliation may occur, affecting the yield considerably.

Management Practices

A single spray of endosulfan or phosphamidon (0.05 per cent) during October-November *i.e.* at the time of emergence of new flush is effective in controlling this pest.

HARVESTING AND YIELD

The tree starts bearing from the third or fourth year, but full bearing is attained at the age of 10th year and continues for another 20-25 years. The tree begins to flower in December and the flowering continues for about three months. The fruits ripen during March to May or early June. Fully mature fruits are collected every day and the nuts are separated. A fully matured tree yields about 25-30 kg of nuts per annum. However, some of the high yielding trees may yield up to 40-50 kg/year. A grafted tree planted at proper spacing yield 5-10 kg from 6th year onwards. Harvesting is done by hand. Generally, ripe fruits fall to the ground and are gathered manually.

PROCESSING OF CASHEWNUTS

Processing of cashewnut involves cleaning, moisture conditioning, roasting, shelling, drying, peeling, grading and packing.

Cleaning and Sizing

Nuts must be cleaned before they are separated and grouped according to size.

Conditioning

Conditioning is done by adding water on the heap of nuts for 10 minutes and then draining. This operation is done several times until the cashewnuts have absorbed the required moisture *i.e.* to bring an optimum moisture level of 15-25 per cent. Conditioning is not required in case if the raw cashewnut is immediately processed without storage.

Roasting

Roasting is carried out at a temperature of 185°C to 190°C to remove the moisture from the nuts. After roasting, the nuts are cooled using cooled water spray and centrifuged in order to remove the excess liquid. Three important methods of roasting are: (1) drum roasting, (2) oil bath roasting, and (3) Steam roasting.

In drum roasting, the nuts are put into a rotating red-hot drum. The drum maintains the burning of the oil oozing out of the nuts. The drum is rotated by hand for about 2-4 minutes. Then the nuts are poured out and covered with wood ash to absorb the oil on the surface. The rate of shelling and turnout of whole kernels are high in this method.

In oil bath roasting, the conditioned raw nuts are passed for 1-3 minutes through a bath of heated cashewnut shell oil maintained at a temperature of 190-200°C by means of screw or belt conveyer. During roasting, the shell gets heated and cell walls get separated releasing oil into the bath. The roasted nuts are then conveyed into a centrifuge. The residual oil adhering to the surface of the nuts is removed by centrifuging. The roasted nuts are then mixed with wood ash.

Steam roasting is finding increasing acceptance in all parts of India. In this method, the raw cashewnuts are treated in a cooker filled with steam at a pressure of 100-110 kg/cm^2 for about 15 minutes. The treated raw nuts are spread out on the floor for cooling and then sent to the shelling section next day. The turnout and appearance of whole kernels from raw nuts treated in this method are said to be better than in any other method. The cashewnut shell liquid (CNSL) obtained in this method from the shells is very clear and command a premium price. About 75 per cent of the CNSL can be extracted from the shells.

Shelling

Shelling is the removal of dry roasted shell. By striking the head of the nut, the natural line of cleavage is broken. It is important when shelling the nut that the kernal is not broken as whole nuts command a higher price in the market. This operation is done manually mostly by skilled women. Wood ash is applied to the hands to prevent damage to the hands and kernel.

Drying

The shelled kernel is covered with the testa and to facilitate removal, *i.e.* to peel in order to produce the blanched kernel, and then the shelled kernel is dried in a drier. The most commonly used method for drying is Broma Drier. After drying for about 6-12 hours, the moisture content of the dried samples will be 2 to 4.5 per cent. This also protects the kernel from pest and fungus attack at this vulnerable stage.

Peeling

After drying, testa is removed from the kernel. The process is termed as 'peeling'. At this stage, the testa is loosely attached to the kernel, although a small amount of kernels may have already lost the testa during the previous operations. Manual peeling is done by gentle rubbing of dried kernels with the fingers. Those parts still attached to the kernel are removed by the use of a bamboo knife.

Grading

The grading operation is important as it is the last opportunity for quality control on the kernels. With the exception of a few grading aids, all grading is done by hand. Use of power driven rotary sieves is one mechanical method, another being two outwardly rotating rubber rollers aligned at a diverging angle. For large operations looking towards export markets, it is necessary to grade the kernels to an international level.

Rehumidification

Before the kernels are packed, it is necessary to ensure that their moisture content rises from 3 per cent to around 5 per cent. This is to make the kernels less fragile, thus lessening the risk of breakage during transport. In humid climates, the kernels may absorb enough moisture during peeling and grading to make a further rehumidification process unnecessary.

Packing

The normal packaging for export of kernels is in air-tight tins of 25 lbs in weight. The packing needs to be impermeable as cashew kernels are subject to rancidity and go stale very quickly. The tin container is familiar to most tropical countries as it is a replica of the four gallon kerosene or paraffin oil tin.

Byproducts of Cashewnut

Cashew Kernel Flour

Lower grade kernels are processed into cashew flour, which contains high protein and is easily digestible.

Cashew Kernel Butter

Kernel residue after extraction of kernel oil is used to produce cashew kernel butter, which is similar to peanut butter. The cake remaining after extraction of oil is used as animal feed.

Cashew Apple

After removing the nuts, cashew apples are wasted and practically not used industrially in any of the cashew growing states in India, except Goa. Cashew juice has 12-14 °Brix, containing 10.1-12.5 per cent sugars (mostly reducing) and about 0.35 per cent malic acid. It contains high amount of vitamin C, up to five times that in citrus fruits. Cashew apple can be used for preparing juice, vinegar, candy and jam. Cashew apple can also be used for making pickles and chutney.

Alcoholic beverages are prepared by fermenting the cashew apple juice. In Goa, a sort of brandy called 'fenni' is prepared from cashew apple.

Cashew Syrup

Extraction of juice and removal of astringency is done by the pre-treatment of juice. For this, add sugar at the rate of 1 to 1.25 kg for every litre of juice. Similary, twenty-to-twenty two gram citric acid per litre and 0.08 per cent as sodium benzoate are added to the juice. Dissolve sodium benzoate in a small quantity of water before adding to the mixture. Mix all ingredients thoroughly and keep it as such for three-to-five hours so that clear syrup forms a separate layer, which can be easily shiphoned. Syrup is then bottled.

Cashew Apple Jam

Cashew apple must be thoroughly cleaned by washing with water. Immerse the apple in 3 per cent salt solution for three days to reduce the tannin content, after which, the fruits are steamed for 15 to 20 minutes at 0.7 to 1.05 kg steam pressure. Then, the apples are crushed and mixed with 750 g sugar per kg of apple and boil it. A pinch of citric acid is added towards the end of the cooling process to improve the taste. Store it in well-sterilized jam bottles.

Cashew Nut Shell Liquid (CNSL)

CNSL is a naturally occurring phenol, which is present between the shell and kernel of cashewnut. It is obtained during the processing of nuts by isolation of kernel by kiln method or by solvent extraction. CNSL is toxic to human skin on contact. CNSL and derivatives have been employed for various preparations for use as resins and polymers. It is used in large quantities in brake-lining and in the preparation of antioxidants, lubricants, disinfectants, bactericides, fungicides, pesticides, herbicides, drugs, etc. It is also used with rice husk boards as binder for false roofing and as insulation panel.

Products from Bark, Stem and Leaves

Bark contains an acrid sap of brown resin, which is used as an indelible ink and printing lines and cottons. It is used as varnish. Since it contains plenty of tannin, it is used for tanning.

Stem yields amber coloured gum, which is used for book binding. The gum has insect repellent properties. The bark is used as firewood, and for charcoal. The pulp from the wood is used to fabricate corrugated and hard board boxes. The ash is rich in potassium and is used as a manure. Cashew wood can withstand sea water and is thus used for building fishing boats. The wood is also used for making furniture, false ceiling, interior decoration, etc.

Export of Cashew Kernels and CNSL

Standard specifications for Indian cashew kernels for purposes of export have been laid down by the Government of India under the Export (Quality Control and Inspection) Act 1963. The Government Act prescribes 33 different grades of cashew kernels. Of these, only 26 grades are commercially available and exported. They are White wholes (W-180, W-210, W-240, W-320, W-450 and W-500), Scorched wholes (SW, SW-180, SW-210, SW-240, SW-320, SW-450 and SW-500), Desert Wholes (SSW, Scorched Wholes Seconds and DW, Dessert Wholes), White pieces (B, Butts; S, Splits; LWP, Large White Pieces; SWP, Small White Pieces and BB, Baby Bits), Scorched Pieces (SB, Scorched Butts; SS, Scorched Splits; SP, Scorched Pieces and SSP, Scorched Small Pieces) and Dessert Pieces (SPS, Scorched Pieces Seconds and DP, Dessert Pieces). India imports raw cashew and also exports them all over the world as cashew kernels and cashew shell liquid (Table 13.5) and earns huge foreign exchange annually.

Table 13.5: Export of Cashew Kernels and Cashew Shell Liquid from India

Year	Cashew Kernel		CNSL		Raw Nut	
	Quantity	Value	Quantity	Value	Quantity	Value
2005-2006	1,14,143	2,51,486	6,405	709	5,65,400	2,16,295
2006-2007	1,18,540	2,45,515	5,589	920	5,92,604	1,81,162
2007-2008	1,14,340	2,28,890	7,813	1,197	6,05,970	1,74,680
2008-2009	1,08,131	2,95,024	6,976	1,679	6,05,654	2,63,178
2009-2010	1,12,322	3,35,235	7,575	1,772	6,16,258	28,54,225
2010-2011	1,28,131	34,25,024	7,972	1,872	6,25,552	3,25,215

Source: NHB Database-2011.

CROP IMPROVEMENT

Objectives

Although cashew is an important dollar earning crop of India, it has remained neglected since its introduction about four centuries ago. It was mainly due to poor genetic yield potential of the existing types and prolonged harvesting season, and low cashew productivity in India. The improvement of genetic yield potential of cashew and also the size of the nut are two most important factors.

Table 13.6: Cashewnut Varieties Developed through Selection from Germplasm at different Cashew Research Stations in India

Research Station	Variety	Source of Germplasm	Yield Potential (kg/tree)	Nut Wt. (g)	Kernel Wt. (g)	Shelling (per cent)	Kernel Grade (W)
National Research Centre for Cashew, Puttur, Karnataka	Selection-1	VTH 107/3 I (3/8 Simhachalam)	10.0	7.6	2.2	28.8	210
	Section-2	VTH 40/1 I (2/9 Dicheria)	9.0	9.2	2.5	28.6	210
Agricultural Research Station, Ullal, Mangalore, Karnataka	Ullal-1	8/46 Taliparamba	16.0	6.7	2.1	30.7	210
	Ullal-2	3/67 Guntur	9.0	6.0	1.8	30.5	240
	Ullal-3	5/37 Manjeri	14.7	7.0	2.1	30.0	210
	Ullal-4	2/77 Tuni	9.5	7.2	2.2	31.0	210
Agricultural Research Station, Chintamani, Karnataka	Chintamani-1	8/46 Taliparamba	7.2	6.9	2.1	31.0	210
Cashew Research Station, Bapatla, Andhra Pradesh	BPP-3	3/3 Simhachalam	11.0	4.8	1.3	28.1	400
	BPP-4	9/8 Epurupalem	10.5	6.0	1.3	23.0	400
	BPP-5	Tree No. 1	11.0	5.2	1.2	24.0	400
	BPP-6	Tree No. 56	10.5	5.2	1.2	24.0	400
Cashew Research Station, Bhubaneshwar, Odisha	Bhubaneshwar-1	WBDC-V (Ven. 36/3)	10.5	4.6	1.4	32.0	320
Regional Research Station, Jhargram, West Bengal	Jhargram-1	T. No. 16 of Bapatla	8.5	5.0	1.5	30.0	320
Cashew Research Station, Anakkayam, Kerala	Anakkayam-1	BLA 139-1 (T. No. 139 of Bapatla)	12.0	6.0	1.7	28.0	280
Cashew Research Station, Madakkathara/Anakkayam, Kerala	Madakkathara-1	BLA 39-4 (T. No. 39 of Bapatla)	13.8	6.2	1.6	26.8	280
	Madakkathara-2	NDR 2-1 (Neduvellur 2-1)	17.0	7.3	2.0	26.2	240
	K 22-1	Kottarakkara 22	13.2	6.2	1.6	26.5	280
Regional Fruit Research Station, Vengurla, Maharashtra	Vengurla-1	Ansur 1	19.0	6.2	1.9	31.0	240
	Vengurla-2	WBDC-VI (Ven. 37/3)	24.0	4.3	1.4	32.0	320
Regional Research Station, Vridhachalam, Tamil Nadu	VRI-1	M 10/4 (Vazhisodhanaipalayam)	7.2	5.1	1.4	28.0	320
	VRI-2	M 44/3 (T. No. 1668 of Kattupalli)	7.4	7.2	1.4	28.3	320
	VRI-3	M 26/2 (Edayanchavadi)	10.0	5.0	2.1	29.1	210

Table 13.7: Chief Characteristics of Cashewnut Hybrids Developed at different Research Stations in India

Research Station	Hybrids	Source of Germplasm	Yield Potential (kg/tree)	Nut Wt. (g)	Kernel Wt. (g)	Shelling (per cent)	Kernel Grade (W)
Cashew Research Station, Bapatla, Andhra Pradesh	BPP-1 (H 2/11)	T. No. 1 x T. No. 273	10.0	5.0	1.3	27.5	400
	BPP-2 (H 2/12)	T. No. 1 x T. No. 273	11.0	4.0	1.0	25.7	450
	BPP-3 (H 2/16)	T. No. 1 x T. No. 39	14.5	8.2	2.3	29.0	210
Cashew Research Station, Madakkathara, Kerala	Dhana (H 1608)	ALGD 1 x K 30-1	17.5	9.5	2.2	28.0	210
	Kanaka (H 1598)	BLA 139-1 x H 3-13	19.0	6.8	2.1	31.0	210
	Priyanka (H 1591)	BLA 139-1 x K 30-1	16.9	10.8	2.8	26.5	180
Regional Fruit Research Station, Vengurla, Maharashtra	Vengurla-3	Ansur 1 x Vetore 56	14.4	9.1	2.4	27.0	210
	Vengurla-4	Midnapore Red x Vetore 56	17.2	7.7	2.4	31.0	210
	Vengurla-5	Ansur Early x Mysore Kotekar 1-61	16.6	4.5	1.3	30.0	400
	Vengurla-6	Vetore 56 x Ansur 1	13.8	8.0	2.2	28.0	210
	Vengurla-7 (H 255)	Vengurla 3 x M 10/4	18.5	10.0	2.9	30.5	180

From commercial point of view, the ideal cashew plant should have dwarf and compact canopy with intensive branching habit, short flowering and fruiting phase, more than 20 per cent perfect flowers, 8-10 fruits/panicle, medium-to-bold nuts (8-10 g each) with shelling of more than 28 per cent, high yield potential (more than 20 kg/tree/year) and tolerance/resistance to major diseases and pests. In order to develop such an ideotype, efforts are being made at various research stations through introduction, selection, hybridization and mutation breeding.

Breeding Methods

Selection

A vast variation in cashew exists in India as a result of introduction of a large quantity of seeds over a number of years. Intensive surveys have been made to select elite cashew clones. This approach has led to significant breakthrough in developing improved cashew cultivars in India. As a result, 23 promising clones have been selected by this method. Characteristic features of these clones are given in Table 13.6.

Hybridization

Hybridization work is in progress in seven research station in India. As a result, eleven hybrids have been released. The salient features of important hybrids are given in Table 13.7.

Mutation Breeding

Irradiation of cashew budsticks with 1 and 2 Kr of gamma rays led to 100 per cent sprouting and variation in phenotypic characters *e.g.* leaf shape, leaf thickness and leaf venation. However, no mutant has been released.

14

Ber

Ber is one of the most suitable fruit trees for arid and semi-arid regions. It is also known as Indian jujube, Chinese date, Chinese fig or poor man's fruit. Ber can do well even under marginal growing conditions and provides quality yields at low-cost. Cultivation of ber under the harsh conditions of the Thar desert of India (West Rajasthan) demonstrates its adaptation to desert conditions. Flowering and fruiting of ber coincide with the availability of maximum rain water (monsoon rains), during July-September. It has a long tap-root and can withstand high temperatures during the summer. During the dry hot summer (April-June), it undergoes dormancy by shedding its leaves, thus evading the injury of drought. The fruits are rich in sugars, vitamin C, A and B complex. Ber leaves contain 5.6 per cent digestible crude protein and 49.7 per cent total digestible nutrients and are thus a nutritive fodder for animals. The tree propagates freely and greatly resists stress conditions in regions experiencing recurrent droughts. It is thus an important tree suitable for integration into the agro-forestry systems in the warm desert eco-regions. The tree can provide economic sustenance to the region and insurance against ecological degradation.

ORIGIN, HISTORY AND DISTRIBUTION

Ber (Ziziphus mauritiana Lamk.) is said to be an indigenous fruit crop. Besides, Z. nummularia (syn. Z. rotundifolia Larnk.), Z. oenoplia Mill., Z. rugosa Larnk. and Z. sativa Gaertn. are also native of India and yield edible fruits. Ziziphus jujuba Mill. (Chinese jujube) also produces large fruits which is grown in China, Korea, Russia and some South European countries but is not common in India. Several Ziziphus species are found in North and Latin America, the Middle East and South East Asia. According to De Condolle (1886), the centre of origin of ber is Central Asia, where it is found under varying climatic conditions. *Ber* is grown in India traditionally from ancient times where it has been in use for almost 4,000 years. It is one of the most suitable fruit trees for arid and semi-arid regions. The *ber* is distributed world-wide including Indian sub-continent, South-East Asia, Australia, China, Africa, Mediterranean region and American centre but its cultivation is confined over drier part of the globe and main cultivation occurs in India. In India, it is most widely cultivated in the states of Punjab, Haryana, Rajasthan, Uttar Pradesh, Madhya Pradesh, Bihar, Gujarat, Maharashtra etc.

COMPOSITION AND USES

Ber is known as poor man's fruit. In some aspects, it is better than apple (Table 14.1). The dehydrated fruits are kept for a long time and are consumed in the off season. The ripe fruits have cooling effect, laxative; remove burning sensation, thirst and blood impurities, whereas dried fruits are laxative and an appetizer. Among processed products, *ber* preserve, candy, squash, jam, nectar etc. can be prepared. Stem bark, root and leaves also have some medicinal values. Leaves of *ber* are used as fodder in dry regions. *Ziziphus xylopyrus* is a host plant for rearing the *Tachardia lacca*, a lac insect. The pruned wood is used as fuel-wood.

Table 14.1: Nutritional Composition of Ripe *Ber* Fruit

Component	Per 100 g Edible Portion	Component	Per 100 g Edible Portion
Total soluble solids	16-23 °Brix	Phosphorus	0.01- 0.02 g
Total sugars	4.9-12.4 g	Iron	0.5-1.0 mg
Reducing sugars	2.0-5.8 g	Ascorbic acid	66-133 mg
Calcium	0.03-0.04 g		

TAXONOMY AND BOTANICAL DESCRIPTION

Taxonomy

Ber or jujube belongs to genus *Ziziphus* of family Rhamnaceae, which has about 50 genera and 600 species. The genus *Ziziphus* consists of more than 100 species, of which, about 40 are grown in India. Several wild forms like *Z. rugosa, Z.rotundifolia, Z. xylocarpa, Z. nummularia, Z. ocenoplia, Z. sativa, Z. lotus,* and *Z. mistol* etc. grow in India and China. Most cultivated types belong to *Z. jujube* (Chinese ber) and *Z. mauritiana* (Indian ber). Indian ber has a chromosome number of 2n = 48, some varieties being tetraploid (2n = 96, Umran), and the Chinese ber has 2n= 24 with polyploidy variant with 2n = 48 and 2n =96. Further, Chinese ber is thornless whereas Indian ber is thorny. Similarly, Indian ber sheds leaves in summer but Chinese ber is evergreen.

There are two species of *Ziziphus*, which are still cultivated on a small scale. These are:

Z. spina-christi (L.) Desf.

Shrub, often with intertwined branches, or small tree up to 10-15 m tall. Dark deeply furrowed and scaly, white-brown to pale-grey. Branchlets densely pubescent white when young. Fruiting branchlets are not deciduous. Tends to produce a very deep tap-root. Mostly spinous with paired spines, unequal in length, one recurved; rarely narmed.

Z. *lotus* (L.) Lam.

This species is a spiny shrub, growing up to 1.5 m tall and resembling Z. *jujuba*. However, fruiting branchlets are not deciduous and twigs are grey. Internodes on branchlets are less than 1 cm long. Leaves are sub-orbicular or broadly elliptic-to-ovate, shallowly glandular-crenate, pubescent beneath and less so above; size (0.5-)1.2(-1.5) cm long × (0.4-)1(-1.3) cm broad.

In the drier parts of northeren-western India, Z. *nummularia* (Syn. Z. *rotundifolia* Lam.) is found. It is also a useful rootstock. The species is a thorny shrub producing red, edible fruits. Three other species need mention since they are sometimes used as rootstocks. Z. *xylopyra* is an erect, small tree frequently unarmed and producing a woody, inedible fruits. It is a species of South India and Sri Lanka. Z. *rugosa* is a straggly bush tending to have solitary spines and edible fruit, found in the central Hills and eastern parts of India. Z. *oenoplia* is a scrambling shrub with spines and small black fruits often used for tanning and found in South India, Sri Lanka and Myanmar.

Botany

The plant is a vigorous grower and has a rapidly-developing taproot. It may be a bushy shrub, 4 to 6 ft (1.2-1.8 m) high, or a tree 10 to 30 or even 40 ft (3-9 or 12 m) tall; erect or wide-spreading, with gracefully drooping branches and downy, zigzag branchlets, thornless or with short, sharp straight or hooked spines. It may be evergreen, or leafless for several weeks in hot summers. The leaves are alternate, ovate- or oblong-elliptic, 1 to 2½ inch (2.5-6.25 cm) long, 3/4 to 1½ inch (2-4 cm) wide; distinguished from those of the Chinese jujube by the dense, silky, whitish or brownish hairs on the underside and the short, downy petioles. On the upper surface, they are very glossy, dark-green, with 3 conspicuous, depressed, longitudinal veins, and there are very fine teeth on the margins.

The 5-petalled flowers are yellow, tiny, in 2's or 3's in the leaf axils. The fruit of wild trees is ½ to 1 inch (1.25-2.5 cm) long. With sophisticated cultivation, the fruit reaches 2½ inch (6.25 cm) in length and 1¾ inch (4.5 cm) in width. The form may be oval, obovate, round or oblong; the skin is smooth or rough, glossy, thin but tough, turns from light-green to yellow, later becomes partially or wholly burnt-orange or red-brown or all-red. When slightly underripe, the flesh is white, crisp, juicy, acid or sub-acid to sweet, somewhat astringent, much like that of a crab apple. Fully ripe fruits are less crisp and somewhat mealy; overripe fruits are wrinkled, the flesh buff-coloured, soft, spongy and musky. At first, the aroma is apple like and pleasant but it becomes peculiarly musky as the fruit ages. There is a single, hard, oval or oblate, rough central stone, which contains 2 elliptic, brown seeds, 1/4 inch (6 mm) long.

SOIL AND CLIMATIC REQUIREMENTS

Soil

Ber provides a good scope for cultivation on soils, which have so far been considered marginal or even unsuitable for growing other fruits as it is not particular in its soil requirement and can grow on a wide variety of soils, including saline and alkali patches. Once established, *ber* can withstand even high salinity (21 dSm^{-1}) in soil and can flourish even in soils with pH as high as 9.2.

Climate

The *ber* is grown on a variety of climates up to 1,000 m above sea level. It can withstand extremely hot conditions but young plants are susceptible to frost. The trees shed leaves and enter into dormancy during summer. Under moderate climate of South India, however, the trees continue to grow throughout the year. *Ber* tree prefers atmospheric dryness for development of good quality in fruits.

IMPORTANT CULTIVARS

Very wide variability with respect to plant and fruit characters exists in *ber* genotypes. In India, more than 300 varieties have been listed but only a few, such as Umran, Banarasi Karaka, Sanaur-2, Mundia, Gola, Seb and Kaithali, are commercially important.

Umran

This variety is cultivated on a large scale in Punjab and Haryana. The fruit is large, oval in shape with a roundish apex and has an attractive golden-yellow colour, which turns into chocolate-brown at full maturity. The fruit is sweet, with 14-19 per cent TSS and has pleasant flavour and excellent dessert quality. It is a prolific cropping variety, yielding 150-200 kg of fruit per tree. The fruits ripen late from second fortnight of March to mid-April and have a good keeping quality. It is susceptible to powdery mildew. Clonal selection form Umran cultivar, Goma Kirti matures about 3 weeks earlier and has similar physical and chemical characters.

Sanaur-2

This is a selection from Sanaur – a small town near Patiala (Punjab), which is known for *ber* cultivation. The fruit is large and oblong with a roundish apex. On ripening, fruits attain light-yellow colour and TSS of 18-19 per cent. Like Umran, it is also a prolofic bearer, yielding about 150 kg fruit per tree. It is a mid-season variety, ripening during second fortnight of March under Punjab conditions and has been found fairly resistant to powdery mildew disease.

Gola

Gola is an early maturing cultivar having spreading growth habit and is suitable for rainfed and saline sodic areas. Its golden yellow, glossy round fruit has 80 per cent edible portion and semi-soft, white and juicy flesh. Average fruit weight ranges from 25 to 40 g and TSS from 17 to 23 °Brix. A well developed tree of Gola produces 40-60 kg fruit under rainfed and 100-110 kg fruit under irrigated conditions.

Mundia

It is an early to mid-season cultivar. Its tree has semi-erect growth habit. It bears bell-shaped fruits of nearly 40 g weight having soft flesh and about 20 °Brix TSS. Average fruit yield under irrigated conditions is about 100-120 kg/tree.

Kaithli

This variety is a selection from Kaithal in Haryana. The fruit is medium in size, oval in shape and has a tapering apex. Fruit pulp is soft and sweet with TSS of 14-16 per cent. Fruits ripen in the second fortnight of March to first week of April. The average yield is 120 kg fruit per tree. This is an excellent variety but appears to be more susceptible to powdery mildew disease.

Banarasi Karaka

It is also a mid-season cultivar and produces yellow coloured oval fruits. Its tree has semi-erect growth habit. Average fruit weight ranges from 24-30 g. Fruit flesh is creamy, soft and has 16-18 °Brix TSS.

Seb

It is a late-maturing cultivar with upright growth habit. The fruits are apple-shaped of about 25 g in weight and the flesh has 17.5 °Brix TSS. Its slightly rough fruit surface remains greenish-yellow even after maturity. The cultivar produces little lower yield (40-50 kg/plant) under rainfed while under irrigated conditions, the yield is equal to that of Gola.

Z.G.-2

The fruit is medium in size and roundish in shape with smooth skin. The fruit pulp is soft with an excellent sugar-acid blend. When ripe, the fruits attain light yellow colour and TSS of 15-16 per cent. The average yield amounts to 150 kg fruit per tree and the ripening time extends from second fortnight of March to first week of April. This variety is recommended for growing in local markets only. It is less susceptible to powdery mildew.

Cultivars such as Gola, Seb and Mundia are suitable for extremely dry areas, Banarasi Kadaka, Kaithali and Meharun for the dry regions and Sanaur 2, Meharun and Umran for comparatively humid regions. In northern India, Gola is earliest to ripen, Kaithali and Mundia are mid-season and Umran and Seb are late cultivars. An early maturing selection from population of Umran, trees named 'Goma Kirti', has been identified at Godhra. Cultivar Gola has also been found tolerant to saline soils.

In Assam, 5 wild or cultivated types, collected from various parts of the state, have been described by S. Dutta:

1. 'Var. 1'–a very thorny wild shrub, with small, round, inferior fruits; grown as a fence to protect crops.
2. 'Var. 2'–a wild, thorny tree to 30 ft (9 m) with red-brown, tough-skinned fruit having slimy, acid-sweet pulp. Much eaten by children and rural folk. Commonly used in cooking and preserving.

3. *'Var. 3'*–a very thorny, spreading tree. Fruit dark-red or brown, with sour pulp. Bears heavily. Planted for shade.

4. *'Var. 4'* ('Bali Bogri')–a wild, thornless tree, with greenish-yellow fruits blushed with red; pulp slightly slimy, mealy, sweet-and-acid, of good flavour. Bears heavily.

5. *'Var. 5'* ('Tenga-Mitha-Bogri')–A wild, thorny tree, with oblong, brownish fruit; pulp slightly slimy, sweet-and-acid, with very pleasant flavour. Bears heavily. A choice jujube recommended for vegetative propagation and commercial cultivation.

FLOWERING, POLLINATION AND FRUIT SET

Flowering in *ber* starts in August and continue upto September-October. Pollen of the *ber* is thick and heavy. It is not airborne but is transferred from flower-to-flower by honeybees (*Apis* spp.), a yellow wasp (*Polister hebraeus*), and the house fly (*Musca domestica*). The cultivars 'Banarasi Karaka', 'Banarasi Pewandi' and 'Thornless' are self-incompatible. 'Banarasi Karaka' and 'Thornless' are reciprocally cross-incompatible. Initial fruit set is very high but 14-16 per cent fruit set is considred good for commercial yield.

PLANT PROPAGATION AND ROOTSTOCK

Seeds of *Katha ber* or *boradi* (*Zizyphus mauritiana* Lamk) are generally used for raising rootstock, which are easily available and possess the qualities of a good rootstock. The *ber* plants should be budded on *Ziziphus mauritiana* (Elongated Dehradun) for higher fruit yield. *Ber* plants raised on semi-vigorous rootstock, *Ziziphus mauritiana* (Coimbatore) can profitably be planted at a closer spacing of 6 × 6 m. Therefore, about 50 per cent more plants/ha can be accommodated with over 20 per cent increase in yield of equally good quality fruit. Seeds of Mallah *ber* (*Ziziphus nummularia*) can also be used as rootstock. The seedlings of Mallah *ber* are slow growing and become buddable after longer period than the seedlings of Katha *ber*.

The *ber* is usually propagated by budding. The most common methods are modified ring, patch and 'I' or 'T' (shield) budding. Rootstock seedlings are raised by sowing seed kernels extracted by breaking the stones, which germinate in about one week. Seed stones collected from dropped fruits contain 50-70 percent non-viable seeds. Seeds should be dipped into a salt solution of 17-18 percent concentration for 24 hours before sowing. The floating seeds should not be sown as these are generally non-viable. If the seed stones are sown as such, germination takes nearly one month. Germination of seed stones can be improved by soaking them for 48 hours in water or by treatment with sulphuric acid. The *ber* seedlings are raised from Katha *ber* stones, which are sown during March-April, after fresh extraction, in well-prepared nursery field at a distance of 15 cm in rows and 30 cm apart. However, seeds of other *Ziziphus* species such as *Z. nummularia* can also be used for raising rootstocks. Germination starts in about 3-4 weeks and seedlings make a rapid growth. The seedlings should be trained to a single stem. Nearly one-fourth of the seedlings attain buddable size of a lead pencil by August, while the rest are ready for budding by next April. To minimize root damage during transplanting, seeds can be sown in 300 guage polythene tubes of 25 cm length and 10 cm diameter, filled with a mixture of FYM, sand and clay in 1: 1: 1 ratio.

When raised in polytubes, the budlings become ready for transplanting in 1-2 months after budding during August-September. Plants raised by polytube technique have been observed to retain deep rooting tendency and thus prove the most suitable under the low rainfall areas. The stock seedlings should be healthy and vigorous and allowed to grow as a single stem only. The budding is done when the stock stem has attained the thickness of a lead pencil. *Ber* orchard can also be raised by *in situ* budding after transplanting the polytube raised seedlings during the monsoon for budding them during the next year. To ensure maximum success, 2-3 shoots emerging from ground level are budded in the field so that later, one of them can be selected to form the canopy. Sprouts emerging from the rootstock portion should be regularly removed. Budding operation should be done when there is a proper flow of sap in the stock to be budded. Shield-budding is done during March-April or August-September, but it has been found that August-September budding gives a far better success. The ring-budding is preferable during June-July when the new growth starts. Shield-budding done during August-September has given success of 75-81 per cent, whereas budding in April has given a little success. The highest budding success is also achieved in June.

At Punjab Agricultural University, Ludhiana, horticulturists have experimented with stooling as a means of propagation. They transplanted one-year-old seedlings into stool beds, cut them back to 4 inch (10 cm) and found that the shoots produce roots only if ringed and treated with IBA, preferably at 12,000 ppm. Likewise, they also recommended that air-layers will root if treated with IBA and NAA at 5,000 to 7,500 ppm and given 100 ppm boron. Cuttings of mature wood at least 2-year-old can be rooted and result in better yields than those taken at a younger stage.

Selection of Stock and Scion

The stock seedlings should be healthy and vigorous and is allowed to grow as a single stem only. The budding is done when the stock stem attains the thickness of a lead pencil. Collection of budsticks from selected trees, which are known for bearing a heavy fruit crop of good quality is inevitable to ensure the quality of planting materials. In addition, the mother plant should be healthy, vigorous, free from diseases and insect-pests and should be true-to-type. About 30 cm long, two-to-three months old shoots with plump buds should be selected as budsticks. For sending to distant places, about 20 cm long budsticks should be taken. The leaves of the budsticks should be cut away, keeping the leaf stalks attached to the buds. These should be tied in small bundles and wrapped in moist piece of cloth. The budsticks should be kept moist till they are utilized for budding. Through such arrangements, the budsticks can be kept for 2-3 days in good condition.

PLANTING AND ORCHARD ESTABLISHMENT

Transplanting of budlings or rootstock seedlings is best done with onset of monsoon. For this, pits of 60 × 60 × 60 cm are dug either in a square or rectangular system during summer at a spacing of 6 × 6 m in low rainfall areas and at 8 × 8 m in irrigated and higher rainfall regions. While filling the pits in sandy areas, a layer of bentonite clay can be placed at the bottom and sides of the pit to reduce moisture losses by infiltration. Each pit, is refilled with top-soil mixed with 15-20 kg FYM alongwith 50 g of chloropyriphos dust to protect damage by termite. In rainfed areas, the interspaces between the tree rows can be shaped to provide 5 per cent slope towards the plant. This would help to accumulate runoff water during monsoon near the tree roots, thereby resulting in higher establishment success. In irrigated areas, *ber* plants can be transplanted during January-March also.

INTERCULTURAL OPERATIONS

Training and Pruning

The pruning of *ber* trees is highly desirable to maintain their vigour and productivity as well as to improve fruit size and quality. Pruning also saves the fruit from being affected by the powdery mildew disease and strong winds. *Ber* has a characteristic growth form producing branches, usually starting from the sixth or ninth node from the base and subsequently at regular intervals of three nodes. These form the first order sylleptic branches or secondaries. Sylleptic branches are produced on proleptic branches, which are laterals from a terminal meristem with some intervening period of the rest of the lateral meristem. The *ber* tree remains young upto 30 years, if proper pruning is done regularly. During the first 2-3 years after planting, *ber* trees are trained to develop a strong frame. After planting during July, the main shoot is headed back during March keeping 1-2 nodes above the bud union to induce vigorous new growth from which one upright vigorous shoot is retained to develop into the main trunk. This is kept clean of side shoots up to 30 cm height from ground level and thereafter, 3-4 well spaced shoots are allowed to grow. The main trunk is then headed back. During the spring of the second year, the secondaries are pruned to retain their basal buds as these basal buds produce vigorous shoots. More than one shoot may emerge from the basal bud, but only one of these is retained, while rest are removed. These form the main branches of the main tree. On these main branches also, 3-4 upward growing well spaced shoots are allowed to grow. The process is continued to develop tertiary branches. Care is taken to allow only one upright growing shoot at each node to avoid narrow crotches and thereby develop a strong frame. This basic frame of the tree is maintained by removal of water sprouts from time-to-time.

In unpruned *ber* tree, the canopies of the trees get un-necessarily enlarged, the growth and branchlets become weak and both fruit size and quality get impaired. Ultimately, such trees become economically unproductive besides occupying large orchard space. Thinning out of some branches of *ber* trees is also necessary to avoid too much crowding so as to admit adequate sunlight and facilitate proper aeration. Moreover, fruit bearing in *ber* on current season's growth and the fruit quality depends on the vigour of shoots. Therefore, annual pruning is done to induce maximum number of healthy new shoots. The best time for pruning is during the hot and dry season when the tree sheds leaves and enters into dormancy. The time of pruning differs in different parts of the country. In Tamil Nadu, pruning is done during January to April, in Maharashtra by the end of April and in North India, it is done during the last week of May. In general, the light pruning, *i.e.* heading back of 25 per cent of the previous year's growth (branchlets, shoots, etc) or of the past season's main shoot at 25 buds is desirable to obtain heavy yield, good fruit size and better quality. However, under moderate climatic conditions, pruning is done at 15-20 buds and then, all the secondary shoots are completely removed. In order to avoid the occurrence of long unfruitful basal portions of branches as a result of light pruning for several years, half the number of past season's shoots are pruned to 20 buds and the remaining half to the basal 1 or 2 nodes. Spray of 3 per cent thiourea or potassium nitrate, 2 days before pruning helps to induce bud sprouting from maximum number of nodes.

Nutrient Management

Ber orchards are seldom manured. However, productivity of trees can be improved by nutrient application. The doses of manures and fertilizers depend on the fertility status of the soil age of tree, fertilizer to be used and nutrient status of soil. Proper nutrition of *ber* tree is necessary to get good crop over the years. The fruit becomes large and attractive and get decent price in the market. In sandy soils at Jobner (Rajasthan), application of 750 g N per tree gave the highest yield, while at Rahuri (Maharashtra), 250 g N and 250 g P_2O_5 increased fruit yield. Potash application does not give any response. Besides, an yearly dose of 40-50 kg well rotten FYM per tree is also applied. However, when no recommendation is available, 20 kg farmyard manure and 100 g nitrogen (400 g CAN) can be applied for one-year-old *ber* tree. Similar amount of farmyard manure and nitrogen should be increased every year up to the age of five years. The quantity of farmyard manure and nitrogen should be stabilized at 100 kg and 500 g (2 kg CAN), respectively, after the age of five years. Half of the phosphatic fertilizers may be applied during rainy season (July-August) and the other half at the time of fruit-set (October-November). The fertilizer should be evenly spread in the basins of trees upto the periphery. After adding the fertilizer, light hoeing with spade or *khurpa* should be given to the basins to mix it thoroughly with the soil. After this, a light irrigation should be given. In sodic soils, having pH over 8.5 and ESP over 21 per cent, *ber* can be successfully planted if amendments such as gypsum and pond soil are added to each pit at the time of filling.

Weed Management

The *ber* tree begins to bear after one year of its planting in the field. To develop the tree properly, it is advisable that no fruit should be taken at least for the first two-three years. Intercropping can be successfully practised on the vacant land in the young orchard during the first four years. Only leguminous crops of short stature like gram, moong and *urd* can be grown to get some income from the land in these initial years. These crops also enrich the soil by fixing nitrogen. Covers cropping with stylosanthes and moth bean have been found to improve fertility and moisture status of the soil in *ber* orchards. The other exhaustive and tall-growing crops should not be grown in the *ber* orchard as they deplete the soil of its nutrients to a greater extent and compete for light with the trees.

Intercropping

Weeds compete for nutrients and water and act as alternate host for diseases. Therefore, timely removal of weeds is essential. Pre-emergence application of hexuron 80 WP (diuron) @1.2 kg/acre can be made during the first fortnight of August when field is free from growing weeds and stubbles. Glyphosate @ 1.2 litres/acre or paraquat @ 1.2 litres/acre as post-emergence should be sprayed when the weeds are growing actively preferably before weeds flower and attain a height of 15-20 cm. Spray of glycel should be done during the calm day to avoid spray drift to the foilage of the fruit trees.

Water Management

For establishment of young plants, 10 irrigations are considered essential during the first year. Irrigation is also essential during the development of fruit, *i.e.* from October-to-February at intervals of 3 or 4 weeks depending upon the weather. Trees will continue to bear even if no irrigation is applied during this period but the yield is substantially reduced because of heavy fruit drops and smaller size of the remaining fruit. Irrigation should be stopped in March as fruits on the branches lying on the ground get damaged and their ripening is delayed. During dormancy, *ber* need little or no irrigation; therefore, if irrigation is applied during this period, the trees would continue to putforth growth haphazardly, which is not desirable. In rainfed orchards, arrangement for *in situ* water harvesting should be made. For this giving 5 per cent slope to the inter-row spaces towards the trees has been found be useful. Mulching by use of black polythene conserves soil moisture and improves growth of the *ber* trees. Sprays of antitranspirants like 0.1 per cent power oil and 7.5 per cent kaolin have also been found effective to reduce transpiration and conserve water in the *ber* trees.

CROP PROTECTION

A. Major Diseases and their Management

Powdery mildew

This is the most serious disease of *ber*, and if not controlled/managmed in time, it may spoil whole crop. It is incited by a fungus, *Oidium erysiphoides,* which starts producing symptoms in November and lasts upto April. With the rise in temperature after February, the disease comparatively subsides. Powdery mildew of *ber* caused by *Microsphaera alphitoides* f. sp. *ziziphi* has also been reported from India.

Symptoms

The disease appears in October-November, when the temperature comes down the weather is cloudy (humid). Initially white spots of the fungus appear on leaves and developing fruits. In case of heavy infestation, powdery mass of the fungus spreads all over the surface and leaves of the fruits. The fruits drop off pre-maturely or become shrunken, misshapen and corky. The disease, if not controlled properly, may remain in the orchard until February-March, when temperature starts rising, thereby causing great losses.

Management Practices

Powdery mildew can be controlled effectively by following measures:

1. Give two sprays either of karathane (0.2 per cent), calixin (0.02 per cent) or wettable sulphur (0.2 per cent) at fortnightly interval well before the expected attack of powdery mildew. The disease can also be effectively controlled by two foliar applications of bayleton or karathane (0.1 per cent) initiating it with the first appearance of the disease, however, an alternate spray schedule of triadimefon (0.1 per cent) and wetsul (0.3 per cent) was also found to be most suitable and economical for controlling the disease.

2. *Ber* cultivars having field resistance to powdery mildew have also been reported. These are Chhuhara, Safeda Selected, Nazuk, Glory, Sanaur-2, ZG 2, ZG 3, ZG 4, Chinese, BS-1, Kathaphal, Kismish, Narna and Sonari No.3. The genotypes Darakhi-1, Darakhi-2, Seedless and *Zizhyphus rugosa* exhibited immune reaction to powdery mildew both in leaves and fruits.

Alternaria Leaf Spot

Several species of *Alternaria* have been reported to be associated with this disease of *ber*.

Symptoms

The disease is characterized by the formation of small irregular brown spots on the upper surface of the leaves. On lower surface, dark-brown to black spots are formed. The spots coalesce to form big patches. The affected leaves may dry and drop off from the trees.

Management Practices

The disease can be controlled effectively by spraying 0.2 per cent difolatan or dithane Z-78.

Sooty Mould or Black Spot

Sooty mould usually develop on the honeydew secreted by some sap sucking pests like aphids, hopper etc. However, in *ber*, it is caused by a fungus, *Sariopsis indica*.

Symptoms

The disease appears during October-November in the form of tuft like circular to irregular black spots on the lower surface of leaves. In advanced stages, the entire lower and upper leaf surface may be covered by sooty mould. The affected leaves may drop down.

Management Practices

Pathologists have recommended the adoption of the following measures for the control of sooty mould effectively:

1. Grow resistant varieties like, ZG-3, Safeda Rohtak, Mundia Murhera, Sanaur-1, Jhajjar Selection etc.

2. Spray bavistin (0.2 per cent) or difolatan (0.2 per cent) or bengard (0.1 per cent) or dithane Z-78 (0.3 per cent).

3. A mycoparasite, *Hansfordia pulvinata* keeps the disease under control.

Cercospora Leaf Spot

Two species of *Cercospora viz., C. zizyphi* and *C. jujubae* cause this disease.

Symptoms

Very minute, circular-to-oval yellowish spots, which are surrounded by dark-brown margins appear on the leaves. Later, these spots turn to brown colour. Affected leaves may wither and die.

Management Practices

Growers should adopt the following measures for the effective control of this disease:

1. Grow resistant varieties like, Safeda Rohtak, ZG-3, Kakrola Gola, Bahadurgarhia, Reshmi etc.

2. Spray of dithane M-45 (0.2 per cent) gives a very good control of this disease.

Fruit Rot

It is caused by many fungi *i.e., Phoma* sp., *Colletotrichum* sp., *Alternaria* sp. and *Pestalotia versicolour*. Light gray-brown spots appear on the lower end of the fruits and later whole area of the fruit is covered and the fruit turns dark-brown in colour. For control, collect all the affected fruit and destroy. Spray diathane M-45 or diathane Z-78 @ 0.2 per cent at the early stage of infection and repeat the spray after 15 days for effective control of rotting in fruits.

B. Major Insect-Pests and their Management

As many as 80 insect species have been reported to attack *ber*, but fruit fly and *ber* beetle are its most serious and destructive pests. The other pests, like bark eating caterpillar, leaf roller, shoot and fruit borer are also accountable for damage, but are generally regarded a minor pests. A brief description of the major pests of *ber* is given below:

Ber Fruit Fly (*Carpomyia vesuviana*)

It is the most serious pest of *ber* and sometimes the damage caused by it goes up to 80 per cent, rendering the *ber* cultivation unprofitable. The fly infestation hastens ripening and dropping of fruits, and such fruits are prone to bird attack. Off-season infested fruits exhibit more unevenness in shape than seasonal ones. The damage is caused only by the larvae, while the adult flies punctures the fruits for egg laying, affecting the growth of fruit in the vicinity of puncture, which results in deformity of fruits. The larvae feed on the fruit pulp and make galleries towards the center. The larval burrows are filled with brownish or dirty-white maggoty frass. The attacked fruits are rotten near the stones and emit a strong smell. Maximum damage is caused during February-March.

Management Practices

1. Collect fallen/damaged fruits twice a week, destroy or burry or feed to the animals.
2. Destroy the wild *ber* species from the vicinity of orchard.
3. Regular ploughing, digging or raking of soil around trees is necessary.
4. Left over, off-season and early set fruits on wild *ber* should be destroyed.
5. Harvest the fruit at maturity when colour starts changing.
6. Do not allow the fruits to over ripe.
7. Cuelure, methyl eugenol, trimedlure, terpinyl acetate, vertlure can be used to catch the adult flies.
8. Apply insecticides as 600 ml oxydemeton methyl 25 EC or 500 ml dimethoate 30 EC in 500 litre water/acre when fruits are of pea size. If necessary, repeat after 30 to 45 days.
9. At ripening, spray 500 ml malathion 50 EC + 5 kg gur or sugar in 500 litre water/acre.
10. Bait sprays having brewery waste and malathion and hydrolyzed protein for bait (20 g solids/litre solution) are also effective.
11. Methyl parathion 2D or fenvalerate 0.4 D @ 10 kg/acre provides good control.

Ber Chaffer Beetle (*Adoretus pallens*)

This scarabaeid beetle is predominant in northern India having preference for *ber* and grapes. The adults, commonly known as chafer beetles or May-June beetles are harmful to fruit and several forest trees, whereas the white grubs chiefly destroy the roots of several field crops. Only the adult beetles cause damage by cutting round holes in the leaves during night. In case of heavy attack, the entire tree is defoliated, adversely affecting the fruit bearing capacity of the tree. Swarms of adults appear early in spring, but their activity is maximum during the summer, which declines with heavy showers of monsoon.

Management Practices

1. Light traps are quite effective in trapping the adult beetles.
2. Raking of the soil around the tree is useful in exposing the hibernating grubs and killing them.
3. Spray carbaryl (0.02 per cent) in the evening as soon as the damage starts. If the damage persists, repeat the spraying at weekly intervals.

Ber Leaf Webber (*Synclera univocalis*)

The leaf webbers are more destructive to tender leaves of new flush especially on young plants and trees receiving frequent irrigation and nitrogenous fertilizers. The newly hatched caterpillar attacks the unopened or partially opened leaf and the later instars keep the leaf folded along the long axis with the help of silken threads. The larva consumes the green matter by scrapping, leaving behind the papery epidermis. Sometimes the advanced stage larva may join the two adjacent terminal leaves together longitudinally.

Management Practices

1. To control these leaf rollers, hand-picking and destruction of caterpillars is suggested.

2. On appearance of the caterpillars on new vegetative flush, spray 25 ml of fenvalerate 20 EC or 200 ml of quinalphos 25EC or 25 ml of cypermethrin 25 EC or 200 g of carbaryl 50 WP in 100 litre of water.

3. Repeat the spray after 3-4 weeks, if there is recurrence of pest infestation.

Hairy Caterpillar (*Euproctis fraterna*)

Hairy caterpillars are polyphagous and sporadic pests of several fruit trees, which are widely distributed in Africa, Australia and the Indian subcontinent. The caterpillars cause considerable damage to *ber*, and several other fruits. The caterpillars feed gregariously on ventral surface of tender leaves skeletonising the leaves completely. After third instars, the caterpillars segregate and feed voraciously, resulting in defoliation of twigs and branches and sometimes the whole tree. The caterpillar also nibbles the epidermis of fruits, resulting in corky and deformed surface. At harvest, body contact with hairs of caterpillarm may lead to skin irritation and itching to workers.

Management Practices

1. The freshly hatched larvae feed gregariously, so hand picking and mechanical destruction of egg masses and congregating larvae is quite effective on a small scale.

2. In case of severe infestation, spray the affected trees with 100 ml monocrotophos 36 per cent or 200 ml of endosulfan 35 EC or quinalphos 25 EC or 300 g carbaryl 50 WP or 25 ml of fenvalerate 20 EC in 100 litre of water. Repeat these sprays after 2-3 weeks, if necessary.

HARVESTING AND YIELD

The *ber* tree grows quickly and and the first crop can be harvested within 2-3 years of planting. In North India, flowering in *ber* takes place during August-September when flower clusters of 15-22 emerge from the leaf axils and are pollinated by honeybees and houseflies. The fruit itself requires about 22-26 weeks to mature after fruit-setting. The peak season of harvesting in North India is in mid-March to mid-April but some early varieties may ripen by end-February. *Ber* is a non-climacteric fruit and all the fruits on the tree do not ripen at one time. Therefore, the fruit should always be picked at the right stage of maturity, *i.e.* when it is neither under-ripe nor over-ripe. It should be picked when it has acquired normal size and characteristic colour of the variety, *e.g.* golden yellow colour in Umran. Picking should be done in the forenoon. Pre-harvest spray of 750 ppm 2-chloroethyl phosphonic acid (ethephon) at colour turning stage induces early and uniform ripening and reduces number of pickings.

Time of harvesting depends on cultivar and agroclimatic conditions. In South India, the fruits are harvested during October-November, in Gujarat during December-March, in Rajasthan during January-March and in Haryana, Punjab and Uttar Pradesh these are harvested during February-April. Early maturing cultivars ripen during middle of February, mid-season cultivars during March and the late cultivars ripen by the end of March to mid of April.

The average yield in different varieties during the prime bearing period (10-20 years) ranges between 80-200 kg per tree. Under rainfed conditions, 50-80 kg fruit per tree can be obtained.

POSTHARVEST TECHNOLOGY

After harvest, sorting is done to discard the damaged, overripe, unripe and misshapen fruits. Fruits are graded into large, medium and small size groups. For local market, fruits are generally packed in cloth sheets or in gunny bags but for long distance transport, packing should be done according to grades. While' A' grade fruits are packed in perforated cardboard cartons of 10 kg capacity with paper cuttings as cushioning material, the lower grade fruits are packed in baskets or gunny bags. *Ber* fruits can be stored for 8-12 days after packing in perforated polythene bags at room temperature. Storage life of fruits gets prolonged until 30-40 days if stored at 3°C and 85-90 per cent humidity. Pre-cooling of fruits at 10°C immediately after harvest increases the shelf life at room temperature. Similary pre-harvest spray of 0.1 per cent calcium nitrate and dipping the fruits in 1 per cent bavistin before storage also improves the shelf life. *Ber* fruits can be processed to prepare *murabba*, candy, dehydrated *ber*, pulp, jam, ready-to-serve beverage, etc.

CROP IMPROVEMENT

Objectives

There has been a wide cultivar diversity in *ber* in India. However, only a few cultivars are commercially important, which lack resistance to insect-pests and diseases. Good productivity along with ability to withstand transport and storage are the basic requirements, one should look for in a *ber* variety. In addition, it should have high TSS, good sugar-acid blend, crisp flesh and good flavour besides resistance to insect-pests like fruitfly and diseases like powdery mildew and sooty mould. Earliness is a desirable trait in cultivars meant for dry, rainfed regions or to be grown under irrigated conditions.

Breeding Methods

Selection

A wide range of variability exists in *ber* in India for all the characters suggesting substantial scope for improvement. Most of the present-day cultivars have been evolved through selection. Major emphasis has been laid on clonal selection and an early maturing clone named 'Early Umran' has been identified from normally late maturing Umran cultivar. This, has been named as 'Goma Kirti'.

Hybridization

Hybridization to evolve superior quality, high yielding, early and drought tolerant cultivars was taken up at HAU, Hisar and for inducing long shelf life and resistance to fruitfly at CAZRI, Jodhpur. Evaluation of hybrids is in progress.

Constraints in Breeding

Ber cultivars have polyploidy and incompatibility'which are serious bottlenecks in its successful breeding programme. Most of the cultivated *ber* cultivars are tetraploids, and show much segregation. Accordingly, many *ber* cultivars have been reported to be reciprocally cross-incompatible, while several cultivars including Umran are self-incompatible.

15

Date Palm

Date palm is probably the most ancient cultivated tree in the world. The botanical name of the date palm, *Phoenix dactylifera* L., is presumably derived from a Phoenician name 'phoenix', which means date palm, and 'dactylifera' derived from a Greek word 'daktulos' meaning a finger, illustrating the fruit's form. Another source refers this botanical name to the legendary Egyptian bird, 'Phoenix', which lived to be 500 years old, and cast itself into a fire from which it rose with renewed growth (Pliny, 1489; Van Zyl, 1983). This resemblance to the date palm, which can also re-grow after fire damage, makes the bird and the date palm share this name, while 'dactylifera' originates from the Hebrew word 'dachel' which describes the fruit's shape. *Phoenix sylvestris* (L.) Roxb. is a related wild species'which produces fruits of inferior quality and is widely used in India as a source of sugar.

Besides date palm and *Phoenix sylvestris*, five other species bear edible fruit (*P. atlantica* chev., *P. reclinata* Jacq., *P. farinifera* Roxb., *P. humilis* Royle., and *P. acaulis* Roxb.). Most of the *Phoenix* species are well known as ornamentals, the most highly valued is *P. canariensis* Chabeaud, commonly called the Canary Island Palm. The plant is not affected by frost and requires intense heat in summer for the development and ripening of the fruits. The maxim 'head in fire and feet in water' indicates the conditions that are ideal for date-cultivation.

COMPOSITION AND USES

Date fruits are nutritious having high calorific value. One kilogram of fully ripe fresh date fruits provide approximately 3,150 calories. Date fruits are highly rich in protein, iron, B complex vitamins and folic acid (Table 15.1). Date fruits can be consumed fresh (*doka* fruits of non-astringent varieties and *dang* and *pind* of all the varieties) as well as after prepartion of *chhuhara* and other products. In Iraq, a liquor known as *arak* and date juice'*dibbis* are preparared. Besides fruits, through the centuries the use of palm products in the date producing areas was diffused in all sectors of the economy from agriculture, transport and construction, to domestic use and reaching out also into the urban centres. Date palm trunks are exploited for use as poles, beams, rafters, lintels, girders, pillars, jetties and light foot bridges. Sawn into coarse planks'they are made up into doors, shutters and staircases for houses. Whole leaves are used in fencing, roofing or in partitioning in houses and enclosures of terraces providing privacy but keeping a certain ventilation. The midrib is used for making crates and furniture. Leaflets are mainly used in plait, which are sewn together in a wide array of baskets and sacks, but also mats and smaller articles like fans and hats. Leaflets are also used for making cord.

Table 15.1: Nutritional Composition of Dates

Component	Per 100 g Edible Portion	Component	Per 100 g Edible Portion
Moisture	7.3-9.5 per cent	Iron	5-10.7 mg
Protein	2.16-2.78	Thiamine	80-130 mg
Fat	0.32-0.51	Riboflavin	135-173 mg
Crude-fibre	1.72-2.56	Biotin	4-5 mg
Minerals	1.80-2.12	Folic acid	43-70 mg
Total sugars	86-87 per cent	Ascorbic acid	2-18 mg
Reducing sugars	73-83 per cent		

ORIGIN, HISTORY AND DISTRIBUTION

Date palm is considered to be native to Iraq. Date palm is found in both the Old World (Near East and North Africa) and the New World (American continent) where it is grown commercially in large quantities. The date belt stretches from the Indus Valley in the East to the Atlantic Ocean in the West. In India, date palm is being grown in dry tracts of Rajasthan, Haryana, Abohar (Punjab) and Kutchh of Gujarat.

Date palm was introduced into the western hemisphere in the late 18th or early 19th century by the Spanish missionaries. Dates were also introduced in the Atacama Desert and other parts of South America, Kalahari Desert of South Africa and the Great Central Desert of Australia. At present, the principal date growing countries of the world are Iraq, Saudi Arabia, Algeria, Iran, Egypt, Libya, Pakistan, Morocco, Tunisia, Sudan, USA and Spain. Of these countries, Iraq alone produces about 1/3rd of total world production of dates.

In the Indus valley, date palm was believed to have been introduced by the soldiers of Alexander the Great in the 4th century BC and also by the Moslem invaders at the beginning of the 8th century AD. Almost all of these areas are now in Pakistan. There is no commercial plantation of good dates in India and most of them are wild groves of seedling date palms. Introduction of some commercial cultivars of date palm from the USA, Pakistan and the Middle East countries were made at Regional Fruit Research Station, Abohar, under the aegis of the Indian Council of Agricultural Research. Further importation of date suckers was started since 1978 under the All India Coordinated Research Project on Arid Zone Fruits of the ICAR at Bikaner, Jodhpur and Jaisalmer in Rajasthan, Abohar in Punjab, Hisar in Haryana and Mundra in Gujarat.

TAXONOMY AND BOTANICAL DESCRIPTION

Taxonomy

The cultivated date palm is botanically called as *Phoenix dactylifera* L., and belongs to family Palmae.This name was given by Theophratus after Phoenica, where the Greeks first see it. The word *'dactylus'* means fingers, and *'ferre'* means to bear. The genus *Phoenix* has about 12 species. The important species are *P. sylvestris, P. humilis* and *P. farinifera,* which are grown in wild form throughout the India. These wild forms are used for the production of *Nira,* a drink. These species don't produce offshoots. The literature on chromosome number of date palm is confusing and the number of chromosomes reported to range from 2n=26 to 2n=36, in addition to the possible occurrence of polyploidy and aneuploidy. As early as 1910, Nemec examined young embryos of *P. dactylifera* and reported the presence of 2n=28 chromosomes. Beal (1937), however, observed 18 pairs of chromosomes (18 bivalents) and 2n=36 chromosomes in root tips of 6 varieties of both male and female plants. Beal also observed in 2 other species of the genus *Phoenix i.e., P. canariensis* and *P. sylvastris* 2n=36 chromosomes. Now, it is established that chromosome number of date palm, 2n = 36.

Botany

Being a monocotyledonous plant, date palm has no taproot. Its root system is fasciculated and roots are fibrous, similar to a maize plant. Secondary roots appear on the primary root, which develop directly from the seed. All date palm roots present pneumatics, which are respiratory organs. The date palm trunk, also called stem or stipe is vertical, cylindrical and columnar of the same girth all the way up. The girth does not increase once the canopy of fronds has fully developed. It is brown in colour, lignified and without any ramification. Being a monocotyledon, date palm does not have a cambium layer. Vertical growth of date palm is ensured by its terminal bud, called 'phyllophor', and its height can reach 20 metres. Horizontal or lateral growth is ensured by an extra fascicular cambium, which soon disappears, and which results in a constant and uniform trunk width during the palm's entire life. Depending on variety, age of a palm and environmental conditions, leaves of date palm are 3 to 6 m long (4 m average) and have a normal life of 3 to 7 years. The frond midrib or petiole is bare of spines for a short distance but full of spines on both sides thereafter. Intermediate zones have spine-like leaflets, also called leaflet-like spines. Leaflets or pinnae are between 120 to 240 per frond, entirely lanceolate, folded longitudinally and oblically attached to the petiole.

Date palm is a dioecious species with male and female flowers being produced in clusters on separate palms. These flowering clusters are produced in the axils of leaves of the previous year's growth. The unisexual flowers are pistillate (female) and staminate (male) in character; they are borne in a big cluster (inflorescence) called 'spadix' or 'spike', which consists of a central stem called 'rachis' and several strands or spikelets (usually 50 – 150 lateral branches). The inflorescence, also called as flower cluster, in its early stages, is enclosed in a hard covering/envelope known as 'spathe', which splits open as the flowers mature, exposing the entire inflorescence for pollination purposes. The male spathes are shorter and wider than the female ones. Each spikelet carries a large number of tiny flowers as many as 8,000 to 10,000 in female and more in male inflorescence. The male flower is sweet-scented and normally has six stamens, surrounded by waxy scale-like petals and sepals (3 each). Each stamen is composed of two little yellowish pollen sacs. The female flower has a diameter of about 3 to 4 mm and has rudimentary stamens and three carpels closely pressed together and the ovary is superior (hypogynous). Only one ovule per flower is fertilised, leading to the development of one carpel, which in turn gives a fruit called a date; the other ovules get aborted. The date fruit is a single, oblong, terette, one-seeded berry, with a terminal stigma, a fleshy pericarp and a membranous endocarp (between the seed and the flesh).

GROWTH AND DEVELOPMENT OF FRUIT

The growth and development of date fruit involves several external and internal changes. These changes are often classified on the basis of change in colour and chemical composition of the fruit. Five distinct stages of fruit development, known as Hababouk, Kimri, Khalal, Rutab and Tamar have been reported (Table 15.2).

Table 15.2 : Fruit Developmental and Maturity Stages in Date Palm

Growth Stage		Duration	Characteristics
Arabic Name	Indian Name		
Hababouk	-	Pollination to 4 weeks	Loss of unfertilized carpels.
Kimri	Gandora	9 weeks	Hard, mature, green fruits.
Khalal	Doka	4-5 weeks	Hard, yellow or pink or red, 10-15 g in size, TSS 30-45 per cent, moisture content 50 to 65 per cent, astringency present or absent depending on cultivar. Edible stage.
Rutab	Dang	4 weeks	Softening starts at tip of fruit, astringency disappears, weight lose and moisture content 30-45 per cent, Edible stage.
Tamar	Pind	One week	Fully ripe fruit dark red, weight lose 60-84 per cent TSS. Edible stage.

a) Hababouk Stage

This stage starts soon after fertilization and continues until the beginning of the *kimri* stage. It usually takes four-to-five weeks to complete and is characterized by the loss of two unfertilised carpels; a very slow growth rate is another characteristic. Fruit at this stage is immature and is completely covered by the calyx and only the sharp end of the ovary is visible. Its average weight is one gram and the size is about that of a pea.

b) Kimri Stage

At this stage the fruit is quite hard, its colour is apple green and is not suitable for eating. This stage lasts from a small green berry to an almost full sized green date. It is the longest stage of growth and development of dates, which lasts a total of nine to fourteen weeks, depending on the variety.

During the first 4 to 5 weeks, there is an average relative weekly growth of 90 per cent, while during the second period of *kimri* stage, there is only about 22 per cent growth. The first phase is characterized by a rapid increase in weight and volume, rapid accumulation of reducing sugars, low but increasing rate of accumulation of total sugars and total solids, highest active acidity, high moisture content though slightly less than that of the second phase.

c) Khalal Stage

It is also called as colour stage. The fruit is physiologically mature, hard ripe and the colour changes completely from for green-to-greenish yellow, yellow, pink, red or scarlet, depending on the variety. It lasts three-to-five weeks, depending on the variety, with a low average relative weekly increase in fruit weight (3 to 4 per cent). At the end of this stage, date fruit reaches its maximum weight and size, but sugar concentration (saccharose), total sugars and active acidity have a rapid increase associated with a decrease in water content (around 50-85 per cent). It is to be noted that date fruits accumulate most of their sugar, both the sucrose type and the reducing sugar type, as sucrose during the *khalal* stage. At this stage, colour of the seed changes at the end from white-to-brown.

Some varieties such as Barhee, Halaby, Hayani and Zaghloul are consumed in this stage, as they are very sweet, juicy and fibrous but not sour. However, *khalal* dates must be eaten immediately after harvesting as they will keep for only a few days without cold storage (7°C for one week or 0-1 °C for longer periods) due to their high sugar and water content, which cause fermentation during hot weather. If supply and demand are in equilibrium, the *khalal* season will last for a couple of weeks.

Varieties harvested and marketed at *khalal* stage present the following advantages: minimum infestation, possibility of cutting the whole bunch, easy handling and packing, high yield and consequently high income.

d) Rutab Stage

It is also called as 'soft ripe' stage. At this stage, the tip of fruit at the apex starts ripening, changes in colour to brown or black and becomes soft. It begins to lose its astringency and starts acquiring a darker and less attractive colour from the previous stage. However, some varieties such as Khadraoui (Iraq) and Bousekri (Morocco) turn green at this stage.

At this stage, which lasts for 2 to 4 weeks, there is a continuous decrease in fresh fruit weight, mainly due to loss of moisture. The average weekly decrease in fresh fruit weight is 10 per cent during the last week of the *rutab* stage.

An increase in reducing sugar, a rapidly increasing rate of conversion of sucrose, a gain of total sugars and total solids also characterize this stage. It has already been observed in respect of the reducing sugar type date, *i.e.* Barhee, in which all the sucrose accumulated during the previous, *khalal* stage, inverts and there is a continuous decrease in active acidity and decrease also in moisture content (average 30-45 per cent). With softening, the remaining tannins under the skin are precipitated in an insoluble form, so that the fruit loses any astringency that may have remained in the *khalal* stage from the *kimri* stage.

It is a very good stage for consumption as a hard ripe date. With the exception of a few varieties, fruit at this stage is very sweet. It is, however, very important to harvest and market the fruit at this stage. Unless they are coldstored, the fruits quickly turn sour, become unacceptable with of no commercial value. For dessert purposes, most people prefer dates after they have passed the *rutab* stage.

e) Tamar Stage

It is also called as 'full ripe' stage or final stage in the ripening. This is the stage when the dates are fully ripe, and they completely change the colour from yellow to dull-brown or almost black. The texture of the flesh is soft. The skin in most varieties adheres to the flesh, and wrinkles as the flesh shrinks. The colour of the skin and underlying flesh darkens with time.

At this stage, the date contains the maximum total soluble solids and has lost most of its water to such an extent (below 25 per cent down to 10 per cent and less) that it makes the sugar water proportion sufficiently high to prevent fermentation. This is the best condition for storage. The average relative decrease in fruit weight during this stage is 35 per cent. The loss in fruit weight continues if fruits are left on the palm. This stage is equivalent to that of the raisin in the grape and the dried prune in the prune type of plum. At the tamar stage, all the fruits on a bunch do not ripen simultaneously, but over almost a month. Hence, three-to-four harvest times are necessary.

In India, the four development stages are called *gandora, doka, dang, pind*. These stages, respectively, correspond to *kimri, khalal, rutab* and *tamar* in the Arabs. Dates can be eaten right from *doka* or *khalal* stage of ripening. At *doka* stage, the fruits may be sweet or astringent depending on the cultivar. Sweet fruits are preferred for dessert. At the next *dang* or *rutab* stage, the fruits become softened and lose most of the astringency but still have high moisture content to make them unsuitable for long storage. These fruits can either be eaten fresh or used for processing. This stage reaches in India only in selected locations. The final *pind* or *tamar* stage (when the soft fruits lose extra moisture) rarely reaches on the tree under Indian agro-climatic conditions. However, in Mediterranean region, *pind* khajoors are harvested because in this region, rains are rare during the ripening period of dates. In India, *doka* or *khalal* fruits, having about 30-35 per cent TSS, are harvested since if left on the tree, the fruits will be damaged by rains.

The home of date palm is in the Mediterranean region, where practically no rain occurs during February-to-September and thus, its fruit matures upto *pind* or *tamar* stage in about 180 days. There is no location in India, which meets this requirement, although fairly dry conditions are available in north-western India. The number of days available for fruit development in Indian districts is cut short to 122-170 days (Table 15.3), as the growers are obliged to harvest early to protect fruit spoilage owing to rainfall and high humidity, even though the fruits are not yet fully mature.

Table 15.3: Spathe Opening and Fruit Development Periods in Date Palm in North-West State of India

Location	Spathe Opening	Doka Stage	Dang Stage	Pind Stage	No. of Days (Pollination to Harvest)
Abohar	Mar–Apr	III week July-I week Aug	–	–	122
Bikaner	Feb–Mar	June–July	July–Aug	–	150
Jaisalmer	26 Feb–12 Mar	5 May–3 July	1 July–7 Aug	Aug	170
Jodhpur	20 Feb–26 Mar	4 June–6 July	26 June–8 Aug	–	145
Hissar	Mar–April	July-Aug	–	–	110
Kutchh	Mid–Feb	III week June, II week July	–	–	151

SOIL AND CLIMATIC REQUIREMETS

Soil

The date palm can be grown in soils containing more alkali or salts. It can tolerate such conditions better than many other fruit plants. Sandyloam soil with good water holding capacity and good drainage is the best for date palm cultivation. The soil profile should be free from stones or calcium carbonate concretions and hard pan at least upto 2 m depth. Date palm can tolerate a pH 8 to 10 and upto 3-4 per cent sodicity (alkalinity) in the soil. If the water-table is within 3 to 3.5 meters from the surface, mature trees need no surface irrigation.

Climate

The date palm is a tree, which flourishes well under varied sets of climate. However, for proper maturity of fruit, the date requires prolonged summer heat without rain or high humidity during the ripening period. The ideal mean temperatures for flowering and fruit ripening in date palm have been found to be 25 and 40 °C, respectively. A total of 3,000 degree day heat units (base 10 °C) are required for full maturity of date palm berries. Rains cause damage to fruits at ripening. A wet season during flowering and fruit-ripening is a limiting factor for the successful cultivation of the date-palm. South-western Punjab, northern Rajasthan and Kutchh offer suitable conditions for date growing.

IMPORTANT CULTIVARS

More than 1,000 cultivars of date palm are known but only a few of them are commercially important, *e.g.*, Halawy, Khadrawy, Sayar, Barhee, Chip Chap, Brem,

Table 15.4: Commercial Cultivars of Date Palm in different Countries

Country	Cultivars	Country	Cultivars
Akgerua	Ghars, Deglet Noor	Libya	Bikraari, Taasfirt, Murzabad, Saidy
Baharain	Murzaban, Khanezi	Somalia	Sucotari, Succari
UAE	Angal	Oman	Mabsaly, Fardh
Egypt	Haiyani, Saidy, Zagloul, Samani	Mauritania	Tingerguel
PDR Yemen	Hamraiya	Morocco	Bufaguus, Medjool
Saudi Arabia	Irzeiz	Sudan	Barakaavi, Mishrig-Khatib
Iraq	Zahidi, Halawy, Khadrawy, Sayer	Tunisia	Fatumi, Deglet Noor
Iran	Ustaumran (Sayer)	Pakistan	Mozawati, Begam Jangi, Dhakki, Assi, Halawy

Shakar and Zahidi of Iraq, Deglet Noor, Medjool of north Africa and Dhakki, Begam Jungi of Pakistan (Table 15.4). Fruits of cultivars such as Barhee, Halawy, Khalas and Khunezi having non- astringent berries at *doka* stage are also eaten raw. These cultivars have good prospects of cultivation in North-West India since these can be harvested before the onset of monsoon and used at *doka* stage. Fruits of varieties such as Medjool, Zahidi, Shamran and Khadrawy have astringent fruits at *'doka'* stage. These can also be harvested for processing to prepare dry dates (*chhuhara*). Berries of the red coloured varieties such as Zagloul and Hayani are suitable for the preparation of date juice (RTS) and other products like jam and *chutney*.

Clasification of Cultivars on the Basis of Flesh Consistency

On the basis of flesh consistency, cultivars of dates have been divied into three groups, such as, soft, semi-dry and dry. In soft cultivars (invert sugar dates) like Halawy, Khadrawy, Shamran and Hayani, almost all cane sugar is converted to invert sugars during ripening, while in dry and semi-dry dates (cane sugar dates), a significant proportion of cane sugar is retained even after full ripening. In bread type date ('Thoory') as much as one-third of total sugars may be sucrose. In India, several cultivars of dates have been introduced at different research insitutes/centres. Based on experimental results by three centres, the cultivars like Khadrawy, Shamran, Barhee, Halawy, Zahidi, Medjool and Khalas have been reported to be promising. Characteristics of some of these cultivars as recorded in north-west India are given below:

Halawy

It is a soft date from Iraq. This variety is relished both in 'doka' and 'dang' stages of maturity. Total soluble solids range between 28 to 42 per cent and astringency in the fruit at 'doka' stage is low or almost absent. It is an early variety yielding good crops and is somewhat more tolerant to rains. Dry dates (*Chhuhara*) of good quality and cured soft dates (*Khajoor*) can be prepared. Its fruit at 'dang' stage (fresh form) is very delicious. Average yield ranges from 50 to 80 kg per palm.

Barhee

It is a soft date from Iraq. The variety has proved extremly good for table use in fresh form (doka stage). The doka fruit contains about 32 per cent TSS, has golden-yellow colour and has a very pleasant taste because of low astringency and high pulp content. The trees of this variety are prolific in bearing and the fruit is ready for eating late in the season *i.e.* first fortnight of August. Its yield ranges from 60 to 110 kg per palm.

Medjool

The variety originated in Morocco and has large size fruits. It is late in ripening and has proved particularly good for preparing dry dates of attractive bold size and good quality. Its yiled is 35 to 50 kg per palm.

Zahidi

It is a mid-season variety, slightly tolerant to rains or high humidity. Its fruit is small-to-medium, obovate and yellow at doka stage.

Khadrawy

It also originated from Iraq. This variety proved successful under Abohar (Punjab) conditions. The trees of this variety are comparatively less tall and and yield good crops. It can be used both for dry dates and soft dates. The yield ranges from 40 to 70 kg per palm.

Khalas

The fruits are yellow and sweet at full *doka* stage, with average fruit weight of 15.2 g and 25 °Brix TSS. It is a mid-season cultivar and is suitable for raw eating and for processing.

Khuneizi

The fruits of this variety at *doka* stage are purple coloured and non-astringent with average fruit weight of 18.7 g and 47 °Brix TSS. It is an early variety and the fruits are suitable for fresh eating.

Some seedling palms selected from the coastal belt of Kutchh district of Gujarat bear sweet *doka* fruits of large size and yellow or red colour.

PLANT PROPAGATION

The date palms are always propagated through suckers (offshoots) for commercial plantation. The suckers usually arise near the ground around the trunk. The offshoots of about 10 kg weight with well-developed roots are separated from the mother palm during March or August-September. About 3 to 5 offshoots can be obtained every year from a mother palm starting from the fourth year after planting. But after tenth year of age, the palms do not produce any offshoots. Thus, a total 15-20 offshoots can be obtained from one mother palm in its whole life. Prior to the removal of offshoots, the field is irrigated and the outer leaves of the offshoots are cut back to two-third of their length and the inner leaves to half. The tender young unopened leaves near the central bud and parts of the bare stalks of the old leaves necessary to protect the bud, should be kept on the offshoots. The outer whorl of leaves may be fastened to afford protection from heat and cold to the central bud. It also facilitates detachment and transport. The offshoot should be separated by cutting at its connection with the mother palm by a sharp chisel in such a way that no injury is caused to the mother palm. The cut end of the offshoot should be painted with copper fungicidal paste.

The rapid propagation of date palm as well as propagation from a mature specimen is impossible due to the limited number of offshoots produced and the fact that offshoot production is limited to a certain period in the palm's life span. While, seed propagation of date clones and cultivars is impractical. Therefore, date palm can also be propagated by tissue culture technique. The *in vitro* propagation technique of date palm has been standardized at Anand Agriculture University in Gujarat. Tissue culture assures true-to-type multiplication of plants with known sex, production potential and quality, leading to increased production and productivity. The stages involved in production of tissue culture plants of date palm are callus initiation, followed by somatic embryo induction, development, maturation, germination and finally *in vitro* plantlet development. For further growth, the plantlets are then transferred to a greenhouse with controlled temperature and humidity conditions. These plants are then gradually acclimatised/hardened by decreasing the humidity and increasing the temperature. These plants are later transferred to polythene bags containing soil based potting mixture for their further growth and development. The plants with one compound leaf are fit to be transferred to the field for cultivation.

PLANTING AND ORCHARD ESTABLISHMENT

Planting is generally done at 6-7 m distance in a square system to facilitate intercultural operations and proper development of the palms. In this way, 202 to 275 plants/hectare will be accomodated. If transported from long distance, the offshoots may be kept in the straight position. After planting, the soil around the offshoots should be pressed firmly and the field is irrigated immediately after planting. Thereafter, frequent light irrigations are given to keep the soil always moist. For supply of adequate pollen for pollination, 10 per cent of the population in the orchard should consist of male palms.

For planting, pits of $1 \times 1 \times 1$ m size are dug and refilled with FYM and soil mixture. The pit should be treated with chloropyriphos or endosulfan to protect the palms from termites and with captan or carbendazim for the control of soil-borne pathogens while refilling them. Before plantation, the offshoot's root portion should be dipped in 0.2 per cent carbendazim solution. The suitable time for planting the offshoots in the field is either rainy season (July-September) or spring season (February-March).

INTERCULTURAL OPERATIONS

Water Management

Date palm is known as drought resistant fruit tree, which can survive for long periods without irrigation. After establishment, plants should be irrigated once a week during winter and twice a week in summer. Though, date palm is highly tolerant to excessive irrigation and floods, yet continuous stagnation of water or waterlogged conditions is injurious for its growth. One irrigation should be given 15 days before flowering and then regularly from fruit set until fruit ripening. The deficit of water during this stage may cause pre-harvest fruit drop and reduction in fruit quality. Irrigation is withheld when ripening starts to facilitate harvesting, hasten fruit ripening and to reduce fruit drop caused by high humidity.

Mulching with black polythene or by spreading 10 cm thick layer of organic materials such as date palm leaves or weeds or other wastes in the basin around the main trunk helps in conservation of soil moisture. Locally available weeds such as *'bui'* (*Aerva persica*), organic meterials or black polythene can also be used as mulch material.

Intercropping

Intercropping in date palm with suitable crops bring good income to the grower and also improves the fertility of the soil. During the first few years, intercropping can be practised with no shortage of irrigation. Crops such as gram, lentil, pea and *guar*, mustard, berseem, vegetables and fruits such as pomegranate, *phalsa*, papaya and citrus can be grown between date palms. Under such management, the additional requirement of nutrients and water for the intercrop should be provided.

Weed Management

For the control of monocot and dicot weeds, mechanical weed control and post emergent application of diuron (5 kg/ha) followed by glyphosate (3.75 l/ha) after 20 days of diuron application has been found effective. Manual weeding or hoeing is also very effective for reducing weed population.

Nutrient Management

Generally, the farmers do not apply manures to the date palm. The trees certainly respond to manuring as indicated by increased vigour and growth. However, for optimum yield of fruits, full grown palms should be given 20-40 kg farmyard or sheep or goat dung manure every year during September-December. Besides, inorganic fertilizers containing 0.5-1.0 kg each of nitrogen and phosphorus and 0.25-0.50 kg of potash should be applied. Full quantity of phosphorus and potash and 50 per cent of nitrogen should be applied three weeks before flowering and the remaining dose of nitrogen should be applied during March-April.

Leaf Pruning

Keeping about 100 healthy leaves on a palm is considered desirable for good fruiting. The old, photosynthetically less efficient, dried and damaged leaves should be removed. Pruning of leaves is done during January-February before spathe emergence. Leaf pruning operation can be combined with dethroning of leaf stalks adjoining the tree crown and the spathes. This facilitates the operations of pollination, bending, harvesting, etc. To maintain proper vigour and productivity of the mother palms, offshoots should be removed as soon as they attain proper size and develop roots. The leaf stalks should be removed from as close to the stem as possible to provide smooth trunk surface.

Bending and Fruit Thinning

After fruit set, during April-May, the main stalks bearing the bunches are twisted downwards so that these hang straight below the leaves without touching the leaf midribs. This operation on one hand minimizes the risk of breaking of the stalks due to fruit load and also reduces damage to the berries from rubbing against the leaf midribs. Excess load of fruit may cause shrivelling of berries, breaking of spathe stalks, more damage due to rain and humidity, delay in ripening and alternate bearing. It also reduces fruit size and produce poor quality of fruits. It is, therefore, necessary to keep only optimum number of fruits and thin out the rest. This is usually accomplished either by reducing the number of fruits on each bunch and or by removing some of the bunches. Depending upon variety, age and vigour of the palm, 5 to 10 bunches or 1,300 and 1,600 fruits per palm should be retained on a palm. Then, one-third strands from the centre of each bunch should be cut to hasten ripening and improve fruit quality. The per cent thinning is generally done 40 – 50 in Khadrawy, 50 – 55 in Halawi, 50 – 60 in Zahidi and Barhee. Ethephon @ 100 – 400 ppm after 10 to 30 days from fruit set is quite effective in fruit thinning of cv. Hayani. The biennial bearing habit of the treated palms can be reduced by ethephon treatment. It also advances the ripening of fruit.

FLOWERING, POLLINATION AND FRUIT RIPENING

Flowering in date palm takes place during March-April. Male and female flowers are borne on separate palms. For natural pollination by wind and insects, nearly half the population should be of male palms. But by resorting to hand pollination, the requirement of male palms can be reduced to about 5-10 per cent. Manual pollination is done by inserting about three mature male strands in inverted position in the centre of each female spathe. Previously stored or extracted pollen can also be dusted on the female spathes during morning hours with the help of a cotton swab or brush. However, this has to be continuously done for 2-3 days.

In date palm, selection of good polliniser is important because the pollen affects (metazinia) the size of the fruit and the time of ripening. Thus, identification of good polliniser is of a great significance under Indian climatic conditions to advance ripening of fruits so as to avoid rain damage. A good polliniser should have the following characters:

1. It should produce plenty of pollen.
2. Its pollen should be viable.
3. Flowering time of a polliniser should match with the flowering time of female cultivar.

4. It should produce overlapping flushes of flowers.

5. Pollen produced by the polliniser should neither be too light or too heavy in weight.

Phoneix species like *P. sylvestris, P. reclinata, P. rupicola, P. lourairli, P. robellini, P. padulosa, P. canariensis* could also be exploited as pollenisers.

Five stages are recognized in the development of date palm fruits, which had been discussed earlier in this chapter. The changes associated with ripening and the period during which the fruit may be consumed extend from the peak of the *khalal* stage, when the fruit has its most intense red or yellow colour and maximum weight, to the final *tamar* stage, when it has lost the greater part of its moisture content and will keep well without special attention to storage. Most of the people like to eat the fruit in *khalal* stage. At least two varieties of dates *i.e.,* Halawy and Barhee, are liked most for eating at *khalal* stage. At first stage, the fruits are not edible while, depending upon cultivar, one or all the three later stages are edible. Dates are hand-picked at the stage of maturity. All the dates in the same bunch do not ripen at same time; it has been the practice to make several pickings to harvest the fruit during a season. The fruit ripening can be hastened by a spray of 1,000 ppm ethrel on fruit bunches at early *doka/khalal* stage.

PLANT PROTECTION

A. Major Diseases and their Management

The major diseases of date palm observed in India are *Graphiola* leaf spot or false smut, crown rot of offshoot and fruit rot, which have described briefly hereunder:

Graphiola Leaf Spot

Graphiola leaf spot is also called as 'false smut' and is the most common disease of dates in India. It is caused by *Graphiola phoenicus*. It is more serious in high humidity conditions.

Symptoms

The fungus attacks the leaves, forming numerous dark-brown black pustules, which are full of yellow spores. In case of severe infestation, the whole leaf may be covered by the pustules, which may result in the death of leaves.

Management Practices

Though, it is very difficult to control this disease, if the infestation has already occurred. However, the following measures should be adopted for its effective control:

1. Grow resistant varieties like, Khadrawy etc.

2. Collect and burn the affected/fallen leaves to avoid further spread of the inoculum.

3. Spray carbendazim (0.2 per cent) as the infection is noticed.

Fruit Rot

This disease usually causes serious loss when humid weather occurs during the ripening time. Under such conditions, various fungi may develop on the fruits which causes spotting, dropping and rotting of fruits. The damage can be reduced by better ventilations of the branches and protecting the fruit from rain. In addition, spray of the bunches with a mixture of 5 per cent fahana (ferlic dimethyl dithiocarbamate) in sulphur.

B. Major Insect-Pests and their Management

About 13 insect-pest species have been reported to cause damage to date palm. Of these, black headed caterpillar, red palm weevil, scale insects are major pests, whereas termites and aphids are of minor importance. A brief description about the major pests and their management has been given below:

Black Headed Caterpillar (*Opisina arenosella*)

It is one of the most serious pests causing heavy losses to the date palms. In India, it is commonly found during January-May. It is a polyphagous pest attacking coconut and almost all other palms including date palm. Outbreaks of the pest occur during dry months and the pest population declines during rainy season. The caterpillars cause damage, which feed gregariously on lower surface of the leaves, hiding inside the tunnels in the leaf folds. The affected fronds turn brown and dry up. In case of severe attack, the leaflets are reduced to papery tissues and the photosynthetic activity of the fronds is adversely affected. With too many fronds affected, the tree dries up and loses its fruit bearing capacity. Infested trees are easily recognized by dried up patches in the fronds.

Management Practices

1. Remove and burn the infested plant parts along with silken galleries and larvae as and when located.
2. Spray the infested trees with 100 ml monocrotophos (Monosil 36 WSC), 200 ml endosulfan (Thiodan 35 EC) or 200 g carbaryl (Sevin 50 WP) in 100 litre water during March.
3. Release of some parasitoids such as *Bracon brevicornis*, *Elasmus nephantidis*, *Trichospilus pupivora* and *Perisieola nephantidis* in areas where infestation occurs regularly is very useful in controlling this pest.

Red Palm Weevil (*Rhynchophorus ferrugineus*)

The palm weevil is an important pest of date palm, although its preferred hosts are coconut and oil palm. It is widely distributed in Indian sub-continent and generally young palms are preferred. The grub is the chief damaging stage and is a voracious feeder. They bore into the stem near the ground and make tunnels in all the directions. Once inside, the grubs do not come out and hence, it is difficult to detect the infestation early in the season. The larvae burrow into the hard woody portion. As a result of tunnelling, palm gets riddled, affecting the supply of water or nutrients to the other parts. As the galleries become more extensive, the trunk weakens and the trees may be easily decapitated by strong wind. The outer leaves turn chlorotic and die and the symptoms soon spread to innermost leaves. The trunk is completely hollowed out and filled with decaying rubbish. Viscous plant sap, unpleasant smell and chewed frassy material are thrown out from the holes in the stem. Growing points of infested young plants wither, dry, rot and produce offensive smell. In older plants, top portion of trunk bends and breaks just below the crown.

Management Practices

1. Orchard sanitation, especially prompt destruction of infested, dying and dead palms at initial stage of attack suppresses the pest development.
2. Avoid the occurrence of the cuts, wounds, scars, etc. on the plants as female weevils search such sites actively for oviposition.
3. The cut portions or holes of the trunk should be painted either with lime or coal tar, which prevents the female weevils from egg laying.
4. The pest is very difficult to control in advance stage of infestation. Hence, the affected plants should be destroyed promptly to prevent further spread.
5. Kill the grubs by inserting wire in the holes.
6. Fumigate the holes with diesel/petrol or with some insecticide and plug them with mud.

Rhinoceros Beetle (*Oryctes rhinoceros*)

Rhinoceros beetle is considered as a serious pest of date palm in certain localities. Adults of rhinoceros beetles are very harmful, which cause damage during night. Young plantations are more prone to attack than the older ones. It is also a serious pest of coconut and maximum damage is caused during monsoon. Grubs feed on decaying organic matter and are harmless. Isolated trees of irregular heights are more prone to attack. Only beetles cause damage, which bore into unopened tender fronds and feeds on the growing point, which is damaged and palm tree dries up. Infested fronds when open, give rise to leaflets having V-shaped cuts. The beetle throws out fibrous mass while feeding in the burrows made in young plants, which die due to destruction of growing points. The growth and yield of older palms suffer badly. Damaged palms are often infested secondarily by fungal pathogens.

Management Practices

1. Orchard sanitation checks the pest development. Burn all dead palm trunks, leaves and logs.
2. Compost pits should not be kept near the plantations.
3. Kill the beetles in the bore hole with a pointed wire or iron spoke during rainy season.
4. Inject some fumigant like, formalin, petrol or carbon bisulphide in the holes and plug the holes with mud.
5. Destroy all potential breeding sites (manure/compost pits, refuse dumps, decaying material etc.) in and around the plantation. All rubbish and rotten vegetation should be dried and burnt. Spraying of insecticides is not practicable on large scale and is uneconomical.

Date Palm Scale (*Parlatoria blanchardi*)

This scale insect seems to be one of the most injurious pests. It is specific to palms, especially *Phoenix dactylifera* L. Closely planted, frequently irrigated, young plants grown under shade suffer the most and dry climate is favourable for this insect. The adult females and their nymphs cause the damage by sucking sap and feeding on succulent tissues of the leaf stalk, mostly during summer months. On severe infestation, fruits are also affected. There is formation of a continuous crust of yellowish-brown scales on the underside of leaves, which interfere with photosynthesis, respiration and transpiration of the fronds. The growth, fruit yield and quality is drastically reduced in the affected palms. Attacked fruits become dry and deformed.

Management Practices

1. Plant healthy offshoots free from scales.
2. Remove and burn all the infested leaflets and fronds in the initial stage of infestation.
3. Sanitation and good crop management practices check pest development and favour the coccinellid predators.
4. Spray 125 ml monocrotophos (Monosil 36 WSC), 150 ml dimethoate (Rogor 30 EC) or 200 ml diazinon (Basudin 25 EC) in 100 litre water. Repeat the spray after 3 weeks, if necessary.

Termites

Offshoots and fully grown trees are often damaged by termites, which feed on the roots, make gallaries inside the main trunk and tunnel upwards. Under severe attack, the offshoots wither and die. For the control of termite, 2-3 ml chlorpyriphos is applied with irrigation water to each basin and also on the trunk at monthly interval.

Birds and their Control

Considerable damage of fruits at *doka, dang* and *pind* stages is caused by mynas, bulbuls, parrots, crows and peacocks. Covering the fruit bunches with wire gauge (3 × 3 mm mesh) just after colour turning stage helps to protect them from birds. Although the initial cost of these covers is high but considering that these can be reused for 6-7 years, these are cost-effective. Covering of bunches by 500 gauge polythene bags having holes for air circulation is also effective in protecting the bunches from bird damage.

C. Physiological Disorders and their Management

Black Nose

Black nose applies to the abnormally shrivelled and darkened tip of dates. Deglet Noor and Hayani seem to be the most susceptible varieties to this physiological disorder. Black nose results from excessive checking of the epidermis, especially in the form of numerous small, transverse checks or breaks at the stylar end of the fruit.

Pronounced shriveling and darkening occur in proportion to the abundance of the checks and are related to humid weather at the *khalal* stage.

The conditions to be avoided include excessive soil moisture and the presence of intercrops and weeds, especially at the susceptible stage of fruit development. Bagging the fruits in brown wrapping paper inhibits the occurrence of black nose. Over thinning can also increase the incidence of checking and subsequent development of black nose.

Crosscuts

Crosscuts is a physiological disorder of fruit stalks and fronds reported from the United States, Pakistan and a few Middle East date growing countries such as Israel and Iraq. Crosscuts, or V- cuts, are clean breaks in the tissues of the fruit stalk bases and on fronds. It consists of a slight to deep notch, similar to a cut artificially done by a knife. Fruits borne on strands in line with the break wither and fail to mature properly. Crosscuts result from an anatomical defect in the fruit stalks and fronds involving internal, sterile cavities leading to mechanical breaks during elongation of the stalk or the fronds. Crosscuts are commonly found in varieties having crowded leaf bases and its incidence increases as the palms get older. Sayer and Khadrawy varieties are especially susceptible to this disorder, and are no longer propagated in some countries.

Crop losses caused by crosscuts may be avoided by using non-susceptible varieties, or by reducing the number of fruit stalks in susceptible varieties.

White Nose

White nose disorder is commonly found in Iraq, Libya and Morocco. Dry and prolonged wind in the early *rutab* stage causes rapid maturation and desiccation of the fruit resulting in whitish drying at the calyx end of the fruit. The affected fruit becomes very dry, hard and has high sugar content. Hydration may correct this condition in harvested fruits.

Barhee Disorder

Barhee disorder is characterised by an unusual bending of the crown of Barhee variety. The disease was first reported in California (USA) and later in Al Basra (Iraq). Affected palms bend mostly to the South and sometimes to the South-West.

Neither the cause nor the control of this disorder is known. However, bunch handling is proposed to correct such an abnormality.

Black Scald

Black scald, different from blacknose, is a minor disorder of date palm of unknown cause occurring in the United States. It consists of a blackened and sunken area with a definite line of demarcation. The disorder usually appears on the tip or the sides of the fruit, and affected tissues have a bitter taste. It is suggested that the appearance of the disorder is due to exposure of fruits to high temperature, but the exact cause is not definitely known.

Bastard Offshoot

This is a deformed growth of date palm vegetative buds especially of offshoots fronds. The bastard condition could be due to infestation of date palm by bud mite, *Makiella phoenicis*. It may also be due to reduction in growth caused by an inequilibrium of growth regulators. Conrol of mite by suitable insecticide also reduces the incidence of this disorder.

HARVESTING AND YIELD

In India, owing to monsoon rains during July-September, it is not possible to allow complete ripening of fruits on the palms and the bunches must be harvested at *doka* stage or at best at early *dang* stage. Fruits for raw eating and for preparation of dry dates (*chhuhara*), these are harvested at full *doka* stage. Yield of date palm depends on age of the plant and cultivars. On an average, 50 kg *doka* fruits are produced from each palm of 10 years age, which increases to 75 kg at the age of 15 years. Fruit yield of some cultivars under Indian condition has been given in Table 15.5.

Table 15.5: Fruit Yield of different Cultivars Under Indian Conditions

Cultivar	Fruits/Bunch	Fruits/kg	Yield/Palm (kg)
Degana	800-850	100	50-60
Sikkori	1,100-1,300	160	40-60
Bintaisha	450-500	85	30-35
Gondilia	750-800	95	40-50
Bartamonde	400-450	150	35-40

POSTHARVEST TECHNOLOGY

Individual bunch of date palm is harvested with the help of a sharp knife and the *doka* berries are packed in gunny bags for local markets. The *doka* berries are highly perishable and cannot be stored for more than 24 hrs at room temperature. Berries from non-astringent varieties are consumed fresh. *Doka* fruits are also processed to prepare *chhuhara* (dry dates). Scientists working at Abohar (Punjab) have developed a technique to transform satisfactorily the date fruit at *doka* stage into dry dates (*Chhuhara*) of good quality. Four varieties Halawy, Khadrawy, Shamran and Medjool have yielded very good product. The technique developed involves immersion of frut at *doka* stage in boiling water for 6 to 8 minutes and then drying either in temperature-controlled oven (air-circulation type) for 80 to 120 hours at 48° to 50° C or in the sun for 10-15 days if weather is dry. Thus, an average of 45 per cent fruit product is obtained. Fruits at advanced *doka* stage or when they attain one-fourth, one-half or full *dang* (the berries become mellow and soft starting from the distal end) can be converted into soft dates (*khazoor*) of good quality by drying either in the oven at 40° C or in the sun. Thus, for soft dates, only drying the berries at partial or full *dang* stage is required with no other treatment and this way, a final product of soft *Khazoor*, ranging from 50 60 per cent is obtained.

Artificial ripening of fruit at *doka* stage to transform it into *dang* stage was also attempted at Abohar. For these studies, berries were treated with 0.5 per cent to 2.0 per cent common salt (sodium chloride) and similar concentrations of acetic acid in combination with 1.0 per cent salt. The fruits to be treated with salt are spread on polythene sheet and requisite quantity of salt is applied by rubbing and smearing uniformly on the berries. The fruits, which are given acetic acid plus salt treatments, are first dipped in solutions of desired acetic acid concentrations for 2 minutes followed salt application by the method described above. Each treated lot is packed into wooden boxes lined with old newspapers and packed in laboratory at room temperature. The boxes are opened after 24 hours. Though, the *dang* obtained by this treatment are not as good in taste as that of naturally ripened berries on the tree, but still these are edible and generally acceptable on account of having lost the astringency. But such products cannot be stored more than 24 hours and, as such, should be consumed to as early as possible.

References

Atwal, A.S. and Dhaliwal, G.S. (1997). Agricultural pests of South East Asia and their management. Kalyani Publishers, Ludhaina, India.

Bal, J.S. (2006). Fruit growing. Klayani Publishers, Ludhiana, India.

Bose, T.K. and Mitra, S.K. (1990). Fruits: Tropical and Sub-tropical. Naya Prokash, Kolkata, India.

Bose, T.K., Mitra, S.K. and Sanyal, D. (2001). Fruits: Tropical and Sub-tropical. Naya Udyog, Kolkata, India.

Bose, T.K., Sanyal, D. and Sadhu, M.K. (1998). Propagation of Horticultural Crops. Naya Prokash Publishers, Kolkata, India.

Chadha, K.L. (2001) Handbook of Horticulture, DKMA, ICAR, New Delhi, India.

Chadha, K.L. and Awasthi, R.P. (2005). The apple. Malhotra Publishing House, New Delhi, India.

Chadha, K.L. and Pareek, O.P. (1991). Advances in Horticulture fruit crops (volume 1-4). Malhotra Publishing House, New Delhi, India.

Chadha, K.L. and Shikhamany, S.D. (1990). The grape. Malhotra Publishing House, New Delhi, India.

Chadha, T.R. (2002). Textbook of Temperate Fruits. DKMA, ICAR, New Delhi, India.

Chattopadhyay, T.K. (2012). A text book on pomology. Vol. II: Tropical fruits. Kalyani Publishers, Ludhiana, India.

Chattopadhyay, T.K. (2012). A text book on pomology. Vol. III: Sub-tropical fruits. Kalyani Publishers, Ludhiana, India.

Chattopadhyay, T.K. (2012). A text book on pomology. Vol. IV: Temperate fruits. Kalyani Publishers, Ludhiana, India.

Creasy, G.L. and Creasy, L.L. (2009). Grapes, CABI Publishing, U.K.

Edmond, J.B., Senn, T.L., Andrews, F.S. and Halfacre, R.G. (1977). Fundamentals of horticulture. Tata Mac-Graw-Hill Publishing Company Ltd., New Delhi, India

Garner, V.R. (1986). Principles of horticultural production. Michigam State University Press, East Lansing, U.K.

Gupta, V.K. and Paul Y.S. (2012). Diseases of vegetable crops. Kalyani Publishers, Ludhaina, India.

Gupta, V.K. and Sharma, S. (2012). Fungi and plant diseases. Kalyani Publishers, Ludhaina, India.

Gupta, V.K., Sharma, S. and Y.S. Paul. (2008. Diseases of fruit crops. Kalyani Publishers, Ludhaina, India.

Hartman, H. T. and Kester, D.E. (1986). Plant propagation. Prentice Hall of India Pvt. Limited, New Delhi.

Jones, D.R. (2000). Diseases of banana, Abaca and Ensete. CABI Publishing, New York, USA.

Kader, A.A., Kasmire, R.E., Mitchell, F.G., Reid, M.S., Sommer, N.F., Thompson, J.F. (1999). Postharvest Technology of Horticultural crops. University of California, USA.

Krishna, H. (2012). Fruit Physiology. Stadium Press, U.K.

Lal, G., Siddappa, G.S. and Tandon, G.L. (1998). Preservation of Fruits and Vegetables. DKMA, ICAR, New Delhi, India.

Litz, R.E. (1998). The mango. CABI Publishing, New York, USA.

Luis, H.I. (2002). Coconut, the wonder palm. Hi-Tech Coconut Corporation, Nagercoil, Tamil Nadu, India.

Mitra, S.K. Bose, T.K and Rathjore, D.S. (1991). Temperate fruits. Hort. and Allied Publishers, Kolkata, India.

Mukherjee, S.K. and Majumder, P.K. (1986). Propagation of tropical and Sub-tropical fruit crops. ICAR, New Delhi, India.

Nakasone, H.Y. and Paull, R.E. (1998). Tropical Fruits. CABI Publishing, New York, USA.

Nath, V., Kumar, D. and Pandey, V. (2008). Fruit: Volume-1. Satish Serial Publishing House, New Delhi, India.

Pandey, R.M. and Pandey, S.N. (1990). The grapes. DKMA, ICAR, New Delhi, India.

Pandey, R.M. and Sharma, H.C. (1988). The litchi. DKMA, ICAR, New Delhi, India.

Pandey, S.N. (2012). The mango. DKMA, ICAR, New Delhi, India.

Patil, R.T., Dingh, D.B. Gupta, R.K. (2008). Postharvest management of Horticultural Produce. Daya Publishing House, New Delhi, India.

Pena, J.E., Sharpa, J.L. and Wysoki, M. (2002) Tropical fruit pests and pollination, CABI publishing, New York USA.

Peter, K.V. (2005). Plantation crops. National Book Trust, New Delhi, India.

Rajput, C.B.S and Babu, S.H. (1999). Citriculture. Kalyani Publishers, Ludhiana, India.

Ray, P.K. (2004). Breeding Tropical and Subtropical Fruits. Narosa Publishing House, New Delhi, India.

Samson, J.A. (1980). Tropical fruits. Longman, London, U.K.

Sharma, R.R. (2006). Fruit production: problems and solutions. International Book Distributing Company, Lucknow, India.

Sharma, R.R. (2013). Propagation of horticultural crops. Kalyani Publishers, Ludhiana, India.

Sharma, R.R. and Krishna, H. (2013). Textbook on plant propagation and nursery management. International Book Distributing Company, Lucknow, India.

Sharma, R.R. and Srivastav, M. (2004). Plant propagation and nursery management. International Book Distributing Company, Lucknow, India.

Singh, A. (2009). Fruit physiology and production. Kalyani Publishers, Ludhiana, India.

Singh, H.P. and Mustaffa, M.M. (2009). Banana: new Innovations. Westville Publication House, New Delhi, India.

Singh, R. and Saxena, S, K. (2008). Fruits. National Book Trust, New Delhi, India.

Singh, R. N. (1990). The mango. DKMA, ICAR, New Delhi, India.

Singh, R.S. (2000). Diseases of fruit crops. Oxford and IBH Publishing Company Pvt. Ltd., New Delhi.

Singh, S.J. (1996). Advances in diseases of fruit crops in India. Kalyani Publishers, Ludhiana, India.

Singh, S.S. and Naqvi, S.A.M.H. (2001). Citrus. International Book Distributing Company, Lucknow, India.

Singh, Z. and Dhillon, B.S. (1993). Mango Malformation. **In**: Advances in Horticulture, Vol. 4. Fruit crops. (Eds., Chadha, K.L. and Pareek, O.P.). Malhotra Publishing House, New Delhi, India.

Srivastava, R.P. (1998). Mango cultivation. International Book Distributing Co., Lucknow, India.

Srivastava, R.P. (1998). Mango insect pest management. International Book Distributing Co., Lucknow, India.

Srivastava, R.P. and Kumar, S (2001). Fruit and Vegetable Preservation: Principles and Practices. International Book Distributing Co., Lucknow, India.

Westwood,M.N. (1993). Temperate zone pomology. W.H. Freeman and Co., San Francisco, USA.

Wills,R.B.H., McGlasson, W.B., Graham, D. and Joyce, D.C. (2002). Postharvest. University Press, U.K.

A. Important Varieties and Hybrids

Bright-N-Early Apple

Mollie's Delicious Apple

Royal Delicious Apple

Oregon Spur Apple

Crab Apples

Pusa Nanha Papaya

Pusa Delicious Papaya

A Bunch of Basrai Banana

Ambika Mango

Pusa Arunima Mango

Pusa Urvashi Grapes

Fruit Production

Calcuttia Litchi

Mridula Pomegranates

Halawy Dates

Umran Ber

Nagpuri Mandarin

Perlette Grapes

A View of Profuse Bearing in *Ber*

Mallika Mango

Pusa Surya Mango

B. Propagation and Cultural Practices

Papaya Seedlings

Stooling in Guava

Side-Veneer Grafting in Mango

Profuse Bearing in Papaya High Density Orchard

A View of High Density Planting in Banana

Amrapali Mango in High Density

Flowering in High Density Apple Orchard

Frost Affected Banana Plants

Micropropagation Protocol for Grape

Harvested Apples in *Kilta*, being taken to Packing Shed

C. Diseases of Fruits

Anthracnose in Mango

Cigar End Rot in Banana

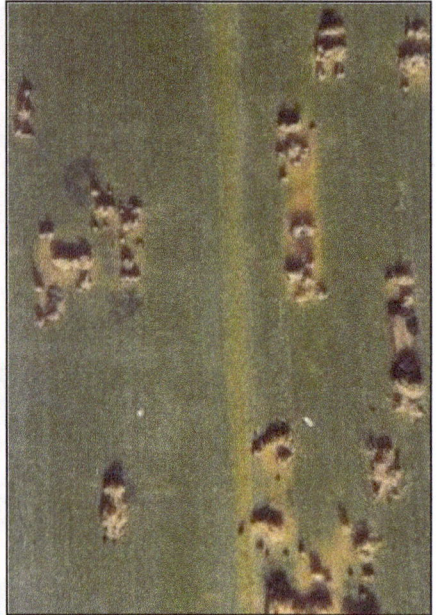

Graphiola Leat Spot in Date

Leaf Curl in Papaya

Scab in Apples

Pre-Mature Leaf Fall in Apple

Penicillium **Rot in Citrus**

Canker in Lemon

Floral Mango Malformation

Powdery Mildew in *Ber*

Bacterial Blight in Pomegranate

Bunchy Top of Banana

D. Physiological Disorders of Fruits

Cracked Lemons

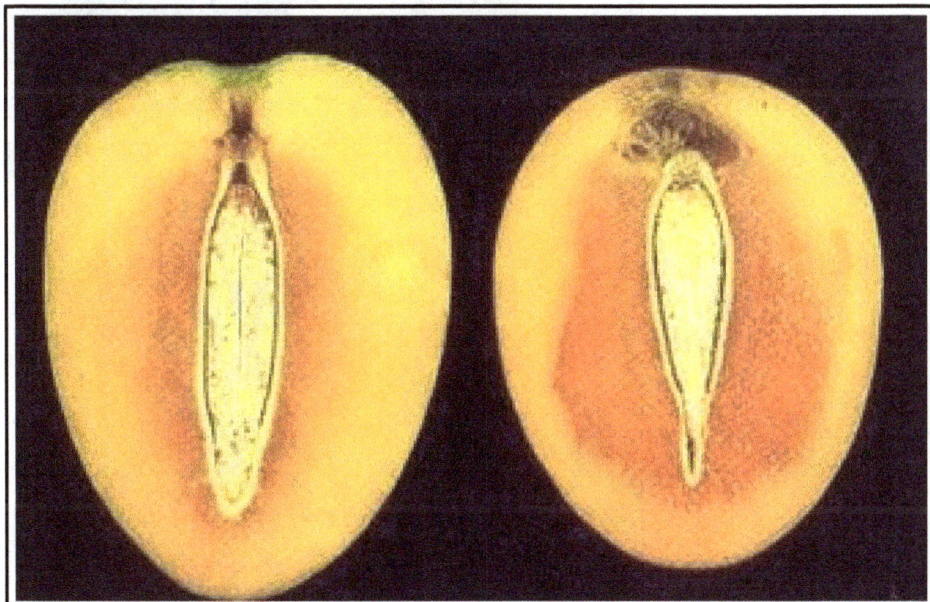

Jelly Seed in Mango

Fruit Production

Pitting in Mango Fruits

Pink Berry Formation in Grapes

Bitter Pit in Apple

Fruit Cracking in Litchi

Fruit Cracking in Pomegranates

E. Insect-Pests of Fruits

Woolly Apple Aphid

Sanjose Scale on Apples

Damage Caused by Anar Butter Fly

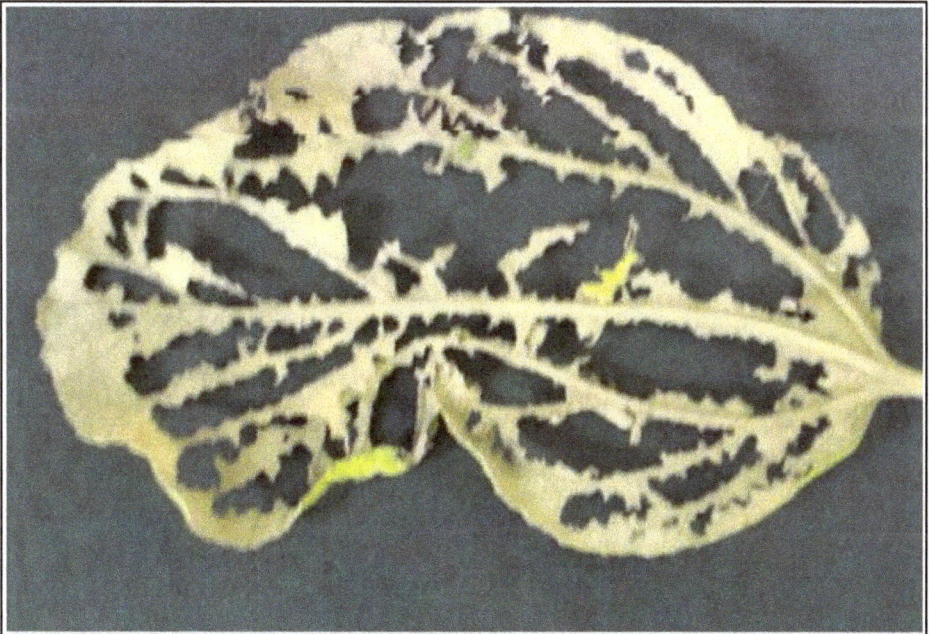

Chafer Beetle Damage in *Ber*